Railway Transportation Systems

Railway Transportation Systems

Design, Construction and Operation

Second Edition

Christos N. Pyrgidis

CRC Press
Taylor & Francis Group
Boca Raton London New York

CRC Press is an imprint of the
Taylor & Francis Group, an **informa** business

Second edition published 2022
by CRC Press
6000 Broken Sound Parkway NW, Suite 300, Boca Raton, FL 33487-2742

and by CRC Press
2 Park Square, Milton Park, Abingdon, Oxon, OX14 4RN

© 2022 Taylor & Francis Group, LLC

First edition published by CRC Press 2018

CRC Press is an imprint of Taylor & Francis Group, LLC

Library of Congress Cataloging-in-Publication Data

Names: Pyrgidis, Christos N., author.
Title: Railway transportation systems : design, construction and operation
/ Christos N. Pyrgidis.
Description: Second edition. | Boca Raton : CRC Press, 2022. | Includes
bibliographical references and index.
Identifiers: LCCN 2021020721 (print) | LCCN 2021020722 (ebook) | ISBN
9780367494216 (hbk) | ISBN 9780367494230 (pbk) | ISBN 9781003046073 (ebk)
Subjects: LCSH: Railroads--Design and construction. | Railroad engineering.
| Transportation engineering.
Classification: LCC TF200 .P97 2022 (print) | LCC TF200 (ebook) | DDC 625.1--dc23
LC record available at https://lccn.loc.gov/2021020721
LC ebook record available at https://lccn.loc.gov/2021020722

ISBN: 978-0-367-49421-6 (hbk)
ISBN: 978-0-367-49423-0 (pbk)
ISBN: 978-1-003-04607-3 (ebk)

DOI: 10.1201/9781003046073

Typeset in Sabon
by Deanta Global Publishing Services, Chennai, India

To my wife, Maria, and my son, Nikos

Contents

19 Railway and the natural environment   449

Preface to the Second Edition

This book constitutes the second edition of the book entitled *Railway Transportation Systems – Design, Construction, and Operation* that was published from the same house, Taylor & Francis Group, LLC, which I wholeheartedly thank for the collaboration.

The term 'railway transportation systems' includes all means of transport whose rolling systems involve at least one iron component (steel wheels on rails or rubber-tyred wheels on steel guideway). Because of this, this book, just like in the first edition, examines only transport systems that have this particular characteristic in common.

> *The development of railway technology reached an early peak and then was found to be in question. During the last twenty years, it has managed to not only rise again but also be on the cutting edge of technology in many countries.*

In recent years, there has been a continuous and rapid increase in the number of urban/ suburban railway transportation systems put in revenue service worldwide. As new suburban, metro, tramway, and monorail systems are being put into circulation, they are supplemented by innovations in rolling stock, infrastructure, and operational equipment/ procedures. The same upward trend is also observed in airport rail links. As a result, all these systems provide more efficient public transport to more people.

Over the last years, the rapid advances in technologies have led to a rapid and continuous increase in the length of high-speed railway tracks worldwide, accompanied by a corresponding increase in running speeds. As a result, they provide a higher service level to the passengers while a distinction should now be made for both the infrastructure and the offered railway services between high and very high speeds.

The transportation of goods is one of the most lucrative functions of railway transport systems. This profitability, besides the volume of transported goods, the distances over which they are transported, and the tariffs that are imposed, is also depended upon other characteristics that permeate freight railway transportation. The choice of these characteristics is not an easy task for railway companies since, on the one hand, their impact (positive or negative) is not clearly and definitively known and, on the other hand, their choice is not dependent solely on parameters that may be affected solely by railway companies (e.g., What is more lucrative for a railway organisation, to circulate passenger and freight trains on the same line or to separate passenger and freight traffic?). In order to answer these dilemmas and in an effort to improve the technical and operational characteristics of railway systems so as to make them more efficient for operators, safe and economical for users, and friendlier to the environment, new technologies that are related to the three railway constituents (infrastructure, rolling stock, and operation) are explored. The railway in an effort to be more competitive and to take over a larger share of the freight transport market is evolving through the use of such technologies. Technologies, choices, and development strategies are, in the end, directly linked.

Finally, and due to climate change, extreme weather conditions and natural phenomena occur that have an impact on both the rolling stock and the infrastructure and, thus, the operation and safety of railway transportation systems. Their occurrence is likely to increase in the future and, thus, existing railway systems will increasingly have to operate under conditions beyond their initial design specifications.

All the above support the fact that there is an imperative need to regularly update relevant data, to adjust them to new needs that arise, and to complement them with additional information. These are the main reasons that led me to continue with a new edition of my book. At the same time, corrections have been made regarding some errors in the previous edition (mostly in some figures) that were noticed after careful reading by both me and various readers of the book.

This new version contains 21 chapters. The book presents a comprehensive overview of passenger and freight railway transport systems, from design through to construction and operation. It covers the range of railway passenger systems, from conventional and high and very high-speed intercity systems (Chapters 11, 12, 13, and 14) through suburban, regional, urban, and rail transport systems for steep gradients (Chapters 4, 5, 6, 7, 8, 9, and 10). Moreover, it thoroughly covers freight railway systems transporting conventional loads, heavy loads, and dangerous goods (Chapters 15 and 16). For each system it provides a definition, a brief overview of its evolution and examples of good practice, the main design, construction and operational characteristics, and the preconditions for its selection.

This book also contains seven chapters that provide cross-cutting information to the reader that concerns all railway systems. Specifically, these issues are related to safety (Chapter 18), the interfaces with the environment (Chapter 19), cutting-edge technologies (Chapter 20), the loads applied on track (Chapter 2), the techniques that govern the stability and guidance of railway vehicles on track (Chapter 3), the operation of railway systems under extreme weather conditions and natural phenomena (Chapter 17), and the methodology and the steps required to verify the pre-feasibility of its implementation (Chapter 21). Finally, an introductory chapter is included that defines the railway as a transport system by describing the basic constituents of the system and by classifying railway systems into categories (Chapter 1). All of these chapters incorporate the author's 30 years of involvement in teaching, research, and experience in railway engineering.

Until recently, knowledge of railway technology was shared only among railway organisations. Many of the organisations' executives changed job positions in order to broaden their vision and knowledge. In recent years, an increasing number of people have become involved in the field of rail transport worldwide (engineers, consultants, manufacturers, transport companies, etc.).

This book provides additional information for those interested in learning about railway transportation systems. It can be used as a decision-making tool for both designers and operators of railway systems. In addition, it attempts to educate young railway engineers to enable them to deal with rail issues that may be assigned to them during the course of their careers.

All the data recorded and analysed in this book relate to the end of the year 2019.

I hope that this new edition of my book will support and promote the scientific dialogue, research, and education in the field of railway transportation by complementing and updating the already existing bibliography in a scientific and industrial field that is becoming more and more relevant and crucial worldwide.

Professor Christos N. Pyrgidis
Aristotle University of Thessaloniki, Greece

Acknowledgements

I thank again everyone who has helped me to complete the first edition of this work. Specifically, I wish to thank:

- My colleagues: Nikolaos Asmeniadis, Alexandros Panagiotopoulos, Rafael Katkadigkas, civil engineers, Dr. Evangelos Christogiannis, Dr. Constantina Anastasiadou, Dr. Alexandros Deloukas, Dr. Nikolaos Demiridis, Assist.Professor Nikolaos Gavanas, Dr. Anastasios Lambropoulos, Dr. Georgios Leoutsakos, Professor Antonios Stathopoulos, and Professor François Lacote. All of them agreed with pleasure to read through chapters of my book and provided useful comments.
- Ifigeneia Balampekou, transport engineer, for assisting with the translation and text editing.
- Artemis Klonos, professional photographer, who provided most of the pictures in this book.
- Ntia Regka, architect, for editing the figures.
- My dear friend Alexandros Kolokotronis, who helped in the selection and design of the book's cover.
- All my graduate and postgraduate students, who contributed in different ways to the final result.

I acknowledge all organisations, institutions, private companies, and individuals for providing and permitting me to use tables, figures, and photographs. Specifically, I thank: edilon)(sedra, Siemens, SNCF, SNCF Médiathèque, Posco ICT Co. Ltd, DCC Doppelmayr Cable Car, Innsbrucker Nordkettenbahnen Betriebs GmbH, Voestalpine Nortrak Ltd., Talgo, Vossloh Fastening Systems GmbH, Mermecgroup, BONATRANS GROUP a.s., Peer plus, Struktonrail bv, ANSALDO STS, Bombardier Transport, Alstom Transportation, UIC, ABB, Railway Gazette International, RATP, UITP, LEA+ELLIOTT Inc., Eyrolles Editions, MRT-Productions, Transportation Safety Board of Canada, CERTU, LEITNER AG/SpA, U.S. Department of Transportation Federal Highway Administration, Urbanaut, Hitachi Ltd, Professor C. Hass-Klau, Ev. Krussev, K. Pedersen, A. Takahashi, Br. Finn, Dr. K. Giannakos, L. Zlateva, Professor S. Iwnicki, Professor V. Profillidis, Professor K. Vogiatzis, Professor Coenraad Esveld, D. Oikonomidis, Ben Betgen, Harald Loy, and S. Dale.

I also thank everyone who has helped me to prepare this second edition of my book, and specifically, I wish to thank:

- My colleague Alexandros Dolianitis, civil engineer, MSc, for assisting with the translation as well as text and figure editing.

- Mr. Carlo Borghini, Executive Director of the Shift2Rail Joint Undertaking, who agreed with pleasure to read Chapter 20 of my book and provided useful comments.
- Mr. Michele Barbagli, Ing. MSc, PMP Assessor, who agreed with pleasure to read through the chapters of my book and provided useful comments.
- My colleague Stella Poravou, civil engineer, who helped in the selection and design of the book's cover.
- All my graduate and postgraduate students, who contributed in different ways to the final result.
- All the readers of the first edition who contacted me in order to bring to my attention the (thankfully few) errors that they found, as well to provide useful comments and suggestion for updating the book.

Last but not least, I thank Taylor & Francis Group for editing and publishing both editions of my book and for the excellent collaboration throughout the entire process.

Author

Christos N. Pyrgidis is a professor in railway engineering at the Aristotle University of Thessaloniki (AUTh), Greece. He earned a diploma in civil engineering (AUTh, 1981). He specialised for five years at the Ecole Nationale des Ponts et Chaussées, Paris, France, in transportation infrastructure (CES), transport economics (DEA), and railway engineering (PhD). From 2004 to 2007, he served as the Greek representative to the Administrative Board of the European Railway Agency. Since 2014, he has been a member of the Scientific Committee of SHIFT2RAIL.

Symbols and Abbreviations

A	track category in accordance with UIC (based on the permitted axle load)
A_r	rail cross section
A_p	bogies pivot centre
A_w	parameter depending on the characteristics on the rolling stock and representing the rolling resistances
AC	Alternative Current
AFC	Automatic Fare Collection system
AGT	Automated Guideway Transit system
APMs	Automated People Movers
APS	Alimentation Par le Sol – ground power supply system for tramways
APT	Advanced Passenger Train
ATO	Automatic Train Operation system
ATP	Automatic Train Protection system
ATS	Automatic Train Supervision system
ATSM	Automatic Train Stopping Management
a_b	parameter that depends on the classification of the track in the UIC classes
AVE	Alta Velocidad Española
a_d	lateral distance of the noise barrier from the track centre
2a	bogie wheelbase
a_i	coefficient that indicates the uneven distribution of the centrifugal force among the two axles of a bogie
B	track category in accordance with UIC (based on the permitted axle load)
B_r	rail's weight per metre
B_t	vehicle's weight
B_{tr}	train's weight
B_{ty}, B_{tz}	lateral and vertical components of the vehicle weight-motion in curves
B_w	parameter depending on the characteristics of the rolling stock and representing the mechanical resistances
BODu	Biochemical Oxygen Demand
BRT	Bus Rapid Transit
BTU	British Thermal Unit
b_b	parameter that depends on the sleepers' length and material
b_{cp}	width of a centre (island) platform

b_{em}	width needed for the installation of electrification masts
b_{lp}	width of a side platform
b_{sw}	width of separator
$b1, b2$	width of two intersected roads (1 and 2) (tramway network)
C	track category in accordance with UIC (based on the permitted axle load)
C_{hmin}	constructional height in the middle of the catenary opening
C_o, C_o'	track maintenance cost
C_{thr}	centre throw
C_v	transport capacity of a passenger train or vehicle
C_{vph}	transport capacity of a passenger train or vehicle in peak hours
C_{vnph}	transport capacity of a passenger train or vehicle in non-peak hours
C_w	parameter depending on the characteristics of the rolling stock and representing the aerodynamic resistance
$\bar{C}_x, \bar{C}_y, \bar{C}_z$	damping coefficients of the secondary suspension dampers in the three directions, respectively
C_ρ	damping torsional torque
C_φ	damping coefficient
CBTC	Communications-Based Train Control systems
CCAs	Cross-Cutting Activities
CCTV	Closed-Circuit TeleVision
CIM	Convention Internationale Marchandises par chemin de fer (contract of international carriage of goods by rail)
Classes A, B, C, D, E	tramway corridors categories
COTIF	COnvention Transports Internationaux Ferroviaires (convention concerning international carriage by rail)
CWR	Continuous Welded Rails
c_b	parameter that depends on the volume of the required track maintenance work
c_{ij}	coefficients of Kalker
c_{11}	longitudinal creep coefficient of Kalker
c_{22}	transversal creep coefficient of Kalker
c_{23}, c_{33}	spin coefficients of Kalker
D	track category in accordance with UIC (based on the permitted axle load)
D_m	dead mileage' (distance between the entrance of the depot and the nearest terminal station of the tramway line)
D_{mmax}	maximum permissible value for 'dead mileage'
D_o	minimum wheel diameter of trains running along the line
DC	Direct Current
DFS	Direct Fastening System
DMU	Diesel Multiple Unit
DOT	Department of Transportation (USA)
DRNT	Design Rail Neutral Temperature
DTO	Driverless Train Operation system
DVT	Driving Van Trailer
d_b	parameter that depends on the maximum axle load
d_o	vibration displacement

$d_o^{'}$	reference vibration displacement
2d	lateral distance between springs and dampers of the primary suspension
$2d_a$	back-to-back wheel distance (inside gauge)
d_i	maximum distance between the point of contact of the total vertical loads that are imposed on the rail, from the edge of the foot of the rail
E	steel elasticity modulus
E_b	parameter that depends on the quality category of the soil and the bearing capacity of the substructure
E_c	track cant excess
E_{cmax}	maximum track cant excess
E_d	total required ground plan area of a depot
E_a	total available ground plan area of a depot
E_{thr}	end throw
EMU	Electrical Multiple Unit
ERA	European Railway Agency
ERTMS	European Rail Traffic Management System
ERRAC	European Rail Research Advisory Council
ESS	Energy Storage Systems
ETCS	European Train Control System
EU	European Union
EXCA	EXceeded CApacity
e_b	ballast layer thickness
e_{bt}	total thickness of ballast and sub-ballast layers
e_{sb}	sub-ballast layer thickness
2e	track gauge
$2e_a$	outer flange edge-to-edge distance (flange gauge)
$2e_o$	theoretical distance between the running surfaces of the right and the left wheels when centred ≈ distance between the vertical axis of symmetry of the two rails
F	guidance force (effort) exerted from the wheel to the rail
F_{cf}	centrifugal force
F_{cfy}, F_{cfz}	lateral and vertical components of the vehicle's centrifugal force – motion in curves
F_{ij}	guidance forces exerted from the four wheels of a 2-axle bogie to the rails (i = 1,2 front and rear wheelsets, respectively; j = 1,2 left and right wheels, respectively, in the direction of movement)
F_j	guidance force exerted from one of the wheels of a wheelset to the respective rail (j = 1,2 left and right wheels, respectively, in the direction of movement)
F_l	fleet size – number of vehicles needed for the daily service
F_{lt}	fleet size – total number of vehicles needed for the operation of the system (reserve vehicles included)
F_{nc}	residual centrifugal force
F_{res}	lateral forces of springs of the primary suspension
FRA	Federal Railroad Administration (USA)
F_t	traction effort developed on the driving wheel treads
F_{tr}	traction effort acting on the axles

f	frequency of oscillation
f_b	parameter that depends on the track design speed and the bearing capacity of the substructure
f_d	wheel flange thickness
f_{ph}	train headway in peak hours
f_{nph}	train headway in non-peak hours
G	geometrical centre of a railway wheelset
G'	centre of gravity of the vehicle
GoA	Grade of Automation (metro systems)
GPS	Global Positioning Satellite system
GRT	Group Rapid Transit system
GSM-R	Global System for Mobile Communications – Railway
g	gravity acceleration
g_{dv}	dynamic gauge width of tram vehicle
g_i	maximum for track twist
g_v	static gauge width of tram vehicle
g_{imax}	highest permitted value for track twist
H	total transversal force transmitted from the vehicle to the rail
H_R	lateral track resistance
H_w	cross or side wind force
h	height clearance under civil engineering structures
h_f	wheel flange height
h_{fc}	height of the catenary contact wire
h_{KB}	distance between the vehicle's centre of gravity and the rail rolling surface
h_o	track lifting after maintenance work
h_r	rail height
h_v	vehicle height
I	track cant deficiency
I_{max}	maximum track cant deficiency
I_1	isolation distance of wire-grounded structures
IEC	International Electro-Technical Commission
IFEU	Institute for Energy and Environmental research (Institut für Energie- und Umweltforschung)
IPs	Innovation Programmes
IRR	Internal Rate of Return
ITDs	Integrated Technology Demonstrators
ITS	Intelligent Transport Systems
i	track longitudinal gradient (or slope)
i_d	track longitudinal gradient in depot
i_{dmax}	maximum permissible track longitudinal gradient in depot
i_{max}	maximum track longitudinal gradient (or slope)
i_{min}	minimum track longitudinal gradient (or slope)
JTTRI	Japan Transport and Tourism Research Institute
$j = 1, 2$	indicator relative to the two wheels of the same wheelset
K',K	factors of increase of the aerodynamic load exerted to noise barriers
K_b	angular stiffness of the link between the two wheelsets of the bogie (bogies with self-steering wheelsets)

K_{bt}	overall longitudinal stiffness of the primary suspension (bogies with self-steering wheelsets)
K_d	derailment factor (due to wheel climb)
K_{dyn}	vertical dynamic stiffness of elastomers
K_m	coefficient with values varying between 1.15 and 1.45
K_o	bogie-yaw dampers stiffness
KP	Kilometric Point
K_{st}	overall lateral stiffness of the primary suspension (bogies with self-steering wheelsets)
K_{stat}	vertical static stiffness of elastomers
K_t	coefficient that depends on the rolling conditions of the power vehicle axles on the track
K_x	longitudinal stiffness of the primary suspension (springs)
K_y	lateral stiffness of the primary suspension (springs)
$\bar{K}_z$	vertical stiffness of the secondary suspension (springs)
K_ς	lateral stiffness of the link between the two wheelsets of the bogie (bogies with self-steering wheelsets)
K_1	Front wind force coefficient (parameter depending on the shapes of the 'nose' and the 'tail' of the train)
K_2	Side-wind force coefficient (parameter depending on the lateral external surface of a train)
k	vertical track stiffness
k_v	track design speed coefficient
L_{den}	day-evening-night equivalent noise level
L_{dn}	day-night equivalent noise level
$L_{eq,T}$	equivalent energy noise level
L_h	length over headstock
L_i	Length of inclined sections of the track
L_{kmin}	minimum allowed length for a transition curve
L_{max}	maximum noise level
L_{st}	distance between two successive stops/stations
L_T	civil engineering structure width
L_{tr}	train's length
L_v	vehicle's length (tram)
L_w	oscillation wave length (hunting of railway wheelset)
LC	LoComotive
LCD	Liquid Crystal Display screens
LCL	Less-than-Car-Load
LED	Light Emitting Diode
LIM	Linear Induction Motors
LRC	Laser Railhead Cleaner
LRTs	Light Rail Transport systems
LRVs	Light Rail Vehicles
LTL	Less-than-TruckLoad
l_A	expansion zone length of rail
l_o	initial rail length
l_T	civil engineering structure length
M	spin moment on wheels
M′	mass of one bogie
$\bar{M}$	car body mass

M_t	total mass of the vehicle
M_1, M_2	spin moments in the left and right wheels, respectively, in the direction of movement of a railway wheelset
MC	Motor Car
M_{RLC}	Traffic moment of a Railway Level Crossing (RLC)
MU	Multiple Unit
m	mass of one railway wheelset (axle + wheels + axle boxes)
N	temperature force
N_{ac}	acceleration force
N_{br}	braking force
N_1	reaction force in the wheel-rail contact surface (wheel 1)
NATM	New Austrian Tunnelling Method
NPV	Net Present Value
n_b	total number of bogies of a train formation
n_p	coefficient of the probability augmentation of the mean square value of standard deviations of vertical dynamic forces of a vehicle
n_{ph}	number of peak hours within the 24-h day
n_{nph}	number of non-peak hours within the 24-h day
n_s	number of intermediate stations/stops
OCS	Overhead power supply (Catenary) System
OESS	Overhead Energy Storage System
ORE	Office de Recherches et d'Essais de l' UIC (office for research and experiments of the UIC)
OTMs	On Track Machines
P_d	total number of passengers expected to be transported along a specific connection (passengers/hour/direction or daily-potential transport volume)
P_d'	passenger transport capacity of the system (passengers/hour/direction)
P_{dph}	total number of passengers expected to be transported along a specific route during peak hours
P_{dyn}	transversal force due to vehicle oscillations
P_{dyear}	yearly predicted passenger transport volume
P_f	fishplate force
P_t	net or useful power of motors
P_{4w}	power consumed at the level of the four wheels of the bogie
PDP	Plasma Display Panel
PPV	Peak Particle Velocity
PR	single railcar
PRT	Personal Rapid Transit systems
PSD	Platform Screen Doors
PSE	Paris-Sud-Est
p	the perimeter that encloses the rolling stock laterally, up to rail level (rolling stock outline)
p_o	mean noise pressure
p_o'	the relative mean reference pressure
ppl	population of a city
Q	axle load

Q_d	design vertical wheel load
Q_{Do}	maximum passing axle load (wheels of diameter D_o)
Q_{dyn}	dynamic vertical wheel load
Q_{dyn1}	dynamic vertical wheel load due to the vehicle's sprung masses
Q_{dyn2}	dynamic vertical wheel load due to the vehicle's semi-sprung masses
Q_{dyn3}	dynamic vertical wheel load due to the vehicle's unsprung masses
Q_{dyn4}	dynamic vertical wheel load due to the oscillations of the elastic parts of the rail-sleeper fixing system
Q_H	quasi-static vertical wheel load
Q_{max}	maximum axle load
Q_{nc}	vertical wheel load due to residual centrifugal force
Q_o	wheel load (= Q/2)
Q_t	total vertical wheel load
Q_w	vertical wheel load due to crosswinds
q	uniform load applied to noise barriers
q_o	vertical distance between the geometrical centre of the lateral surface of the car body and the rail rolling surface (height of centre of the crosswind force)
q_r	flange cross-dimension (the horizontal distance between the intersection point of the joint geometric level with the flange face and the intersection point of a reference line at a distance of 2 mm from the flange tip with the flange face) (critical lateral dimension of the wheel)
$q^*_o = 1.25\, q_o$	compensated value of the nominal height of application point of the crosswind force, as taken from the rolling surface of the rails
R	curvature radius of the wheel tread
R′	radius of curvature of the rolling surface of the railhead
R_c	radius of curvature in the horizontal alignment
R_{cmin}	minimum radius of curvature in the horizontal alignment
R_{co}	horizontal alignment radius as it derives from simulation models
R_g	Switch turnout radius of curvature
R_s	sound-insulating capacity index of the construction material of noise barriers
R_v	radius of curvature in the vertical alignment
R_{vmin}	minimum radius of curvature in the vertical alignment
RID	Regulation International de transport des produits Dangereux par chemin de fer (international carriage of dangerous goods by rail)
RLCs	Railway Level Crossings
RNT	Rail Neutral Temperature
RSSB	Rail Safety Standards Board
r_o	rolling radius of the wheels in the central equilibrium position
r_1, r_2	rolling radii of the left and the right wheels (in the direction of movement) of a railway wheelset, in case of lateral displacement from its central equilibrium position
$2r_o$	wheel diameter
S	route, link, line, track, or connection length
S_A, S_B, S_C, S_D, S_E	tramway corridor length for corridor categories A, B, C, D, E, respectively

S_1	front surface cross section area of the train (frontal external affected surface of a train)
S_2	lateral surface area of the vehicle (lateral external affected surface of a train)
S_{max}	maximum route or link or connection length
S_{min}	minimum route or link or connection length
S_p	total gravitational force, restoration force, or gravitational stiffness
S_{p1}, S_{p2}	gravitational forces exerted on the left and the right wheels when the wheelset is displaced from its central equilibrium position
S_u	useful cross section area of the tunnel
S_v, S_m	coefficients with values depending on the speed of passenger (with the highest speed) and freight (with the lowest speed) trains, respectively, running on the track
S_{w1}, S_{w2}	paths covered by the two wheels 1 and 2 of a wheelset in curves, during a time interval t
SEL	Sound Exposure Level
SIL4	Safety Integrity Level 4
SNCF	Société Nationale des Chemin de Fer Français (national company of French railways)
SPDs	Suspended Particle Devices
SPL	Sound Pression Level
SSEM	Supervised Speed Envelope Management
STI	Sustainable Technologies Initiatives
STO	Semi-automatic Train Operation
SRIA	Strategic Research and Innovation Agenda
SW	loading model in railway bridges (heavy loads)
SWL	Single Wagon Load services
T	lateral creep force applied on the wheel
T_f	total daily traffic load
T_{fr}	friction forces between rails and sleepers and between sleepers and ballast
T_g	daily traffic load of freight trains
T_{ij}	lateral creep forces exerted to the four wheels of a 2-axle bogie (i = 1,2 front and rear wheelset, respectively, j = 1,2 left and right wheels, respectively, in the direction of movement)
T_m	average daily traffic load of freight wagons
T_p	daily traffic load of passenger trains
T_{tm}	average daily traffic load of freight trains' power vehicles
T_t	total traffic load
T_{tv}	average daily traffic load of passenger trains' power vehicles
T_v	average daily traffic load of trailer passenger cars
T_1, T_2	lateral creep forces exerted to left and right wheels (in the direction of movement) of a railway wheelset
TBM	Tunnel Boring Machine
TC	Trailer vehicle (Car) of a train
TC'	Trailer vehicle (Car) of a railcar
TDs	Technology Demonstrators
TFEU	Treaty on the Functioning of the European Union
TGV	Train Grande Vitesse (high-speed train [French technology])

TGV-A	TGV Atlantique (TGV Atlantic)
TGV (PSE)	TGV (Paris-Sud-Est) (Paris South East)
TL	Train Load services
TMS	Traffic Management System
TOFC	Trailer Of Flat Cars
TPs	Transforming Projects
TRL	Technology Readiness Level
TSIs	Technical Specifications for Interoperability
TSP	Total Suspended Particles
TT	single tramway track
TTROW	Total Tramway infrastructure Right-Of-Way
TTROWC	Total Tramway infrastructure Right-Of-Way in Curves
TTROWS	Total Tramway infrastructure Right-Of-Way in Straight path
TTROWST	Total Tramway Right-Of-Way in STops' areas
TTSM	TimeTable Speed Management
t	run time (or time interval)
t'	year of change of the corridor's operating frame
t_{AB}	run time from a point A (origin) to a point B (destination)
t_{fin}	year of the end of the economic life of a project
$t_{re} - t_{in}$	actual (recorded) temperature minus initial temperature of the rail
t_s	dwell or waiting time at stations/stops
t_{tot}	total travel time (door-to-door)
t_{ts}	dwell time (waiting time) of trains at the two terminal stations of a route
U	track (normal) cant
U_{max}	maximum (normal) track cant
U_{th}	theoretical track cant
U_{thvmax}	theoretical track cant for the maximum running speed
UAE	United Arab Emirates
UIC	Union International des Chemins de Fer (international union of railways)
UIC 1, 2, 3, 4, 5, 6	track categorisation in accordance with UIC (based on the total daily traffic load)
USA	United States of America
USM	UnSprung Masses of the vehicle (one wheelset)
UTO	Unattended Train Operation system
UTS	Ultimate Tensile Strength
V	train or vehicle or wheelset running or transit or forward speed
V_{amaxtr}	average permissible track speed
V_{ar}	average running speed
V_c	commercial speed
$V_{CA}, V_{CB}, V_{CC}, V_{CD}, V_{CE}$	commercial speed of tramways running on corridor categories A, B, C, D, E, respectively
V_{cmax}	maximum commercial speed
V_{cr}	vehicle critical speed
V_d	track design speed
$V_{der.dis}$	speed over which derailment due to lateral displacement of the track occurs
$V_{der.ov}$	speed over which derailment due to overturning occurs

$V_{der.wcl}$	speed over which derailment due to wheel climb occurs
V_{dmax}	maximum track design speed
V_{max}	train maximum running speed
V_{maxtr}	permissible track speed
V_{min}	running speed of the slowest trains circulating along a line – minimum running speed
V_p	train passage speed
V_{pas}	maximum speed of passenger trains
V_{fr}	maximum speed of freight trains
V_{pmax}	maximum train passage speed
V_{rs}	rolling stock design speed
V_t	train instant speed
V_w	wind speed
V_1, V_2	relative velocities of the left and right wheels (in the direction of movement) of a railway wheelset
VAL	Vibration Acceleration Level
VDSC	Vehicle Detection and PRIMOVE Segment Control cable
VN	Visual Nuisance
VNPs	Visual Nuisance Points
VPF/VPC	Value of Preventing a Fatality/Casualty
VPI	Value of Preventing an Injury
VVL	Vibration Velocity Level
v_o	vibration velocity
v'_o	reference level of vibration velocity
W	total train resistance
W_{ac}	acceleration resistance
W_B	basic resistance
W_i	(track) gradient resistance
W_m	movement resistance
W_{Rc}	(track) curve resistance (drag)
W_{tr}	(total) track resistance
W_α	air resistance or aerodynamic resistance or aerodynamic drag
WHO	World Health Organization
w_{wh}	wheel width
w_i	weighting factor that defines the level of influence every structural element i of a railway system has on the visual nuisance caused by the system as a whole
WILD	Wheel Impact Load Detector
X	longitudinal creep force applied on the wheel
X_{ij}	longitudinal creep forces exerted to the four wheels of a 2-axle bogie (i = 1,2 front and rear wheelset, respectively; j = 1,2 left and right wheels, respectively, in the direction of movement)
X_{vr}	reference values of yearly passenger transport volumes of tramway systems
X_1, X_2	longitudinal creep forces exerted to left and right wheels (in the direction of movement) of a railway wheelset
x	longitudinal displacement of the wheelset
x'	derivative of the longitudinal displacement x of a railway wheelset

Y_2	transversal force acted on the horizontal level and exerted in the point of contact of wheel 2, in case of contact of wheel 1 with the rail
y	lateral displacement of the wheelset in relation to its central equilibrium position
y_i	lateral displacements of the two wheelsets of a bogie (i = 1,2 front and rear wheelset, respectively)
y_o	lateral displacement of the wheelset in case of its radial positioning in curves (wheelset lateral offset)
y_w	oscillation wave amplitude (hunting of railway wheelset)
y'	derivative of the lateral displacement y of a railway wheelset
y''_{max}	maximum lateral acceleration of a railway wheelset
yy	derailment resultant force axis
σ	flangeway clearance
$\sigma\ (Q_{dyn1},\ Q_{dyn2})$	typical deviation of the dynamic vertical forces of the sprung and semi-sprung masses of the vehicle
$\sigma\ (Q_{dyn3})$	typical deviation of the dynamic vertical forces of the unsprung masses of the vehicle
ω	angular velocity of the two wheels of a conventional wheelset
$\omega_1,\ \omega_2$	angular velocities of the left and the right wheels (in the direction of movement) of a railway axle equipped with independently rotating wheels
γ_o	inclination of the tangent plane at the contact point between rail and wheel when the wheelset is in central position
$\gamma_1,\ \gamma_2$	angles formed by the horizontal plane, and the tangent planes at the contact points I_1 and I_2, respectively, when the railway wheelset is displaced from its central equilibrium position
γ_{nc}	lateral residual acceleration
γ_{ncmax}	maximum permitted lateral residual acceleration
γ_e	equivalent (or effective) conicity of the wheel
α	yaw angle of the wheelset
α_i	yaw angle of the two wheelsets of a bogie (i = 1,2 front and rear wheelset, respectively)
α'	derivative of the yaw angle α of a railway wheelset
α_{br}	coefficient of the static vertical loads applied on railway bridges
α_o	vibration acceleration
α'_o	reference level of vibration acceleration
α_s	sound-absorption coefficient
α_t	steel thermal expansion coefficient
$2\alpha_f$	angle of vertical displacement of the joint (sum of the angles that are formed by the two rails and the horizon) (rad)
ΣQ	overall train weight
Π	adhesion force
π	constant equal to 3.14
μ	wheel–rail friction coefficient (adhesion coefficient, Coulomb coefficient)
Δ	distance between track centres (double track)
Δ_t	temperature change
$\Delta I_{max}/\Delta t$	maximum rate of change of cant deficiency

Δl	variation of the length of the rail
ΔP_{max}	maximum permissible change in pressure generated inside the tunnels
δ_p	angle of cant
φ	angle of rotation of the wheels and the axle (conventional bogies)
φ'	derivative of the angle of rotation φ of the wheels and the axle (conventional bogies)
φ_{bri}	dynamic coefficient for the loading of railway bridges (i = 2 or 3)
φ_o	road intersection angle
φ_t	tilting angle of car body
φ_1, φ_2	angles of rotation of the two wheels of the same wheelset (axle with independently rotating wheels)
β_1	rail–wheel flange contact angle (wheel 1 under derailment)
ν	exponent with values between 3 and 4
ρ	air density
ρ_1	angle of wheel-rail friction of wheel 1
ρ_2	angle of wheel-rail friction of wheel 2

Chapter 1

The railway as a transport system

1.1 DEFINITION

The 'railway' or 'rail' is a terrestrial-guided mass transport system. Trains move on their own (diesel traction) or through remotely transmitted power (electrical traction) using steel wheels[1] rolling on a dedicated steel guideway defined by two parallel rails.

The railway transports passengers and freight. Its capability can extend to cover any distance in any environment (urban, suburban, periurban, regional, and interurban). Its range for passengers' transportation is usually suited to approximately 1,500 km[2], while for freight the distances can be much greater[3].

From a transport system point of view, it is by default considered to comprise three constituents:

- Railway infrastructure.
- Rolling stock.
- Railway operation.

1.2 CONSTITUENTS

1.2.1 Railway infrastructure

The term 'railway infrastructure' describes the railway track and all the civil engineering structures and systems/premises that ensure the railway traffic (Figure 1.1).

The railway track consists of a series of components of varying stiffness that transfer the static and dynamic traffic loads to the foundation. Hence, the railway track comprises successively from top to bottom the rails, the sleepers, the ballast, the sub-ballast, the formation layer, and the subgrade (Figures 1.2 and 1.3) (Giannakos, 2002; Profillidis, 2014).

The rails are mounted on the sleepers on top of elastic pads to which they are attached by means of a rail hold-down assembly called the rail fastening (Figure 1.4).

Rails, sleepers, fastenings, elastic pads, ballast, and sub-ballast constitute the 'track superstructure', while the subgrade and the formation layer constitute the 'track substructure' (Figure 1.2).

The upper section of the track superstructure that comprises the rails, the sleepers, the fastenings, and the rail pads forms what could be commonly called the 'track panel'. Switches

---

[1] For a small number of metro lines and for many cases of driverless railway systems (cable-propelled and self-propelled) of low/medium transport capacity, rubber-tyred wheels are also used.

[2] The longest railway route worldwide has a length of 2,439 km (Hong Kong–Beijing Xi, China) (Hartill, 2019).

[3] The longest freight train routes connect Europe and Asia (distances in the range of more than 10,000 km).

DOI: 10.1201/9781003046073-1

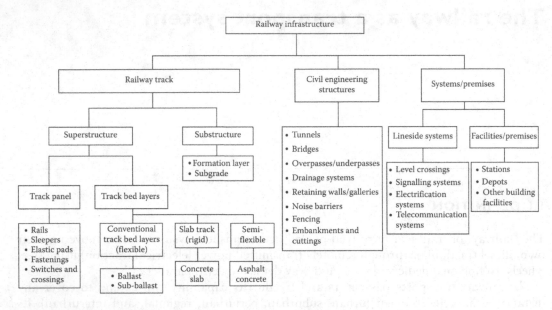

Figure 1.1 Constituents and components of the railway infrastructure.

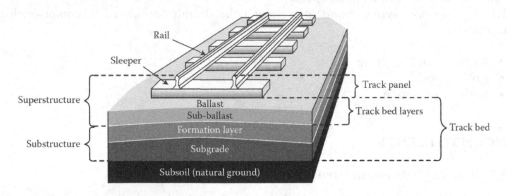

Figure 1.2 Railway track. (Adapted from Giannakos, K. 2002, *Actions in Railways*, Papazisi (in Greek), Athens.)

and crossings by means of which the convergence, cross section, separation, and joining of tracks at specific points of the network are accomplished are also considered to be part of it (Figure 1.5).

The lower part of the track superstructure that comprises the ballast and its sublayers is called 'track bed layers'. The track bed layers and the track subgrade, considered as a whole, are called 'track bed'.

Apart from the ballasted track bed (conventional or flexible track bed), a concrete track bed (slab track or rigid track bed) is more and more frequently used. The latter solution has proven to be very efficient in the case of underground track sections, where maintenance requirements are greatly restricted (Figure 1.6).

A third track bed system seldom applied is the 'asphalt concrete track bed', or otherwise called the 'semiflexible track bed'. This solution is used on certain occasions in Italy and Japan for the construction of new high-speed lines. It is also extensively used in North America for the restoration of short lengths in critical segments of the track (tunnels, switches and crossings sections, and transition zones before or after major civil engineering

Figure 1.3 Railway track; ballasted track superstructure, Athens-E. Venizelos Airport suburban line, Greece. (Photo: A. Klonos.)

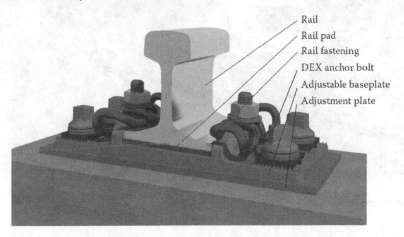

Rail
Rail pad
Rail fastening
DEX anchor bolt
Adjustable baseplate
Adjustment plate

Figure 1.4 DFS (Direct Fastening System) with adjustable steel baseplate. (From edilon)(sedra, 2015.)

works). Finally, this system is an alternative method for the improvement of the mechanical strength of the existing infrastructure (Buananno and Mele, 1996; Schoch, 2001).

The civil engineering structures comprise the underground sections of the track, the bridges, the overpasses/underpasses, the embankments and cuttings, the drainage systems, the soil retaining walls, the galleries, the noise barriers, and the fences.

Figure 1.5 Switches and crossings configurations, Zurich, Switzerland. (Photo: A. Klonos.)

Figure 1.6 Slab track, Tempi tunnel, Greece. (Photo: A. Klonos.)

The track systems and premises are separated into:

- Lineside systems that comprise the level crossings as well as the electrification, signalling, and telecommunication systems.
- Facilities and premises that comprise the stations, the depots, and other building facilities (administration buildings, warehouses, etc.).

Two special terms are usually used to describe the track structural characteristics along its length:

- 'Plain' track (line): A segment of a railway track that does not have any switches and crossings configurations on it (turnovers, crossovers, etc.).
- 'Open' track: A segment of a railway track that does not have any tunnels, bridges, overpasses, high embankments, deep cuttings, and stations/stops on it.

1.2.2 Rolling stock

'Rolling stock' is the term employed to describe all railway vehicles, both powered and hauled, used as power, trailer, or engineering vehicles (Figure 1.7).

The power vehicles are self-propelled, that is, they are equipped with traction motors. These vehicles may:

- Serve the sole purpose of hauling the trailer vehicles and are then called 'locomotives' (or 'traction units' or 'engines').
- Transport a number of passengers and are then called either 'single railcars' or 'rail-buses' (when they have a driver's cab at one or both ends) or 'motor cars' (when they are remote controlled from other vehicles).
- Be used for shunting, hence they are called 'shunting locomotives' or 'shunters' or 'switchers'.

Locomotives and single railcars, depending on the type of traction power they utilise, are classified into: (a) steam, (b) diesel, (c) electric, (d) hybrid, (e) gas turbine, and (f) fuel cell locomotives/railcars. All these categories may be further classified based on a series of criteria specific to each category (Pyrgidis and Dolianitis, 2021).

The trailer vehicles are not self-propelled. They serve the purpose of transporting people or goods. They may be classified into three basic categories depending on their use (Metzler, 1985), namely:

- 'Passenger vehicles' (or 'passenger cars' or 'coaches' or 'carriages') intended to transport passengers.
- 'General-use freight vehicles' (or 'freight cars' or 'good's wagons' or 'trucks') intended to transport goods.
- 'Specific-use freight wagons' intended for the transportation of certain types of freight only.

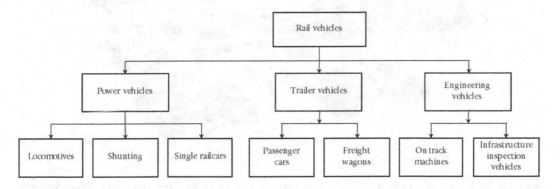

Figure 1.7 Categories of rolling stock.

The 'rolling stock' also encompasses the various engineering vehicles that are used to carry out track panel installation works as well as the various track inspection and maintenance works. These types of vehicles, due to their special use, may on their own be considered as a special category (Figure 1.7). These vehicles are further distinguished into two main categories, specifically into (Biasin et al., 2008; Pyrgidis and Dolianitis, 2021):

- 'On Track Machines' (OTMs), which include vehicles specifically designed for the construction and maintenance of the track. OTMs may be used in different modes, namely, working mode and transport mode (either as a self-propelled vehicle or as a hauled vehicle).
- 'Infrastructure inspection vehicles' or 'track recording cars' or 'diagnostic cars', which are utilised to monitor the condition of the infrastructure. For such vehicles, there is no distinction between transport and working modes; instead, they are operated in the same way as freight or passenger vehicles.

Every railway vehicle, either trailer or power, consists of three basic parts (Figure 1.8):

- The car body (body shell).
- The bogies (trucks).
- The wheelsets (axle + 2 wheels).

Several specific characteristics of these parts may be used to classify railway vehicles further (Pyrgidis and Dolianitis, 2021).

A 'train' or 'trainset' or 'consist' is an operational formation consisting of one or more units, while a unit may, in turn, be composed of several vehicles.

The combination of locomotives and trailer vehicles forms the loco-hauled passenger or freight trains depending on the category of the trailer vehicles (Figure 1.9). When two

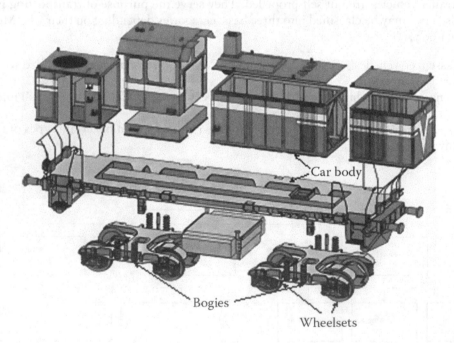

Figure 1.8 Main parts of a railway vehicle (locomotive): car body, bogies, and wheelsets. (From Siemens, 2015.)

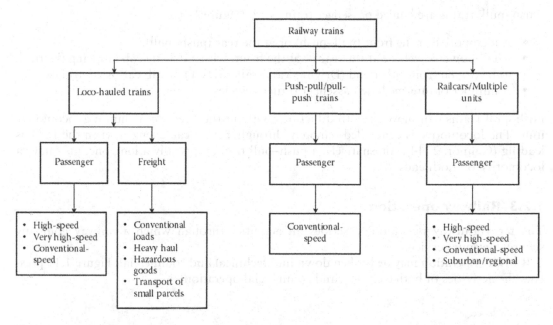

Figure 1.9 Types of trains.

locomotives are included in the same train formation then the operation is called '2-loco operation'.

The combination of single railcars with motor cars and/or trailer vehicles forms the 'railcars'.

The railcars can move in both directions without the need for a shunting locomotive in contrast with loco-hauled trains, which need a shunting locomotive.

The term 'Multiple Unit (MU)', either Electric (EMU) or Diesel (DMU), refers to a train-set of more than one unit in which all units are capable of carrying a payload (passengers, luggage/mail, or freight) (Biasin et al., 2008). Specifically, they exhibit the following characteristics (Connor, 2014):

- Units are railcars formed by single railcars, motor cars, and/or trailer vehicles, which are semi-permanently coupled.
- A driving cab is provided at each end of the unit. Drivers just change ends at the terminus.
- The train length can be varied by adding or subtracting units.
- Power equipment is distributed along the whole train (only motor cars and single rail-cars have power equipment).

Example formations of Multiple Units are:

- PR + TC + PR + PR + TC + PR.
- PR + MC + MC + PR.
- PR + PR + PR + PR.

    where:
    PR: Single railcar.
    TC: Trailer vehicle (Car).
    MC: Motor Car.

'Push–pull' trains are hauled passenger trains with (Figure 1.10):

- A locomotive at the front (pull–push) or at the rear (push–pull).
- An unpowered vehicle at the rear or at the front, with a driving cab allowing the train to be driven from either end (Driving Van Trailer (DVT) or cab car or control car).
- A number of intermediate passenger trailer vehicles.

Push–pull trains can move in both directions without the need for a shunting locomotive unit. The locomotive is controlled remotely through a train cable length when the DVT is leading (Connor, 2014). Alternatively, a push–pull train, especially a long one, may have a locomotive on both ends.

1.2.3 Railway operation

The term 'railway operation' describes all activities through which a railway company secures revenue service.

Railway operation may be broken down into technical and commercial. Figure 1.11 presents the activities of both technical and commercial operation.

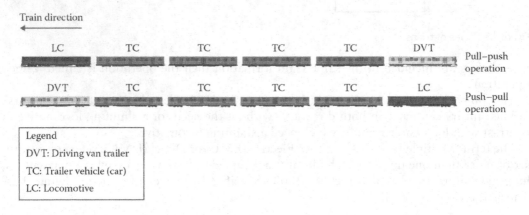

Figure 1.10 Pull–push/Push–pull train formations.

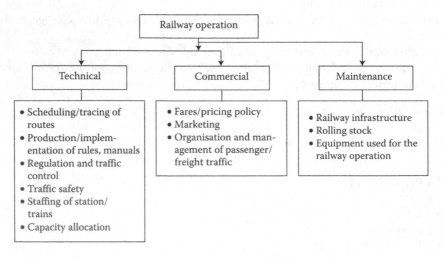

Figure 1.11 Technical and commercial railway operation activities.

Sound maintenance is a prerequisite for the smooth operation of the railway system. Maintenance is characterised as a 'horizontal activity', since it applies to all three constituents of the railway system.

Subsequently, definitions for several terms, which are directly related to the operation of a railway system and will reappear in later sections, are provided. The definition of these terms at this point is consistent with the structure of the book and is aimed at facilitating the reader.

'Track': Refers to the railway permanent way, which is defined by two parallel rails. The term 'track' is mostly used when one refers to the infrastructure (superstructure, substructure, civil engineering structures, and systems/premises) of a railway system. A track links two stations. It is comprised of track sections. The permanent way may include one track (single track), two tracks (double track), or more than two tracks.

(Railway) 'Line': Refers to the railway permanent way, which is defined by two parallel rails. The term 'line' is mostly used when one refers to the geographical integration and the operation of a railway system. A line usually connects two stations of importance. A line may include one track (single track line), two tracks (double track line), or more than two tracks.

'Railway network': Denotes all tracks (lines) that are located within a specific geographical area (continent, country, region, and county) as a whole.

'Main line' or 'main track': Refers to all the lines/tracks used by the trains servicing the routes included in the current 'timetable manual' of a railway network.

(Scheduled) 'Railway route': Refers to a specific link defined by an origin and a destination station and is included in the current 'timetable manual' of a railway network.

'Railway corridor': Denotes the line(s) that connect(s) two wider geographical locations without necessarily constituting a railway route (e.g., the railway corridor X Thessaloniki–Salzburg and Europe–Asia railway corridor).

'Track section': This term is used in the process of calculating track capacity (UIC, 1983). A track section connects two overtaking or crossing stations (in general terms, non-neighbouring stations), or two turnovers under the condition that the number and the proportion of different types of passing trains do not alternate more than 10%. It is comprised of more than one track subsection. In a wider sense, the term may simply refer to a segment of track.

'Track subsection': This term is also used in the process of calculating track capacity. A track subsection connects two neighbouring overtaking or crossing stations as well as one overtaking or crossing station with a turnover (UIC, 1983).

1.3 THE RAILWAY SYSTEM TECHNIQUE

1.3.1 Description of the system

The two basic 'technical units' securing the 'transport' with railway means are the vehicle's wheelset and the rails (Figures 1.12 and 1.13).

The wheelset consists of three basic parts:

- The axle.
- The wheels.
- The axle boxes.

The wheels consist of:

- The wheel tread, being the outer section of the wheels that allows the rolling on the rails.
- The wheel body.

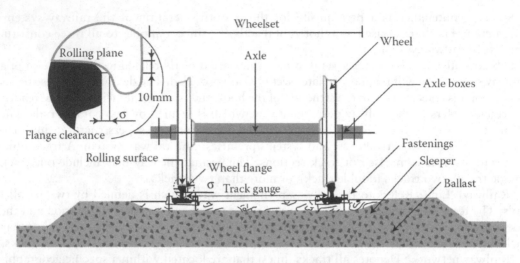

Figure 1.12 Rolling of a conventional railway wheelset on rails. (Adapted from Petit, J. M. 1989, *Conception des Bogies Modernes*, Revue ALSTHOM, Code APE 2811, France.)

Figure 1.13 Railway wheelsets, OSE, Piraeus factory, Greece. (Photo: A. Klonos.)

Over the last 35 years, all freight and passenger vehicles running at speeds V > 160 km/h have been equipped with cast wheels (meaning that the wheel rim and the body are a monobloc).

The wheel flanges (one on each wheel) are characteristic of the inner section of the wheels; their mission is to prevent derailment in case the wheelset lateral displacement exceeds the limits set by the track gauge (Figures 1.12 and 1.14). Meanwhile, the flanges support the wheelsets' self-guidance when passing through switches and crossings configurations (Figure 1.15).

The wheel cross section (profile) is not orthogonal as it is in the case of road vehicle tyres. It features a slight variable conicity that results in larger wheel diameters towards the inner part of the track (Figures 1.16 and 1.17).

Figure 1.14 Railway wheel rolling on plain track. (Adapted from Vignal, B. 1982, SNCF Médiathèque, France.)

Figure 1.15 Railway wheel rolling on crossing's frog area. (Photo: A. Panagiotopoulos.)

The two wheels are rigidly connected through a cylindrical rod (axle) resulting in the rotation of both the wheels and the axle at the same angular velocity ω.

The aforementioned 'wheel-axle' system is called conventional (or classic); the wheelset runs a steel guideway consisting of two parallel rails set at a fixed distance between them (rail inside face), commonly called the 'track gauge' (Figure 1.12).

The rail consists of three main parts (Figure 1.18):

- The head.
- The web.
- The foot.

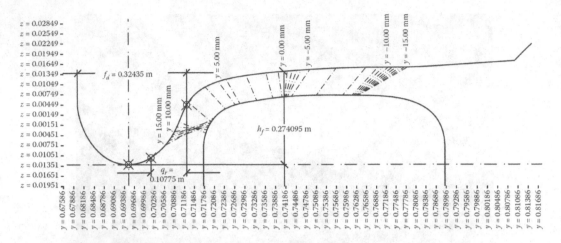

Figure 1.16 Wheel–rail contact surface geometry (indicative).

Figure 1.17 Geometry of rolling surface of the wheel.

The upper surface of the railhead is curved and, hence, forms the surface over which the wheel treads run.

Rails are mounted on sleepers at a certain angle called rail inclination angle (usually 1:20 or 1:40) (Figure 1.19). This layout improves the transversal stability of the vehicles in a straight path.

1.3.2 Fundamental functional principles

During wheel rolling, elastic forces develop on the contact surface (creep, gravitational forces). Under smooth running conditions (good track ride quality, allowable speed limits, rolling stock in good condition), these forces guarantee the stability and guidance of vehicles on straight paths and in curves (see Chapters 2 and 3).

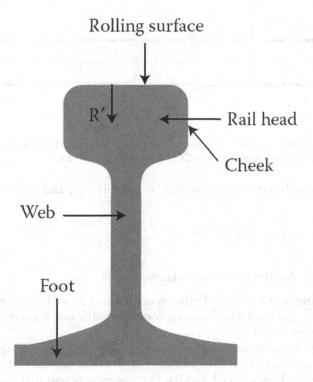

Figure 1.18 Main rail parts.

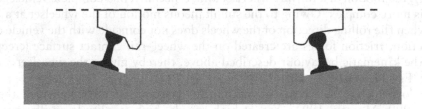

Figure 1.19 Inclined rails on sleepers.

The generation of these forces comes about through:

- The specific profile of the wheels.
- The rigid connection of the wheels to the axle.
- The geometry of the upper outer part of the railhead surface.
- Creep phenomena.

1.3.2.1 Running on a straight path

We consider a conventional railway wheelset centred on the track, running at a constant speed, V, on a straight path. If for any reason (track geometry defects, wheel asymmetry, etc.), the axle is displaced transversally with regard to the initial equilibrium position, then, due to their profile, the two wheels roll with different radii ($r_1 \neq r_2$).

In the case of Figure 1.20, $r_2 > r_o > r_1$ applies (where r_o indicates the rolling radius of the two wheels in the initial equilibrium position). Owing to the rigid connection of the wheels

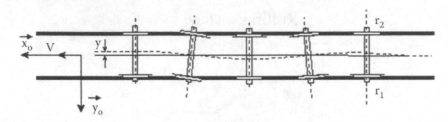

Figure 1.20 Sinusoidal motion of a railway wheelset (hunting). (Adapted from Petit, J. M. 1989, *Conception des Bogies Modernes*, Revue ALSTHOM, Code APE 2811, France.)

to the axle, both wheels have the same angular velocity, 'ω', and consequently Expression 1.1 applies:

$$\omega r_2 > \omega r_o > \omega r_1 \Rightarrow V_2 > V > V \tag{1.1}$$

where:

V_1, V_2: Relative velocities of the two wheels.

The wheel running at a relatively higher speed (wheel 2) will overtake the other wheel (wheel 1), and due to the rigid wheel connection, this will cause a rotation of the axle with regard to the transversal axis $\bar{y}_o$.

Owing to the simultaneous forward rolling of the wheelset, the rolling radius r_1 of wheel 1 constantly increases while the rolling radius r_2 of wheel 2 constantly decreases. When $r_1 > r_2$, wheel 1 starts overtaking wheel 2 and the phenomenon is repeated.

This motion of the wheelset is known as 'hunting' oscillation.

In reality, the motion of a railway wheelset and, especially, of a complete vehicle (car body + bogies) is more complex. Owing to the simultaneous motion of the wheelset at a constant speed V, when the rolling direction of the wheels does not coincide with the vehicle displacement direction, friction forces are created on the wheel–rail contact surface (creep forces) and alter the kinematic behaviour described above, thereby giving the wheelset a dynamic behaviour (Esveld, 2001).

At very low speeds, this physical mechanism guarantees the stability of the vehicle (Pyrgidis, 1990; Moreau, 1992). Yet, at high speeds, high-amplitude oscillations are created and the motion becomes unstable. In such a case, vehicle stability is secured, thanks to the longitudinal elastic connection between the bogies and the wheelsets (primary suspension), which limits the amplitude of oscillations. Should the wheel transversal displacements exceed the anticipated flangeway clearance, the rolling of the wheels on the rails is secured by the presence of wheel flanges (Figure 1.12).

1.3.2.2 Running in curves

Let us examine the layout of Figure 1.21. Upon entering the curve, the wheelset is displaced by 'y' with regard to its outer face. Owing to the conic profile of the wheels, the initial rolling radius r_o of the two wheels changes into r_1 and r_2 for the outer wheel and the inner wheel, respectively.

The rolling radius of the outer wheel is larger. The inequity $r_1 > r_2$ applies, and by extension, $V_1 > V_2$ also applies.

Owing to the rigid connection of the wheels, the wheelset tends to rotate by itself towards the inner face of the curvature, displaced by 'y_o', seeking for a radial positioning inside the curve (the two wheels cover unequal paths).

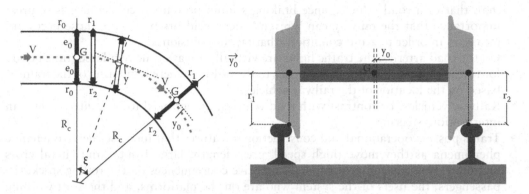

Figure 1.21 Movement of a single railway wheelset in a curvature of track.

As in the case of running on straight sections of track, when the wheelset transversal off-set 'y_0' exceeds the existing flangeway clearance 'σ', the running of the wheels on the rails is secured by the presence of flanges.

The motion described above concerns a single isolated wheelset. The bogie negotiation in curves is more complex, and the axle's positioning is affected by the motions of both the bogies and the car body. However, the wheelset inscription mechanism in curves remains the same.

Since the birth of the railway (1825), the system described above has represented the basic unit materialising the physical mechanism of guidance of the railway vehicles on a straight path and in curves. Unlike other means of transport, railway vehicles do not require human intervention (steering wheel operation) or complicated mechanisms.

At this point, the following classification of curves is proposed, based on the range of the horizontal curve radii values R_c:

• $R_c \geq 5{,}000$ m	Very large curve radii
• $2{,}000$ m $\leq R_c < 5{,}000$ m	Large curve radii
• 500 m $\leq R_c < 2{,}000$ m	Medium curve radii
• 250 m $\leq R_c < 500$ m	Small curve radii
• 100 m $\leq R_c < 250$ m	Very small curve radii
• 20 m $\leq R_c < 100$ m	Tramway network curve radii

1.3.3 Distinctive features of railway systems compared to road means of transport

Being a terrestrial-guided mass transportation system, the railway differs from the road means vis-à-vis its three constituents, namely, the railway infrastructure, the rolling stock, and the railway operation. Indicatively:

* The railway has only one degree of freedom. This single degree of freedom facilitates the automation of a range of operations such as driving, signalling, braking, and electrification. Conversely, unlike road vehicles, the railway cannot provide 'door to door' services (rigid system).
* Owing to the low adherence between wheel and rail (steel/steel contact) and the greater braking weight, the braking distance of a train is, for the same speed, much greater

than that of a road vehicle; since braking seldom prevents a collision, it is of great importance that the railway can 'prevent' such accidents by taking those necessary measures in order to avoid conditions that favour collisions.

- On the road arteries, the traffic lights are virtually always time regulated (time period for traffic signals). The opposite applies to the railway where regulating of the trains is based on the location of the railway vehicles.
- Railway vehicles, by contrast with road vehicles, do not need to be guided by human intervention (steering).
- Trains possess operational and constructional features that increase the aerodynamic phenomena as they move (high speed, great length, large frontal and lateral cross section). These phenomena may have negative consequences on the rolling stock, the passengers, the users of the system who are on the platforms, and the staff working near the track.

1.4 CLASSIFICATION OF RAILWAY SYSTEMS

Railway systems can be classified in many ways. This section defines the term 'speed' in railway engineering and attempts a classification of railway systems based on the speed, the functionality/provided services, the track gauge, and the traffic composition.

1.4.1 Speed in railway engineering: design and operational considerations

1.4.1.1 Definitions

The term 'speed' in a railway context may be defined in various ways, depending on the technical and/or operational context being considered. The following definitions are commonly used:

- *Track design speed* (V_d), which is defined as the speed for which the track alignment and corresponding railway infrastructure as a whole (superstructure, substructure, civil engineering structures, systems/premises) have been designed and constructed. This speed is not related to any operational or track capacity constraints and it is regarded as the maximum speed at which a train can safely and comfortably operate on a given track.
- *Permissible track speed* (V_{maxtr}), which is defined as the maximum speed that may be developed on a railway track section at the time a given rolling stock is commissioned. This speed is determined by the infrastructure manager of a railway network taking into consideration the track ride quality as well as other performance aspects at the moment. The permissible track speed is directly related to the maintenance level of the track.
- *Maximum running speed* (V_{max}), which is defined as the maximum speed developed by a particular train type on a given line, while performing a scheduled route. This speed may either refer to a small segment of the line, or it may develop at the biggest part of the route.
- *Passage speed* (V_p), which is defined as the constant speed with which a train passes from a particular, characteristic segment of the line which is of small length (e.g., passing through a tunnel and passing through stations).

- *Instant speed* (V_t), which is defined as the speed with which a train passes from a specific kilometric point at a specific time.
- *Commercial speed* (V_c), which is defined as the ratio of the length of a railway route (usually between the two terminals or between two important intermediate stations) to the time it takes to cover it, including halt times at all intermediate stations and delays. Commercial speed always refers to a particular type of train and a given route.
- *Average running speed* (V_{ar}), which is defined as the quotient of the length of a line segment (usually between two successive stations), to the time taken to pass this segment, considering normal traffic conditions (e.g., no unforeseen delays). The average running speed always refers to a particular train type and a given line segment.
- *Rolling stock design speed* (V_{rs}), which is defined as the maximum speed that, according to the manufacturer, can be developed by a particular type of locomotive, or with which a trailer vehicle can move, or, finally, the maximum speed that can be developed by multiple units of a given formation taking into consideration the traction system (diesel or electric power), the hauled weight, and the track geometry alignment design and considering the track to be of very good ride quality.

The mathematical Expression 1.2 generally applies:

$$V_{ar} \leq V_{maxtr} \leq V_d \qquad (1.2)$$

As regards the quality of the railway infrastructure, it is secured when:

- V_{maxtr} at individual track segments coincides with the track design speed V_d which, however, corresponds to a particular traction system.
- The average running speed V_{ar} is nearly equal to V_{maxtr} (in case this refers to a segment between two successive stations, due to the train's acceleration and deceleration while starting and stopping, these two speed values cannot coincide and, in this occasion, the value of V_{maxtr} is greater).

As regards the combination of the track and rolling stock, the design speed of the rolling stock V_{rs} must be slightly greater than V_d or at least equal to V_d.

Finally, regarding the level of service and, more specifically, the run times, the maximum running speed V_{max} must be achieved for the biggest part of the route.

Figure 1.22 illustrates the graphical representation of the speeds V_d, V_{maxtr}, V_{ar}, and V_c, for a route AB with an intermediate stop, considering that the speeds V_d and V_{maxtr} are the same for the entire route length S.

It should be noted, at this point, that:

- The 'track design speed' is a structural parameter of the system. It concerns designers and construction contractors of railway systems and subsequently railway infrastructure managers. The track design speed should be the same on all track sections of a railway line.
- 'Commercial speed' is an operational parameter of the system. It is directly associated with run time and concerns the users of railway systems and subsequently railway system operators.

In cases of construction of a new railway line or upgrading of an existing connection, an issue, which usually is of concern to railway companies, is the track design speed that should be selected.

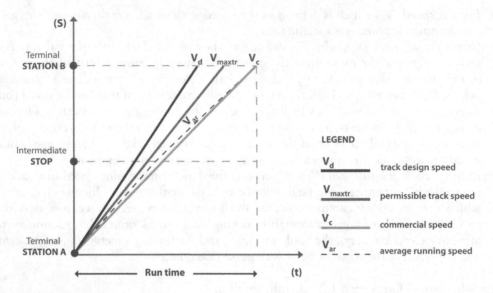

Figure 1.22 Speed in railway engineering.

An approach that is usually adopted indicates that the selected track design speed V_d must ensure a desired run time t (target time) between the two terminal stations of the connection or, preferably, a desired total travel time (t_{tot}) between two specific critical points in the wider areas of both ends of the connection (door-to-door service).

On the other hand, the target time should be set directly based on the run/travel time that is required by the existing competitive transportation means (e.g., bus, private car, and aeroplane). Under such a condition, the choice of a target time requires a decision of the responsible authority for the construction of the project (e.g., Department of Transportation).

The following steps are proposed in order to calculate the track design speed V_d:

- The definition of the desired run time t (target time) for the connection length S (taking into account the total travel time and particularly the travel time that is required by competitive transportation means).
- The definition of the number of intermediate stops n_s and the dwell time t_s at each stop.
- The calculation of the average running speed V_{ar} of trains that is required in order to achieve the target time via the use of Equation 1.3.

$$t = \frac{S}{V_{ar}} + n_s \cdot t_s \tag{1.3}$$

The initial track design speed V_d is calculated as a percentage of V_{ar}, as follows:

$$V_d = V_{ar} \cdot k_v \tag{1.4}$$

The value of the track design speed coefficient k_v, ranges from 1.15 to 1.25, and is the result of trials within simulation models applied on existing connections.

The final value of V_d is calculated by rounding up the result of Equation 1.4 and by selecting values of V_d usually equal to 120, 140, 150, 160, 200, 250, 300, and 350 km/h.

1.4.1.2 Case study

Two cities are planned to be connected via a double, standard gauge, railway track. The length of the connection is to be S = 800 km. Intermediate stops are provisioned every 200 km with a waiting time at each stop of t_s = 1.5 min.

Currently, air travel service is offered for the same connection with a total travel time (door-to-door) of t_{tot} = 6 h.

The desired total travel time via rail (door-to-door) should be equal to that of air travel. Transportation to the train origin station requires 30 min, where waiting time at the platform and boarding time amount to another 20 min. Similarly, alighting the train and exiting the destination station amount to 20 min, while transportation from the train destination station to the final destination requires another 35 min.

In order to select an appropriate track design speed V_d for the new line, the following calculations are required:

$$n_s = \frac{800\,km}{200\,km} - 1 = 3\,\text{intermediate stops}$$

$$t_{tot} = 6h = t + 30\,min + 20\,min + 20\,min + 35\,min$$

$$= \frac{S}{V_{ar}} + n_s \cdot t_s + 105\,min$$

$$6h = 6 \times 60\ min = \frac{800\,km}{V_{ar}} + 3 \times 1.5\,min + 105\,min$$

$$V_{ar} = 191.616\,km/h$$

It is assumed that k_v = 1.20.

$$V_d = V_{ar} \cdot k_v = 191.616\,km/h \times 1.20 = 229.99\,km/h$$

Finally, a track design speed V_d = 250 km/h is selected.

1.4.2 Classification of railway systems based on speed

Based on its speed a railway system may be placed into one of the following three categories:

- Conventional-speed rail.
- High-speed rail.
- Very high-speed rail or super-fast rail.

As of yet, there exists no universally accepted definition of distinguishing between conventional, high, and very high-speed rail. On the contrary, various approaches may be found in the relevant literature as proposed by different railway organisations or researchers (EC, 1996; Peterman et al., 2009; Pyrgidis and Demiridis, 2012; UIC, 2014).

In this section, after presenting the existing definitions, a new definition is proposed (Pyrgidis et al., 2017; Pyrgidis, 2018, Pyrgidis et al., 2020):

Existing definitions (indicatively):

Definition 1: The speed of V = 200 km/h was initially established as a limit of distinction between a train running at 'conventional' speeds and a train running at 'high' speeds. The main reasons for the adoption of the above limit were the following:
- In most upgraded tracks, the curvature radii in the horizontal alignment of the layout had been selected for maximum passage speed V_{pmax} = 200 km/h.
- Just over this speed, the impacts from the geometric track defects are intensified, while some of the train functions become troublesome and need special handling (e.g. an increase of the braking distance and the aerodynamic resistances of the trains and the train driver's inability to identify the indications of the track-side signalling).

Definition 2: In the Technical Specifications for Interoperability (TSIs), the Trans-European High-Speed Network lines are classified in the following three categories (EC, 1996; UIC, 2014):
- *Category I:* New tracks which are specially constructed for high speeds and are suitably equipped so that running speeds V_{max} ≥ 250 km/h can be reached.
   In some sections of these tracks, where, for technical reasons, the maximum speed provided for the interoperable trains may not be reached, it is possible for a lower permissible track speed to be imposed.
- *Category II:* Existing tracks which are specially upgraded for high speeds and are suitably equipped so that running speeds in the region of V_{max} = 200 km/h can be reached.
- *Category III:* Tracks that are specially upgraded for high speeds (V_{max} = 200 km/h), with special specifications, due to the limitations/enforcements imposed by the landscape or the compulsory passage through the urban environment, resulting in speed adjustment, depending on the case.

Definition 3: In the Technical Specifications for Interoperability (TSIs) for the 'Infrastructure' subsystem, the performance levels for 'TSI categories of line' (passenger and freight) are defined. For different types of passenger traffic, this document defines railway line specifications as regards the parameters loading gauge, axle load, line speed, and usable length of station platforms, as shown in Table 1.1.

Definition 4: In the United States, the Department of Transportation (DOT) and the Federal Railroad Administration (FRA) have defined high-speed rail as a service that is competitive to air and/or road travel for routes ranging from 160 km to 800 km (100–500 miles). As stated by the FRA, this definition arises from the recognition that when choosing a transport mode, passengers are more interested in the overall travel time than in the running speed (run time) of a particular transport mode (Peterman et al., 2009).

Table 1.1 Performance parameters for different codes of traffic (Commission Regulation, 2014)

Traffic code	Type of loading gauge	Axle load (t)	Line speed (km/h)	Usable platform length (m)
P1	GC	17	250–350	400
P2	GB	20	200–250	200–400
P3	DE3	22.5	120–200	200–400
P4	GB	22.5	120–200	200–400
P5	GA	20	80–120	50–200
P6	GI	12	n.a.	n.a.
P1520	S	22.5	80–160	35–400
P1600	IRL1	22.5	80–160	75–240

Proposed definition:

A specific procedure is proposed for making such a distinction and concerns passenger intercity and regional trains (definition of and further information on these systems are provided in Section 1.4.3 as well as in Chapters 8 and 11) (Pyrgidis et al., 2020).

The distinction is based on the characteristics of the three constituents of the railway system (infrastructure, rolling stock, operation). Initially, based on speed, the railway infrastructure of, the rolling stock of, and the services provided by passenger intercity and regional railway systems are classified into the following three categories:

- Conventional-speed.
- High-speed.
- Very high-speed or super-fast.

The same classification can be applied to the system as a whole.

Subsequently, separate criteria for each of the three constituents of a railway system are formulated. They correspond and characterise to an extent the needs of the actors that manage these constituents, namely, the railway infrastructure managers (infrastructure) and the railway system operators (rolling stock, operation).

It is very important that all actors that are involved in the railway domain (railway infrastructure managers, railway operators, regulating organisations, railway industry, financing organisations, travel agents, users, etc.) 'speak the same language' when referring to such systems, meaning that they all understand the same structural and operational characteristics.

- *Definition as regards to the railway infrastructure, and in a broader sense, the railway infrastructure managers (and possibly other actors such as financing organisations)*

The rail infrastructure manager is interested in whether a track/track section can be characterised as a 'high-speed track/track section'. In this regard, a single criterion is proposed, specifically, the value of the maximum permissible track speed V_{maxtr} along a specific track or track section.

If the infrastructure is maintained at optimal conditions, then this value will coincide with the one of track design speed ($V_{maxtr} = V_d$).

For this distinction of the infrastructure based on speed, the following V_{maxtr} (V_d) limits are adopted:

1. Conventional-speed tracks: V_{maxtr} (V_d) < 200 km/h   (1.5)
2. High-speed tracks: 200 km/h $\leq V_{maxtr}$ (V_d) < 250 km/h   (1.6)
3. Very high-speed tracks: V_{maxtr} (V_d) $\geq$ 250 km/h   (1.7)

- *Definition as regards to the operation, and in a broader sense, the system operator (as well as other actors such as travel agents and the users of the system)*

The railway operator is interested in whether a train service that is offered on a specific route can be characterised as a 'high-speed service', which depends on whether or not, on a specific route, the rail transport mode offers competitive travel times (door-to-door services), as compared to other transport modes. In this regard:

(a) For conventional-speed services, a single criterion is proposed. This criterion may be expressed by either the commercial speed V_c or by the average running speed V_{ar}. For this distinction of railway services based on speed, the following V_c (V_{ar}) limit is adopted:

$$\text{Conventional-speed services } V_c \, (V_{ar}) < 150 \text{ km/h} \qquad (1.8)$$

(b) For high-speed and very high-speed services, three criteria, which must be concurrently fulfilled, are proposed. Specifically, the length of the route S, the maximum running speed V_{max}, and the commercial V_c or average running speed V_{ar}. This distinction essentially reflects whether a railway service offered along a specific route may be characterised as a 'high/very high-speed service'. This characterisation is based, to a large extent, on whether, for this specific route, railways offer run times and, more importantly, door-to-door travel times that are competitive to other modes of transport, namely, private cars and aeroplanes.

The limit values vary for high- and very high-speed services. More specifically, for high-speed services, two cases may be identified:

The first case refers to a service that covers the whole length S of a railway line (whether or not intermediate stops exist). In this case, three criteria, with the following limit values, are adopted:

1. $V_{max} \geq 200$ km/h
2. $S \geq 150$ km $\qquad\qquad\qquad\qquad\qquad\qquad\qquad\qquad\qquad (1.9)$
3. $V_c \geq 150$ km/h

Secondly, there is the case of a service that runs over only a part L_{st} of a railway line of length S including intermediate stops. In this case, the adopted criteria and limit values are as follows:

1. $V_{max} \geq 200$ km/h
2. $L_{st} \geq 150$ km $\qquad\qquad\qquad\qquad\qquad\qquad\qquad\qquad\qquad (1.10)$
3. $V_{ar} \geq 150$ km/h

High-speed railway services compete with air transportation and, mostly, with relatively short-distance travel via private cars.

On the other hand, for very high-speed services, the following limit values apply:

1. $V_{max} \geq 250$ km/h
2. $S \geq 400$ km $\qquad\qquad\qquad\qquad\qquad\qquad\qquad\qquad\qquad (1.11)$
3. $V_c \geq 180 \, (200)$ km/h

Such services typically involve long distances and compete primarily with air transportation.

The scientific/methodological background for extracting Equations 1.5–1.11 and the associating limit values are taken from and presented in relevant published articles (Pyrgidis et al., 2017; Pyrgidis, 2018, Pyrgidis et al., 2020).

- *Definition as regards to the rolling stock, and in a broader sense, the railway industry (as well as other actors such as the railway operators)*

A single criterion is proposed. This criterion is the rolling stock design speed V_{rs}. The following limit values are adopted (Pyrgidis et al., 2020):

1. Conventional-speed trains: $V_{rs} < 200$ km/h $\qquad\qquad\qquad\qquad\qquad (1.12)$
2. High-speed trains (fast trains): 200 km/h $\leq V_{rs} \leq 250$ km/h $\qquad (1.13)$
3. Very high-speed trains (super-fast trains): $V_{rs} > 250$ km/h $\qquad\quad (1.14)$

1.4.3 Classification of railway systems based on functionality/provided services

In general, the railway systems fall under the category of terrestrial-guided transport modes (moving along a dedicated corridor – 'fixed permanent way'). Depending on the permanent way they use, these guided transport modes are distinguished into railway (Figure 1.23), aerial (i.e. aerial cable car), road (e.g. BRT, trolley), and magnetic levitation systems.

The term 'railway system' may include all transport means whose rolling system involves at least one iron component (steel wheels on rails or rubber-tyred wheels on a steel guideway). In this context, this book examines only those means of transport that have this particular characteristic in common.

Railway systems transport passengers (passenger railway systems) or goods (freight trains).

On the basis of the geographic/urban environment in which they operate, and generally on their functionality/provided services, passenger railway systems are distinguished into (Figures 1.24–1.37):

- Intercity systems.
- Suburban/regional systems.
- Urban systems.
- Steep gradient railway systems.

The *intercity railway* serves trips greater than 150 km and usually links major urban centres. Terminal stations are usually located in large urban centres. It includes high-speed trains, very high-speed trains (Figure 1.24), and conventional-speed trains (Figure 1.25).

The suburban/regional railway (Figures 1.26 and 1.27) is a railway means of transport with characteristics adapted to commuter services within the limits of the influence area of major urban areas (suburbs and satellite regional centres). Its range can exceed 100 km and may even reach up to 150 km. The nomenclature varies. The length of the route and the frequency of the service are usually used to distinguish three subsystems. When covering distances of 10–40 km, it is designated as a suburban railway. The track is usually electrified, and the operation is characterised by very high-frequency services (usually trains run every 15–30 min). When covering distances of 30–50 km, it is designated as commuter or periurban or urban rail. The track is usually electrified and the operation is defined by relatively high-frequency services (usually trains run every 20–60 min). Finally, when it covers greater lengths (50–150 km), it is designated as regional railway (usually trains run every 1–3 h). One of the two terminal stations is usually located in a small or medium sized urban centre.

Urban railway systems include:

- The metro (Figures 1.28 and 1.29).
- The light metro (Figure 1.30).
- The tramway (Figure 1.31).
- The monorail (Figures 1.32 and 1.33).
- The driverless railway systems of low/medium transport capacity (Figures 1.34 and 1.35).

Out of the above systems, the first three serve trips that are performed exclusively within a city (urban transport), whereas the latter two are mainly used for trips with a different character. More specifically:

- In essence, metros move underground and are characterised by great transport capacity and high implementation cost.

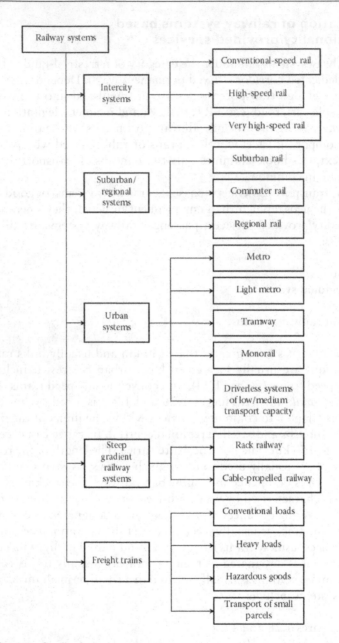

Figure 1.23 Classification of railway transport systems.

- Trams are integrated into the road arteries of the city, using a specialised track superstructure.
- The light metro is, based on its construction and operation features, a system somewhere between the tram and the metro. Light metro and tram belong to the so-called 'Light Rail Transport Systems' (Light Rail Vehicles – LRVs or Light Rail Transport systems – LRTs).
- The monorail moves using a system of rubber tyres (this is the most common type) on an elevated guideway comprising a single beam made of concrete or steel. It serves short distances mainly within the urban environment and is particularly suitable for

Figure 1.24 High-speed intercity railway (THALYS). (Photo: A. Klonos.)

Figure 1.25 Conventional-speed intercity train, Helsinki, Finland (hauled electric passenger train). (Photo: A. Klonos.)

trips within recreation areas (thematic parks, zoo parks, etc.), as well as for connecting the city centre to the airport. In recent years, monorails are increasingly used as a means to circumvent land scarcity issues in congested cities.

- Finally, driverless systems of low/medium transport capacity move on an exclusive transport corridor using either single vehicles with a transport capacity of 3–25 persons (Figure 1.34) or trains of low and medium transport capacity. They are either cable-propelled (Figure 1.35) or self-propelled electric systems, and they belong in

Figure 1.26 Suburban railway, Berlin, Germany (double-deck electric railcar). (Photo: A. Klonos.)

Figure 1.27 Suburban railway, Greece (diesel railcar). (Photo: A. Klonos.)

the category of Automated People Movers (APMs). In the urban environment, such systems may serve as feeders for heavy rail transport systems. However, they usually operate for the service of trips within airports, large hotels, casinos, congress centres and health centres, educational institutions, and big companies' premises.

The *steep gradient railway* serves small-distance connections with an important difference of altitude between the two edges of the railway line. They are separated into rack railways (Figure 1.36) and cable-propelled railway systems (Figure 1.37).

Figure 1.28 Metro (with driver), Athens, Greece. (Photo: A. Klonos.)

Figure 1.29 Metro (driverless), Paris, France. (Photo: A. Klonos.)

The rack (or cog) railway is mainly used to approach remote mountain developments and tourist resorts on tracks with longitudinal slopes usually exceeding 50–70‰. Apart from the two classical rails, the cog railway superstructure includes a special toothed rack rail mounted between the two conventional running rails. The wheelsets of the power vehicles are fitted with one or more cog wheels that bind in the rack rail. The required supplementary traction force is achieved through the engagement of the rack rail teeth with the locomotive pinion teeth.

Cable-propelled railway systems for steep gradients use vehicles that are hauled via cables. On the basis of the technique that is used for their traction, they are divided into funicular

Figure 1.30 Driverless light metro, Copenhagen, Denmark. (Photo: A. Panagiotopoulos.)

Figure 1.31 Modern tramway, Athens, Greece. (Photo: A. Klonos.)

(non-detachable, cable-propelled vehicles for steep gradients), cable railway (detachable, cable-propelled vehicles for steep gradients), and inclined elevator.

The funicular (Figure 1.37) operates using two vehicles that move on rails with the aid of a cable; one of the vehicles is ascending while the other one is descending. The cable rolls over pulleys which are mounted on the track superstructure. The vehicles are permanently connected to both ends of the cable and start and stop simultaneously. The ascending vehicle

Figure 1.32 Monorail (suspended system), Memphis, Tennessee, USA. (Adapted from Nightryder84 on the English Wikipedia, 2005, available online at: https://commons.wikimedia.org/wiki/File:Memph is_front_view.jpg.)

Figure 1.33 Monorail (straddled system), Sydney, Australia. (Adapted from Hpeterswald, 2013, available online at: https://commons.wikimedia.org/wiki/File:Metro_Monorail_Pitt_Street.jpg [accessed 8 August 2015].)

uses the gravitational force of the descending one (counterbalance system). The system usually connects distances of less than 5 km, with constant longitudinal gradients of around 300–500‰ (max value recorded in practice is 1,100‰, see Chapter 10).

The cable railway also uses vehicles that run on conventional rails, using a cable which moves constantly and at a constant speed. The difference between the two systems lies in

Figure 1.34 Driverless self-propelled railway systems of low/medium transport capacity. (From POSCO ICT Co., Ltd., 2015.)

Figure 1.35 Driverless cable-propelled railway systems of medium transport capacity, Birmingham airport system, UK. (Adapted from Doppelmayr Cable Car, 2008, available online at: https://en.wikipedia.org/wiki/Cable Liner [accessed 7 August 2015].)

the fact that for the cable railway, the vehicles are not permanently connected to the cable. The vehicles can stop independently, disconnecting from the cable, and may start again, reconnecting to the cable. This process may occur either automatically or manually (San Francisco system, USA).

The inclined elevator or inclined lift or inclinator is a variant of the funicular. It operates using a single vehicle which is either winched up at the station on the top of the inclined

Figure 1.36 Cog railway, Arth Goldau, Switzerland. (Photo: A. Klonos.)

section where the cable is wound on a winch drum or the weight of the single vehicle is balanced by a counterweight so that the system operates as a funicular. It usually connects distances of less than 1.5 km, with constant longitudinal gradients of around 500–700‰. This system can serve extremely steep gradients.

Freight trains, as seen in Figures 1.9 and 1.23, are distinguished into trains transporting:

- Conventional loads (axle load Q ≤ 25).
- Heavy loads (axle load Q > 25).
- Dangerous (hazardous) goods.
- Small parcels.

1.4.4 Classification of railway systems based on track gauge

The track gauge (2e) is the distance between the inner edges of the heads of the two rails measured at 14–16 mm below the rolling surface plane (Figure 1.38).

The track gauge is not the same in all countries. In many countries it varies from region to region.

On the basis of the gauge, railway lines are divided into five categories (Esveld, 2001):

- *Standard tracks or standard track gauge*: This category mainly comprises the 1,435 mm gauge. This distance (4 ft and 8 in.) was established by the British engineer George Stephenson (1781–1848).

Figure 1.37 Funicular, Graz, Austria. (Adapted from Riemer, J.F., 2007, available online at: http://en.wikip edia.org/wiki/Schlossbergbahn_(Graz) [accessed 7 August 2015].)

- *Broad tracks or broad track gauge*: This category mainly comprises the following gauges: 1,520/1,524 mm (former Soviet countries), 1,600 mm (Irish gauge), 1,665 mm, and 1,667 mm.
- *Metre tracks or metre track gauge*: This category mainly comprises the following gauges: 914 mm, 950 mm, 1,000 mm (metre), 1,050 mm, and 1,067 mm (Cape gauge).
- *Narrow tracks or narrow track gauge*: This category comprises gauges from 600 mm to 900 mm and mainly 600 mm (Decauville), 700 mm, 750 mm, and 760 mm (Bosnian gauge). These gauges are usually used for secondary lines (e.g., industrial areas, factories, and mine service lines). Metre and narrow tracks are also termed as 'small gauge tracks'.
- *Mixed gauge tracks*: This category comprises tracks on which trains of different gauge category may run simultaneously.

Regardless of the track gauge category, the distance between the rails remains constant throughout the network length apart from the curved alignment sections with small curvature radii ($R_c < 150$–200 m), where, in many cases, a widening of the track gauge is permitted to facilitate the inscription of vehicle wheelsets (gauge widening) (Pyrgidis, 2005).

Figure 1.38 illustrates the layout of the rails and the wheels that roll on them, for the case of standard track gauge (indicatively).

Mathematical Equation 1.15 applies:

$$2\sigma = 2e - (2d_a + 2f_d) \tag{1.15}$$

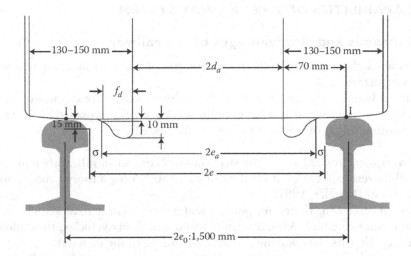

$2e$:	Track gauge	$=1{,}432{-}1{,}470$ (mm) $+0$
$2e_a$:	Outer flange edge-to-edge distance (flange gauge)	$=1{,}426$ (mm) -16 $+3$
$2d_a$:	Back-to-back wheel distance (inside gauge)	$=1{,}360$ (mm) -3
$2e_0$:	Theoretical distance between the running surfaces of the right and the left wheel when centred	$\approx 1{,}500$ (mm)
f_d:	Flange thickness	$= 33{-}25$ (mm) (wear limit $= 25$ mm)
σ:	Flange way clearance	$= (2e - 2e_a)/2$ (mm)
2σ:	Total flange way clearance	$= 2e - 2e_a$ (mm)

Figure 1.38 Railway wheels on rails – track of standard gauge – geometrical and constructional dimensions (track-centred wheelset). (Adapted from Alias, J., 1977, *La voie ferrée*, Eyrolles, Paris.)

1.4.5 Classification of railway systems based on traffic composition

The number and percentage of each category of trains (railway traffic composition) using a specific railway infrastructure can be directly used in classifying railway systems.

On the basis of their traffic composition, railway network/corridors can be classified into five categories as follows (Christogiannis, 2012; Christogiannis and Pyrgidis, 2013):

1. Exclusively used by freight trains (freight-dedicated network/corridor).
2. Mainly used by freight trains.
3. Network/corridor with mixed traffic operation.
4. Mainly used by passenger trains.
5. Exclusively used by passenger trains (passenger-dedicated network/corridor).

The exact limits that are used in order to distinguish networks/corridors based on their traffic composition are set in (Christogiannis and Pyrgidis, 2013).

The term *mixed traffic operation* commonly describes the routing of both freight and passenger trains on the same track.

The term *exclusively used or dedicated* describes the exclusive routing of either passenger or freight trains on the track.

1.5 THE CAPABILITIES OF THE RAILWAY SYSTEM

1.5.1 Advantages and disadvantages of the railway

Table 1.2 presents the advantages and disadvantages of the railway compared with other means of transportation.

Some of the advantages/disadvantages of the railway system are discussed below, while Section 1.5.3 presents a comparison between the level of service provided by the railway and that of other competing transportation systems. Specifically:

- *High transportation capacity*: The steel-on-steel contact significantly reduces the specific rolling resistance (15 N/t for the railway, 150 N/t for a tourist coach, 300 N/t for a road truck) (Metzler, 1981).

  A locomotive can, therefore, pull a greater load than a road vehicle by applying the same tractive effort. Also, a train is formed of many vehicles, thus allowing it to increase or alter its transporting capacity according to the demand.

  For example, in order to transport 700 passengers, a train of length between 280 m and 300 m is needed. For the same number of passengers to be transported by road, it would require:
  - Fifteen coaches with 44 seats covering a length of road of 1,050 m.
  - Hundred and seventy private cars with 4 seats in each, covering a road distance of 11,900 m including a necessary safety distance.
- *High-speed capability*: Nowadays, technical advances in the areas of rolling stock and the track allow a train to move safely on a straight track of good rolling conditions at running speeds $V_{max} > 300$ km/h ($V_{max} = 350$ km/h in China). China holds the record for the fastest average running speed between two successive intermediate stops at $V_{ar} = 317.7$ km/h (Beijing Nan–Nanjing Nan, S = 1,021.9 km) and for the fastest commercial speed at $V_c = 304.1$ km/h (Beijing Nan–Shanghai, S = 1,307.6 km, one intermediate stop) based on 2019 data (Hartill, 2019). France retains the world speed record with a test train (574.8 km/h) recorded on 3 April 2007.

  With regard to passenger transportation, trains in many countries run on conventional lines at speeds in excess of 160 km/h, while there are 19 countries globally operating very high-speed lines ($V_{dmax} \geq 250$ km/h) (Pyrgidis et al., 2020) (see Chapter 12).

  As far as freight transport is concerned, numerous countries operate trains moving at speeds between 100 km/h and 120 km/h. Table 1.3 shows values for characteristic speeds which apply for the railway globally (at the end of 2019).
- *Rail services regardless of weather conditions* (Figure 1.39): Safety in train operation and train movement is generally not affected by extreme weather conditions (fog, snow, ice, strong winds), and cancellation of scheduled services due to weather conditions is

Table 1.2 Advantages and disadvantages of the railway

Advantages	Disadvantages
• High transportation capacity • High-speed capability • Travel safety • Rail services regardless of weather conditions (regularity of services) • Environmentally friendly transport • One degree of freedom (automation of many operations) • Passenger comfort/relaxed state of mind • Small land take (right-of-way)	• Increased requirements in track geometry design (horizontal, longitudinal alignment) • Low wheel–rail adhesion coefficient • One degree of freedom (no door-to-door services) • Hard (noisy) rolling • Low network density

Table 1.3 Railways – characteristic speeds (2019 data)

Characteristic speeds	Maximum value (km/h)
Rolling stock design speed (V_{rs})	400
Average running speed between successive stops (V_{ar})	317.7 (China)
Commercial speed (V_c)	304.1 (China)
Running speed (V_{max})	320 (Europe, Japan), 350 (China)
Track design speed (V_d)	400
Speed record	574.8 (France)

Figure 1.39 Rail services regardless of weather conditions. (Photo: A. Klonos.)

seldom needed (see Chapter 17). Given this fact, the railway ensures regularity in its services, a quality of great importance to its users.

- *Passenger comfort/relaxed state of mind*: Provided that it offers a satisfactory level of service, the railway is generally viewed in a positive light in comparison with road and air transport, as the passenger:
  - Has greater comfort in terms of space when the train is moving; he/she can move about more, visit the restaurant car, work on the train. It should be noted that for very high-speed trains (i.e., trains running with $V \geq 250$ km/h), there is no obligation for passengers to wear a seatbelt.[4]
  - Can enjoy the view throughout the whole journey.
  - Is transported 'on terra firma' without having to drive him/herself, which psychologically is more comforting.
- *Occupies a small space*: A double-standard gauge track occupies a space of about 1/3 of that of a two-way highway with three lanes per direction (Figures 1.40 and 1.41). Indicatively, for 1 km of high-speed railway line, 3.2 hectares of land are needed, while for the same amount of highway length, 9.3 hectares of land are needed.

---

[4] http://www.railwaygazette.com/news/single-view/view/study says no to seat belts on trains.html

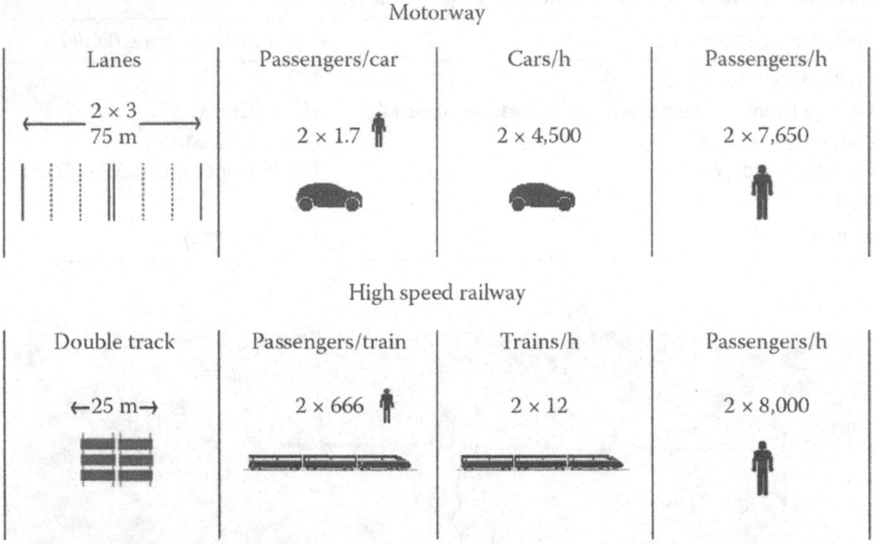

Motorway

Lanes	Passengers/car	Cars/h	Passengers/h
←— 2 × 3 75 m —→	2 × 1.7	2 × 4,500	2 × 7,650

High speed railway

Double track	Passengers/train	Trains/h	Passengers/h
←25 m→	2 × 666	2 × 12	2 × 8,000

Figure 1.40 Comparison between occupied space (right-of-way) of a double-standard gauge track and a highway for approximately the same transport capacity. (Adapted from UIC, 2010, High speed rail-fast track to sustainable mobility, Paris, available online at: www.uic.org/download.php/… /521E.pdf [accessed 20 March 2015].)

Figure 1.41 Right-of-way for a double high-speed railway track and for a highway (2 × 3), Mundener bridge, Koln-Frankfurt, Germany. (Photo: A. Klonos.)

- *Increased requirements in track geometry alignment*: The design of a railway line is more demanding both in terms of the horizontal alignment and in terms of its longitudinal alignment compared with that of a road. Regarding its horizontal plane, the curve radii for the intercity and suburban/regional railway must be greater than $R_c \geq$ 250–300 m in 'open' track sections (outside the area of stations).

In terms of its longitudinal alignment, the effective operation of a railway network sets the gradients for the intercity and suburban/regional railway at i_{max} = 3–4% with usual gradients at less than 2–2.5%. In the case of roadworks, the corresponding values lie between 8% and 10%.

Table 1.4 shows the characteristic gradient values for various means of transport and networks.

* *Low wheel–rail adhesion coefficient*: In railways, the contact surface between the rail and the wheel features a small adhesion coefficient due to the nature of the materials in contact (steel on steel). In road transport, this coefficient is approximately three times greater (Metzler, 1981).

The small adhesion coefficient acts negatively on two basic operations, namely, braking and starting the train. The greatest braking distance required to stop the train automatically sets a maximum speed limit as well as a maximum longitudinal gradient for the railway.

Furthermore, the wear on the wheel–rail contact surface created by the friction between the wheel and the brakes is a major financial burden for the maintenance and operation of a railway network.

As a result of the lower adhesion coefficient, the necessary braking length for the same speed and weight is greater for a train than it is for a private car.

Table 1.5 presents the braking distance for various means of transport and transport conditions. Table 1.6 presents the proportional relations.

Table 1.4 Characteristic values of longitudinal gradients for different means of transport and network cases

Means of transport/network	Longitudinal gradient (%)
Road transport	8 (A road category, V = 60 km/h)
	15–20 (maximum values)
Cog railway	5–48
Cable-propelled systems (funicular)	9–110
Tramway	7–8 (maximum values)
Metro	5 (maximum value)
Monorail	10 (maximum value, 20% gradient)
High and very high-speed network	3.5–4.0 (maximum values)
Conventional-speed intercity/suburban/ regional train	3–3.5 (maximum values)

Table 1.5 Braking distance for different means of transport and transport conditions

Transport means	Braking distance
Boeing 747, landing speed V = 200 km/h	1,500 m
Road vehicle, V = 120 km/h	
Dry road surface	95 m
Wet road surface	142.5 m
Lorry, V = 80 km/h	
Dry road surface	60 m
Wet road surface	90 m
Freight train, V = 80 km/h	700 m (emergency brake)
TGV (PSE), V = 270 km/h	3,000 m (emergency brake)
TGV – A, V = 300 km/h	3,200–3,500 m (emergency brake)

Table 1.6 Comparative braking distances among different modes

	Aeroplane/very high-speed train	Road vehicle/conventional-speed train	Road truck/freight train
Braking distance ratios	1:2	1:10	1:10

- *Hard (noisy) rolling*: The hard steel-on-steel rolling, which characterises the wheel–rail system, increases rolling noise and vibrations, resulting in the requirement for mitigation measures, regarding both the source of the noise (i.e., rolling stock and track) and the receptor.
- *Low network density*: It is possible at specific points of the railway network to converge, cross, split, and join tracks. This is achieved via the use of various switches and crossings configurations provided as part of the superstructure along the section in question.

    However, it is both technically difficult (if not impossible) and economically unprofitable to develop a rail network that has the same level of density as a road network.

1.5.2 Comparison of the characteristics of railway systems

Table 1.7 presents the basic technical and operational characteristics of passenger railway transportation systems.

1.5.3 Comparison of the capabilities of different transportation systems

The advantages and disadvantages of each transport system are usually compared by quantifying the parameters of one system and contrasting them with the corresponding ones of the other modes. The difficulties inherent to such an approach are numerous, since (a) one must compare competitive modes (with similar functionality) in order for the results to be comparable, (b) in the literature, one comes across statistical data comparing the transport systems at various levels with regard to the geographical presence of the means at a global level, at a continental level, within a country or a group of countries (e.g., EU member states), and so on, (c) for the same evaluation indicator, there is a wide variation in the values compared depending on the literature; this shortcoming, which is particularly strong, requires cross-checking data from numerous and various sources in order for the final results to be as reliable as possible.

Within this framework and in the next sections, the two cases below are compared:

Case 1: Long-distance trips (S = 500–1,500 km). Aeroplanes and high-speed trains are compared.
Case 2: Urban trips. Metro, tram, urban bus, and private car are compared.

The comparison does not include parameters related to the natural environment, which are comprehensively examined in Chapter 19.

1.5.3.1 Comparison of air and high-speed train transport

In order to compare high-speed trains with aeroplanes, eight specific connections in Europe served by both means of transportation (Rome–Naples, Rome–Florence, Madrid–Barcelona,

Table 1.7 Main technical and operational characteristics of passenger railway transportation systems

Technical and operational characteristics	High-speed intercity rail	Conventional-speed intercity rail	Suburban/urban/regional rail	Metro	Tramway	Monorail	Cog railway (pure rack system)	Funicular
Route length (km)	> 150 (300–400 – very high-speed rail)	> 150	10–40/30–50/50–150	10–40	5–20	1.5–12 (much longer if urban service)	4–20 (max 19.11, usually 4.5–6)	Usually < 1.2 (max 4.827)
Track gauge	Normal/Broad	All gauges	All gauges	Normal/Metric	Normal/Metric	Beam of 2.30–3.00 m width	All gauges (usually metric)	Various gauges (usually metric)
Number of tracks	Double	Double Single	Double (suburban/urban rail)	Double	Double	Usually two beams at sufficient distance	Usually single	Single with passing loop. Double with or without passing loop
Traction system	Electric	Electric Diesel	Usually electric (suburban/urban rail)	Electric	Electric (overhead wires, through the ground, with energy storage systems)	Electric motors setting a system of rubber-tyred wheels to roll	Usually electric, diesel, bio-diesel, and steam	Via pulled electric cables placed on the track superstructure (non-detachable vehicles – counterbalance system)
Distance between successive stops	150–250 km	50–150 km	Normally 2,000–3,000 m (suburban) 10–30 km (regional)	500–1,000 m	400–600 m	800–1,500 m	Rarely, 1–2 intermediate stations	Without intermediate stops. 1–2 stops located anti-symmetrically for the two running vehicles
Commercial speed (km/h)	≥ 150	< 150	45–65/50–70/ < 100	30–40	15–25	15–40	7.5–20	Usually < 20 (5.5–50)
Longitudinal gradient (%)	0–4	0–3.5	0–3.5	0–5	0–8	0–10 (max 20)	> 5 (maximum 48, usually 20–25)	> 10 (max 110, usually 30–50)
Frequency (headway)	Depending on demand	Depending on demand	Suburban < 60 min (5–30 min. Min headway: 90 sec)	< 15 min (2–8 min. Min headway: 60 sec)	< 20 min (5–15 min. Min headway: 90 sec)	< 20 min (3–15 min. Min headway: 60 sec)	Depending on demand	Frequent services depending on length of trip

(Continued)

Table 1.7 (Continued) Main technical and operational characteristics of passenger railway transportation systems

Technical and operational characteristics	High-speed intercity rail	Conventional-speed intercity rail	Suburban/urban/regional rail	Metro	Tramway	Monorail	Cog railway (pure rack system)	Funicular
Track superstructure	Ballasted and/or slab track. CWR, UIC 60. Concrete sleepers	Ballasted track. CWR, UIC 54. Preferably concrete or wooden sleepers	Usually ballasted track	Usually slab track	Usually embedded in the pavement	Single beam in orthogonal profile or of type I from concrete and rarely from steel	Ballasted track with additional cogged bar in the centre and parallel with the main rails	Ballasted and/or slab track with pulleys along the length of the line to support cable
Maximum transportation work (passengers/hour/direction)	High demand for trips between terminal stations	Demand for passenger transport between terminal stations as well as between intermediate stations	60,000 (suburban) No seasonal fluctuations High demand for trips between intermediate stops of the route	45,000	15,000	Small systems: 2,000 Large systems: 12,500 Compact systems: 4,800	Low/medium transportation capacity	Low transportation capacity (500–2,000)
Integration relative to ground surface	Mainly at grade	Mainly at grade	At grade and underground in small parts	Underground integration for the biggest part of the route	At grade for the biggest part of the route	Mainly over ground integration	At grade (very rarely with underground sections)	At grade Rarely underground
Train formation	4–10 vehicles	4–10 vehicles	2–8 vehicles Push–pull trains	4–10 vehicles	Articulated trains	Articulated trains (2–6 and rarely 8 vehicles)	Usually a single railcar	Usually a single car
Rolling stock	EMU and special loco-hauled trains	MU and loco-hauled trains	Usually MUs	Special rolling stock	Special rolling stock. Low (usually) height floor	Special rolling stock	Electric and diesel railcars	Cabin vehicles (one ascending, one descending)

(Continued)

Table 1.7 (Continued) Main technical and operational characteristics of passenger railway transportation systems

Technical and operational characteristics	High-speed intercity rail	Conventional-speed intercity rail	Suburban/urban/regional rail	Metro	Tramway	Monorail	Cog railway (pure rack system)	Funicular
Signalling	Cab signalling	Electric side/Automatic Train Protection system	Electric side/Automatic Train Protection system	Cab signalling	Electric side signalling	Non-existent	Usually mechanical signalling	-
Level crossings	Prohibited	Permitted	Automatic barriers with warning light and sound signals	Non-existent	Mandatory	Non-existent	Permitted	Non-existent
Environment	Interurban	Interurban	Suburban Periurban Regional	Urban	Urban	Urban. Connections in recreational parks and zoos. Connection of airports with city centres	Mountain. Rarely urban. Connecting locations with great altitude difference	Tourist connections. Urban connections with great altitude differences
Maximum running speed (km/h)	200–350	< 200	120–160	90–100 (120)	80–90	60–100	40 (usually 15–25)	50 (constant speed, usually < 20)
Implementation cost (€ M/ track-km) (2014 data)	10–40 (usually 15–25) (infrastructure only, double track)	8–12 (infrastructure only, double track)	10–20 (infrastructure only, double track)	70–150 (infrastructure and rolling stock, double track)	15–35 (infrastructure and rolling stock, double track)	30–90 (infrastructure and rolling stock)	10–15 (infrastructure only)	15–25 (infrastructure and rolling stock)
Peculiarities	Very high speed. High implementation cost	Passenger and freight trains usually sharing the same infrastructure	Very high transportation capacity. Vehicles with many passenger seats	Underground corridor High transportation capacity Very high implementation cost	Specific superstructure Surface integration into urban areas Very sharp horizontal curve radii	Elevated permanent way Specific superstructure Panoramic view	Steep gradient Rolling stock with effective braking system Specific superstructure Light vehicles	Very steep gradient Short-length connection Movement via cables Limited, low transportation capacity

Madrid–Seville, London–Paris, Amsterdam–Paris, Brussels–Amsterdam, Paris–Lyon) have been taken into consideration (Pyrgidis and Karlaftis, 2010). These routes either concern domestic connections or international connections. The routes concern the 2009–2010 period.

After analysing the data, the following may be reasonably concluded:

- Regarding run times, aeroplane is the fastest mode (ratio 1:1.7 for short distances (250 km) and 1:3 for long distances (500 km)).
- Regarding travel times, the aeroplane prevails only for long and very long distances.
- The ratio of the number of daily services between the aeroplane and the high-speed train is 1:4.
- Transport capacity is calculated by multiplying the number of journeys carried out daily in the eight connections examined by the number of passengers who can be carried every day by each mode. Specifically, regarding air travel, an average aeroplane load is equal to 247 passengers. For connections over 400 km, the conclusion to be drawn is that the transport capacity ratio between aeroplane and high-speed train is 1:3.

1.5.3.2 Comparison of urban systems

Table 1.8 shows the comparison between different urban transport systems, in terms of transportation capacity, commercial speed, fares, frequency, and space occupied.

1.6 HISTORICAL OVERVIEW OF THE RAILWAY AND FUTURE PERSPECTIVES

The development of the railway is directly related to the use of steam as a source of energy and the exploitation of coal and iron mines (Profillidis, 2014). Table 1.9 shows the milestones in the history of railways.

The year 1825 is considered to be the starting point in the history of railway, and George Stephenson, an English engineer, is considered as its pioneering figure.

The railway is perhaps the only technology that during its course of development reached an early peak, then found itself being in question and during the last decade it not only managed to rise again but also to be at the cutting edge of technology in many countries.

In short, the railway dominated terrestrial transportation for over 100 years (1830–1950). During this period, it made an enormous contribution to transportation and civilisation. The railway is considered to be the mode of transport that laid down the foundations for

Table 1.8 Comparison of urban transport systems

Comparable parameters	Comparable systems		
Max number of passengers transported per hour per direction	Metro/tram 2.5:1	Metro/urban bus 13:1	Tram/urban bus 5:1
Commercial speed	Urban bus/tram/private car/metro 1:1.5:2.0:2.5		
Fare	Metro/tram/urban bus 1:1:1 Metro/private (a) car peak hours 1:15 (b) off-peak hours 1:5		
Headway between vehicles	Metro/tram/urban bus 1:2:3		
m² of road occupied per transported passenger	Tram/urban bus/private car 1.2:1.9:23.7–40		

Table 1.9 Important milestones in the history of the railway

1800	Discovery of the steam engine (Watt)
1822	Operation of the first factory to construct steam engines (Stephenson, Newcastle, England)
1825	First commercial steam-powered railway journey (Stockton–Darlington line, England)
1830	Operation of first passenger steam-powered railway (Liverpool–Manchester)
1830	First flat-bottom rails (Stevens, USA)
1858	First steel wheels (Bessemer)
1879	Unveiling of first electric locomotive (Siemens–Halske, Germany)
1938	First appearance of diesel traction
1964	Operation of first high-speed train (V_{max} = 210 km/h, Japan, Tokyo–Osaka line)
1981	Operation of first high-speed train in Europe (V_{max} = 260 km/h, France, TGV PSE)
1989	Operation of speed train 300 km/h (France, TGV-A)
1990	First conventional train to achieve a speed in excess of 500 km/h, (515.3 km/h, France)
2007	The most recent record speed for a conventional train (574.8 km/h, France)
2009	Operation of super-fast trains V_{max} = 350 km/h (China)

inland development on all continents. If it were not for the railway, the coastal towns would have become powerful, as they would be the only ones able to support the growing demand for transport of goods by using the basic means of mass transportation, that is, ships. The 1950s saw the start of the struggle between the railway on the one hand and the aeroplane and car on the other, a struggle which was to intensify from the mid-1960s onwards. At the dawn of the 1970s, not only had the railway begun to lose a worrying amount of ground in terms of its share of the transport market, but also railway organisations had begun to suffer financially (loans, deficits) and have become significantly dependent on state budgets, having problematic cross-border services.

According to the European Union's institutions that are responsible for the drawing up and implementation of transport policies, 'competition' has been the solution to overcome the economic impasse afflicting the rail sector in recent years. For this purpose, important amounts were and are still being spent for the construction of new railway infrastructures and modern rolling stock; the research for increasing running speeds was intensified and crowned with success, whereas significant efforts were made to improve the services offered in both freight and passenger transport. Automation constitutes one of the most recent significant technological achievements, which will be a milestone for the future operation of railway transportation systems. Automation has already been developed and adopted in metros and in monorails, which sustain common operational characteristics. In recent years, research has been intensified so as to apply automation to intercity and regional railway systems of conventional and high speeds as well as to passenger and freight systems and in general to incorporate automation in a mixed traffic network (see Chapter 20).

In parallel, while maintaining all its intrinsic advantages (safety, great transport capacity, environmental friendliness, etc.), rail transport had to try operating with more flexible organisational structures, and the monopolies had to be lifted.

Within this framework, the European Commission published a series of directives calling for the revitalisation of the railway sector and the increase of its competitiveness.

The first attempt to draw up railway legislation was made in 1991 with the adoption of Directive 91/440 by the Council of Ministers. This Directive, concerning the development of the railway in the community, is the first attempt to open up railway transportation to competition. For the first time, it introduced the right of free access both for international railway consortiums and companies carrying out combined transportation to the rail infrastructure. The provisions of the Directive also imposed the accounting separation between infrastructure activities and operation activities.

Since 2010, the entire rail services within the European Community Railway Network are open.

Over the last 35 years, the improvement in the quality of life in the large cities, the dramatic rise in road and airport congestion, the intensification of air and noise pollution, as well as the continuing energy crisis, have all created a massive ecological issue. Thus, the railway has made a comeback since it is an ecologically friendly mode of transport and has become more up-to-date and can move at very high speeds. The use of rail transportation is judged more and more to be imperative, both for movement within urban and suburban environments and, also, in order to serve the need for long-distance travelling.

REFERENCES

Alias, J. 1977, *La voie ferrée*, Eyrolles, Paris.

Biasin, D., Lavogiez, H. and Pichant, J.C. 2008, *Trans-European Conventional Rail System - Subsystem Rolling Stock*, European Railway Agency (ERA), Valenciennes.

Buananno, A. and Mele, R. 1996, New method of track formation rehabilitation in use at Italian State Railways, *Rail Engineering International*, No. 1, Netherlands, pp. 17–20.

Christogiannis, E. 2012, *Investigation of the Impact of Traffic Composition on Economic Profitability of a Railway Corridor – Fundamental Principles and Mathematical Simulation for the Selection of Operational Scenario of a Railway Corridor*, PhD thesis (in Greek), Aristotle University of Thessaloniki, Thessaloniki, Greece.

Christogiannis, E. and Pyrgidis, C. 2013, An investigation into the relationship between the traffic composition of a railway network and its economic profitability, *Rail Engineering International*, No. 1, pp. 13–16.

Commission regulation (EU) No 1299/2014 of 18 November 2014 on the technical specifications for interoperability relating to the 'infrastructure' subsystem of the rail system in the European Union, *Official Journal of the European Union*, 12/12/2014, L356.

Connor, P. 2014, *Mass Rapid Transit Operations: From Tram to High Capacity Metro, PRC Rail Consulting Limited*, Lectures notes, Master in Railway Systems Engineering and Integration, University of Birmingham, UK.

Doppelmayr Cable Car. 2008, Available online at: https://en.wikipedia.org/wiki/Cable Liner (accessed 7 August 2015).

EC. 1996, Council directive 96/48/EC of 23 July 1996 on the interoperability of the trans-European high-speed rail system, *Official Journal L 235, 17/09/1996 P. 0006–0024.*

Esveld, C. 2001, *Modern Railway Track*, 2nd edition, MRT-Productions, Duisburg.

Giannakos, K. 2002, *Actions in Railways*, Papazisi (in Greek), Athens.

Hartill, J. 2019, China powers ahead as new entrants' clock in, *Railway Gazette International*, July 2019, pp. 25–29.

Hpeterswald. 2013, Available online at: https://commons.wikimedia.org/wiki/File:Metro_Mono rail_Pitt_Street.jpg (accessed 8 August 2015).

http://www.railwaygazette.com/news/single-view/view/study says no to seat belts on trains.html (accessed 20 March 2015).

Metzler, J.M. 1981, *Généralités sur la traction*, Lecture Notes, ENPC, Paris.

Metzler, J.M. 1985, *Le matériel à voyageurs*, Lecture Notes, ENPC, Paris.

Moreau, A. 1992, Characteristics of wheel/rail contact, *Rail Engineering International*, No. 3, Netherlands, pp. 15–22.

Nightryder84 on the English Wikipedia, 2005, Available online at: https://commons.wikimedia.org/ wiki/File:Memphis_front_view.jpg.

Peterman, D.R., Frittelli, J. and Mallet, W. 2009, *High Speed Rail (HSR) in the United States*, Congressional Research Service (www.crs.org), 7-5700, R40973, 8 December 2009.

Petit, J.M. 1989, *Conception des Bogies Modernes*, Revue ALSTHOM, Code APE 2811, France.

Profillidis, V.A. 2014, *Railway Management and Engineering*, Ashgate, Farnham.

Pyrgidis, C. 1990, *Etude de la stabilité transversale d'un véhicule ferroviaire en alignement et en courbe – Nouvelles technologies des bogies – Etude comparative*, Thèse de Doctorat de l' ENPC, Paris.

Pyrgidis, C. 2005, Calculation of the gauge widening of a track with the aid of mathematical models, *8th International Congress Railway Engineering –2005*, 29–30/06/2005, London.

Pyrgidis, C. and Karlaftis, M. 2010, A level-of-service comparison of rail, air and road transport systems. A qualitative and quantitative approach, *International Conference 'Railways and Environment'*, Delft University, The Netherlands, 16/12/2010, pp. 11–15.

Pyrgidis, C. and Demiridis, N. 2012, An overview of high-speed railway lines in revenue service around the world at the end of 2010 and new links envisaged, *Rail Engineering International*, No. 1, pp. 13–16.

Pyrgidis, C., Anagnostopoulos, E. and Dimitriadis, P. 2017, Defining high-speed rail: A proposal that endeavors to meet the requirements of both rail infrastructure manager and railway operator, *Rail Engineering International*, No. 2, pp. 14–16.

Pyrgidis, C. 2018, A proposal to define high-speed rail, *8th International Symposium on Speed-Up and Sustainable Technology for Railway and Maglev Systems (STECH2018)*, Barcelona, Spain, 3–7/09/2018.

Pyrgidis, C., Savvas, S. and Dolianitis, A. 2020, Classification of intercity and regional passenger railway systems based on speed – A worldwide overview of very high-speed infrastructure, rolling stock and services, *Ingegneria Ferroviaria*, No. 9, September 2020, pp. 635.

Pyrgidis, C. and Dolianitis, A. 2021, *Rail Vehicle Classification, Encyclopedia of Transportation*, Amsterdam.

Riemer, J.F. 2007, Available online at: http://en.wikipedia.org/wiki/Schlossbergbahn_(Graz) (accessed 7 August 2015).

Schoch, W. 2001, Rail maintenance as a contribution to railway track optimization, *Rail Engineering International*, No. 1, Netherlands, pp. 11–13.

UIC. 1983, *Méthode Destinée à Determiner la Capacité des Lignes*, fiche 405 – 1, January 1983. UIC, Paris.

UIC. 2010, *High Speed Rail-Fast Track to Sustainable Mobility*, Paris, available online at: www.uic .org/download.php/.../521E.pdf (accessed 20 March 2015).

UIC. 2014, *General Definitions for High Speed*, International Union of Railways, available online at: http://www.uic.org/spip.php?article3229 (accessed 28 July 2014).

Vignal, B. 1982, *SNCF Médiathèque*, SNCF, Paris.

Chapter 2

Loads on track

2.1 CLASSIFICATION OF LOADS

The railway track is subjected to loads that are vertical, transversal, and longitudinal (Figure 2.1). Apart from the forces that may be exerted in case of an earthquake, all other forces are generated by the rolling stock which is running on the track (traffic loads).[1]

The vertical loads are exerted on the rail rolling surface and are transferred to the subgrade through the various components of the track. During their transfer, the surface area of exertion of the internal forces increases, while the developing stresses decrease (Esveld, 2001; Lichtberger, 2005).

The transversal loads are first transferred by the wheels to the rails, either solely through the rail rolling surface (when there is no flange contact) or both through the rail rolling surface and mainly through the wheels' flanges (when there is flange contact). Further on, the loads are transferred through the components of the track panel (fastenings, elastic pads, sleepers) to the track bed layers.

The longitudinal loads are exerted on the rail rolling surface and, similar to the transversal loads, they are transferred to the track bed layers. They are distributed to a larger number of sleepers compared with the vertical loads.

Depending on their nature, track loads are classified as follows:

- *Static loads:* as a result of the gross weight of the rolling stock. They are exerted on the track by the rolling stock permanently, whether the rolling stock is immobilised or running.
- *Semi-static or quasi-static loads*: They are exerted on the rolling stock through which they are transmitted to the track for a given period of time. Afterwards, and as soon as the cause provoking such loads stops existing, they vanish. Loads developed as a result of the residual centrifugal force and as a result of crosswinds are examples of semi-static loads (wind forces are considered as forces external to the railway system).
- *Dynamic loads*: They are caused as a result of:
  - Track defects and the heterogeneous vertical stiffness of the track.
  - Discontinuities of the rolling surface (at joints, switches, etc.).
  - Wear on the rail rolling surface and on the wheel treads.
  - The suspension system of the vehicles and the asymmetries of the rolling stock.

They vary in relation to time and cause oscillations of different parts of the vehicle.

---

[1] *Wind forces are exerted on the rolling stock and are subsequently transmitted to the track.*

DOI: 10.1201/9781003046073-2

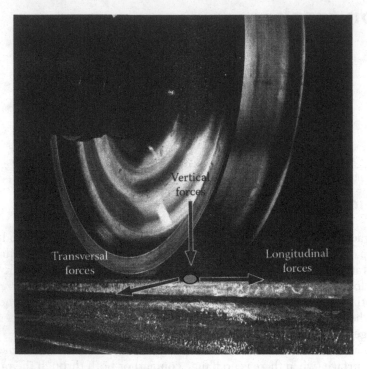

Figure 2.1 Forces acting on the rail in all three directions. (Adapted from Urtado, M., 1993, SNCF Médiathèque, France.)

Table 2.1 presents the classification of loads applied on the track according to the direction in which they act upon the track and provides information regarding their origin and their magnitude (Pyrgidis and Iwnicki, 2006; Pyrgidis, 2009).

2.2 VERTICAL LOADS ON TRACK

Vertical loads play a decisive role in the design, construction, operation, and maintenance of the track. They are a part of the mathematical expressions of all lateral and longitudinal forces acting on the track, either directly or indirectly. More specifically, the vertical strain will determine the selection of the type of the rail, the material of the sleepers and the distance between them, the fastenings, as well as the dimensioning of the elastic pads and the track bed layers.

Last but not least, vertical loads play an important role in the growth rate of deterioration of the track and the damage of the rolling stock.

As already outlined in Section 2.1, the vertical loads on track are divided into static, quasi-static, and dynamic. This classification is consistent in all international literature (Alias, 1977; Esveld, 2001). Apart from these three categories, it is also useful to include and examine separately an additional category of vertical loads which can be called 'characteristic loads'. Included in this category are:

- The axle load Q.
- The total daily traffic load T_f.
- The design vertical wheel load Q_d.
- The design loads of bridges.

Table 2.1 Loads applied on track

Direction of forces as to the rail rolling surface plane	Name – symbol of loads	Type of effort	Cause of generation	Analytical expression ($j = 1$, wheel 1) ($j = 2$, wheel 2)
Vertical loads	Vertical axle load Q (characteristic load)	Static	Vehicle mass	$Q = \left(\dfrac{\bar{M}}{4} + \dfrac{M'}{2} + m\right) \cdot g$ (vehicle with 2-axle bogies)
	Vertical wheel load Q_o	Static	Vehicle mass	$Q_o = \dfrac{1}{2} \cdot \left(\dfrac{\bar{M}}{4} + \dfrac{M'}{2} + m\right) \cdot g$ (vehicle with 2-axle bogies)
	Vertical wheel load due to crosswinds Q_w	Quasi-static	Crosswinds	$\pm Q_{wj} = H_w \cdot \dfrac{q_o}{2e_o}$
	Vertical wheel load due to residual centrifugal force Q_{nc}	Quasi-static	Cant deficiency at the curvatures of track in the horizontal alignment	$\pm Q_{ncj} = \dfrac{F_{nc} \cdot h_{KB}}{2e_o} = \dfrac{Q \cdot l \cdot h_{KB}}{4e_o^2}$
	Dynamic vertical wheel load Q_{dyn}	Dynamic	Track defects, wear of the rail rolling surface, discontinuities on the running rolling surface, asymmetries in rolling stock	$Q_{dynj} = Q_{dyn1j} + Q_{dyn2j} + Q_{dyn3j} + Q_{dyn4j}$
	Total vertical wheel load Q_t	Static, quasi-static, dynamic	Total amount of vertical loads applied	$Q_{tj} = Q_{oj} \pm Q_{wj} \pm Q_{ncj} + Q_{dynj}$
	Design vertical wheel load Q_d (characteristic load)	Static, quasi-static, dynamic	Augmented total vertical wheel load Q_t	$Q_{dj} = Q_{oj} \pm Q_{ncj} \pm Q_{wj} +$ $5 \cdot \left[\sigma\left(Q_{dyn3j}\right)^2 + \sigma\left(Q_{dyn1j} + Q_{dyn2j}\right)^2\right]^{\frac{1}{2}}$

(Continued)

Table 2.1 (Continued) Loads applied on track

Direction of forces as to the rail rolling surface plane	Name – symbol of loads	Type of effort	Cause of generation	Analytical expression (j = 1, wheel 1) (j = 2, wheel 2)
	Design loads of bridges (characteristic loads)	Static and dynamic	Static and dynamic loads	Loading model 71, load models SW/0 and SW/2 (heavy traffic)
	Total daily traffic load T_f (characteristic load)	Static	Daily load of all passing traffic on the line	$T_f = T_p \cdot \dfrac{V_{max}}{100} + T_g \cdot \dfrac{Q_{Do}}{18 \cdot D_o}$
Transversal loads	Total gravitational force (or restoring force or gravitational stiffness) S_p	Dynamic	Wheel profile, railhead profile	$S_p = 2Q_o \cdot \dfrac{1}{R-R'} \cdot y = 2Q_o \cdot \dfrac{\gamma_e \cdot y}{R \cdot \gamma_o}$
	Lateral creep forces T	Dynamic	Wheel profile, rigid wheel linkage, lateral creep phenomena, when the rolling direction of the wheelset forms an angle with its direction of displacement	Straight path $T_1 = T_2 = -c_{22} \cdot \left(\dfrac{y'}{v} - \alpha \right)$ Curves $T_1 = T_2 = -c_{22} \cdot (-\alpha)$
	Longitudinal creep forces X	Dynamic	Wheel profile, rigid wheel linkage, longitudinal creep phenomena, when the rolling direction of the wheelset forms an angle with its direction of displacement	Straight path $X_1 = -c_{11} \cdot \left(\dfrac{x'}{v} - \dfrac{e_o}{v} \cdot \alpha' - \dfrac{\gamma_e}{r_o} \cdot y \right)$ $X_2 = -c_{11} \cdot \left(\dfrac{x'}{v} + \dfrac{e_o}{v} \cdot \alpha' + \dfrac{\gamma_e}{r_o} \cdot y \right)$ Curves $X_1 = -c_{11} \cdot \left(-\dfrac{\gamma_e}{r_o} \cdot y + \dfrac{e_o}{R_c} \right)$ $X_2 = -c_{11} \cdot \left(+\dfrac{\gamma_e}{r_o} \cdot y + \dfrac{e_o}{R_c} \right)$

(Continued)

Table 2.1 (Continued) Loads applied on track

Direction of forces as to the rail rolling surface plane	Name – symbol of loads	Type of effort	Cause of generation	Analytical expression (j = 1, wheel 1) (j = 2, wheel 2)
	Crosswind force H_w	Quasi-static	Crosswinds	$H_w = \frac{1}{2} \cdot \rho \cdot S_2 \cdot V_w^2 \cdot K_2$
	Residual centrifugal force F_{nc}	Quasi-static	Cant deficiency at the curvatures of track in the horizontal alignment	$F_{nc} = \frac{Q}{g} \cdot \left(\frac{V^2}{R_c} - g \cdot \frac{U}{2e_o} \right) = \frac{Q \cdot I}{2e_o}$
	Guidance force F	Dynamic	Flange contact (y = σ)	F_j
	Force due to vehicle oscillations P_{dyn}	Dynamic	Oscillation of car body and bogies	P_{dyn}
	Total transversal force H (meaning the force transmitted from the vehicle to the rail)	Quasi-static + Dynamic		$H = a_i \cdot \left(\frac{Q \cdot I}{1,500} \right) + \left(\frac{Q \cdot V}{1,000} \right)$
Longitudinal (horizontal) loads	Rail creep forces	Dynamic	Trains moving downhill	
	Adhesion force Π	Static	Wheel–rail contact	$\Pi = \mu \cdot Q_o$ (power wheels)
	Traction effort on the treads F_t	Static	Train traction	$F_t = P_t \cdot V$
	Temperature force N	Static	Temperature changes	$N = \pm E \cdot A_r \cdot \alpha_t \cdot \Delta_t$
	Braking force N_{br}	Static	Vehicle braking	$N_{br} = 0.25 \cdot \Sigma Q$
	Acceleration forces N_{ac}	Static	Vehicle acceleration	N_{ac}
	Fishplate force P_f	Dynamic	Passage of trains	$P_f = Q_o + 2\alpha_f \cdot V_P \cdot \sqrt{k} \cdot USM$

Source: Adapted from Pyrgidis, C.,, and Iwnicki, S., 2006, International Seminar Notes, EURNEX-HIT, Thessaloniki; Pyrgidis, C. 2009, Transversal and longitudinal forces exerted on the track – Problems and solutions, 10th International Congress Railway Engineering – 2009, 24–25 June, 2009, London, Congress Proceedings, CD.

A special characteristic of all the above loads, which can be static or dynamic, is that they determine the dimensioning of the railway track as well as the maintenance policy to be followed to a great extent.

During recent years, a continuous increase of the value in the vertical rail loads has been observed. This appears as an increase of the axle load, of the length and, consequently, of the weight of the trains, of the daily traffic load, and finally, as a result of the significant increase of the train speeds (resulting in an increase of the dynamic vertical loads) (Riesberger, 2008).

2.2.1 Static vertical loads

2.2.1.1 Axle load

The term 'axle weight' or 'axle load' describes the static load Q which is individually transferred by each axle of a vehicle, and in general of a train, through the wheels to the rails. The axle load is classified as 'characteristic load'.

Considering a symmetric loading of the various vehicle parts, the axle load substantially expresses the quotient of the total vehicle weight divided by the total number of axles.

For example, in the case of a vehicle with 2-axle bogies, the following mathematical equation applies:

$$Q = \left(\frac{\bar{M}}{4} + \frac{M'}{2} + m \right) \cdot g \tag{2.1}$$

where:
Q: Axle load.
$\bar{M}$: Car body mass.
M': Mass of one bogie.
m: Mass of one railway wheelset (axle + wheels + axle boxes).
g: Gravity acceleration.

The axle load is indirectly or directly involved in the analytical expressions of all the forces applied on the wheel–rail contact surface and affects the behaviour of both the rolling stock and the track. Especially for the track used by very heavy vehicles, an increase in the number and the size of track defects is observed as well as the fatigue of the track superstructure materials leading to an increase in the track maintenance needs and cost.

The International Union of Railways (UIC) classifies the tracks depending on the maximum permitted axle load into four categories: A, B, C, and D (Table 2.2).

Each category is divided into subcategories depending on the distributed load per metre of length (total vehicle load divided by the free length between buffers).

Table 2.2 Categories of tracks according to the permitted axle load of trains in traffic (in accordance with UIC)

Track category	Axle load (t)
A	16
B	18
C	20
D	22.5

Source: Adapted from UIC. 1989, Fiche 714R, Classification des voies des lignes au point de vue de la maintenance de la voie.

The maximum axle load a track can support is a function of all the parameters involved in the construction of the track superstructure and substructure. The maximum axle load differs from country to country and, in most countries, from track to track.

The increase of the track gauge allows a significant increase in the axle load.

Axle loads Q > 16– 17 t are considered prohibitive for the development of very high speeds (V ≥ 250 km/h).

The axle load is characterised as an important issue in order to ensure railway interoperability.

2.2.1.2 Wheel weight

The term 'wheel weight' or 'wheel load' refers to the static load Q_o which is individually transferred by each wheel of the vehicle to the corresponding rail.

Considering a symmetrical loading of the vehicle, the following mathematical equation applies:

$$Q_o = \frac{Q}{2} \tag{2.2}$$

In practice, the loads of both wheels of each axle, especially in the case of running in curves, are not equal to each other.

The wheel weight and, in particular, the weight distribution to the two wheels are directly linked to the phenomena of derailment and overturning of vehicles.

2.2.1.3 Daily traffic load

The qualitative and quantitative assessment of track traffic is usually expressed by the total daily traffic load T_f (in t). The load T_f is often referred to as one of the 'characteristic loads'.

Based on the value of the total daily traffic load, the tracks are classified into categories for the ultimate purpose of standardising the track dimension and maintenance.

To calculate T_f, the following two mathematical equations have been suggested by the UIC:

$$T_f = T_p \cdot \frac{V_{max}}{100} + T_g \cdot \frac{Q_{D_o}}{18 \cdot D_o} \tag{2.3}$$

where:

T_f: Total daily traffic load (in t).
T_p: Daily traffic load of passenger trains (in t).
T_g: Daily traffic load of freight trains (in t).
V_{max}: Maximum running speed (in km/h).
D_o: Minimum wheel diameter of running trains along the line (in m).
Q_{D_o}: Maximum passing axle load (wheels of diameter D_o) (in t).

On the basis of Equation 2.3 the tracks are classified into four categories, as shown in (Table 2.3).

$$T_f = S_V \cdot \left(T_V + K_t \cdot T_{tv}\right) + S_m \cdot \left(K_m \cdot T_m + K_t \cdot T_{tm}\right) \tag{2.4}$$

where:

T_f: Total daily traffic load (in t).
T_V: Average daily traffic load of trailer passenger cars (in t).

Table 2.3 Classification of tracks based on Equation 2.3

Track category	Total daily traffic load (t)
I	$T_f > 40,000$
II	$40,000 \geq T_f > 20,000$
III	$20,000 \geq T_f > 10,000$
IV	$10,000 \geq T_f$

Table 2.4 Classification of tracks based on Equation 2.4

Track category	Total daily traffic load (T_f)
UIC 1	$130,000\ t < T_f$
UIC 2	$80,000\ t < T_f \leq 130,000\ t$
UIC 3	$40,000\ t < T_f \leq 80,000\ t$
UIC 4	$20,000\ t < T_f \leq 40,000\ t$
UIC 5	$5,000\ t < T_f \leq 20,000\ t$
UIC 6	$T_f \leq 5,000\ t$

Source: Adapted from UIC. 1989, Fiche 714R, Classification des voies des lignes au point de vue de la maintenance de la voie.

T_m: Average daily traffic load of freight wagons (in t).

T_{tv}: Average daily traffic load of passenger trains' power vehicles (in t).

T_{tm}: Average daily traffic load of freight trains' power vehicles (in t).

K_m: Coefficient with values varying between 1.15 (standard value) and 1.45 (when > 50% of the traffic takes place with vehicles of axle load Q = 22.5 t or when 75% of the traffic takes place with vehicles of axle load Q ≥ 20 t).

K_t: Coefficient that depends on the rolling conditions of the power vehicle axles on the track. It is usually equal to 1.40.

S_v, S_m: Coefficients with values depending on the speed of passenger (with the highest speed) and freight (with the lowest speed) trains, respectively, running on the track.

Based on Equation 2.4 and according to the (UIC, 1989), the tracks are classified into the six groups, as shown in Table 2.4.

2.2.2 Quasi-static vertical loads

2.2.2.1 Vertical wheel load due to crosswinds

The vertical load due to crosswinds Q_w is given by the following mathematical equation (Figure 2.2) (Esveld, 2001):

$$\pm Q_w = H_w \cdot \frac{q_o}{2e_o} \tag{2.5}$$

where:

H_w: Crosswind force applied on the geometrical centre of the lateral surface of the car body.

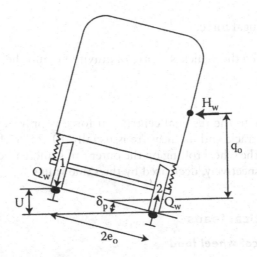

Figure 2.2 Vertical wheel load due to crosswinds ($\pm Q_w$) – motion in curve. (Adapted from Esveld, C., 2001, *Modern Railway Track*, 2nd edition, MRT-Productions, West Germany.)

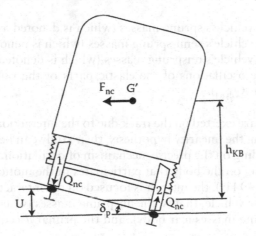

Figure 2.3 Vertical wheel load $\pm Q_{nc}$ due to residual centrifugal force. (Adapted from Esveld, C., 2001, *Modern Railway Track*, 2nd edition, MRT-Productions, West Germany.)

q_o: Vertical distance between the geometrical centre of the lateral surface of the car body and the rail rolling surface.

$2e_o$: Distance between the vertical axis of symmetry of the two rails.

When the force H_w is directed from wheel 2 towards wheel 1, then wheel 1 load increases by Q_w, while wheel 2 load decreases by the same value.

As a result of crosswinds, the load is applied during motion, both on a straight path and in curves, and stops once the wind loads disappear.

2.2.2.2 Vertical wheel load due to residual centrifugal force

The vertical load Q_{nc}, which is due to residual centrifugal force F_{nc}, is expressed as (Figure 2.3):

$$\pm Q_{nc} = \frac{F_{nc} \cdot h_{KB}}{2e_0} = \frac{Q \cdot I \cdot h_{KB}}{4e_0^2} \tag{2.6}$$

where:
    F_{nc}: Residual centrifugal force.
    I: Cant deficiency.
    h_{KB}: Distance between the vehicle's centre of gravity G′ and the rail rolling surface.
    U: Track cant.
    δ_p: Angle of cant.

The load which is due to the residual centrifugal force is only applied during motion at curved segments of the track and its measure is usually 10–25% of the value of the static wheel load. The load of the wheel rolling on the outer rail is increased by Q_{nc}, while the load of the inner wheel is, respectively, decreased by the same value.

2.2.3 Dynamic vertical loads

2.2.3.1 Dynamic vertical wheel load

The total dynamic vertical wheel load Q_{dyn} is calculated as the sum of the four separate dynamic loads, specifically:

- The one due to the vehicle's sprung masses (which is denoted as Q_{dyn1}).
- The one due to the vehicle's semi-sprung masses (which is denoted as Q_{dyn2}).
- The one due to the vehicle's unsprung masses (which is denoted as Q_{dyn3}).
- The one due to the oscillations of the elastic parts of the rail-sleeper fixing system (which is denoted as Q_{dyn4}).

The dynamic forces that are exerted on the track due to the interaction with the rolling stock are random. By employing the linearity hypothesis, their study can be conducted in different frequency ranges, depending on the precise mechanism of oscillation, the causes of the oscillation, and also depending on the body that participates in the motion.

At low frequencies (0–40 Hz), the interest is focused on the interaction of the total of inert and elastic elements of the vehicle, that is, the unsprung masses (wheelsets), the semi-sprung masses (bogies), the sprung masses (car body), and the primary/secondary suspension with the railway track.

At this frequency range, the track appears to be particularly rigid in relation to the vehicle. As a result, it is not taken into consideration during the dynamic study of the mechanical system. The generic mechanism of the oscillations at this particular frequency range mainly consists of the medium and long wavelength track defects L_w, that is, 3–25 m and 25–70 m or 25–120 m, respectively, depending on the speed of the vehicle, a parameter which determines the frequency of oscillation f ($f = V/L_w$, in Hz).

Similarly, at medium and high frequencies (40–400 Hz and 400–2,000 Hz), the interest is focused on the interaction between the unsprung masses of the vehicle (wheelsets) and the track. In this case, the vehicle's suspension isolates its total of masses that are capable of contributing to the motion, with the exception of the unsprung masses. As concerns the railway track, when examining the system, all components are taken into consideration.

The rails' geometric defects as well as any discontinuities of their running surface are considered to be the generic mechanism for the above frequency range (Giannakos, 2002).

- *Forces at a frequency range of 0.5–5 Hz*
  Such forces are due to the sprung masses of the vehicle (car body).

The vertical accelerations of the car body increase less rapidly than the speed. As a result, the increase of the static vertical wheel load caused by these forces is relatively small for both low and high speeds.

These loads may be reduced by reducing the natural frequency of the car body or by an improvement of the track's quality.

- *Forces at a frequency range of 5–20 Hz*
  Such forces are due to the semi-sprung masses of the vehicle (bogies).
  The vertical accelerations of the bogies increase with speed, yet at a quicker rate than that of the car body.
  In this case, the increase of the static vertical load is greater.
  Q_{dyn2} loads may be reduced by lightening the bogies, by reducing the vertical stiffness of the primary suspension springs and by increasing the bogie-axles damping coefficient.
  The presence of continuous welded rails limits the oscillations of the semi-sprung masses.
- *Forces at a frequency range of 20–200 Hz*
  Such forces are due to unsprung masses of the vehicle (wheelsets) and the rails.
  The reduction of the axle mass, the reduction of the track stiffness, and the increase of the ballast thickness contribute to the reduction of the dynamic vertical forces.
- *Forces at a frequency range of 200–2,000 Hz*

Such forces are due to the oscillations of the elastic parts of the rail-sleeper fixing system (elastic pads, fastenings). They generate noise and wear on the rail rolling surface. They are the main cause of short pitch corrugation on the rails.

Thus, the total dynamic vertical wheel load Q_{dyn} is the result of adding all the above forces:

$$Q_{dynj} = Q_{dyn1j} + Q_{dyn2j} + Q_{dyn3j} + Q_{dyn4j} \tag{2.7}$$

where:

j = 1,2: Index related to the two wheels of the same wheelset.

The very high speeds, the continuously increasing weight, and the stiffness of the track's components result in an increase of the effect of the dynamic phenomena and an increase of loads exerted on the track's superstructure, the subgrade, and the vehicle.

The calculation of the dynamic forces remains extremely complex and, in some instances, it is not possible at all. Most analyses are restricted to quasi-static speculations. In most cases, a purely empirical approach based on measurements is adopted (Giannakos, 2014, 2016).

The additional dynamic loading may reach values of up to 50% of the static wheel load (Alias, 1977). According to relevant literature (Profillidis, 1995; Zicha, 1989), for speeds up to 200 km/h, the dynamic impact factor varies from 1.35 to 1.6. Thus, for speeds up to 200 km/h, a dynamic impact factor of 1.5 is suggested. For speeds greater than 200 km/h, an analytical survey could be conducted based on experimental data.

2.2.3.2 Total vertical wheel load

The total vertical wheel load Q_t is calculated as the sum of all the static, quasi-static, and dynamic vertical loads transferred to each wheel by the rolling stock:

$$Q_{tj} = Q_{oj} \pm Q_{wj} \pm Q_{ncj} \pm Q_{dynj} \tag{2.8}$$

2.2.3.3 Design vertical wheel load

The term 'design vertical wheel load' Q_d refers to the characteristic value of the wheel's vertical load exerted on the railway track, which covers the maximum possible theoretical probability of not exceeding such load during the life cycle of the railway track. This probability allows for taking into consideration exceptional track loading conditions, such as anchors forgotten on the running surface of the rails, rail fragmentation, discontinuities of the rolling surface of the rails, and flattening of the wheel treads that exceed the accepted tolerances.

Given the random nature of the load mechanism of the railway track, the probability of the appearance of extreme, maximum wheel loads is generally equal to values around 10^{-6}, that is, 1 in 1,000,000.

From all the above, it becomes clear that having an accurate value for the dynamic design vertical load of the wheel is of great significance for the railway track. The probability approach adopted for this calculation is generally based on the increase of the mean value of the total vertical wheel load, which allows achieving the desired statistical level of safety.

The design vertical wheel load Q_d is considered to be equal to the sum of the static wheel load, the quasi-static wheel load, and the mean of the square deviation of the standard deviations of the dynamic forces of the unsprung, sprung/semi-sprung vehicle masses. The mean is increased so as to cover the statistical probability of not exceeding the calculated load in real conditions (Esveld, 2001; Giannakos, 2002).

Thus, the following mathematical equation applies:

$$Q_{dj} = Q_{oj} + Q_{Hj} + n_p \cdot \left[\sigma\left(Q_{dyn3j}\right)^2 + \sigma\left(Q_{dyn1j} + Q_{dyn2j}\right)^2 \right]^{\frac{1}{2}} \tag{2.9}$$

where:

Q_d: Design vertical wheel load.

Q_o: Static vertical wheel load.

Q_H: Quasi-static vertical wheel load.

$\sigma\left(Q_{dyn3}\right)$: Typical deviation of the vertical dynamic forces of the unsprung masses of the vehicle.

$\sigma\left(Q_{dyn1}, Q_{dyn2}\right)$: Typical deviation of the vertical dynamic forces of the sprung and semi-sprung masses of the vehicle.

n_p: Coefficient of the probability augmentation of the mean square value of standard deviations of vertical dynamic forces of a vehicle, taking a value equal to 5.00 (Demiridis and Pyrgidis, 2010).

$j = 1,2$: Index related to the two wheels of the same wheelset.

2.2.3.4 Design loads of bridges

The design loads of bridges are classified as vertical, longitudinal, and transversal. Taking into account that vertical loads are the most important ones, all three categories are taken into consideration when studying the vertical forces.

To calculate the design loads of bridges two approaches may be adopted (UIC, 1979, 2003; EN 1991–2, 2003; EN 1993–2, 2006; Gerard, 2003; Tschumi, 2008):

- The static analysis.
- The dynamic analysis.

2.2.3.4.1 Vertical static loads

The vertical loads can be defined with the aid of two loading models, of which one represents the conventional traffic loads (loading model 71) and the other represents the heavy loads (SW loading model).

Figure 2.4 illustrates the loading model 71 and the respective characteristic values of the vertical loads. In the case where trains running on the track are heavier or lighter than the loading model 71, these loads are multiplied by a coefficient α_{br}. The values of the coefficient α_{br} are decided by the competent authority and are 0.75, 0.83, 0.91, 1.00, 1.10, 1.21, and 1.33. Loads of the loading model SW concerning bridges with continuous spans, centrifugal forces, acceleration and braking forces, as well as random forces, should also be multiplied by the same coefficient.

Figure 2.5 illustrates the arrangement of the SW loading model while Table 2.5 summarises the respective characteristic values of vertical loads.

2.2.3.4.2 Vertical dynamic loads

Traffic loads cause dynamic oscillation which results in increased stresses and deformations. The main factors that influence the vertical dynamic behaviour are the following:

- Natural frequency of the structure.
- Distance between the railway axles.
- Running speed on the bridge.
- Damping of the construction.
- Arrangement of the bridge's deck.
- Wheel and rail defects, etc.

To take into consideration the influence of these factors on the dynamic behaviour of bridges, regulations define the dynamic coefficient φ_{bri} (i = 2 or 3). This coefficient applies only for speeds V ≤ 220 km/h and for natural frequencies which lay within the limits that are set by

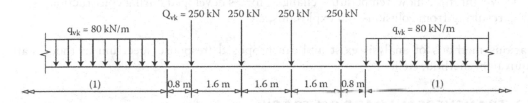

Figure 2.4 Loading model 71. (Adapted from UIC, 2003, 702 loading diagram to be taken into consideration for the calculation of rail carrying structures on lines used by international services, Technical Report, International Union of Railways, Leaflet.)

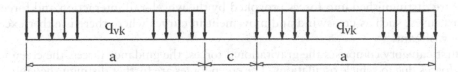

Figure 2.5 Loading model SW. (Adapted from UIC, 2003, 702 loading diagram to be taken into consideration for the calculation of rail carrying structures on lines used by international services, Technical Report, International Union of Railways, Leaflet.)

Table 2.5 Characteristic values of vertical loads for SW loading models

Load classification	q_{vk} (kN/m)	a (m)	c (m)
SW/0	133	15.0	5.3
SW/2	150	25.0	7.0

the regulations. Dynamic coefficient φ_{bri} increases the static stresses and deformations that are caused by traffic loads.

2.2.3.4.3 Transverse loads

- *Centrifugal forces*: Centrifugal forces act at a level of 1.80 m above the rolling surface. These forces are not multiplied by the coefficient φ_{bri}.
- *Total transverse load*: The characteristic value of the lateral force must be considered equal to 100 kN. This force is not multiplied by the coefficients α_{br} and φ_{bri} and it must always be combined with the vertical load.

2.2.3.4.4 Longitudinal loads

- *Acceleration and braking forces*: Acceleration and braking forces act on the rail rolling surface in the longitudinal direction of the track and are considered to be uniformly distributed over their length of influence.

These forces must be combined with the respective vertical loads.

2.2.3.4.5 Other loads

- Aerodynamic effect due to train circulation.
- Random effects – derailment.
- Earthquake.
- Weight, fire, snow, temperature changes, forces developed during construction, forces resulting from collisions and explosions, etc.

Various methods of analysis exist and are proposed (response spectrum method, static equivalent method, etc.).

2.3 TRANSVERSAL LOADS ON TRACK

The transversal forces are directly linked to the safety of the train traffic and the dynamic comfort of the passengers. Under certain circumstances, they may cause the phenomenon of derailment.

They are distinguished into forces provoked by the wheel–rail interaction and forces due to other causes, such as crosswind and movement in curves where there is either excess or deficiency of cant.

The first category comprises the gravitational forces, the guidance forces, the creep forces, and the forces due to vehicle oscillations. The creep forces are further distinguished into transversal and longitudinal forces. Given that they are activated simultaneously and that they affect the transversal behaviour of the vehicles, they are both examined under this section.

The second category comprises the residual centrifugal force and the crosswind force.

2.3.1 Gravitational forces

At the wheel–rail contact surface of each wheel, the reaction force R_o is analysed into two components, namely, Q_o and S_{po} (Figure 2.6). The transversal component S_{po} is defined as 'gravitational' or 'restoring force' or 'gravitational stiffness'. It is exclusively due to the conicity of the wheels, and it acts via the axle on the rail rolling surface.

It is considered as a dynamic force and is equal to:

$$Q_o \cdot \tan \gamma_o$$

where:

Q_o: Static wheel load.

γ_o: Angle between tangent plane and horizontal wheelset in central position.

In the case of a single wheelset, there are two restoring forces, one for each wheel (S_{pj}, j = 1,2) (Figure 2.7).

If 'y' symbolises the transversal displacement of the wheelset, then according to Figure 2.7, the following mathematical equations apply:

$$S_{p1} = Q_1 \cdot \tan \gamma_1 = Q_1 \cdot \gamma_1 \tag{2.10}$$

$$S_{p2} = Q_2 \cdot \tan \gamma_2 = Q_2 \cdot \gamma_2 \tag{2.11}$$

where:

Q_1, Q_2: Vertical components of the reactions R_1 and R_2 at contact points I_1 and I_2, respectively.

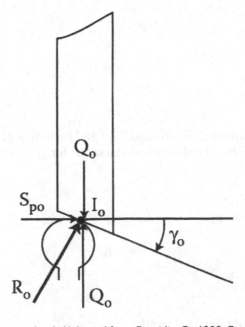

Figure 2.6 Gravitational force per wheel. (Adapted from Pyrgidis, C., 1990, Etude de la stabilité transversale d'un véhicule ferroviaire en alignement et en courbe – Nouvelles technologies des bogies – Etude comparative, Thèse de Doctorat de l' ENPC, Paris.)

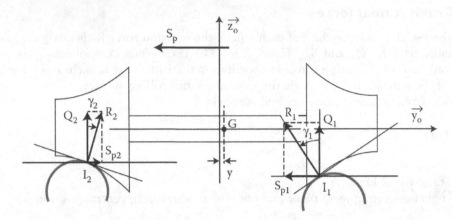

Figure 2.7 Total gravitational force S_p. (Adapted from Pyrgidis, C., 1990, Etude de la stabilité transversale d'un véhicule ferroviaire en alignement et en courbe – Nouvelles technologies des bogies – Etude comparative, Thèse de Doctorat de l' ENPC, Paris.)

γ_1, γ_2: Angles formed by the horizontal plane and the tangent planes at contact points I_1 and I_2, respectively (as γ_1, γ_2 are very small quantities, $\tan \gamma_1 = \gamma_1$ and $\tan \gamma_2 = \gamma_2$ applies).

From the mathematical resolution of the geometry of the wheel–rail contact and assuming that the wheels are of conical shape with curved slants of constant radius while the heads of the rails are spherical, the following linear equations are derived (Pyrgidis, 1990):

$$\gamma_1 = \gamma_o + \frac{\gamma_e}{R \cdot \gamma_o} y \tag{2.12}$$

$$\gamma_2 = -\gamma_o + \frac{\gamma_e}{R \cdot \gamma_o} y \tag{2.13}$$

$$\gamma_e = \frac{R \cdot \gamma_o}{R - R'} \tag{2.14}$$

Taking into account Equations 2.10 through 2.14 and assuming equal load distribution on each wheel, the following two mathematical equations for the total gravitational force are derived:

$$S_p = S_{p1} + S_{p2} = 2Q_o \cdot \frac{\gamma_e \cdot y}{R \cdot \gamma_o} \tag{2.15}$$

$$S_p = 2Q_o \cdot \frac{1}{(R - R')} \cdot y \tag{2.16}$$

where:
R: Curvature radius of the wheel tread.
γ_e: Equivalent (effective) conicity of the wheel.
R′: Radius of curvature of the rolling surface of the railhead.

Considering Equations 2.15 and 2.16, we can conclude that the total gravitational force:

- Is proportional to the displacement 'y' of the axle's centre of gravity. This means that if, for any reason, the axle is displaced laterally, then the gravitational force tends to reinstate it to its initial equilibrium position (for $y = 0$, $S_p = 0$).
- Is inversely proportional to the curvature radius R of the wheel tread and proportional to the equivalent conicity γ_e. Thus, worn wheels (small R values/big γ_e values) automatically generate a greater gravitational force.

The gravitational force S_p is due to the special profile of the rail wheels (variable conicity; see Section 2.38).

The presence of the gravitational force does not create problems for any component of the railway system. On the contrary, it plays a balancing role with regard to the forces that tend to disturb the stability of the vehicle wheelsets.

Regarding the track, no wear is caused on the rolling surface of the rails on which it acts.

As regards the rolling stock, the total gravitational force S_p is always desired since it assists in the centring of the railway wheelsets on the track.

Especially in the case where bogies with independently rotating wheels are used (rolling stock of tramway networks, Talgo trains), while the longitudinal creep forces become null, the gravitational forces are the only forces enabling the centring of the wheelset on the track (Joly and Pyrgidis, 1996; Pyrgidis, 2004).

On the basis of the above, the increase of the value of the gravitational force leads to an increase of the critical speed[2] of the vehicle in a straight path while it allows for a better geometric positioning of the wheelsets in curves.

2.3.2 Creep forces

2.3.2.1 Running on straight path

In the case of conventional axle running on a straight path, the analytical expressions of the creep forces resulting from the application of Kalker linear theory are given by Equations 2.17 through 2.22 (for the motion parameters, the sign convention adopted in Figure 2.8b applies):

$$X_1 = -c_{11} \cdot \left(\frac{X'}{V} - \frac{e_o}{V} \cdot \alpha' - \frac{\gamma_e}{r_o} \cdot y \right) \tag{2.17}$$

$$X_2 = -c_{11} \cdot \left(\frac{X'}{V} + \frac{e_o}{V} \cdot \alpha' + \frac{\gamma_e}{r_o} \cdot y \right) \tag{2.18}$$

$$T_1 = -c_{22} \cdot \left(\frac{y'}{V} - \alpha \right) - c_{23} \cdot \left(\frac{\alpha'}{V} - \frac{\gamma_o}{r_o} - \frac{\gamma_e \cdot y}{R \cdot \gamma_o \cdot r_o} \right) \tag{2.19}$$

$$T_2 = -c_{22} \cdot \left(\frac{y'}{V} - \alpha \right) - c_{23} \cdot \left(\frac{\alpha'}{V} + \frac{\gamma_o}{r_o} - \frac{\gamma_e \cdot y}{R \cdot \gamma_o \cdot r_o} \right) \tag{2.20}$$

---

[2] The critical speed V_{cr} of a railway vehicle on a straight path is the speed beyond which the motions of the vehicle bogies become 'unstable' (the sinusoidal motion of the wheelset is no more dampened).

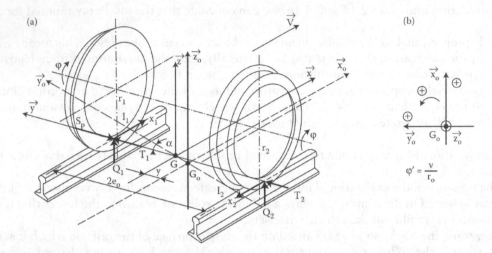

Figure 2.8 Forces applied on the rail wheelset in random position on the track: (a) conventional axle running on straight path and (b) sign convention adopted for the motion parameters. (Adapted from Pyrgidis, C., 2009, Transversal and longitudinal forces exerted on the track – Problems and solutions, *10th International Congress Railway Engineering – 2009*, 24–25 June 2009, London, Congress proceedings, CD; Pyrgidis, C., 1990, Etude de la stabilité transversale d'un véhicule ferroviaire en alignement et en courbe – Nouvelles technologies des bogies – Etude comparative, Thèse de Doctorat de l' ENPC, Paris.)

$$M_1 = c_{23} \cdot \left(\frac{y'}{V} - \alpha \right) - c_{33} \cdot \left(\frac{\alpha'}{V} - \frac{\gamma_o}{r_o} - \frac{\gamma_e \cdot y}{R \cdot \gamma_o \cdot r_o} \right) \tag{2.21}$$

$$M_2 = -c_{23} \cdot \left(\frac{y'}{V} - \alpha \right) - c_{33} \cdot \left(\frac{\alpha'}{V} + \frac{\gamma_o}{r_o} - \frac{\gamma_e \cdot y}{R \cdot \gamma_o \cdot r_o} \right) \tag{2.22}$$

where:
X_1, X_2: Longitudinal creep forces applied on both wheels.
T_1, T_2: Lateral creep forces applied on both wheels.
M_1, M_2: Spin moment on both wheels.
x: Longitudinal displacement of the wheelset.
y: Lateral displacement of the wheelset.
α: Yaw angle of the wheelset.
φ: Angle of rotation of the wheels and of the wheelset.
x', y', α', φ': Derivative of displacements x, y, of the yaw angle α, and of angle of rotation φ.
c_{11}: Longitudinal coefficient of Kalker.
c_{22}: Transversal coefficient of Kalker.
c_{23}, c_{33}: Spin coefficients of Kalker.

From Equations 2.17 through 2.20, the following can be concluded:

- The longitudinal creep forces X_1 and X_2 applied on both wheels create a pair of forces which tend to rotate the railway axle around axis $\vec{z}_o$ (Figure 2.8a). For x = 0, the forces X_1 and X_2 are equal in terms of magnitude and opposite in terms of direction.
- The lateral forces applied on each wheel when C_{23}, $C_{33} = 0$ (spin) are equal and act in the same direction.

- The increase of the displacement velocity V reduces the damping terms (x'/V, α'/V, y'/V) that tend to stabilise the axle.
- The increase of the equivalent conicity γ_e and the decrease of the wheel rolling radius r_o lead to an increase in the value of the longitudinal creep forces.

2.3.2.2 Running in curves

In the case of a conventional axle running on a curvature of the track, the analytical expressions of the creep forces resulting from the application of Kalker linear theory are given by the following equations:

$$X_1 = -c_{11} \cdot \left(-\frac{\gamma_e}{r_o} \cdot y - \frac{e_o}{V} \cdot \alpha' + \frac{e_o}{R_c} \right) \tag{2.23}$$

$$X_2 = -c_{11} \cdot \left(+\frac{\gamma_e}{r_o} \cdot y + \frac{e_o}{V} \cdot \alpha' - \frac{e_o}{R_c} \right) \tag{2.24}$$

$$T_1 = -c_{22} \cdot \left(\frac{y'}{V} - \alpha \right) - c_{23} \cdot \left(\frac{\alpha'}{V} - \frac{1}{R_c} - \frac{\gamma_o}{r_o} - \frac{\gamma_e \cdot y}{R \cdot \gamma_o \cdot r_o} \right) \tag{2.25}$$

$$T_2 = -c_{22} \cdot \left(\frac{y'}{V} - \alpha \right) - c_{23} \cdot \left(\frac{\alpha'}{V} - \frac{1}{R_c} + \frac{\gamma_o}{r_o} - \frac{\gamma_e \cdot y}{R \cdot \gamma_o \cdot r_o} \right) \tag{2.26}$$

$$M_1 = c_{23} \cdot \left(\frac{y'}{V} - \alpha \right) - c_{33} \cdot \left(\frac{\alpha'}{V} - \frac{\gamma_o}{r_o} - \frac{1}{R_c} - \frac{\gamma_e \cdot y}{R \cdot \gamma_o \cdot r_o} \right) \tag{2.27}$$

$$M_2 = -c_{23} \cdot \left(\frac{y'}{V} - \alpha \right) - c_{33} \cdot \left(\frac{\alpha'}{V} + \frac{\gamma_o}{r_o} - \frac{1}{R_c} - \frac{\gamma_e \cdot y}{R \cdot \gamma_o \cdot r_o} \right) \tag{2.28}$$

In the case of movement in curves of small radius, with a displacement velocity that is approximately equal to the equilibrium speed, the inertia and damping forces may be disregarded when compared with the elastic forces. Moreover, by ignoring the spin impact, the following mathematical equations apply:

$$X_1 = -c_{11} \cdot \left(-\frac{\gamma_e}{r_o} \cdot y + \frac{e_o}{R_c} \right) \tag{2.29}$$

$$X_2 = -c_{11} \cdot \left(+\frac{\gamma_e}{r_o} \cdot y - \frac{e_o}{R_c} \right) \tag{2.30}$$

$$T_1 = T_2 = -c_{22} \cdot (-\pm) \tag{2.31}$$

The longitudinal creep forces result in the wear of the wheel and rail rolling surface, the fatigue of the contact materials, and noise.

As shown in Figure 2.8, the longitudinal creep forces result in the horizontal rotation of the axle and, along with the lateral creep forces, they activate the sinusoidal movement (hunting) of the bogie wheelsets, thereby causing oscillations.

The creep forces appear when there is a deviation between the rolling direction of the wheels and the wheelset's direction of displacement. This occurs when there is a transversal displacement 'y' or a yaw angle 'α' of the axle from the initial equilibrium position. Therefore, in order to address creep forces, it is required to focus on the parameters that cause such forces.

These factors are the track defects which either pre-exist due to poor construction of the track or arise during track use. Correcting such track defects essentially involves intervention into the source which produces those creep forces.

Other measures that help to reduce the creep forces are the wheel turning at regular intervals, the appropriate choice of the constructional characteristics of the bogies (wheel profile, wheel diameter, stiffness of the primary suspension, bogie wheelbase), and the suitable choice of the bogie technology for the operability of the network.

The presence or absence of the longitudinal and lateral component of the creep forces depends on how the wheels are connected to the axle while the presence or absence of spin moment depends on the angle formed by the wheel rolling surface plane and the rotation angle.

Table 2.6 presents the forces that are developed on the wheel–rail contact surface for various technologies of railway wheelsets (already manufactured and theoretical) (Frederich, 1985).

To avoid the hunting of the wheelsets, it is essential that the rigid link between the two wheels and the axle be broken. Thus, the two wheels will be able to rotate at different angular velocities, while at the same time, the following mathematical equation shall remain applicable:

$$\omega_1 \cdot r_1 = \omega_2 \cdot r_2 = V \tag{2.32}$$

Table 2.6 Wheelset technologies: forces exerted on wheels

Technology description	Schematic representation	S_p	T	X	M
Conventional axle – wheels of variable conicity		Yes	Yes	Yes	Yes
Axle with independent rotating wheels – wheels of variable conicity		Yes	Yes	No	Yes
Wheels of variable conicity with articulated body axle		Yes	No	Yes	Yes
Conventional axle – cylindrical wheels		No	Yes	Yes	No
Conventional axle – wheels of constant conicity		No	Yes	Yes	Yes
Independent rotating wheels of variable conicity		Yes	No	No	Yes
Axles with independent rotating inclined wheels of variable conicity		Yes	No	No	No
Conventional axles with inclined wheels of variable conicity		Yes	Yes	Yes	No

Source: Adapted from Frederich, F., 1985, Possibilités inconnues et inutilisées du contact rail-roue, *Rail International*, Brussels, November 1985, pp. 33–40.

where:

ω$_1$, ω$_2$: Angular velocities of the two wheels.

r$_1$, r$_2$: Their rolling radii, respectively.

This ensures the rolling of the two wheels without creep and elimination of longitudinal creep forces. The technology of bogies with independently rotating wheels is based on this logic.

2.3.3 Crosswind forces

In the case of crosswinds, a transversal force H$_w$ is transferred through the axles to the rail rolling surface. This force is considered as quasi-static and its direction depends on the direction of the wind. Its initial point of application is the geometrical centre of the car body lateral surface. Equation 2.33 is applied (Hibino et al., 2010).

$$H_W = \frac{1}{2} \cdot \rho \cdot S_2 \cdot V_W^2 \cdot K_2 \tag{2.33}$$

where:

H$_w$: Cross (or side) wind force (in N).

V$_w$: Wind speed (in m/sec).

ρ: Air density (in kg/m^3).

S$_2$: Lateral surface area of the vehicle (in m^2).

K$_2$: Side wind force coefficient (parameter depending on the lateral external surface of a vehicle).

The crosswind force H$_w$ is undesirable as it causes an increase of the transversal displacement of the wheelsets, and it assists the vehicle's overturning mechanism.

To deal with crosswinds, the following measures can be applied:

• Installation of wind barriers.
• Speed reduction or the interruption of operation in areas subject to strong crosswinds.

2.3.4 Residual centrifugal force

When a vehicle of mass M$_t$ runs at a speed V in a curve where the curvature radius is R$_c$, then the vehicle's centre of gravity generates the centrifugal force F$_{cf}$, which pushes the vehicle towards the outer side of the curve (Figure 2.9):

$$F_{cf} = M_t \cdot \frac{V^2}{R_c} \tag{2.34}$$

Owing to the cant of the track U, the transversal component of the weight B$_{ty}$ is simultaneously applied. B$_{ty}$ acts in a direction opposite to that of the centrifugal force and its value equals:

$$B_{ty} = M_t \cdot g \cdot \delta_p = M_t \cdot g \cdot \frac{U}{2e_o} \tag{2.35}$$

The difference between forces F$_{cf}$ and B$_{ty}$ expresses the residual centrifugal force F$_{nc}$. At the level of wheelsets, and consequently at the rail rolling surface, the following applies:

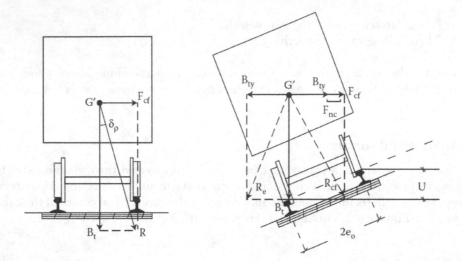

Figure 2.9 Motion in the curvature of the track – centrifugal force F_{cf} and residual centrifugal force F_{nc}. (Adapted from Alias, J., 1977, *La voie ferrée*, Eyrolles, Paris.)

$$F_{nc} = \frac{Q}{g} \cdot \left(\frac{V^2}{R_c} - g \cdot \frac{U}{2e_o} \right) = \frac{Q \cdot I}{2e_o} \qquad (2.36)$$

where:

I: Cant deficiency.

The value:

$$\frac{V^2}{R_c} - g \cdot \frac{U}{2e_o} \qquad (2.37)$$

represents the transversal residual acceleration γ_{nc}.

The increase of the vehicle's speed V as well as the decrease of the radius of curvature R_c and the decrease of the cant U contribute to the increase of the residual centrifugal acceleration.

F_{nc} is considered as a quasi-static force and is always undesirable as it causes not only the displacement of the wheelsets (risk of flange contact), but also problems regarding the transversal dynamic passenger comfort. Moreover, it assists the vehicle's derailment mechanism. However, traffic reasons such as the coexistence of low- and high-speed trains on the same track and the risk of transversal wheelset sliding towards the internal rail render the adoption of a cant, which is smaller than the equilibrium cant (theoretical cant), essential, thereby leading to the appearance of a residual centrifugal acceleration.

For example, when Q = 18 t, V = 150 km/h, R_c = 1,500 m, g = 9.81 m/s², $2e_o$ = 1.50 m, and U = 130 mm, then F_{nc} = 5.6 kN and γ_{nc} = 0.307 m/s².

To reduce the residual centrifugal force, the following measures can be applied:

- Rational choice of the curve's geometrical data (cant deficiency and cant excess).
- Ergonomic seating design.
- Use of tilting trains (see Chapter 13).

2.3.5 Total transversal force transmitted from the vehicle to the rail

According to Alias (1977) and Profillidis (1995), the total transversal force H (in t) transmitted from the vehicle to the rail is calculated by applying the following empirical formula:

$$H = a_i \cdot \left(\frac{Q \cdot I}{1,500}\right) + \left(\frac{Q \cdot V}{1,000}\right) \qquad (2.38)$$

where:

I: Cant deficiency (in case of movement along curved sections of track) or transversal track defect or twist (in case of motion in a straight path) (in mm).

Q: Axle load (in t).

V: Running speed (in km/h).

a_i: Coefficient that indicates the uneven distribution of the centrifugal force among the two axles of a bogie (values 1–1.1) (Montagné, 1975).

The first term of Equation 2.38 refers to the quasi-static forces and specifically to the residual centrifugal force. The second term refers to the random dynamic forces deriving from track alignments irregularities and motions of the vehicle itself, or its bogies, leading to unsteadiness above a critical speed (forces due to vehicle oscillations, creep forces, gravitational forces, – see Equation 2.39) (Montagné, 1975).

2.3.6 Forces due to vehicle oscillations

These transversal dynamic loads P_{dyn} the causes of which are similar to those of vertical dynamic loads confer additional transversal accelerations to the various parts of the vehicle. To address the problem, track defects must be removed and rail grinding must be applied.

2.3.7 Guidance forces

When the transversal displacement 'y' of a railway wheelset is equal to the flange clearance 'σ' between the wheel flange and the rail, the outer side of the wheel flange comes in contact with the inner part of the rails (Figure 2.10).

Transversal dynamic loads are exerted on the contact point and are called guidance forces F_j (j = 1 or 2).

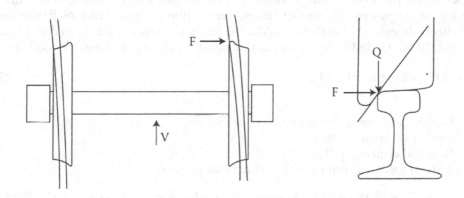

Figure 2.10 Flange contact – guidance force.

The guidance forces create problems not only for the passengers but also for the rolling stock and the track. More specifically:

- They reduce the dynamic passenger comfort (increase of transversal accelerations, jerks).
- They increase the rolling noise considerably.
- They cause wear to the wheels and rails.
- They increase the fatigue of the bogies.
- Under certain circumstances, they may result in a lateral displacement of the track and the derailment of the vehicles (see Section 3.3.3).
- Under certain circumstances, they may provoke the derailment due to wheel climb (see Section 3.3.4).

Therefore, where possible, during railway vehicles' movement the wheel flange contact must be avoided.

To reduce the guidance force, the following measures can be taken:

- Appropriate choice of the constructional characteristics of the bogies (wheel profile, wheel diameter, stiffness of bogie primary suspension, bogie wheelbase).
- Appropriate choice of the bogie technology so as to serve network operability.
- Gauge widening at the narrow curved sections of the track ($R_c < 150$–200 m).
- Lubrication of the inner face of rails at the curved sections of the track.

The guidance force F is stochastic in nature (Profillidis, 2014). Its value is dependent on several parameters. In almost all cases calculating its value is achieved through approximation and through:

(a) In situ measurements along the rail with dynamometers placed on the wheels.
(b) Simulation models.

Several models of this kind exist in the market and are used by both the industry and researchers (SIMPACK, UMLAB, Vampire Pro, Adams/Rail etc.).

Aside from the ones already mentioned, several non-commercial models have been developed by individual researchers or research groups for their own use and are not readily available in the market (Joly and Pyrgidis, 1990; Pyrgidis, 1990; Joly and Pyrgidis, 1996). With these models, it is possible to study in curves the semi-static lateral vehicle behaviour and the effect of the main features of the bogies on the 'geometric' positioning of wheelsets on track and on the wheel rolling conditions (displacements and yaw angles of wheelsets, calculation of the wheel–rail contact forces, verification of appearance of flange contact and of wheels slipping). Using these models, the guidance forces F_j are derived as a result of Equation 2.39 (all the individual forces in this equation are derived from the model).

$$F_j = \pm\left(T_1 + T_2\right) \pm S_p \pm F_{nc} \pm F_{res} \tag{2.39}$$

where:
T_1, T_2: Lateral creep forces applied on both wheels.
S_p: Total gravitational force.
F_{nc}: Residual centrifugal force.
F_{res}: Lateral forces of springs of the primary suspension.

(c) Empirically with the use of Equation 2.40 (see Section 3.3.4) (Figure 2.11) (Profillidis, 1995; Amans and Sauvage, 1969; Iwnicki, 2006).

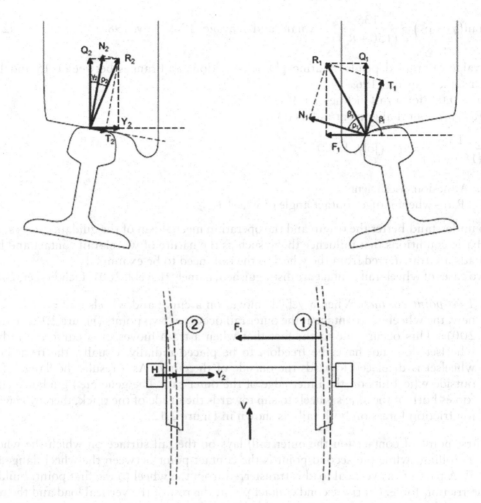

Figure 2.11 Flange contact – guidance force. A case of large yaw angle of the under-derailment wheelset.

Equation 2.40 is applicable when the yaw angle of the under-derailment wheelset is large ($\alpha \geq 5$ rad).

$$F_1 = H + Y_2 \tag{2.40}$$

where (see Figure 2.11):

F_1: Guidance force applied on under-derailment wheel 1.

H: The total lateral force that is transmitted from the vehicle to the rail (at wheelset level).

$$Y_2 = Q_2 \cdot \tan(\gamma_2 + \rho_2) \tag{2.41}$$

Q_2: Static vertical load of wheel 2.

γ_2: Angle between the rolling surface of wheel 2 and the horizontal plane (wheel–rail contact angle).

ρ_2: Angle of wheel–rail friction of wheel 2.

$$\tan(\gamma_2 + \rho_2) = \frac{135}{(150 + R_c)} \quad \text{(Amans and Sauvage, 1969; Joly, 1983)} \tag{2.42}$$

The value of $\tan\rho_2$ depends on atmospheric conditions and ranges between 0.15 and 1.25 (Amans and Sauvage, 1969).

$\gamma_2 = 0.02$ (for a rail incline of 1:40).

R_c: Horizontal alignment radius (in m).

$$\frac{135}{(150 + R_c)} \approx \mu \quad \text{(Joly, 1983)} \tag{2.43}$$

μ: Adhesion coefficient.

β_1: Rail–wheel flange contact angle of wheel 1.

To understand better the origin and the operation mechanism of the guidance forces, certain basic conditions that influence them, such as the nature of wheel–rail contact and how the loads are transferred from the wheel to the rail, need to be examined.

Two cases of wheel–rail contact are distinguished, namely (Esveld, 2001; Lichtberger, 2005):

1. *Two-point contact*: When a vehicle moves on a curve and wheels and rails are both new, the wheel–rail contact at the outer rail occurs at two points (Figure 2.12) (Esveld, 2001). This occurs due to the fact that when a train moves on a curve the railway wheelset does not have the freedom to be placed radially. Usually, the front bogie wheelset is displaced towards the outside of the curve. As a result, the flange of the outside wheel hits on the inner edge of the outer rail. The generated guidance effort forces both of the axle's wheels to slip towards the inside of the track, thereby generating friction forces on both rails as shown in Figure 2.12.

The first point of contact (on the outer rail) lays on the rail surface on which the wheel's tread is rolling, while the second point is the contact point between the wheel flange and the rail. A part of the vertical load is transferred from the wheel to the first point, building up the friction force. On the second contact point, the rest of the vertical load and the total transversal force are transferred.

2. *Single-point contact*: After a certain amount of wear of the wheel, the contact between the wheel and the outer rail occurs at a single point (single-point contact). In

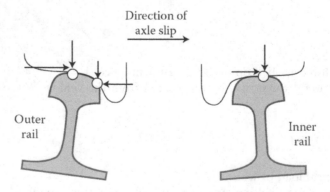

Figure 2.12 Two-point contact. (Adapted from Esveld, C., 2001, *Modern Railway Track*, 2nd edition, MRT-Productions, West Germany.)

this case, the wheel load and the total transversal force are applied at the same point while friction force builds up as a reaction to the wheel slipping towards the inside of the track.

The nature of the wheel–rail contact on a new rail depends on the radius of curvature in the horizontal alignment. The two-point contact is more likely to occur for a radii between $R_c = 1,200$ and $2,000$ m. The single-point contact is less likely to occur for radii $R_c < 1,200$ m. Finally, multiple-point contact is also less likely to occur for radii $R_c > 2,000$ m (Lichtberger, 2005).

2.3.8 Case study

Railway axle is moving with constant speed V at a straight segment of a track. It is assumed that the wheels are of conical shape with curved slants of constant radius R, while the heads of the rails are spherical (R'). The following data are given:

2e: Track gauge = 1,435 mm
f_d: Wheel flange thickness = 30 mm
$2d_a$: Inside gauge = 1,360 mm
r_o : Rolling radius of the wheel in the central equilibrium position = 0.45 m
γ_o: Angle between tangent plane and horizontal wheelset in central position = 0.025
R': Radius of curvature of the rolling surface of the railhead = 0.30 m
R: Curvature radius of the wheel tread = 0.36 m
Q_o: Static wheel load = 7.03 t

You are asked to calculate:

   a. The equivalent conicity γ_e of the wheels.
   b. The flange way clearance 'σ' (in mm) in case of a track-centred wheelset.
   c. The gravitational force S_p in case of wheel flange–rail contact (in t).
   d. The gravitational force S_p when:
        • The wheels have a conical profile (triangular cross section).
        • The wheels are cylindrical (orthogonal cross section).

Solution:

   a. $\gamma_e = R \gamma_o / (R - R') = 0.36 \times 0.025 / (0.36 - 0.30) = 0.15$ (Equation 2.14)
   b. $2\sigma = 2e - (2d_a + 2f_d) = 1,435 - (1,360 + 60) = 15$ mm $\Rightarrow \sigma = 7.5$ mm (Equation 1.15)
   c. $S_p = 2Q_o y / (R - R') = 14.6 \times 7.5 / (360 - 300) = 1.757$ t (Equation 2.16)
   d. In case of conical wheels, the total gravitational force is zero:
        $R = \infty \Rightarrow \gamma_e = \gamma_o \Rightarrow \gamma_1 = -\gamma_2$
        $S_p = S_{p1} + S_{p2} = Q_o (\gamma_1 + \gamma_2) = 0$ (Equations 2.10 and 2.11)
The same applies in case of cylindrical wheels:

$R = \infty, \gamma_o = 0 \Rightarrow -\gamma_e = 0 \Rightarrow \gamma_1 = \gamma_2 = 0$
$S_{p1} = S_{p2} = 0 \Rightarrow S_p = 0$

In Figure 2.13, a random positioning of the aforementioned wheelset on the straight track is given.

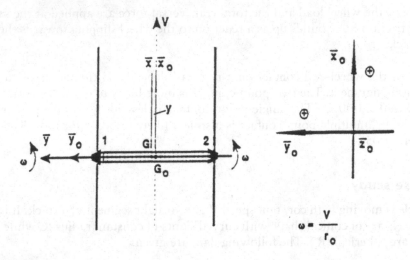

Figure 2.13 Random positioning of a railway wheelset on track – straight path.

You are asked to evaluate/provide:

The values/analytical expressions of all the applied lateral forces acting for this specific case. The following data are given:

Lateral displacement of the wheelset: y = 8 mm
Longitudinal coefficient of Kalker: $C_{11} = 16.85 \times 10^6$ N

The influence of spin is neglected.
Solution:

$X_1 = -c_{11} (-\gamma_e y/r_o) = +c_{11} \gamma_e y/r_o = + 4.49$ t (Longitudinal creep force – wheel 1) (Equation 2.17)

$X_2 = -c_{11} (+\gamma_e y/r_o) = -c_{11} \gamma_e y/r_o = -4.49$ t (Longitudinal creep force – wheel 2) (Equation 2.18)

$T_1 = T_2 = -c_{22} y' / V$ (Lateral creep forces – both wheels) (Equations 2.19 and 2.20)

$S_p = 2Q_o \gamma_e y / R \gamma_o = -1.86$ t (Gravitational force) (Equation 2.15)

2.4 LONGITUDINAL FORCES

2.4.1 Temperature forces

When the rail temperature rises, the rails show a tendency towards an increase or a decrease of their length by a value Δl:

$$\Delta l = \pm\, \alpha_t \cdot \Delta_t \cdot l_o \qquad (2.44)$$

where:

Δl: Variation of the length of the rail (expansion or contraction) (in mm).
α_t: Steel thermal expansion coefficient (in grad^{-1}).
Δ_t: $t_{re} - t_{in}$ = Actual (recorded) temperature – initial temperature (in °C).
l_o: Initial rail length (in mm).

This displacement is countered by the friction forces T_{fr} developed between the rails and sleepers and between the sleepers and ballast (Figure 2.14).

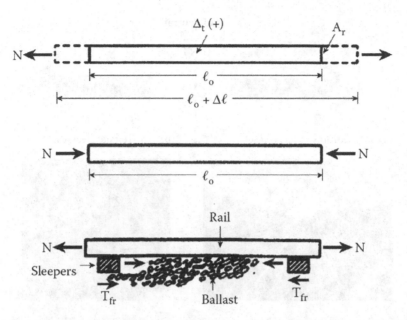

Figure 2.14 Friction forces T_{fr} between the rails and sleepers and between sleepers and ballast. Temperature forces N. (Adapted from Pyrgidis, C., and Iwnicki, S., 2006, *International Seminar Notes*, EURNEX-HIT, Thessaloniki; Profillidis V. A., 2014, *Railway Management and Engineering*, Ashgate, England.)

Fastenings, sleepers, and ballast combined with the rail's weight prevent the rail from expanding or contracting. As a result, the rail is under compressive stress when the temperature increases and under tensile stress when the temperature decreases (Green and Shrubsall, 2002).

In the case of continuous welded rails (CWR), a pair of compressive or tensile forces N applies axially on the rail (as shown in Equation 2.45).

$$N = -E \cdot A_r \cdot \frac{\Delta l}{l_0} = -E \cdot A_r \cdot \alpha_t \cdot \Delta_t \tag{2.45}$$

where:
  A_r: Rail cross section.
  E: Steel elasticity modulus.

These forces are named temperature forces and are considered to be static forces (Lichtberger, 2005). Their values remain constant almost throughout the length of the CWR. At a distance of approximately $l_A = 150$ m from each edge of the CWR (expansion zone), the value of force N is gradually decreased to zero at both the rail's edges.

When $\Delta_t = +40°C$, E = 2.1×10^3 t/cm², $\alpha_t = 1.2 \times 10^{-5}$ grad⁻¹ and rail UIC 50 ($A_r = 63.93$ cm²), then the resulting force is N = 64.5 t.

If the track is subject to excessive compressive stress, it shall buckle. For situations where tensile forces are excessive, the rail shall break (Figure 2.15), most likely at a welding point where the rail is usually weaker. Both phenomena are dangerous; however, buckling is particularly dangerous because the deviation of the track from its 'correct' geometric position causes lateral displacement and yaw angle of the wheelsets, thus favouring the emergence of creep and guidance forces, whose adverse impacts have already been analysed. The rail

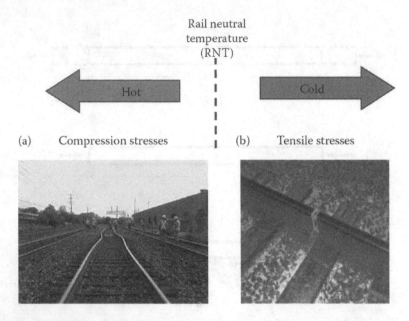

Figure 2.15 Effect of temperature changes on the rails. (a) Adapted from Transportation Safety Board of Canada, 2006, online image available at: http://www.tsb.gc.ca/eng/rapports-reports/rail/2006/r06t0153/r06t0153.asp; (b) Adapted from online image available at: http://www.fuzzyworld3.com/3um/viewtopic.php?f=5&t=4000.)

temperature for which the total longitudinal force on the rail is null is termed Rail Neutral Temperature (RNT). It is often related to the temperature of rail laying.

The temperature at which laying of continuous welded rails happens ensures the development of the lowest possible values of internal stresses (and thus the length of the expansion zone is minimised), which is termed Design Rail Neutral Temperature (DRNT).

Track buckling commonly occurs in the transverse direction (Figure 2.15). The activation of a lateral buckling effect is facilitated by the rails' transversal defects, the temperature difference between the DRNT and the RNT, the high dynamic train loads, and a low lateral track resistance (Lanza Di Scalea, 2012).

For the same track features, at curved segments of the horizontal alignment, a lower temperature increase is required for the transversal buckling effect to appear. This is expected because at curves, the rail is already curved; hence, its further lateral shift is easier to occur.

Buckling is dealt with by increasing the lateral stiffness of the track (elastic fastenings, concrete sleepers, ballast windrow up to rail height outside the sleeper) and more particularly in the case of continuous welded rails, by stress releasing.

Vertical buckling rarely occurs due to the increased weight of the track's panel.

In the case of bridges, the temperature difference Δ_t provokes direct stressing on the bridge as well as on the track's superstructure components. As already mentioned in Section 2.2.3.4, the forces due to temperature changes are taken into consideration in the design. Furthermore, the position where the 'breather switches' (or 'expansion joints') are placed is affected by these temperature changes.

2.4.2 Rail creep forces

Rail creep is defined as the gradual displacement of the rails with regard to the sleepers as well as the gradual displacement of the rails and sleepers with regard to the ballast, in the direction of the train's movement.

This specific phenomenon is due to (Australian Rail Track Corporation LTD, 2014):

- The braking of the train, which 'pushes' the rails in the direction of movement.
- The acceleration of trains, which 'pulls' the rails in the opposite direction.
- The wave motion of the rails caused by the passing trains' wheels, 'pushing' the rails in the direction of movement.
- The weight of the rails, which in the case of longitudinal track gradient, 'pushes' the rails downhill.

For single tracks with bidirectional traffic, the rail creep is smaller. For track sections with a significant gradient, the rail creep follows the direction of the descending gradient of the track, regardless of the direction of train movement.

The rail creep will result in the gradual accumulation of compressive stresses in some segments of the track and tensile stresses in others, causing a deviation of the RNT from the DRNT and rendering different track segments vulnerable to rail buckling or rail breakage.

The impacts of the rail creep are the following (Lanza Di Scalea, 2012):

- In the case of continuous welded rails, an increase of the forces exerted due to temperature changes, and in the case of jointed rails, creation of very large or very small joint gaps.
- Dissimilarities in terms of the degree and extent of the rail creep between the two rails which result in the bending of the sleepers. As a further result, bending moments are developed on the rails.
- Disturbance of the stability of the track's superstructure, resulting from the displacement of the rails and sleepers relative to the ballast layer.

The rail creep can be treated using elastic fastenings, good track ballasting, and adequate maintenance of the track superstructure.

To prevent rail creep, anti-creep devices are positioned on the track (Figure 2.16). These are U-shaped strips that are placed on the rail foot so as to prevent the longitudinal movement of the rail vis-à-vis the sleeper.

2.4.3 Braking forces: acceleration forces

The acceleration forces N_{ac} and the braking forces N_{br} are considered as static loads developed during the acceleration and braking of trains, respectively (Lichtberger, 2005).

More specifically, during acceleration, longitudinal forces build up on the track due to the static friction between the wheel and the rail. In front of the accelerating rail wheelset, tension is developed, while behind the wheelset, compression is developed. The magnitude of these longitudinal forces depends on the vertical wheel load as well as on the adhesion coefficient. The acceleration forces can be considered negligible since they do not exceed 5% of the longitudinal forces that are developed due to temperature changes.

When the vehicles are decelerating, compression tendencies appear in front of the first braking wheelset and tensile ones appear behind it, which is precisely the opposite of what happens during acceleration. Another difference in the above two functionalities is that during braking, all axles are usually contributing. Braking forces may reach up to as much as 15% of the maximum forces due to the variation in temperature. Indicatively, the values of the braking forces for the various vehicles are as follows (Lichtberger, 2005):

- For electric locomotives, they reach values equal to 12–15% of the axle loads.
- For diesel locomotives, they reach values up to 18% of the axle loads.
- For 2-axle freight wagons, they reach values up to 25% of the axle loads.

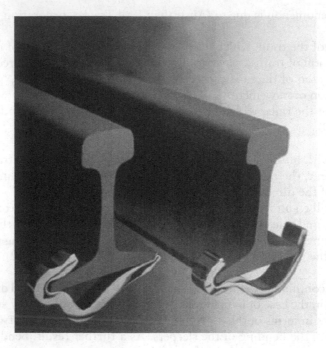

Figure 2.16 Clamp for rail creep protection.

The acceleration and braking forces are taken into account in the design of civil engineering structures (bridges, embankments). The braking force is considered as a distributed moving load.

For many railway networks, it is considered that the total braking force is equal to 25% of the overall train weight.

2.4.4 Traction forces: adhesion forces

Problems regarding railway vehicle traction concern the following two cases (Metzler, 1981):

- Determination of the load which a certain type of power vehicle can pull under certain conditions.
- Identification of the suitable type of power vehicle which can pull a specific load under certain conditions.

These problems can be resolved once the specific features governing the actual transport capacity of the locomotives, and the trains in general, are known.

These features are called 'traction basic elements' and include the following:

- Total train resistance (W).
- Power output of the motor or motors (P_t).
- Traction effort developed on the driving wheel treads (F_t).
- Adhesion force generated on the driving wheels' wheel–rail contact surface (Π).

The fundamental mathematical relation which must apply at all times in order to ensure train movement is the following:

$$\Pi > F_t > W \tag{2.46}$$

In all cases, the traction effort on the treads must be less than the resistance of the couplings.

The force that is derived as a result of the power output of the motors, which acts on the wheel treads (i.e., the contact point between the wheel and the rail, where the motor's torque is converted to force (Riley and Li, 2012)), is called traction effort on the treads (F_t).

The force acting on the axles is called traction effort (F_{tr}) (Figure 2.17).

The traction effort on the treads is not constant but alters with the speed of the locomotive and generally reduces as the speed increases. Thus, we have the following equation:

$$P_t = F_t \cdot V \tag{2.47}$$

where:

P_t: Net (or useful) power.

The maximum traction effort on the treads is developed at start up (Figure 2.18). The increase of the speed results initially in the linear reduction of the traction effort on the treads and to its further excessive reduction thereafter. However, the increase in speed also results in an increase in the train's movement resistance.

It should be noted that an electric locomotive is able to provide for a short time – instantly – traction effort that is greater than the traction effort which it can provide at continuous operation.

The motor's traction effort cannot produce transportation work if it lacks a point of application. On the railway, the point of application is to be found on the wheel tread–rail contact surface. The contact surface, the surface where the traction effort on the treads (F_t) is applied, must produce a certain resistance, which is the adhesion force (Π) (Figure 2.17).

Adhesion force (Π) is defined as the product:

$$\Pi - Q_o \cdot \mu \tag{2.48}$$

where:

Q_o: Vertical wheel load.
μ: Adhesion coefficient.

The adhesion force depends on many parameters that are classified into three categories, namely (Vasic et al., 2003; Cuevas, 2010):

1. Manufacturing features of the vehicle and the track.
2. Construction material of the wheel and the rail.
3. Environmental conditions.

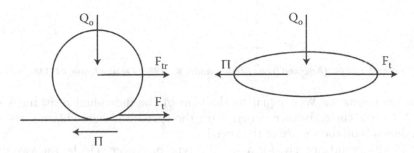

Figure 2.17 Forces appearing during train movement.

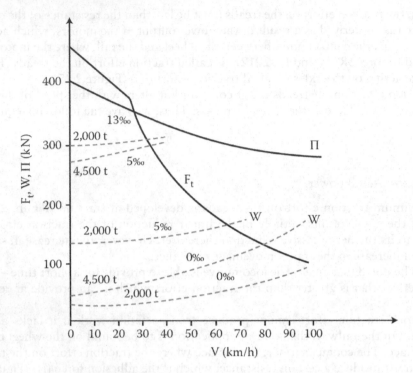

Figure 2.18 Diagram F_t–V, W–V, and Π–V. (Adapted from ABB, 1992, *Traction Vehicle Technic for All Applications*, Information Leaflet, Manheim.)

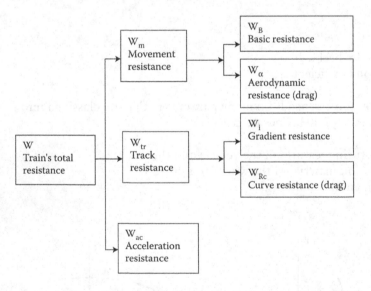

Figure 2.19 Train resistances. (Adapted from Abakoumkin, K., 1986, *Lecture Notes*, NTUA, Athens.)

The total train resistance W is equal to the sum of the individual resistances shown in Figure 2.19. Movement resistance depends on the speed and aerodynamic resistance, in particular, depends on the square of the speed.

Figure 2.18 shows indicatively, for a specific type of power vehicle, the variation of F_t, W, and Π in relation to speed. This diagram allows for the determination of the maximum speed a train can move at, for a specific type of locomotive at a specific gradient along the track and for a specific power vehicle.

The traction effort on the treads is given by the manufacturer of the rolling stock, while the adhesion force is calculated either experimentally by special trains measuring the friction coefficient or with the help of empirical formulae (Bourachot, 1984; Meccanica della Locomozione, 1998) and the train's total resistance is calculated solely by the competent traction engineer based on his network's data.

According to the diagram illustrated in Figure 2.18, the movement of a train that weighs 4,500 t at a gradient of i = 5‰ can be performed at a speed of V = 38 km/h.

2.4.5 Fishplate forces

For rails that are joined by fishplates (Figure 2.20), during the movement of trains, a force P_f is developed on the joint, which can be calculated using the equation:

$$P_f = Q_o + 2\alpha_f \cdot V_P \cdot \sqrt{k \cdot USM} \qquad (2.49)$$

where:

Q_o: Wheel load (during calculations it is increased by 20% so as to take into consideration the track cant) (kN).

$2\alpha_f$: Angle of vertical displacement of the joint (summary of the angles which are formed by the two rails and the horizon) (rad).

V_p: Train passage speed (m/s).

k: Track vertical stiffness (N/m).

USM: Unsprung masses of the vehicle (of one wheelset) (kg).

From Equation 2.49 and considering:

- Wheel loads of 11.25 t (freight trains) and 8 t (passenger trains),
- k = 60 MN/m, USM = 1,200 kg, and

Figure 2.20 Rails jointed using fishplates. (Adapted from Les Chatfield from Brighton, 2006, online image available at: https://commons.wikimedia.org/wiki/File:Fishplate_UK_2006.jpg.)

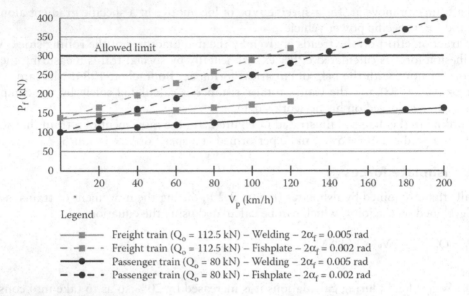

Figure 2.21 Force P_f developed at the rail connection – fishplate joint and welding – passenger and freight trains. (Adapted from Christogiannis, E., 2012, Investigation of the impact of traffic composition on the economic profitability of a railway corridor – Fundamental principles and mathematical simulation for the selection of operational scenario for a railway corridor, PhD thesis, Aristotle University of Thessaloniki, Thessaloniki, Greece.)

- The angle of vertical displacement in the case of a fishplate link $2\alpha_f$ equals 0.02 rad, while in the case of a single welding equals 0.005 rad, respectively (typical values),

the diagram of Figure 2.21 can be derived.

Looking at the same diagram, the following can be concluded (Bona, 2006):

- The fishplate force P_f increases as the speed increases.
- In the case of passenger trains, the force P_f which is developed on the fishplate joint for V_p = 120 km/h is equal to 259 kN, that is, it is about 3 times the wheel load, while for V_p = 200 km/h, the P_f force reaches a value of 378 kN, that is, about 5 times the wheel load.
- In the case of freight trains, the force P_f which is developed on the fishplate joint for V_p = 80 km/h is equal to 244 kN, that is, it is almost double the wheel load, while for V_p = 120 km/h, the P_f force reaches a value of 303 kN, that is, about 2.5 times the wheel load.

From the above, and taking into consideration that the maximum value of the force P_f adopted by different networks is around 300 kN, it can be said that for V_p > 120 km/h, the use of fishplate joints is not suitable, while for speeds V_p < 120 km/h, the use of fishplate joints may be adopted only for freight-dedicated lines. The technique of continuous welding, especially for lines that are aimed to be used by high-speed trains or heavy haul loads, is the only acceptable technique for the joining of rails.

REFERENCES

Abakoumkin, K. 1986, *Lecture Notes*, NTUA, Athens.
ABB. 1992, *Traction Vehicle Technic for All Applications*, Information Leaflet, Manheim.

Alias, J. 1977, *La voie ferrée*, éditions Eyrolles, Paris.

Amans, F. and Sauvage, R. 1969, La stabilité de la voie vis a vis des efforts transversaux exercés par les véhicules, *Annales des Ponts et Chaussées*, Vol. 1.

Australian Rail Track Corporation LTD. 2014, *Managing Track Stability – Concrete Sleepered Track*, ETM-06-06, November 2014, available online at: http://extranet.artc.com.au/docs/eng /track-civil/procedures/track-ls/ETM-06-06.pdf (accessed 26 March 2015).

Bona, M.E. 2006, *The Effect of Straightening and Grinding of Welds on Track Roughness*, PhD thesis, Queensland University of Technology, Brisbane, Australia.

Bourachot, J. 1984, *Calcul de la marche des trains – Principes et Algorithmes*, Lecture Notes, ENPC/ EPFL, Paris, 1984–1985.

Christogiannis, E. 2012, *Investigation of the Impact of Traffic Composition on the Economic Profitability of a Railway Corridor – Fundamental Principles and Mathematical Simulation for the Selection of Operational Scenario for a Railway Corridor*, PhD thesis, Aristotle University of Thessaloniki, Thessaloniki, Greece.

Cuevas, O.A. 2010, *Low Adhesion in the Wheel-Rail Contact, Investigations towards a Better Understanding of the Problem and Its Possible Countermeasures*, PhD thesis, Delft University of Technology, Netherlands.

Demiridis, N. and Pyrgidis, C. 2010, Analytical method for determining the maximum wheel load, *Technika Chronika*, April, 1, pp. 13–32.

EN 1991-2. 2003, *Eurocode 1: Actions on Structures – Part 2: Traffic Loads on Bridges*, European Committee for Standardization, Brussels.

EN 1993-2. 2006, *Eurocode 3: Design of Steel Structures – Part 2: Steel Bridges*, European Committee for Standardization, Brussels.

Esveld, C. 2001, *Modern Railway Track*, 2nd edition, MRT-Productions, Duisburg.

Frederich, F. 1985, Possibilités inconnues et inutilisées du contact rail-roue, *Rail International, Brussels*, November 1985, No. 11, pp. 33–40.

Giannakos, K. 2002, *Actions in Railways*, Papazisi, Athens.

Giannakos, K. 2014, Actions on a railway track, due to an isolated defect, *International Journal of Control and Automation*, Vol. 7, No. 3, pp. 195–212, available online at: http://dx.doi.org/10 .14257/ijca.2014.7.3.19 ISSN: 2005-4297 IJCA Copyright 2014 SERSC.

Giannakos, K. 2016, Modeling the influence of short wavelength defects in a railway track on the dynamic behavior of the non-suspended masses, *Mechanical Systems and Signal Processing*, Volumes 68–69, February 2016, pp. 68–83

Gerard, J. 2003, *Analysis of Traffic Load Effects on Railway Bridges*, PhD, Structural Engineering Division Royal Institute of Technology, Stockholm, Sweden.

Green, J. and Shrubsall, P. 2002, Management of neutral rail temperature, *The American Railway Engineering and Maintenance-of-Way Association AREMA 2002 Annual Conference*, Washington, DC, 22–25 September, available online at: https://www.arema.org/files/library /2002_Conference_Proceedings/00033.pdf (accessed 26 March 2015).

Hibino, Y., Shimomura, T. and Tanifuji, K. 2010, Full scale experiment on the behavior of a railway vehicle being subjected to lateral force, *Journal of Mechanical Systems for Transportation and Logistics*, Vol. 3, No. 1, pp. 35–43.

Iwnicki, S. 2006, *Handbook of Railway Vehicle Dynamics*, Taylor and Francis.

Joly, R. 1983, *Stabilité transversale et confort vibratoire*, Thèse de Doctorat d'Etat, Université de PARIS VI, Paris.

Joly, R. and Pyrgidis, C. 1990, Circulation d'un véhicule ferroviaire en courbe – Efforts de guidage, *Rail International*, No. 12, December 1990, Brussels, pp. 11–28.

Joly, R. and Pyrgidis, C. 1996, Etude de la stabilité transversale d'un véhicule ferroviaire muni des bogies à essieux à pseudo glissement contrôlé et à roues indépendantes, *Rail International, Brussels*, No. 12, pp. 25–33.

Lanza Di Scalea, F. 2012, Rail neutral temperature measurement, *FRA Research & Development Research Review*, 22 May 2012.

Les Chatfield from Brighton, 2006, Online image, available online at: https://commons.wikimedia .org/wiki/File:Fishplate_UK_2006.jpg.

Lichtberger, B. 2005, *Track Compendium-Formation, Permanent Way, Maintenance, Economics*, Eurail Press, Hamburg.

Meccanica della Locomozione. 1998, *Lecture Notes on Railway Engineering*, Universita degli studi di Catania, Catania, 1998–1999.

Metzler, J.M. 1981, *Géneralités sur la traction*, Lecture Notes on Railway Engineering, ENPC, Paris.

Montagné, S. 1975, Permanent way for high speed, *Institution of Mechanical Engineers*, 1975, pp. 35–37.

Online image, available online at: http://www.fuzzyworld3.com/3um/viewtopic.php?f=5&t=4000.

Profillidis, V. 1995, *Railway Engineering*, Farnham.

Profillidis V.A. 2014, *Railway Management and Engineering*, Ashgate, Farnham.

Pyrgidis, C. 1990, *Etude de la stabilité transversale d'un véhicule ferroviaire en alignement et en courbe – Nouvelles technologies des bogies – Etude comparative*, Thèse de Doctorat de l', ENPC, Paris.

Pyrgidis, C. 2004, Il comportamento transversale dei carrelli per veicoli tranviari, *Ingegneria Ferroviaria*, October 2004, Rome, No. 10, pp. 837–847.

Pyrgidis, C. 2009, Transversal and longitudinal forces exerted on the track – Problems and solutions, *10th International Congress Railway Engineering – 2009*, 24–25/06/2009, London.

Pyrgidis, C. and Iwnicki, S. 2006, *International Seminar Notes*, EURNEX-HIT, Thessaloniki.

Riesberger, K. 2008, Wheel/rail interaction on high-speed and heavy-haul railway lines: Reflections, *Rail Engineering International*, Netherlands, No. 1, pp. 11–15.

Riley, J.E. and Li, P. 2012, Tractive effort, *The American Railway Engineering and Maintenance-of-Way Association AREMA 2012 Annual Conference & Exposition*, 16–19 September, Chicago, IL.

Transportation Safety Board of Canada, 2006, Online image, available online at: http://www.tsb.gc .ca/eng/rapports-reports/rail/2006/r06t0153/r06t0153.asp.

Tschumi, M. 2008, Railway actions selected from EN 1991-2 and annex A2 of EN 1990, available online at: http://eurocodes.jrc.ec.europa.eu/doc/WS2008/EN1991_9_Tschumi.pdf (accessed 13 May 2015).

UIC. 1979, 776-1R, *Lads to Be Considered in the Design of Railway Bridges*, Technical Report, International Union of Railways, Amendment.

UIC. 1989, Fiche 714R, *Classification des voies des lignes au point de vue de la maintenance de la voie*, edition 1989, Paris, France.

UIC. 2003, *702 Loading Diagram to Be Taken into Consideration for the Calculation of Rail Carrying Structures on Lines Used by International Services*, Technical Report, International Union of Railways, Leaflet.

Urtado, M. 1993, *SNCF Médiathèque*, Paris.

Vasic, G., Franclin, F.J. and Kapoor, A. 2003, *New Rail Materials and Coatings*, Dissertation, University of Sheffield, July 2003.

Zicha, J.H. 1989, High speed rail track design, *ASCE Journal of Transportation Engineering*, Vol. 115, No. 1, pp. 68–83.

Chapter 3

Behaviour of rolling stock on track

3.1 BEHAVIOUR OF A SINGLE RAILWAY WHEELSET

3.1.1 Movement on straight paths

The movement of a conventional single (isolated) railway wheelset is known as a sinusoidal or hunting motion (Figure 3.1) and was first studied by Klingel in 1883 (Julien and Rocard, 1935).

Klingel simulated the railway wheelset with a bicone. He assumed that the axle moved at a constant speed V in the direction of the track axis while at the same time, at a random moment in time, it is laterally displaced by 'y' and rotated by an angle 'α'.

Klingel proved that the bicone motion is sinusoidal with a difference of phase of the parameters 'y' and 'α' equal to π/2 and with the following characteristics:

Wave amplitude : y_w (3.1)

Wavelength : $L_w = 2\pi \sqrt{\dfrac{r_o \cdot e_o}{\tan\gamma_o}}$ (3.2)

Frequency : $f = \dfrac{V}{2\pi} \sqrt{\dfrac{\tan\gamma_o}{r_o \cdot e_o}}$ (3.3)

Maximum lateral acceleration : $y''_{max} = 4\pi^2 \cdot y_w \cdot \dfrac{V^2}{L_w^2}$ (3.4)

The reduction of conicity γ_o of the wheels, the increase of the rolling radius r_o, and the increase of the width of the railway wheelset $2e_o$ increase the wavelength of the sinusoidal motion and reduce the lateral accelerations of the axle.

Klingel presented a pure kinematic analysis of the phenomenon assuming a harmonious motion without damping and without flange contact. In reality, the motion of a railway wheelset and particularly of a whole vehicle (car body + bogies) is much more complex (Esveld, 2001).

3.1.2 Movement in curves

Let us consider the layout of Figure 1.21. Upon entering the track and trying to achieve equilibrium, the axle is displaced by y_o with regard to the curve's outer face.

DOI: 10.1201/9781003046073-3

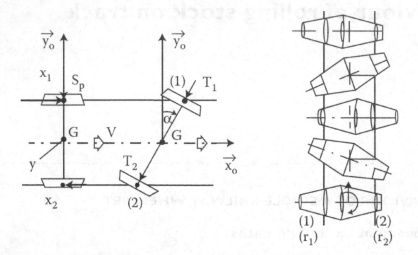

Figure 3.1 Sinusoidal motion of a railway wheelset.

Let us calculate the displacement y_o for the above position. The rolling radii of the two wheels will be:

$$r_1 = r_o + \gamma_e \cdot y_o$$
$$\text{and}$$
$$r_2 = r_o - \gamma_e \cdot y_o \tag{3.5}$$

Let us consider S_{w1} and S_{w2} as the paths covered by the two wheels during the time interval t, we have:

$$S_{w1} = V_1 \cdot t$$
$$\Rightarrow \frac{S_{w1}}{S_{w2}} = \frac{V_1}{V_2} = \frac{\omega \cdot r_1}{\omega \cdot r_2} \tag{3.6}$$
$$S_{w2} = V_2 \cdot t$$

and

$$\frac{S_{w1}}{S_{w2}} = \frac{V_1}{V_2} = \frac{(R_c + e_o) \cdot \xi}{(R_c - e_o) \cdot \xi} = \frac{\omega \cdot (r_o + \xi_e \cdot y_o)}{\omega \cdot (r_o - \xi_e \cdot y_o)} \Rightarrow y_o = \frac{e_0 \cdot r_o}{\gamma_e \cdot R_c} \tag{3.7}$$

According to the mathematical Equation 3.7 the displacement y_o is reversely proportional to the equivalent conicity γ_e and the radius of curvature R_c. On the contrary, the increase of track gauge and the increase of the wheel diameter lead to an increase of the displacement y_o. For $y_o = \sigma$ we have a contact of the flange with the outer rail, where 'σ' denotes the flange way clearance.

3.2 BEHAVIOUR OF A WHOLE VEHICLE

3.2.1 Operational and technical characteristics of bogies

3.2.1.1 Object and purposes of bogies

The term 'bogie' sometimes simply denotes a construction that supports the car body without including the wheelsets. However, and this is usually the correct definition, the term refers to the aggregate 'secondary suspension-bogie frame and primary suspension-wheelsets'.

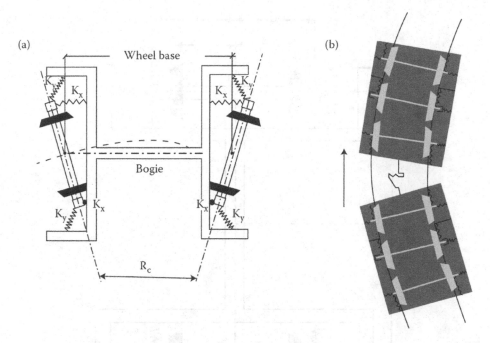

Figure 3.2 (a) Inscription of a two-axle bogie in curves – ideal inscription. (b) Inscription of two, three-axle bogies in curves.

The ability of the inscription of a railway vehicle in curves depends directly on the length of the vehicle. Initially, railway trailer vehicles were relatively short and their inscription in curved sections of the horizontal alignment was achieved through two or three single wheel-sets linked directly to the car body. The evolution of the railway as a means of transport went hand in hand with the increase in the vehicles' transport capacity, a fact which dictated the increase in the vehicles' length. Under these circumstances, the inscription of vehicles could no longer be attained using the technique of the single wheelsets.

Using the bogies, the inscription is achieved essentially via the bogies (wheelbase length $2a < 4.0$ m) while the car body follows their movement (Figure 3.2).

The bogies must:

- Allow the smooth inscription of the wheelsets in curves.
- Assist the optimum transfer of loads from the car body to the rails.
- Provide stability of the vehicles on a straight path (development of very high speeds).
- Provide dynamic comfort to passengers in three directions.
- Have relatively low construction and maintenance cost.

3.2.1.2 Conventional bogies

3.2.1.2.1 Description and operation

Nowadays, 'conventional' or 'classic' bogies are broadly used in trailer and power vehicles. In this technology, the bogies are fitted with wheelsets, the wheels of which are rigidly linked to the axle resulting in the rotation of the wheels and the axle at the same angular velocity (classical wheelsets). The bogie frame is connected to the car body and the wheelset by means of elastic elements and dampers providing the vehicle with two suspension levels, namely the

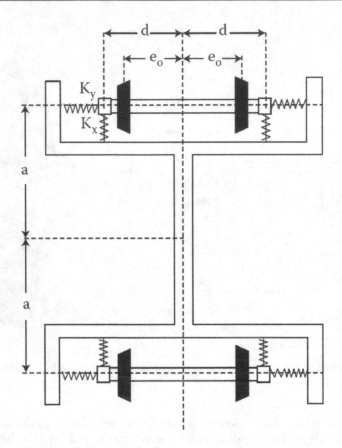

Figure 3.3 Conventional bogie with classical wheelsets.

primary suspension: wheelset-bogie suspension (which usually utilises coil springs and dampers or rubber elements [chevrons]) and secondary suspension: bogie-car body suspension (which usually utilises air suspension or coil springs and dampers) (Figure 3.3).

There are various types of conventional bogies; the choice and design among them depend directly on the functionality of the vehicles on which they will be mounted and on the geometrical characteristics of the track they will run on.

Figure 3.4 shows in detail all the individual parts which form a conventional power bogie (Schneider Jeumont Rail, n.d.).

3.2.1.2.2 Design of the bogies

Good construction of the track superstructure does not guarantee on its own a smooth train ride and the achievement of the desired performances; design and construction of the rolling stock are of equivalent importance. Developing a railway bogie from design to commissioning involves the following main stages:

- Conception of the bogie technology and physical explanation of the bogie behaviour.
- Theoretical study and modelling of its dynamic behaviour using simulation models.
- Design and construction.
- Testing.
- Commissioning and entering into operation.

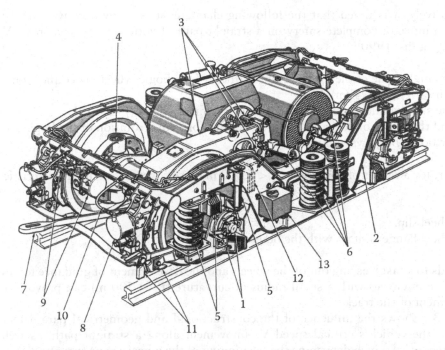

1. Wheelset–axle box	6. Secondary suspension	11. Sand and stone scattering device
2. Bogie frame	7. Brake lever	12. Contact brush
3. Reducer device	8. Air brake	13. Wheel flange lubricator
4. Transmitter	9. Brake pad	
5. Primary suspension	10. Brake pipe	

Figure 3.4 **Conventional bogie – main parts.** (Adapted from Schneider Jeumont Rail., n.d., *Bogie CL93 à Moteurs Asynchrones, Catalogue pièces de rechange*, Le Creusot, France.)

The geometrical and technical characteristics of the bogies that substantially affect the dynamic behaviour of the vehicles are (Joly, 1983):

- The longitudinal (K_x) and lateral (K_y) stiffness of the primary suspension springs.
- The bogie wheelbase (2a) (Figure 3.3).
- The wheel diameter ($2r_o$).
- The mass of the bogie (M') and of the wheelsets (m).
- The equivalent conicity of the wheels (γ_e).

All the above elements are directly related to the lateral behaviour of the bogies that determine the steady motion of the vehicles on straight paths and the good negotiation of curves as well as:

- The car body mass ($\bar{M}$).
- The vertical stiffness ($\bar{K}_z$) of the secondary suspension springs.
- The damping coefficients ($\bar{C}_x, \bar{C}_y,$ and $\bar{C}_z$) of the secondary suspension dampers.

The last three features are directly related to the vertical behaviour of the bogies that characterises the dynamic comfort of the passengers.

Despite the technological advances for the rolling stock and the track equipment, it is not possible to guarantee both high speeds on straight paths and good negotiation of wheelsets in curves.

Indicatively, it is noted that the following characteristics allow a conventional railway vehicle to move in complete safety on a straight path of good quality at a speed V > 350 km/h. (Pyrgidis, 1990):

- The high value of the longitudinal stiffness of the bogie-wheelsets connection (3×10^7 N/m $\geq K_x \geq 10^7$ N/m).
- The small value of the equivalent conicity of the wheels ($\gamma_e < 0.12$).
- And the fixing of devices which restrict the horizontal rotation of the bogie and of the car body (bogie yaw dampers).

However, for a radius of curvature $R_c < 6{,}000$ m these constructional characteristics relate to:

- Wheel slip.
- Wheel flange contact with the outer rail.

This leads to a fast wearing out of the wheels and the development of guidance forces which in the case of curves with a small radius of curvature ($R_c < 500$ m) can provoke a lateral displacement of the track.

Table 3.1 shows the influence of the constructional and geometrical parameters of the bogies at the vehicle's critical speed V_{cr} (movement along a straight path) as well as on parameters (y, α) which determine the positioning of the wheelsets in curves.

An increase in the critical speed increases the stability of the vehicle while moving on a straight path, hence the possibility of achieving higher speeds. An increase in the lateral displacement 'y' and the yaw angle 'α' of the bogie's front wheelset is equivalent to an increase in creep forces, a likely wheel slip and the appearance of guidance forces (flange contact); and in general to an expected poor negotiation of curves (wear on the wheels and the rail, lower speeds, risk of derailment and lateral displacement of the track).

The inability of classical bogies to combine the stable motion of vehicles at high speeds on straight paths with a safe and wear-free negotiation of curves has led to a continuous effort

Table 3.1 Influence of constructional and geometrical characteristics of bogies

Constructional and geometrical characteristics of bogies	Movement along a straight path – change of V_{cr}	Movement in curves – change of y, α
Reduction in equivalent conicity γ_e	Increase	Increase
Increase of the wheelbase 2a	Increase	Increase
Increase in the diameter of the wheels $2r_o$	Increase	Increase
Increase in longitudinal stiffness of the bogie-wheelset linkage K_x	Increase	Increase
Increase in lateral stiffness of the bogie-wheelsets linkage K_y	Increase	Unchanged
Increase in mass of the bogies M' and the wheelsets m	Reduction	Unchanged
Placement of yaw dampers between the bogie and the car body	Increase	Increase

Source: Adapted from Joly, R., 1983, Stabilité Transversale et Confort Vibratoire en Dynamique Ferroviaire, Thèse de Doctorat d'Etat, Université de Paris, Paris; Pyrgidis, C., 1990, Etude de la Stabilité Transversale d'un Véhicule Ferroviaire en Alignement et en courbe – Nouvelles Technologies des Bogies – Etude Comparative, Thèse de Doctorat de l', ENPC, Paris.

Note: Movement along straight path and in curves.

to improve the performance of the wheel-rail system. Within the context of this effort, many improvements have been made regarding the way in which bogies are designed and manufactured (new techniques, new elastic connecting materials, lighter bogies and wheel sets, etc.).

One parameter that restricts the performance of conventional bogies is the equivalent conicity γ_e which characterises the wheel wear. This parameter significantly influences the lateral behaviour of the bogies (Pyrgidis and Bousmalis, 2010). The increase in the number of kilometres run by the bogies translates to increased wheel wear and increased equivalent conicity; this results in a decrease in the vehicle's critical speed for which it was originally designed. To regain the initial performance, the profile of the wheel treads needs to be reshaped at frequent intervals.

3.2.1.3 Bogies with self-steering wheelsets

The behaviour of a bogie in curves is improved when the bogie's wheelsets are placed radially within the curved path. This positioning is not possible with conventional bogies, where generally, bogie-wheelset connections are rigid.

The self-steering (or auto-oriented wheelsets or radial wheelsets or steered bogies) technology allows an almost ideal negotiation of bogies in curves of a small radius of curvature ($100 \text{ m} < R_c < 500 \text{ m}$). For this technology, the two classical wheelsets of a conventional bogie are also connected to each other by means of elastic connections of defined stiffness K_s and K_b (Figure 3.5), where K_s is the lateral stiffness and K_b the angular stiffness.

The bogie frame-wheelset connection is achieved in different ways (Scheffel, 1974; Scheffel and Tournay, 1980; Joly, 1988; Pyrgidis, 1990).

In the case of conventional bogies, the value of the lateral stiffness K_y between the bogie and the wheelsets depends on the value of the longitudinal stiffness K_x ($K_x = \lambda' \cdot K_y$ where $\lambda' < 1$). On the contrary, the technology of self-steering wheelsets allows the manufacturing of springs with independent lateral and angular stiffness values (the angular stiffness K_b plays the same role as the K_x). As the value of the lateral stiffness does not practically influence the

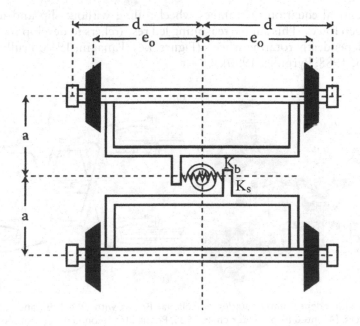

Figure 3.5 Bogie with self-steering wheelsets.

negotiation of bogies in curves, it is possible, using this technology, by reducing the angular stiffness and increasing the lateral stiffness to achieve very satisfactory results in small radii of curvature while securing at the same time average values of speed on straight paths (V = 160–220 km/h).

Moreover, this technology seems to provide the wheelsets with a guidance mechanism that tends, in most cases, to position the wheelset inwardly to the curve in a radial way. This technology was first proposed by H. Scheffel in South Africa and it was afterwards developed in countries with railway networks comprising a significant percentage of curves with a small radius of curvature (Scheffel, 1974). This technology is also applied in the bogies of tilting trains (Pyrgidis and Demiridis, 2006).

3.2.1.4 Bogies with independently rotating wheels

In conventional bogies, during a lateral movement of a wheelset on the track, the two wheels of each axle rotate at different rolling radii due to their conical profile. This results in the appearance of longitudinal creep forces of equal value and opposite direction ($X_1 = -X_2$) on each wheel.

To avoid the sinusoidal movement of a wheelset, it is necessary to dispense the rigid connection of the wheels with the axle in order for the two wheels to be able to rotate at different angular velocities, thus maintaining during their movement the mathematical Equation 2.32:

$$\omega_1 \cdot r_1 = \omega_2 \cdot r_2 = V$$

where:

ω_1, ω_2: Angular velocities of the two wheels.
r_1, r_2: Rolling radii of the two wheels.
V: Forward wheelset speed.

This mathematical equation guarantees wheel rolling without slip and nullification of longitudinal creep forces. This simple reasoning led researchers to develop the technology of bogies with independently rotating wheels (Figure 3.6) (Panagin, 1978; Frullini et al., 1984; Frederich, 1985, 1988; Pyrgidis, 1990).

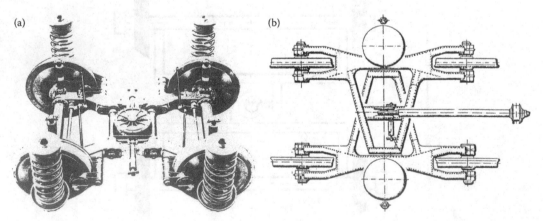

(a)   (b)

Figure 3.6 Bogies with independently rotating wheels. (a) Bogies with wheelsets and (b) bogies without wheelsets. (Adapted from Frederich, F., 1985, Possibilités inconnues et inutilisées du contact rail-roue, *Rail International*, Brussels, Novembre, 33–40.)

Using this technology each bogie has four wheels, which rotate at different angular velocities (freely).

Two techniques of implementation of this technology are distinguished:

- Bogies with wheelsets (Figure 3.6a).
- Bogies without wheelsets (Figure 3.6b).

Irrespective of the technical implementation, this technology theoretically allows for the development of very high critical speeds on straight paths. However, the wheelset is very vulnerable to lateral displacement since it can only apply the gravitational force on the track, which is activated at each lateral displacement of the wheelset due to the variable conical wheel profile (Pyrgidis and Panagiotopoulos, 2012). Conversely in curves, wheelsets with wheels, which rotate independently, cannot be placed on the track in a radial position.

It is possible to improve the positioning of the wheelsets in curves should a high value of equivalent conicity be used. This choice does not cause problems on straight paths (since the sinusoidal wheelset movement is mitigated), while at the same time the value of stabilising gravitational forces increases.

The technology of independently rotating wheels is extensively implemented in tramways (Pyrgidis, 2004; Pyrgidis and Panagiotopoulos, 2012).

3.2.1.5 Bogies with creep-controlled wheelsets

In this technology, each bogie bears four wheels rotating at different angular velocities; hence in this case, the mathematical Equation 2.32 does not apply.

Magnetic coupling of the two wheels (Figure 3.7) generates a damping torsional torque, the value of which is proportional to the difference of the angular velocities of the two wheels (Geuenich et al., 1985):

$$C_\rho = C_\varphi \cdot \left(\omega_1 - \omega_2 \right)$$

where:

C_ρ: Damping torsional torque.
C_φ: Damping coefficient.

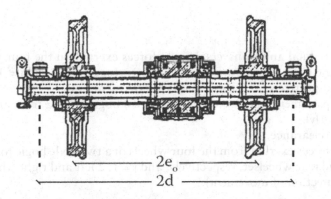

Figure 3.7 Creep-controlled wheelset. (Adapted from Pyrgidis, C., 1990, Etude de la Stabilité Transversale d'un Véhicule Ferroviaire en Alignement et en courbe – Nouvelles Technologies des Bogies – Etude Comparative, Thèse de Doctorat de l' ENPC, Paris; Geuenich, W., Cunther, C. and Leo, R. 1985, Fibre composite bogies has creep controlled wheelsets, *Railway Gazette International*, April, 279–281.)

The torque C_ρ is of the same nature as the one that causes the pair of lateral creep forces, but much smaller.

The bogie wheelset with controlled creeping technology was mainly developed in the USA and Germany. It allows for very high speeds on straight paths without the use of bogie-yaw dampers, thus significantly simplifying the bogie-wheelset connection.

The performance of this technology can be optimised by varying the damping coefficient C_φ as a function of the vehicle forward speed V.

In spite of its positive impact on vehicle behaviour, the development of this technology was abandoned due to its increased implementation cost.

3.2.1.6 Bogies with wheels with mixed behaviour

These bogies are equipped with a special mechanism that allows them to behave like conventional bogies on straight sections of the track, and to behave like bogies with independently rotating wheels in curves. This technology is applied in tramway vehicles (Pyrgidis, 2004).

3.2.2 Wheel rolling conditions and bogies inscription behaviour in curves

In curves, the following wheel rolling and bogie-wheelsets positioning cases may be considered:

1. Rolling of all bogie wheels without flange contact, without slip and without the development of creep forces (pure rolling). This case is purely theoretical and is considered to be perfect since due to the absence of forces, no wear is noticed either on the rolling stock or the track.

In this case, the following mathematical expressions apply:

$$X_{ij} = T_{ij} = 0$$

and

$$|y_i| < \sigma \left(F_{ij} = 0 \right)$$

where:

X_{ij}, T_{ij}: Longitudinal and transversal creep forces exerted on the four wheels of a two-axle bogie (i = 1,2 front and rear wheelset, respectively, and j = 1,2 left and right wheel, respectively, in the direction of movement).

y_i: Lateral displacements of the two wheelsets of a bogie (i = 1,2 front and rear wheelset, respectively).

σ: Flange way clearance.

F_{ij}: Guidance forces exerted from the four wheels of a two-axle bogie to the rails (i = 1, 2 front and rear wheelset, respectively, and j = 1, 2 left and right wheel, respectively, in the direction of movement).

2. Rolling of all bogie wheels without flange contact and without slip. The only forces applied on the wheels are the creep forces. This case is considered ideal and may be met under real service conditions where the negative impact on the rolling stock and on the track is considerably minimised (minimal material fatigue and low-level noise emission).

In this case, the following mathematical expressions apply:

$$\sqrt{X_{ij}^2 + T_{ij}^2} < \mu \cdot Q_o \tag{3.8}$$

and

$$|y_i| < \sigma(F_{ij} = 0)$$

where:
   μ: Wheel-rail friction coefficient.
   Q_o: Wheel load.

3. Rolling featuring contact of the front bogie wheelset with the outer rail (via their outer wheel flange) and, accordingly, no contact of the rear bogie wheelset (the rear wheelset may or may not slip) (Figure 3.8).

This case is frequently encountered in small radii curves. It results in wearing out of the contact wheel and mainly of the outer rail, rolling noise, and fatigue of contact materials. This case of inscription is not desirable; however, it is considered acceptable, provided that the value of the exerted guidance force F_{11} is not very high and, obviously, it is lower than the limits set by derailment and track lateral shift criteria.
   In this case, the following mathematical expressions pertain:

$$F_{11} \neq 0, \quad F_{21}, \quad F_{22}, F_{12} = 0, \quad y_1 = +\sigma, \quad y_2 \neq \pm\sigma$$

4. Rolling of both bogie wheelsets in contact with the outer rail (via their outer wheel flanges) (both wheelsets may or may not slip) (Figure 3.9). This case is more adverse than case 3 as two of the four wheels come in contact (via the flange) with the rail,

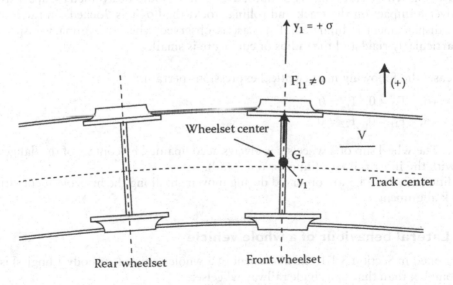

Figure 3.8 Third special rolling condition. Left-wheel flange of front wheelset-outer rail contact – rolling of rear wheelset without wheel flange-rail contact.

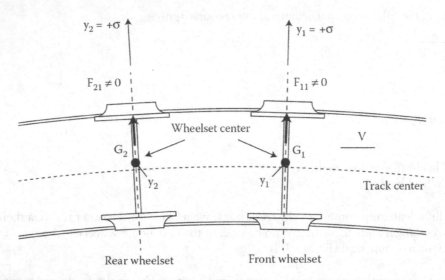

Figure 3.9 Fourth special rolling condition. Left-wheel flange of front and rear wheelset-outer rail contact.

and as a result the adverse impact is increased. However, this case may be acceptable provided that derailment and track lateral shift are within the set limits.

In this case, the following mathematical expressions pertain:

$$y_1 = +\sigma, \ F_{11} \neq 0, \quad F_{12} = 0$$
$$y_2 = +\sigma, \ F_{21} \neq 0, \quad F_{22} = 0$$

5. Rolling with front bogie wheelset-outer rail contact (via the outer wheel flange) and rear bogie wheelset-inner rail contact (via the inner wheel flange) (Figure 3.10). This case, known as 'crabbing', is the most averse and should be avoided as, apart from the adverse impact on the track and rolling stock, the bogie is 'locked' on the track and its displacement is hindered. This case is observed when the primary suspension is particularly rigid and the radius of curvature is small.

In this case, the following mathematical expressions pertain:

$$y_1 = +\sigma, \quad F_{11} \neq 0, \ F_{12} = 0$$
$$y_2 = -\sigma, \quad F_{21} = 0, \ F_{22} \neq 0$$

Remark: The wheel slip of a wheelset is always accompanied by contact of the flange of one wheel with the inner rail face (inner or outer rail).

Conditions 3, 4, and 5 are observed during movement along the horizontal curvatures of the track alignment.

3.2.3 Lateral behaviour of a whole vehicle

As mentioned in Section 3.1.1, the movement of a whole vehicle (car body + bogies) is much more complex than that of a single railway wheelset.

Dynamic railway engineering, a division of the applied engineering sector, allows the development of mathematical models simulating the lateral behaviour of a railway vehicle

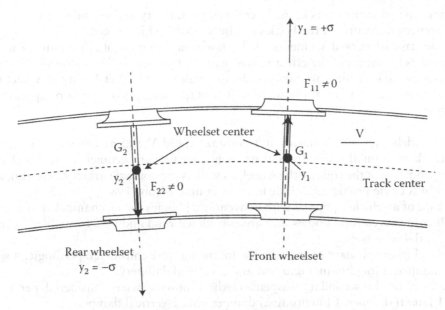

Figure 3.10 Fifth special rolling condition. 'Crabbing rolling': left-wheel flange of front wheelset–outer rail contact. Right-wheel flange of rear wheelset–inner rail contact.

on straight paths and in curves. With the aid of these models, it is possible to study the influence of the construction characteristics of the bogies on the 'geometric' positioning of the wheelsets on track and to determine, for a given speed and bogie construction characteristics, the minimum radius of curvature in the horizontal alignment, which ensures acceptable rolling conditions and inscription of the wheelsets in curved sections of the track (avoidance of slipping, absence of guidance forces).

There are many models available in the market. These models are used both in the industry and in academia and also in research institutes. Whether these models approach reality and to what extent depends on the assumptions and the hypotheses made during their development as well as their mathematical approach. These models are constantly evolving helping to improve traffic safety at all levels and to achieve a lower vehicle construction cost and a lower cost for the maintenance of track infrastructure and rolling stock.

The simulation models that can be acquired from the market are, among others: SIMPACK, UMLAB, Vampire Pro, Adams/Rail, NUCARS, GENSYS, and MEDYNA. All these models take into account the real wheel profile, the track geometric defects, and rolling conditions of the wheels. In straight segments, these models calculate speeds and accelerations (some models only calculate speeds) and in curves they calculate the applied forces/stresses, the contact surface area, and the geometric positioning of the wheelsets on the track (some models only calculate the forces).

Apart from the above models that are available in the market and some other models that are free to use (wheel rail contact calculator) there are some models that have been developed by individual researchers or research groups for their own use, and cannot be found in the market.

A group of such models are described in the literature references (Joly, 1983, 1988; Joly and Pyrgidis, 1990, 1996; Pyrgidis, 1990, 2004; Pyrgidis and Joly, 1993). With these models it is possible to study the following features for five different technologies of bogies (conventional bogies, bogies with self-steering wheelsets, bogies with independently rotating wheels, bogies with creep-controlled wheelsets and bogies with mixed behaviour):

- In the case of tangent track, the lateral vehicle stability and the influence of the main construction features of the bogies on the 'critical' vehicle speed.
- In the case of curved segments of the horizontal alignment, the semi-static lateral vehicle behaviour and the effect of the main features of the bogies on the 'geometric' vehicle behaviour (displacements and yaw angles of wheelsets) and the wheel rolling conditions (calculation of the wheel-rail contact forces, verification of appearance of slipping).

For these models, the vehicle moves at a constant speed V and its movement occurs on a railway track without the longitudinal gradient and without geometric track defects. To study the geometry of the contact the wheel, as well as the contact surface of the rail, is both simulated by a circle profile (i.e., circle to circle contact) (Joly, 1983).

In the case of a vehicle equipped with conventional bogies the mechanical system consists of the following seven solid bodies that are considered rigid and undeformable: 1 car body, 2 bogies, and 4 wheelsets.

At the level of the primary suspension the following were considered per bogie: 4 springs, 1 lateral damper, 1 longitudinal damper, and 2 vertical dampers.

At the level of the secondary suspension the following were considered per bogie: 2 springs, 1 lateral damper, 1 longitudinal damper, and 2 vertical dampers.

This mechanical system illustrates in fact the French passenger vehicles of the 'Corail' type.

Additionally, the following key assumptions/assumptions are made:

At straight segments:
- The creep forces are expressed on the basis of the linear theory of Kalker and the creep coefficients C_{ij} are considered to be reduced by 33%.
- The rails are not taken into consideration, and as a result the guidance of the wheelsets during the movement is ensured by the combined action of the equivalent conicity of the wheels and the creep forces that are exerted on the wheel-rail contact level. At the same time their lateral stiffness, which is bigger than the stiffness of the vehicle's elastic links, is ignored.

In curves:
- The study of the lateral vehicle behaviour refers to the circular segment of the curve.
- The vertical loads are distributed equally on both wheels of the axles.
- For the calculation of the creep forces the nonlinear theory of Johnson–Vermeulen was adopted (Vermeulen and Johnson, 1964).

3.2.3.1 Vehicles with conventional bogies

Figure 3.11 shows the variation in the vehicle critical speed V_{cr} as a function of the longitudinal bogie-wheelsets stiffness K_x for both values of bogie-yaw dampers longitudinal stiffness ($K_o = 3 \times 10^6$ N/m and $K_o = 0$).

There is a zone of values of K_x between 7×10^6 and 1.5×10^7 N/m, where the critical speed reaches its highest values ($V_{cr} = 465–495$ km/h, for $K_o = 3 \times 10^6$ N/m) (Pyrgidis, 1990). This area of K_x values is seen as being the greatest 'safety margin' as regards the stability of vehicles on straight paths for the constructional characteristics of the vehicle illustrated in Figure 3.11.

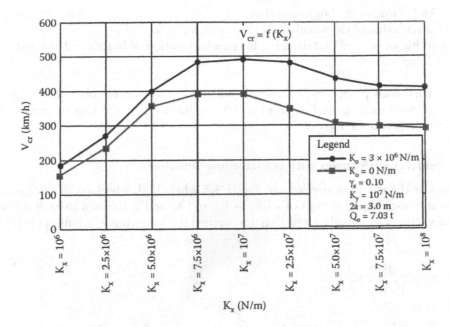

Figure 3.11 Conventional bogies – variation of V_{cr} as a function of K_x. (Adapted from Pyrgidis, C., 1990, Etude de la Stabilité Transversale d'un Véhicule Ferroviaire en Alignement et en courbe – Nouvelles Technologies des Bogies – Etude Comparative, Thèse de Doctorat de l', ENPC, Paris.)

Table 3.2 Performances of vehicles with conventional bogies – running on straight path and in curves

$K_x = 8 \times 10^6 \ N/m$	$2a = 3.0 \ m$	$2e_o = 1.50 \ m$	$\sigma = \pm 10 \ mm$
$K_y = 10^7 \ N/m$	$2r_o = 0.90 \ m$	$Q_o = 7.03 \ t$	$\gamma_{nc} = 0.02 \ g$
Adjustable characteristics	*Straight path (V_{cr})*	*Curves (occurrence of slip)*	*Curves ($R_c = 500 \ m$) ($K_o = 0$)*
$K_o = 3 \times 10^6 \ N/m$ $\gamma_e = 0.05$	658 km/h	$R_c = 6,300 \ m$ (contact)	$F_{11} = 35.1$ kN
$K_o = 3 \times 10^6 \ N/m$ $\gamma_e = 0.10$	482 km/h	$R_c = 4,800 \ m$	$F_{11} = 20.7$ kN
$K_o = 3 \times 10^6 \ N/m$ $\gamma_e = 0.20$	320 km/h	$R_c = 4,300 \ m$	$F_{11} = 0$ (slip)

Source: Adapted from Pyrgidis, C., 1990, Etude de la Stabilité Transversale d'un Véhicule Ferroviaire en Alignement et en courbe – Nouvelles Technologies des Bogies – Etude Comparative, Thèse de Doctorat de l' ENPC, Paris.

For values $K_x > 3.5 \times 10^7$ N/m approximately (and $K_o = 3 \times 10^6$ N/m), the critical speed remains roughly equal to $V_{cr} = 450$ km/h. The absence of bogie-yaw dampers ($K_o = 0$) reduces the critical speed by about 20%.

The value of $K_x = 8 \times 10^6$ N/m is considered as the optimum value for longitudinal stiffness. On the one hand, this specific value is within the 7×10^6–1.5×10^7 limits, and, on the other hand, it is relatively small which facilitates negotiation of curves.

Table 3.2 shows the performances of vehicles with conventional bogies on straight paths and in curves.

Indicatively, it is noted that the following properties allow, in theory, a classical railway vehicle to run on a straight path of good ride quality at speeds V > 650 km/h:

- The high value of the bogie-wheelset longitudinal stiffness ($K_x = 8 \times 10^6$ N/m).
- The small value of the wheel equivalent conicity ($\gamma_e = 0.05$).
- The fixing of devices that limit the horizontal rotation of bogies and car body (bogie-yaw dampers).

However, with such properties, in case of radii $R_c < 6,300$ m we observe contact of the wheel flange with the outer rail (Pyrgidis,1990). It has to be noted that in curves the yaw dumpers are deactivated ($K_o = 0$).

3.2.3.2 Vehicles with bogies with self-steering wheelsets

If we consider the mechanical system in Figure 3.5 where both wheelsets of a bogie are connected to the bogie using springs of stiffness K_x and K_y and connected to each other using springs of stiffness K_b (angular) and K_s (lateral), then the following relations apply (Pyrgidis, 1990):

$$K_{st} = K_s + \frac{d^2 \cdot K_x \cdot K_y}{d^2 \cdot K_x + a^2 \cdot K_y} \tag{3.9}$$

$$K_{bt} = K_b + K_x \cdot d^2 \tag{3.10}$$

where:

K_{st}: Overall lateral stiffness of the primary suspension (mechanical system).
K_{bt}: Overall longitudinal stiffness of the primary suspension (mechanical system).
2a: Bogie wheelbase.
2d: Lateral distance between springs and dampers of the primary suspension.

For d = 1.0 m and 2a = 3.0 m Equations 3.9 and 3.10 become:

$$K_{st} = K_s + \frac{K_x \cdot K_y}{K_x + 2.25K_y} \tag{3.11}$$

$$K_{bt} = K_b + K_x \tag{3.12}$$

and for $K_s = K_b = 0$, Equations 3.11 and 3.12 become:

$$K_{st} = \frac{K_x \cdot K_y}{K_x + 2.25K_y} \tag{3.13}$$

$$K_{bt} = K_x \tag{3.14}$$

From the above equations, the following may be concluded:

- Stiffnesses K_s and K_b increase the total stiffness (lateral and longitudinal) of the system.
- The angular stiffness K_b plays the same role (for d = 1.0 m) as the longitudinal stiffness K_x of a conventional bogie.
- The total lateral stiffness of the primary suspension of a conventional bogie depends as much on K_x as on K_y.

Table 3.3 shows the performance of bogies with self-steering wheelsets on straight paths and in curves. Compared with conventional bogies, a smaller value of the total longitudinal stiffness ($K_{bt} = 2 \times 10^6$ N/m $< 8 \times 10^6$ N/m) and a smaller value of the total lateral stiffness ($K_{st} = 1.3 \times 10^6$ N/m $< 2.62 \times 10^6$ N/m) are observed.

Indicatively, it is noted that bogies with self-steering wheelsets make it possible to combine very good negotiation of small and very small radius curves with a fair value of speed on straight paths.

3.2.3.3 Vehicles with independently rotating wheels

Table 3.4 presents the performances of bogies with independently rotating wheels on straight paths and in curves (Pyrgidis, 1990; Pyrgidis and Joly, 1993; Joly and Pyrgidis, 1996).
Indicatively, it is noted that:

- This technology allows theoretically, without the fixing of bogie-yaw dampers, the development of very high critical speeds on straight paths while eliminating the hunting of wheelsets (absence of longitudinal creep forces). However, the wheelset is very sensitive to lateral displacements.
- A great value of equivalent conicity facilitates both the negotiation of bogies in curves and the motion on straight paths as it increases the value of the gravitational force that tends to centre the wheelset on the track.
- For wheelbase and wheel diameter values, which are the same as those of high and very high-speed conventional bogies, a wheel slip is observed in comparatively much

Table 3.3 Performances of vehicles with bogies with self-steering wheelsets – running on straight path and in curves

$K_b = 10^6$ N/m $K_x = K_y = 10^6$ N/m $K_s = 10^6$ N/m	$2a = 3.0$ m $2r_o = 0.90$ m	$2e_o = 1.50$ m $Q_o = 7.03$ t	$\sigma = \pm 10$ mm $\gamma_{nc} = 0.02$ g	
Adjustable characteristics	Straight path (V_{cr})	Curves (occurrence of slip)	Curves ($R_c = 500$ m)	Curves ($R_c = 200$ m)
$K_o = 3 \times 10^6$ N/m $\gamma_e = 0.10$	323.5 km/h	$R_c = 1,200$ m	$F_{11} = 15.4$ kN ($K_o = 0$)	$F_{11} = 59.6$ kN ($K_o = 0$)
$K_o = 0$ $\gamma_e = 0.20$	198 km/h	$R_c = 250$ m	$F_{11} = 0$	$F_{11} = 14.8$ kN

Source: Adapted from Pyrgidis, C., 1990, Etude de la Stabilité Transversale d'un Véhicule Ferroviaire en Alignement et en courbe – Nouvelles Technologies des Bogies – Etude Comparative, Thèse de Doctorat de l', ENPC, Paris.

Table 3.4 Performances of vehicles with bogies with independently rotating wheels – motion on straight path and in curves

$K_x = 10^8$ N/m $K_y = 10^6$ N/m	$2e_o = 1.50$ m $K_o = 0$	$\sigma = \pm 10$ mm $Q_o = 7.03$ t	$\gamma_e = 0.20$ $\gamma_{nc} = 0.02$ g
Straight path (V_{cr})	Curves (occurrence of slip)	Curves ($R_c = 500$ m)	Curves ($R_c = 100$ m)
$V_{cr} \approx \infty$ Vulnerability in lateral displacements	1,400 m	$F_{11} = 1.7$ kN	$F_{11} = 15.7$ kN

Source: Adapted from Pyrgidis, C., 1990, Etude de la Stabilité Transversale d'un Véhicule Ferroviaire en Alignement et en courbe – Nouvelles Technologies des Bogies – Etude Comparative, Thèse de Doctorat de l' ENPC, Paris.

smaller curvature radii, while the forces exerted in very small radius curves are much smaller.

3.2.3.4 Comparative assessment

In Table 3.5, a comparison of the performances of the three bogie technologies under examination on straight paths and in curves is attempted (where P_{4w} is the power that is consumed at the level of the four wheels of the bogie when rolling occurs without contact between the wheel flange and the inner side of the rail).

3.2.4 Selection of bogie design characteristics based on operational aspects of networks

The selection of bogie design characteristics is directly related to the operational characteristics of the network in which the trains will operate. In this section, technical characteristics for vehicle bogies are suggested considering the operational characteristics of the network. Data are obtained from the application of the mathematical models developed by Joly and Pyrgidis (Joly, 1983, 1988; Joly and Pyrgidis, 1990, 1996; Pyrgidis, 1990, 2004; Pyrgidis and Joly, 1993).

3.2.4.1 High-speed networks

These are characterised by:

Track design speed: $V_d \geq 200$ km/h.

Table 3.5 Performances of examined bogie's technologies in motion on a straight path and in curves

Technology/characteristics of bogies	Straight path	Occurrence of slip	Curves		
			$R_c = 500$ m $K_o = 0$	$R_c = 200$ m $K_o = 0$
Conventional bogies $K_x = 8 \times 10^6$ N/m $K_y = 10^7$ N/m $\gamma_e = 0.10$ $K_o = 3 \times 10^6$ N/m $2a = 3.0$ m, $2r_o = 0.90$ m	482 km/h	$R_c = 4,800$ m	$F_{||} = 20.7$ kN	$F_{||} = 65.9$ kN
Bogies with self-steering wheelsets $K_x = K_y = 10^6$ N/m $K_b = 10^6$ N/m $K_s = 10^6$ N/m $\gamma_e = 0.20$ $K_o = 3 \times 10^6$ N/m $2a = 3.0$ m, $2r_o = 0.90$ m	226 km/h	$R_c = 250$ m	$F_{||} = 0$ $P_{4w} = 1.11$ kW	$F_{||} = 14.9$ kN
Bogies with independently rotating wheels $K_x = K_y = 10^8$ N/m $\gamma_e = 0.30$ $K_o = 0$ $2a = 3.0$ m, $2r_o = 0.90$ m	$V_{cr} = \infty$ Vulnerability in lateral displacements	$R_c = 1,400$ m	$F_{||} = 0$ $P_{4w} = 4.3$ kW	$F_{||} = 0$ $P_{4w} = 12.2$ kW

Source: Adapted from Pyrgidis, C., 1990, Etude de la Stabilité Transversale d'un Véhicule Ferroviaire en Alignement et en courbe – Nouvelles Technologies des Bogies – Etude Comparative, Thèse de Doctorat de l', ENPC, Paris.

Alignment layout: Small percentage of curved sections out of total track length. Large and very large curve radii (R_{cmin} = 2,000–5,000 m, depending mainly on the speed and the stiffness of the primary suspension of bogies, see Section 12.3.1.1).

The following are proposed for the rolling stock:

Bogies: Conventional or with independently rotating wheels (Talgo trains).
Equivalent conicity: Small (e.g., γ_e = 0.05–0.10).
Stiffness of the primary suspension: High (e.g., K_x = 8 × 10⁶ N/m and K_y = 10⁷ N/m).
Bogie wheelbase: High (e.g., 2a = 3.0 m).
Wheel diameter: Big (e.g., $2r_o$ = 0.90 m).
Bogie and wheelset masses: Small.

3.2.4.2 Conventional speed networks

They are characterised by:

Track design speed: 140 km/h $\leq V_d$ < 200 km/h.
Alignment layout: Mainly medium curve radii (R_{cmin} = 500–2,000 m, depending mainly on the speed, see Section 12.3.1.1).

Conventional bogies are proposed. The selection of values of the bogie design characteristics depends on the track design speed and the track geometry alignment.

Remark: If it is desired to improve the performance on an existing track (assuming the track superstructure is in a very good state) tilting trains may be used.

3.2.4.3 Mountainous networks

Characterised by:

Track design speed: V_d < 140 km/h.
Alignment layout: Large percentage of curved sections out of the total track length mainly medium and small horizontal alignment radii (R_c = 250–750 m).

The following proposals are made for the rolling stock:

Bogies: With self-steering wheelsets (or conventional wheelsets).
Equivalent conicity: Medium (e.g., γ_e = 0.20).
Total longitudinal stiffness of primary suspension: Small (e.g., K_{bt} = 2 × 10⁶ N/m).
Total lateral stiffness of primary suspension: Small (e.g., K_{st} = 1.3 × 10⁶ N/m).

3.2.4.4 Metro networks

These are characterised by:

Track design speed: V_d = 90–100 (120) km/h.
Alignment layout: Very large percentage of curved sections out of the total track length. Mainly small curve radii (R_c = 150–300 m).

The following options are proposed for the rolling stock:

Bogies: Conventional.
Equivalent conicity: High (e.g., γ_e = 0.30).

Bogie wheelbase: Small (e.g., 2a = 2.00–2.40 m).
Wheel diameter: Small (e.g., $2r_o$ = 0.70–0.75 m).
Bogie and wheel masses: Small.
Stiffness of primary suspension: Small (e.g., K_x = 4.10^6 N/m and K_y = 10^6 N/m).

3.2.4.5 Tramway networks

These are characterised by:

Track design speed: V_d = 80–90 km/h.
Alignment layout: Very large percentage of curved sections out of the total track length.
    Curve radii mainly in the range of R_c = 25–50 m.

The following options are proposed for the rolling stock:

Bogies: With independently rotating wheels.
Equivalent conicity: Very high.
Bogie wheelbase: Small (e.g., 2a = 1.80 m).
Wheel diameter: Small (e.g., $2r_o$ = 0.65 m).
Bogie and wheelset masses: Small.
Longitudinal stiffness in primary suspension: No effect.
Lateral stiffness in primary suspension: Small (e.g., K_y = 10^5–10^6 N/m).

Apart from the technology of bogies with independently rotating wheels, a mixed system can be used (bogies of the Sirio series tramway). For the accurate operation of this technology, a wheel profile with varying conicity is needed to secure small values for γ_e and high values ('smart profile') of γ_e (for the lateral displacements – that is to say during the motion in curves) (Pyrgidis, 2004; Pyrgidis and Panagiotopoulos, 2012).

3.3 DERAILMENT OF RAILWAY VEHICLES

3.3.1 Definition

The term 'derailment' is used to describe the definite loss of contact of at least one vehicle wheel with the railhead rolling surface (Figures 3.12 and 3.13).
    The derailment of a railway vehicle may occur as a result of:

- Lateral displacement (shift) of the track.
- Overturning/tilting of the vehicle.
- Wheel climb.
- Track gauge widening and rail rollover.

The causes of derailment can be internal to the railway system (high exerted forces from the vehicle to the track, excessive train speed, bad condition and design of rolling stock, poor ride quality of track and of track layout, embankment collapse, etc.) or external to the railway system (human error, crosswind, obstacle on tracks, etc.). The internal causes based on the effect they have on the development of the phenomenon may be characterised as direct and indirect (Figure 3.14).

Figure 3.12 Derailment of railway vehicles. (Photo: A. Klonos.)

Figure 3.13 Derailment. (Photo: A. Klonos.)

3.3.2 Derailment as a result of vehicle overturning

Vehicle derailment due to overturning may occur both during movement along curved segments of the track and movement along a straight path.

3.3.2.1 Check for derailment due to overturning – movement along curved track segments

During the movement or immobilisation of a railway vehicle on curved segments of the horizontal alignment, the vehicle may overturn under certain conditions.

Overturning may occur toward the outside or the inside of the curve.

For the former (Figure 3.15a), the following reasons may pertain (Esveld, 2001):

- Significant cant deficiency in relation to the passage speed V_p and the radius of curvature, which translates to an increased value of the lateral residual centrifugal force F_{nc}.

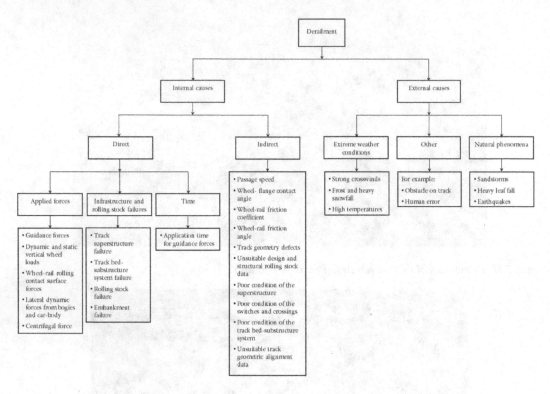

Figure 3.14 Causes of derailment.

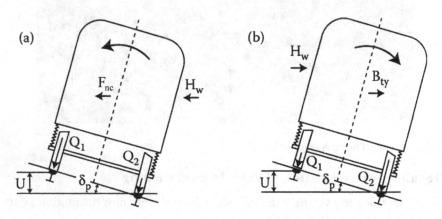

Figure 3.15 Vehicle overturning mechanism in curves: (a) Toward the outside of the curve and (b) Toward the inside of the curve. (Adapted from Esveld, C., 2001, *Modern Railway Track*, 2nd edition, MRT-Productions, West Germany.)

- Crosswind force H_w directed toward the outside of the curve.
- Unequal load distribution on two wheels with lower loading on the inside wheel ($Q_1 > Q_2$).

All the above reasons result in the development of moments, which tend to overturn the vehicle toward the outer rail.

For the latter (Figure 3.15b), the following reasons may pertain, respectively:

- Crosswind force H_w directed toward the inside of the curve.
- Immobilisation of vehicles ($V_p = 0$) on a curved track segment with a high cant U.
- Low axle load.
- Displacement of load towards the inside wheels.

Under these circumstances moments which tend to overturn the vehicle toward the inner rail are developed resulting ultimately in derailment.

The check for derailment due to overturning may be accomplished by:

3.3.2.1.1 Using analytical relations

This check is realised taking into account the moment of forces in regard to the railhead over which the vehicle overturns (Figure 3.16).

Equation 3.15 applies (Rivier, 1984/85):

$$V_{der.ov}^2 = \frac{\left[R_c \cdot g \cdot \left(\frac{U}{2e_o} + \frac{2e_o}{2h_{KB}} \right) \right]}{\left[1 - \left(\frac{U}{2e_o} \cdot \frac{2e_o}{2h_{KB}} \right) \right]} \tag{3.15}$$

where:

$V_{der.ov}$: Speed over which derailment due to overturning occurs.

h_{KB}: Distance between the vehicle's centre of gravity and the rail rolling surface.

g: Gravity acceleration.

$2e_o$: Distance between the vertical axis of symmetry of the two rails.

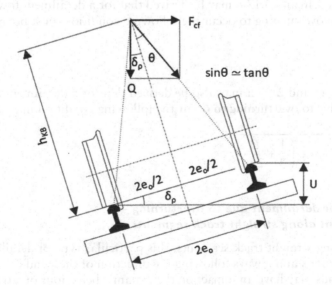

Figure 3.16 Vehicle overturning mechanism. (Adapted from Rivier, R. 1984/85, *Railway Notes*, EPFL.)

3.3.2.1.2 Using semi-empirical formulas

'Kunieda's formula' (Matsumoto et al., 2016; Hibino et al., 2010) is a simplified formula, to calculate the risk against train overturning, considering the centrifugal force of curving, the vehicle vibration (especially lateral) and the crosswind effect. It is a simplified equation based on the static mechanism. It has been certified in many cases and used for the guideline of safety in Japan.

3.3.2.1.3 Using empirical formulas

It is applicable only when $h_{KB} > 2.25$ m and refers only to cases where the overturning occurs towards the outside of the track (Amans and Sauvage, 1969).

For derailment due to overturning to occur the following condition must apply:

$$\gamma_{ncmax} > \frac{g}{3} \qquad (3.16)$$

where:

γ_{ncmax}: Maximum permitted lateral residual acceleration.

Moreover, the following equations apply:

$$\gamma_{nc} = \frac{V^2}{R_c} - \frac{g \cdot U}{2e_o} \qquad (2.37)$$

$$\gamma_{nc} = g \cdot \frac{I}{2e_o} \qquad (3.17)$$

where:

V: Vehicle running speed.
I: Cant deficiency.

From Equations 3.16 and 3.17 it may be derived that for a derailment towards the outside of the track due to overturning to occur the following condition must be true:

$$I > \frac{2e_o}{3} \qquad (3.18)$$

From Equations 3.16 and 2.37 it may also be derived that for a derailment towards the outside of the track due to overturning to occur the following condition must be true:

$$V > V_{der.ov} = \sqrt{R_c \cdot g \cdot \left(\frac{1}{3} + \frac{U}{2e_o}\right)} \qquad (3.19)$$

3.3.2.2 Check for derailment due to overturning – movement along straight track segments

When moving along straight track segments, this particular type of derailment may occur due to high crosswinds and always following the direction of the wind.

Strong crosswinds may have an impact on the dynamic behaviour of a train as a result of lateral wind forces that affects the lateral dynamic vehicle behaviour of the entire train. In that they cause lateral vehicle instability, which may lead to a derailment (see Chapter 17).

The check for derailment due to overturning may be conducted by:

3.3.2.2.1 Using analytical relations

This check is effectuated taking into account the moments (a) of the crosswind force, and (b) of the total weight of the vehicle in relation to the railhead over which the vehicle is overturned. The following equations apply:

$$M_t \cdot g \cdot e_o = q_o^* \cdot H_w \tag{3.20}$$

$$H_w = \frac{1}{2} \cdot \rho \cdot S_2 \cdot V_w^2 \cdot K_2 \tag{2.33}$$

$$V_w = \sqrt{\frac{M_t \cdot g \cdot 2e_o}{q_o^* \cdot \rho \cdot S_2 \cdot K_2}} \tag{3.21}$$

where:

M_t: Total mass of the vehicle (in kg).

H_w: Cross (or side) wind force (in N).

V_w: Wind speed (in m/sec).

ρ: Air density (in kg/m³).

S_2: Lateral surface area of the vehicle (in m²).

K_2: Side wind force coefficient (parameter depending on the lateral external surface of a vehicle).

q_o^* (= $1.25q_o$): Compensated value of the nominal height of centre of the crosswind force, as taken from the rolling surface of the rails (in m).

3.3.2.2.2 Using semi-empirical formulas

Kunieda's formula is also applicable (U = 0, R_c = ∞).

As derived by the information recorded above, regardless of whether we refer to an overturning towards the inside of the curve, the outside of the curve, or at a straight segment, a 'high' centre of gravity of the vehicle, a low weight, and a high value of the lateral wind force (high wind speeds and large surface area of the vehicle) increase the risk of overturning.

3.3.3 Derailment due to track displacement

In the case of derailment through the displacement of track the track panel (rails + sleepers) of a track segment is displaced due to the effect of significant lateral forces, resulting in the derailment of one or more of the train's vehicles.

Derailment through displacement of the track occurs when:

$$H > H_R \tag{3.22}$$

where:

H: Total lateral force, which is transferred from the vehicle to the rail.

H_R: Lateral track resistance.

This type of derailment is solely due to causes internal to the railway system and is the most common type of derailment.

The total lateral force H (in t) may be calculated using the Empirical Formula 2.38.

In regard to the calculation of the lateral track resistance H_R various formulas have been proposed (ORE, 1984; Amans and Sauvage, 1969; Prud'homme, 1967). Indicatively, the following formulas are given (H_R is calculated in t):

3.3.3.1 Prud'homme limit

$$H_R = (0.85)\left(1 + \frac{Q}{3}\right) \tag{3.23}$$

where:
   H_R: Lateral track resistance (in t).
   Q: Vertical axle load (in t).

Equation 3.23 (with the multiplying factor 0.85) takes into consideration the track alignment and the thermal forces applied on rails but supposes a destabilised track.

3.3.3.2 Empirical formulas considering the stabilisation degree of the track and the type of sleepers

- For concrete sleepers

$$H_R = 0.6 \cdot (Q + 6) \cdot \left(1 - 0.4 \cdot e^{-\frac{T_t}{60,000}}\right) \tag{3.24}$$

where:
   Q: Vertical axle load (in t).
   T_t: Total traffic load (in t).

For fully stabilised track ($T_t = \infty$) Equation 3.24 is transformed as follows:

$$H_R = 0.6 \cdot Q + 3.6 \tag{3.25}$$

For fully destabilised track ($T_t = 0$) Equation 3.24 is transformed as follows:

$$H_R = 0.36 \cdot Q + 2.16 \tag{3.26}$$

- For wooden sleepers

$$H_R = 0.5 \cdot (Q + 4) \cdot \left(1 - 0.4 \cdot e^{-\frac{T_t}{60,000}}\right) \tag{3.27}$$

For fully stabilised track ($T_t = \infty$) Equation 3.27 is transformed as follows:

$$H_R = 0.5 \cdot Q + 2 \tag{3.28}$$

For fully destabilised track ($T_t = 0$) Equation 3.27 is transformed as follows:

$$H_R = 0.3 \cdot Q + 1.2 \tag{3.29}$$

Moreover, the following equations apply:

$$U_{thvmax}(mm) = 11.8 \cdot \frac{V_{max}^2 \left(\frac{km}{h}\right)}{R_c(m)} \tag{3.30}$$

$$I = U_{thvmax} - U \tag{3.31}$$

where I, U_{thvmax}, and U are expressed in mm and U_{thvmax} refers to the theoretical track cant for the maximum running speed V_{max}.

Taking into account Equations 3.24, 3.28, 3.30, 3.31, and 2.38 as well as by assuming an even distribution of the centrifugal force among the two axles of the bogie ($a_i = 1$, see Equation 2.38), it may be concluded, based on Equation 3.22, that the speed $V_{der.dis}$ over which derailment due to lateral displacement of the track occurs (for concrete or wooden sleepers over a fully stabilised track) is given by the following equations:

Track with concrete sleepers:

$$\left(11.8 \cdot \frac{Q}{R_c}\right) \cdot V_{der.dis}^2 + 1.5 \cdot Q \cdot V_{der.dis} - Q \cdot (U + 900) - 5,400 = 0 \tag{3.32}$$

Track with wooden sleepers:

$$\left(11.8 \cdot \frac{Q}{R_c}\right) \cdot V_{der.dis}^2 + 1.5 \cdot Q \cdot V_{der.dis} - Q \cdot (U + 750) - 3,000 = 0 \tag{3.33}$$

where Q is expressed in t, R_c in m, U in mm, and $V_{der.dis}$ denotes the speed (in km/h) over which derailment due to lateral displacement of the track occurs.

In straight track segments, due to the absence of cant, in Equation 2.38 the parameter I (cant deficiency) expresses the transversal track defect (or the track twist). In cases when the traversal track defect is null, theoretically there is no risk of derailment due to lateral displacement of the track. If in equation 2.38 we set I = 0, then Equations 3.32 and 3.33 are transformed as follows:

$$V_{der.dis} = 600 + \frac{3,600}{Q} \tag{3.34}$$

$$V_{der.dis} = 500 + \frac{2,000}{Q} \tag{3.35}$$

The application of these equations for Q = 22.5 t requires $V_{der.dis.} \geq 760$ km/h for the case of concrete sleepers and $V_{der.dis.} \geq 588.99$ km/h for the case of wooden sleepers, for a derailment due to lateral displacement of the track to occur.

As derived from the above, to reduce the risk of a lateral displacement of the track, we must reduce the horizontal force H that is transferred from the vehicle to the rail or increase the transversal resistance of the track H_R or both. Regarding H, the parameters that affect it are derived either directly or indirectly from Equation 2.38.

Regarding H_R, the following choices/parameters increase its value:

> Heavy concrete sleepers, heavy continuous welded rails, elastic fastenings, fully stabilised track, slab track, very good condition of the system track bed-substructure, and, finally, in the case of ballasted track: (a) large ballast occupancy width; (b) high ballast super elevation; (c) high degree of compacting and hardness of the ballast; and (d) rare tamping.

3.3.4 Derailment due to wheel climb

For derailment due to wheel climb to occur, firstly flange contact with the inner side of the rail must occur and, therefore, a guidance force (F) be applied. At the contact surface of the wheel flange with the inner side of the rail (Figure 2.11) the wheel imposes on the rail a guidance force F_1 and the vertical load of the wheel Q_1. It receives the vertical reaction N_1 and the lateral creep force T_1 (when slide occurs the force T_1 is equal to the Coulomb force). In practice, derailment through wheel climb occurs when the resultant of the projections of all these forces on the yy axis (derailment force axis) is directed upwards and the application time of this resultant force is long enough for the wheel to climb over the rail (Figure 3.17).

Derailment due to wheel climb can occur when there is a significant unloading of the derailed wheel with simultaneous loading of the non-derailed wheel. This phenomenon can be observed in the case of movement at low speeds in curves with a small radius of curvature and high values of cant and twist.

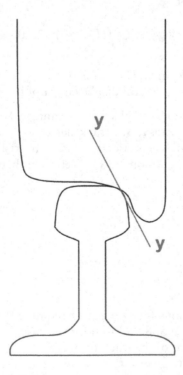

Figure 3.17 Derailment due to wheel climb.

Derailment due to wheel climb also occurs through external to the railway system causes, that is, poor operation and adjustment of switches, etc.

Most derailments occur in areas of switches and crossings due to several causes (see Section 3.4).

It should be noted that derailment due to wheel climb can occur in turnouts, in areas of switches, when the developed centrifugal force is significantly large (in straight turnouts no cant is given to the track) and the lateral track resistance is high (sudden derailments) (Centre for Advanced Maintenance Technology,1998).

The risk of derailment due to wheel climb increases when there is:

- An increase of the value of the guidance force.
- An increase in the application time of guidance force.
- An increase in the value of the wheel-rail friction coefficient (during rain the risk of derailment by rail climbing is smaller).
- An increase in the value of the yaw angle of the wheelset.
- A decrease in the value of the rail-wheel flange contact angle.
- A decrease in the value of the vertical load on the derailed wheel with a simultaneous increase of the vertical load of the non-derailed wheel.

Wheel climb does not occur instantaneously. A certain amount of time is required, and thus the derailing wheel covers some distance on the track, usually some metres. This distance is called 'flange-climbing distance' and is defined as the distance covered from the moment when the total value of the guidance force is applied until the moment on which the wheel-rail contact flange angle reaches 26.6° (Dos Sandos et al., 2010). The shorter this distance is, the faster the derailment will become apparent. The influence of various wheel parameters on the aforementioned distance was examined with the aid of simulation modelling. The results have shown that the flange-climbing distance increases (and thus the appearance of derailment slows down) when (Dos Sandos et al., 2010):

- The yaw angle of the wheelset is smaller.
- The height of the wheel flange is increased. When the wheel is worn out the distance required for rail climbing increases. The positive effect of a high flange is significantly limited when the yaw angle is large.
- The value of q_r increases (see Figures 1.16 and 3.18).

For checking the derailment due to wheel climb the following 'tools' may be used:

3.3.4.1 Criteria that evaluate the F_1 / Q_1 ratio

Derailment is avoided when

$$\frac{F_1}{Q_1} < K_d \qquad (3.36)$$

where:

K_d: Derailment factor (derailment due to wheel climb).
Q_1: Vertical static load of the under derailment wheel (assuming wheel 1).
F_1: Guidance force.

Methods for calculating the guidance force F_1 were given in Section 2.3.7.

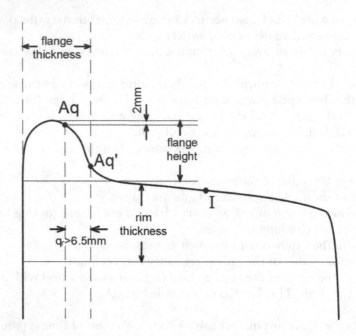

Figure 3.18 Wheel climb derailment check with the use of the q_r criterion.

Such criteria are found in the relevant literature (FP7, 2011; Iwnicki, 2006; Ishida and Matsuo,1999; Alias, 1977; Profillidis, 2005) and indicatively include the following:

- Nadal's criterion (the yaw angle is supposed to be not zero).
- Weinstock's criterion.
- Chartet's criterion (applies for yaw angles of $\alpha > 1°$).
- Derailment criterion for yaw angles of $\alpha > 5$ mrad.

According to relevant literature (FP7, 2011; Iwnicki, 2006; Ishida and Matsuo,1999; Umdrucke zur Grundvorlesung, 2002/2003):

- In Japan and Western Europe, it is assumed that $K_d = 0.8$.
- In South America it is assumed that $K_d = 1.0$.
- In China it is assumed that the limit value of K_d is 1.0 while the risk limit is taken equal to $K_d = 1.2$.

3.3.4.2 Criteria related to the time or distance limits, which are applied to limit the exceeding duration of the F_l / Q_l ratio limit in either time or distance scale

These include, indicatively, the following (Iwnicki, 2006):

- High-speed passenger distance limit (5 ft)-FRA, US.
- CHXI 50 millisecond time limit-Association of American, US.
- F_1 / Q_1 time duration criterion – proposed by Japanese National Railways.
- F_1 / Q_1 time duration criterion – proposed by Electromotive Division of General Motors.
- Wheel climb distance criterion – proposed by Transportation Technology Centre, Inc.

3.3.4.3 Criterion q_r

where:

q_r: Flange cross-dimension (the horizontal distance between the intersection point of the joint geometric level with the flange face and the intersection point of a reference line at a distance of 2 mm from the flange tip with the flange face) as shown in Figures 1.8 and 3.18.

This particular check is conducted with the use of a special verification device for the q_r distance and may be used in the following circumstances:

- Vehicle movement along small radius curves with a large axle yaw angle.
- Vehicle movement through switches and crossings.

For derailment to be avoided the following must be true (as defined by ORE):

$$q_r > 6.5\,\text{mm} \tag{3.37}$$

3.3.4.4 Empirical formula that calculates the speed over, which a vehicle is derailed due to wheel climb

Equation 3.40 is derived from the empirical Formula 3.38 and the analytical Equation 3.39 (Figure 3.19) (Rivier, 1984/85):

$$F_{nc} = \frac{Q}{4} \tag{3.38}$$

$$F_{nc} = \frac{Q}{g} \cdot \left(\frac{V^2}{R_c} - g \cdot \frac{U}{2e_o} \right) \tag{3.39}$$

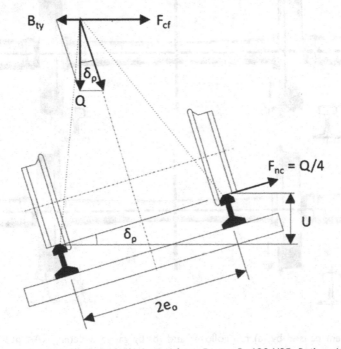

Figure 3.19 Derailment due to wheel climb. (Adapted from Rivier, R. 1984/85, Railway Notes, EPFL.)

$$V_{der.wcl} = \sqrt{R_c \cdot g \cdot \left(\frac{U}{2e_o} + \frac{1}{4} \right)} \qquad (3.40)$$

where:

$V_{der.wcl}$: The speed over which derailment due to wheel climb occurs.

Equations 3.38 and 3.40 are suggested by Professor Rivier of Ecole Polytechnique Fédérale de Lausanne and are based on purely experimental data. They correspond to the least favourable conditions in regard to derailment due to wheel climb.

3.3.5 Derailment caused by gauge widening or rail rollover

In this case, due to large lateral forces acting from the wheels to the rails in curves, both rails may experience significant lateral shift and/or railhead roll which often cause the non-flanging wheel to drop between rails (Figure 3.20) (Iwnicki, 2006; Blader,1990). Rail gauge wear is another cause for gauge widening.

For the check for derailment caused by gauge widening the following formula is used (Figure 3.21):

$$2e \geq 2d_a + f_d + w_{wh} \qquad (3.41)$$

where:

2e: Track gauge.

f_d: Flange thickness.

$2d_a$: Back-to-back wheel distance (inside gauge).

w_{wh}: Wheel width.

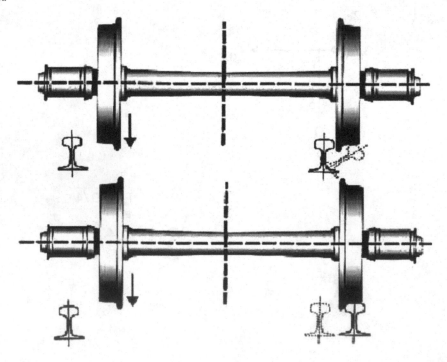

Figure 3.20 Derailment caused by (a) rail rollover and (b) by gauge widening. (Adapted from Iwnicki, S., 2006, *Handbook of Railway Vehicle Dynamics,* Taylor & Francis.)

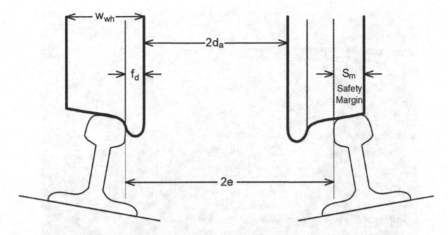

Figure 3.21 Criterion for derailment caused by gauge widening. (Adapted from Iwnicki, S., 2006, *Handbook of Railway Vehicle Dynamics*, Taylor & Francis.)

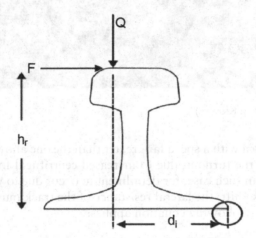

Figure 3.22 Criterion for derailment caused by rail roll. (Adapted from Iwnicki, S., 2006, *Handbook of Railway Vehicle Dynamics*, Taylor and Francis.)

For the check for derailment caused by rail rollover the following formula is used (Iwnicki, 2006) (Figure 3.22):

$$F > Q \cdot \frac{d_i}{h_r} \tag{3.42}$$

where:

h_r: Rail height.

d_i: Maximum distance between the point of contact of the total vertical loads that are imposed on the rail, from the edge of the foot of the rail (Figure 3.22).

3.4 DERAILMENT IN TURNOUTS

In straight turnouts and specifically in the circular part of the turnout, no track cant (U = 0), no track gauge widening, and no transition curve are foreseen (Figure 3.23). In case when a

Figure 3.23 Turnout. (Photo: A. Klonos.)

train enters the switch area with a speed far greater than the one allowed by the radius of the horizontal alignment of the turnout, due to increased centrifugal force, derailment due to wheel climb may occur (in such case, for derailment to occur due to wheel climb and before a derailment due to track shift, the lateral resistance of the track must be high).

In straight turnouts the following equation applies:

$$U = U_{th} - I = 0 \qquad (3.43)$$

where:

$$U_{th}\text{: Theoretical cant} = 11.8 \ V^2 / R_g \ (\text{in mm}). \qquad (3.44)$$

I: Cant deficiency (in mm).
V: Turnout passage speed (in km/h).
U: Track cant (in mm).
R_g: Switch turnout radius of curvature (in m).

By replacing Equation 3.43 in Equation 3.44 it is concluded that:

$$V = 0.29 \cdot \sqrt{R_g \cdot I} \qquad (3.45)$$

Based on the above and by assuming U = 0, Equations 3.15, 3.19, 3.32, 3.33, and 3.40, that reflect the derailment speed for various causes of derailment, are transformed as follows:

- Derailment due to overturning

$$V_{der.ov} > \sqrt{R_c \cdot \frac{g}{3}} \qquad (3.46)$$

$$V_{der.ov} > \sqrt{R_c \cdot g \cdot \frac{e_o}{h_{KB}}} \qquad (3.47)$$

- Derailment due to track displacement
  For a fully stabilised track with concrete sleepers:

$$\left(11.8 \cdot \frac{Q}{R_c}\right) \cdot V_{der.dis}^2 + 1.5 \cdot Q \cdot V_{der.dis} - Q \cdot 900 - 5,400 = 0 \qquad (3.48)$$

For a fully stabilised track with wooden sleepers:

$$\left(11.8 \cdot \frac{Q}{R_c}\right) \cdot V_{der.dis}^2 + 1.5 \cdot Q \cdot V_{der.dis} - Q \cdot 750 - 3,000 = 0 \qquad (3.49)$$

- Derailment due to wheel climb

$$V_{der.wcl} = \sqrt{R_c \cdot \frac{g}{4}} \qquad (3.50)$$

Specifically for turnouts the value of K_d must be reduced. According to literature (Amans and Sauvage, 1969; Profillidis, 1995) it should be taken equal to 0.4. In such a case, however, the vertical load Q_1 should correspond to the dynamic rather than the static load.

Finally, according to literature (Franklin, 2018) observation and experiment indicate that for safety, F/Q in switches and crossings should be limited to be in the order of 0.8.

3.5 CASE STUDY

A train enters with a running speed of V = 140 km/h the diverging branch of a turnout and specifically the curved part of the turnout (switch area). The track is equipped with concrete sleepers and is fully stabilised. The following data are given:

Normal track gauge: $2e_o = 1,500$ mm.
Straight turnout.
Radius of curvature of the turnout: $R_g = 500$ m.
Axle load: Q = 17 t.
Even distribution of the axle load among the wheels: $Q_1 = Q_2 = Q/2 = 8.5$ t
Gravity acceleration: g = 10 m/sec².
Distance of the centre of gravity of the vehicle from the rolling surface of the rails: $h_{KB} = 1,600$ mm.
Side wind speed: $V_w = 0$.

You are asked to:

1. Calculate (in km/h) the maximum allowed running speed along the curved segment of the turnout taking into account that the current Track Regulation imposes a maximum allowed centrifugal residual acceleration of $\gamma_{ncmax} = 0.7$ m/sec^2.
2. Undertake, for a speed of 140 km/h, the appropriate derailment checks.
3. Calculate, for a speed of 140 km/h, the centrifugal force that is imposed on the vehicle.

Solution

1. By replacing the given values in Equation 2.37 and specifically $R_c = R_g = 500$ m, U = 0, $\gamma_{ncmax} = 0.7$ m/sec^2, we get that:

$$\gamma_{ncmax} = \frac{V_{max}^2}{R_c} - g \cdot \frac{U}{2e_o} \Rightarrow V_{max} = 67.34\,km/h$$

2. Derailment checks
   Derailment due to overturning
   a) *Analytical calculations*
      From Equation 3.47 and for $R_g = 500$ m, g = 10 m/sec^2, $e_o = 0.75$ m, $h_{KB} = 1,600$ mm, we have that:

      $V_{der.ov} = 174.278$ km/h > 140 km/h

      Therefore, derailment due to overturning does not occur.
   b) *Calculation with the use of the empirical Equation 3.46 (for $h_{KB} > 2.25$ m)*
      By setting $R_c = R_g = 500$ m and g = 10 m/sec^2 we get that:

      $V_{der.ov} = 146.96$ km/h

      Therefore, derailment due to overturning will not occur since V = 140 km/h < $V_{der.ov}$ = 146.96 km/h.
      The fact that the distance of the centre of gravity of this particular vehicle from the rolling surface is significantly lower than 2.25 m (most favourable case for avoiding vehicle overturning) also allows us to conclude that derailment due to overturning will not occur.

   Derailment due to track shift

      From Equation 3.45 and for V = 140 km/h and $R_g = 500$ m, it is derived that:

      I = 466.111mm

      By setting I = 466.11 mm, Q = 17 t, and V = 140 km/h in Equation 2.38, it is derived that:

      H=7.76t

      By applying Equation 3.25:

      $$H_R = 0.6 \cdot Q + 3.6 = 13.8t$$

$H<H_R$

Therefore, derailment due to lateral displacement of the track does not occur. Using another approach, from Equation 3.48 and for $R_g = 500$ m and $Q = 17$ t it is derived that:

$V_{der.dis} = 197.57$ km/h

$V = 140$ km/h $< V_{der.dis} = 197.57$ km/h

Therefore, derailment due to lateral displacement of the track does not occur.

Derailment due to wheel climb

a) *Based on empirical/experimental criteria that calculate the speed over which derailment due to wheel climb occurs*
    From Equation 3.50 and for $R_g = 500$ m and $g = 10$ m/sec^2, it is derived that:

$V_{der.wcl} = 127.27$ km/h

$V = 140$ km/h $> V_{der.wcl} = 127.27$ km/h

Therefore, derailment due to wheel climb will occur.

b) *Empirically, based on Equation 2.40*
    By considering that $Q_2 = Q/2 = 8.5$ t, $R_g = R_c = 500$ m, $\tan(\gamma_2+\rho_2) = \dfrac{135}{(150 + R_c)}$, and $H = 7.76$ t, it is derived that:

$F_1 = 9.52$ t

and

$$\frac{F_1}{Q_1} = \frac{9.52}{8.5} = 1.12 > 0.8$$

Therefore, derailment due to wheel climb will occur.

3. Calculating the centrifugal force
    Equation 3.39 applies for the remaining centrifugal force F_{nc}

$$F_{nc} = \frac{Q}{g} \cdot \gamma_{ncmax} = \frac{Q}{g} \cdot \left(\frac{V_{max}^2}{R_g} - g \cdot \frac{U}{2e_0} \right)$$

And by setting $R_g = 500$ m, $g = 10$ m/sec^2, $Q = 17$ t, and $U = 0$, it is derived that $F_{nc} = 5.14$ t.

Important remarks:
    The centrifugal force F_{nc} that occurred was 5.14 t $> Q/4 = 4.25$ t. Based on the above, it may be assumed that the speed for which the train would derail could be lower than that of 127.27 km/h.

If the track is considered fully destabilised, then the Prud'homme limit (Equation 3.23) may be applied. In that case:

$$H = 7.76t > H_R = (0.85)\left(1 + \frac{Q}{3}\right) = (5.66)6.66t$$

Therefore, derailment due to track shift occurs before wheel climb.

The same conclusion is obtained if we proceed by estimating the derailment speed. By considering $H = a_i \cdot \left(\frac{Q \cdot I}{1,500}\right) + \left(\frac{Q \cdot V}{1,000}\right) = H_R = 5.66$ t and by setting $I = U_{th} - U = U_{th} = 11.8 \cdot V^2/R_g$, $R_g = 500$ m, $Q = 17$ t, it is derived that $V_{der.dis} = 116.48$ km/h $< V_{der.wcl} = 127.27$ km/h.

Therefore, derailment due to lateral track shift occurs before wheel climb.

REFERENCES

Alias, J. 1977, *La Voie Ferrée*, Eyrolles, Paris.

Amans, F. and Sauvage, R. 1969, La stabilité de la voie vis à vis des efforts transversaux exercés par les véhicules, *Annales des Ponts et Chaussées*, Vol. 1.

Blader, F.B. 1990, A review of literature and methodologies in the study of derailments caused by excessive forces at the wheel/rail interface, *Association of American Railroads Report R - 717*, December 1990.

Centre for Advanced Maintenance Technology. 1998, *A Technical Guide for Derailments*, CAMTECH/M/3, Maharajpur, Gwalior.

Dos Sandos, G.F.M., Lopes, L.A.S., Kina, E.J. and Tunna, J. 2010, The influence of wheel profile on safety index, *Proceedings of IMechE, Vol. 224, Part F: J. Rail and Rapid Transit*, Special issue paper, pp. 429–434.

Esveld, C. 2001, *Modern Railway Track*, 2nd edition, MRT-Productions, Duisburg.

FP7-TPT-2011-RTD-1P. 2011, Current status of studies on derailment (state of the art), European Commission, Brussels.

Franklin, A. 2018, *Wear and Derailment Risk at Facing Switches*, Network Rail, London.

Frederich, F. 1985, Possibilités inconnues et inutilisées du contact rail-roue, *Rail International*, Brussels, November, No. 11, pp. 33–40.

Frederich, F. 1988, A bogie concept for the 1990s, *Railway Gazette International*, September, No. 9, pp. 583–585.

Frullini, R., Casini, C. and Tacci, G. 1984, Simulation du comportement en ligne d'un véhicule à roues indépendantes, comparaison avec les résultats, *Ingegneria Ferroviaria*, January-February, Rome (translation RATP No 84 - 307).

Geuenich, W., Cunther, C. and Leo, R. 1985, Fibre composite bogies has creep controlled wheelsets, *Railway Gazette International*, April, No. 3, pp. 79–281.

Hibino Y., Shimomura, T. and Tanifuji, K. 2010, Full scale experiment on the behavior of a railway vehicle being subjected to lateral force, *Journal of Mechanical Systems for Transportation and Logistics*, Vol. 3, No. 1, pp. 35–43.

Ishida, H. and Matsuo, M. 1999, Safety criteria for evaluation of railway vehicle derailment, *QR of RTRI*, Vol. 40, No. 1, March 1999, pp. 18–25.

Iwnicki, S. 2006, *Handbook of Railway Vehicle Dynamics*, Taylor and Francis.

Joly, R. 1983, Stabilité transversale et confort vibratoire en Dynamique Ferroviaire, Thèse de Doctorat d'Etat, Université de Paris, Paris.

Joly, R. 1988, Circulation d'un véhicule ferroviaire en courbe de faible rayon – Bogies de conception classique/Bogie à essieux auto – orientés, *Rail International*, Brussels, April, No. 3, pp. 31–42.

Joly, R. and Pyrgidis, C. 1990, Circulation d'un véhicule ferroviaire en courbe – Efforts de guidage, *Rail International*, Brussels, No. 12, pp. 11–28.

Joly, R. and Pyrgidis, C. 1996, Stabilité transversale des véhicules en alignement, *Rail International*, No. 12, pp. 25–33.

Julien, M. and Rocard, Y. 1935, *La stabilité de route des locomotives*, HERMANN et Cie Editeurs, Paris.

Matsumoto, A., Michitsuji, Y. and Tobita Y. 2016, Analysis of train-overturn derailments caused by excessive curving speed - Akira, *3rd International Conference of Railway Technology: Research and Maintenance Meeting*, Cagliari, Italy, April 2016.

ORE, C138, RP8. 1984, *Permissible Maximum Values for the Y and Q Forces and Derailment Criteria*, ORE, Utrecht.

Panagin, R. 1978, La technica delle ruota independenti al fine di eliminare l'instabilita laterale nei veicoli ferroviari, *Ingegneria Ferroviaria*, February, Rome, 2, pp. 143–150 (translation SNCF No. 10 -7 9).

Profillidis, V. 1995, *Railway Engineering*, Avebury, Farnham.

Profillidis, V. 2005, *Railway Engineering*, Avebury, Farnham.

Prud'homme, A. 1967, La résistance de la voie aux efforts transversaux exercés par le matériel roulant, *Révue Générale des Chemins de Fer*, January, 1967.

Pyrgidis, C. 1990, *Etude de la stabilité transversale d'un véhicule ferroviaire en alignement et en courbe – Nouvelles technologies des bogies – Etude comparative*, Thèse de Doctorat de l', ENPC, Paris.

Pyrgidis, C. 2004, Il comportamento transversale dei carrelli per veicoli tranviari, *Ingegneria Ferroviaria*, October, Rome, 10, pp. 837–847.

Pyrgidis, C. and Bousmalis, T. 2010, A design procedure of the optimal wheel profile for railway vehicles running at conventional speeds, *5th International Congress for Transport Research*, Volos, 27–28 September 2010.

Pyrgidis, C. and Demiridis, N. 2006, The effects of tilting trains on the track superstructure, *1st International Congress, Railway Conditioning and Monitoring* 2006, IET, 29–30 November, Birmingham, UK, pp. 38–43.

Pyrgidis, C. and Joly, R. 1993, Forces acting in the guidance of a railway vehicle with conventional axles and independently rotating wheels, *Ingegneria Ferroviaria*, August, Rome, pp. 511–529.

Pyrgidis, C. and Panagiotopoulos, A. 2012, An optimization process of the wheel profile of tramway vehicles, *Elsevier Procedia Social and Behavioral Sciences*, Vol. 48, pp. 1130–1142.

Rivier, R. 1984/5, *Railway Notes*, EPFL, Lausanne.

Scheffel, H. 1974, Conceptions nouvelles relatives aux dispositifs de suspension des véhicules ferroviaires, *Rail International*, Brussels, December, pp. 801–817.

Scheffel, H. and Tournay, H.M 1980, The development of an optimum wheel profile for self-steering trucks under heavy axle load conditions, *Winter Annual Meeting*, 16–21 November 1980, Chicago, IL, Manuscript received at ASME, 23 June 1980.

Schneider Jeumont Rail. n.d., Bogie CL93 à moteurs asynchrones, *Catalogue pièces de rechange*, Le Creusot, France.

Vermeulen, P.J. and Johnson, K.L. 1964, Contact of non-spherical elastic bodies transmitting tangential forces, *Journal of Applied Mechanics*, Vol. 86, pp. 338–340.

Umdrucke zur Grundvorlesung. 2002/3, Technische Universitat Munchen, TeilI 5. Semester WS.

Chapter 4

Tramway

4.1 DEFINITION AND DESCRIPTION OF THE SYSTEM

The 'modern tramway' is a steel wheel electric train, running almost exclusively at grade along urban or suburban roads. It either shares the same infrastructure as the rest of the road traffic or moves on a specially built corridor or, finally, on a segregated (protected) lane, placed at one side, at two opposite sides, or in the middle of the roadway.

It serves distances usually in the range of 5–20 km, and it may be integrated into horizontal alignment radii as tight as 20–25 m. It is characterised by commercial speeds in the range of 15–25 km/h, and it is able to carry about 15,000 passengers/h/per direction (Brand and Preston, 2005). It commonly uses two, one-way traffic lines (double track), which are constructed with either grooved rails embedded in the pavement or conventional flat-bottom rails.

As far as technology and operation are concerned, the 'modern tramway' is a newer version of the 'conventional tram' (streetcar), which monopolised the urban public transport of most cities in Europe and the United States in the early decades of the last century (Figures 4.1 and 4.2).

4.2 CLASSIFICATION OF TRAMWAY SYSTEMS

Figure 4.3 summarises the main categories in which tramway systems can be classified based on their operational and constructional characteristics.

4.2.1 Based on physical characteristics of the corridor

Tramway corridors may be classified into five different categories (classes: E, D, C, B, A) (Bieber, 1986).

a. *Common corridor (class E)*: In this case, railway vehicles are mixed with road vehicles and pedestrians (Figure 4.4).

   In order not to hinder the movement of road vehicles, a tram runs on special rails (grooved rails), which are properly embedded in the pavement. The implementation cost of this corridor is relatively low, but the train's commercial speed remains low, similar to the commercial speed of urban buses (12–15 km/h) (Bieber, 1986). Furthermore, priority at traffic lights in relation to road transport cannot be given to the tram.

DOI: 10.1201/9781003046073-4

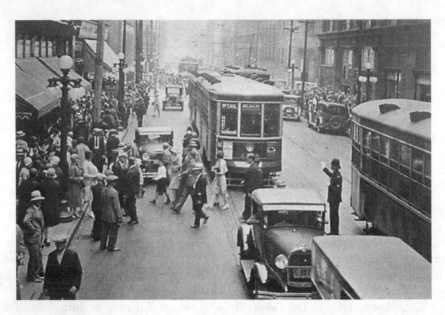

Figure 4.1 Streetcar. (From Collection CERTU, 1999, *Nouveaux Systèmes de Transports Guidés Urbains*, Paris, March 1999.)

Figure 4.2 Modern tramway, Zagreb, Croatia. (Photo: A. Klonos.)

b. *Exclusive separated corridor (class D)*: In this case, grooved rails are also used, but they are separated from the general traffic by means of horizontal lining or obstacles accessible to pedestrians (Figure 4.5).

   The tramway is, theoretically, separated from the rest of the traffic, except at level crossings. Separation of a corridor increases the commercial speed of the trains (16–22.5 km/h) (Bieber, 1986; Pyrgidis and Chatziparaskeva, 2012).

   At road intersections, level crossings are maintained; however, at these locations, priority at traffic lights can be given to the tram.

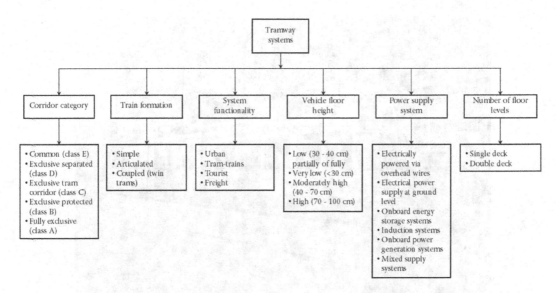

Figure 4.3 Classification of tramway systems based on their different constructional and operational characteristics.

Figure 4.4 Tram operation on 'common corridor' (class E), Oslo, Norway. (Photo: A. Klonos.)

c. *Exclusive tram corridor (class C)*: The existing road is used exclusively for the movement of the tram while the remaining road width is pedestrianised. The solution is applicable for narrow streets or when it is deliberately sought for sole use of the road by the tramway traffic only (for instance, in the case of commercial or historical city centres).

Grooved rails, which are embedded in the pavement, are used for the construction of the corridor, while the segregation of the tram corridor from the pedestrian area is usually achieved with the use of horizontal signalling (Figure 4.6).

Figure 4.5 Tram movement on 'exclusive separated corridor' (class D), Amsterdam, Holland. (Photo: A. Klonos.)

Figure 4.6 'Exclusive tram corridor' (class C), Zurich, Switzerland. (Photo: A. Klonos.)

These three solutions deliver a relatively small improvement in the quality of service in comparison with urban buses or trolleys moving on an exclusive lane, while at the same time allowing the implementation of a tram line with a relatively low construction cost. On the contrary, the environmental impact of such an intervention is very positive.

d. *Exclusive protected corridor (class B)*: In this case, the tramway is completely separated from the circulation of road vehicles and pedestrians. The separation can be achieved by artificial or natural means (trees, plants, railings, walls, etc.), and pedestrian crossings are placed at specific intervals, depending on pedestrian flows (Figure 4.7).

Railway vehicles usually move on rails that are similar to those of classic rail tracks (Vignoles type) but are of lighter construction. However, there is always the possibility of embedding them, as in the example of Figure 4.7. At intersections with roads, level crossings are maintained, but at these locations, priority at traffic signals can be given to the tram.

The movement of trains occurs with no particular problems, and despite the presence of level pedestrian crossings, it is possible to achieve commercial speeds of 20–25 km/h (Bieber, 1986).

e. *Fully exclusive corridor (class A)*: In this case, tramway vehicles move as in the previous case (class B) on flat-bottom rails, at grade, or underground, or elevated section (Figure 4.8).

Any level pedestrian crossings and intersections with roads are removed so that the commercial speed reaches 30 km/h (Bieber, 1986). This type of corridor is also used for the light metro and suburban rail.

The two latter solutions provide a better quality of service, but are comparatively far more expensive, especially when it is required that underground or elevated track sections are built.

4.2.2 Based on functionality/provided services

Based on functional/operational aspects and, more specifically, based on the nature and extent of services they provide, tramway systems may be divided into four categories:

a. *Urban tramways*: Serve passenger movements for relatively short distances (S = 5–20 km) within an urban area; they move at low commercial speeds (V_c = 15–25 km/h).

b. *Long-distance tramways ('tram-trains')*: This technique was first applied in Germany for the Regio Citadis train at Karlsruhe (1992, 1997). Tram-trains usually serve trips

Figure 4.7 Tram movement on 'exclusive protected corridor' (class B), Athens, Greece. (Photo: A. Klonos.)

Figure 4.8 Tram movement on 'fully exclusive corridor' (class A), Paris, France. (Photo: A. Klonos.)

Figure 4.9 Regio Citadis, Hague, Netherlands. (Photo: A. Klonos.)

that are 15–50 km long, connecting city centres to suburban and periurban areas. The maximum running speed that can be developed is $V_{max} = 80$–120 km/h, and the commercial speed is around $V_c = 60$ km/h.

Tram-trains are operated on infrastructure which is used not only by trams but also by other categories of railway systems (suburban, commuter/regional passenger trains, and freight trains) (Figure 4.9). The vehicles are equipped, for example, with two traction systems (dual-mode vehicles, diesel/750 V DC) or with dual-current systems (dual-voltage vehicles, e.g., 15 kV AC and 750 V DC), whilst the vehicle design is such that it allows trains to operate on platforms of different heights. Thus, the need to change mode is eliminated, accessibility is improved, and run times are reduced.

The length of tram-trains range between 26.50 m and 37 m, the width ranges between 2.40 m and 2.65 m, the height of the floor ranges between 350 mm and 450 mm with folding stairs and, finally, the bogies can negotiate curves of horizontal alignment radii up to $R_c = 20–30$ m. The average distance between stops ranges from 500 m up to 5 km.

c. *Tourist tramways or cultural heritage trams*: These systems serve tourist and recreational needs. They have a short connection length and move at low commercial speeds.

d. *Freight trams*: Since the beginning of the twenty-first century, urban tramway systems have been used for freight. The incentive is to reduce air pollution, traffic congestion, and the wear and tear of city centre traffic infrastructure. Urban trams that are able to carry goods are those at Dresden (Figure 4.10), Cologne (Germany), and Zurich (Switzerland). In Amsterdam, such trams were pilot tested, but ultimately not commissioned for revenue service.

4.2.3 Based on floor height of the vehicles

Depending on the distance between the floor of the vehicle and the top of the rail, tramway systems are divided into low floor, very low floor, moderately high floor, and high floor.

4.2.3.1 Low floor

The height between the rail-running table and the floor of the vehicle is 30–40 cm (commonly 35 cm), resulting in passenger access to vehicles without any steps (Figures 4.11 and 4.12). This design improves the tram's accessibility for the public, and particularly for

Figure 4.10 Freight tram in Dresden, Germany. (Adapted from http://www.flickr.com/people/77501394@N00 kaffeeeinstein, 2008.)

Figure 4.11   Wheelchair access at low-floor tram. (Adapted from Lasart75, 2010, available online at http://en.wikipedia.org/wiki/Low-floor_tram (accessed 7 August 2015).)

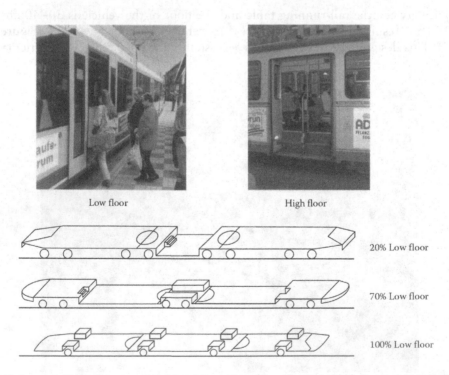

Figure 4.12   Low- and high-floor tram. (Adapted from Hass-Klau, C., et al., 2003, Bus and Light Rail: Making the Right Choice, ETP, Brighton.)

people with disabilities, and allows for the construction of larger windows. However, it also generates some problems such as:

- Difficulty in the installation and maintenance of electrical equipment.
- Discontinuity at floor level where carriages connect to each other.
- Positioning of the equipment at the edges of the car body, thereby increasing the length of the vehicle by 10–20%.

The minimum height of the platforms is considered 0.25 m. If low-floor vehicles are selected, the platforms' height can be reduced or even eliminated.

Depending on the percentage of their length, which is low floor, trams are distinguished (Figure 4.12) into the following categories:

- Totally (100%) low floor (low floor throughout the length of the vehicle).
- Partially low floor (e.g., for 70% of the vehicle length).

The technology of partially low-floor trams is old. On the contrary, the technique of low-floor trams is relatively new and is now an established current state-of-the-art for the urban tram. The Socimi Company in Milan manufactured the first vehicle of this type as a laboratory prototype in 1989. In 1990, the German company M.A.N. GHH presented the prototype of the first totally low-floor vehicle (350 mm) in Bremen.

4.2.3.2 Very low floor

The height between the rail-running table and the floor of the vehicle is < 30 cm. A typical value for that height is 180 mm.

4.2.3.3 Moderately high floor

The height between the rail-running table and the floor of the vehicle is 40–70 cm. For such trams, there are stairs in order for the passengers to access the vehicle floor.

4.2.3.4 High floor

The height between the rail-running table and the floor of the vehicle is 70–100 cm. For these trams, passengers' access to the vehicle floor is achieved by means of steps (Figure 4.12).

4.2.4 Based on power supply system

Based on the power supply system that is used for their movement, modern tramway systems are divided into the following categories (Guerrieri, 2019):

- Electrically powered via overhead wires.
  - Trolley type.
  - Overhead catenary systems (single or dual voltage, e.g., 15 kV AC and 750 V DC).
- Electrical power supply at ground level.
  - Conventional system with a third rail.
  - Conventional system with fourth and fifth rails.
  - APS (Alimentation Par Sol).
  - Tramwave.

- Onboard energy storage systems.
  - Batteries.
  - Supercapacitors.
  - Flywheels.
- Induction systems.
  - Primove.
- Onboard powered generation systems.
  - Hydrogen fuel cell.
  - Microturbines.
  - Diesel engines.
- Mixed supply system.
  - Diesel-powered and electrically powered via overhead wires (dual mode).
  - Electrically powered via overhead wires and power supply at ground level.
  - Electrically powered via overhead wires and onboard energy storage systems (supercapacitors, batteries).
  - Power supply at ground level and onboard energy storage systems (supercapacitors, batteries).

The vast majority of tramway systems are powered via the catenary system, however, the techniques that do not use overhead wires have largely developed over the last few years. These schemes are considered to be cutting-edge technologies for rail and are examined in detail in Chapter 20.

4.2.5 Other classifications

Based on the formation of trains, trams may be divided into simple, articulated (Figure 4.13), and coupled (articulated or simple).

Basing on the number of levels of their floors, vehicles are distinguished as:

- Single deck.
- Double deck.

Figure 4.13 Double articulated tram, Hague, Netherlands. (Photo: A. Klonos.)

Double-decker trams were used extensively in Great Britain until 1950 when they were dismantled. Today, this type of tram is still in operation in Alexandria, Blackpool, and Hong Kong.

Basing on the bogie's technology, modern trams are divided into those using bogies with independently rotating wheels and those using bogies of mixed behaviour (see Section 3.2.1.6).

Finally, based on their construction history, urban tram systems can be divided into the following three categories:

- Category 1: This category includes new systems that were built after 1980.
- Category 2: This category includes systems that were built many years ago. These systems were taken out of circulation and their tracks were dismantled. However, new infrastructure has recently been built for these systems and they were reopened.
- Category 3: This category includes systems that were built many years ago and were modernised and upgraded.

4.3 CONSTRUCTIONAL AND OPERATIONAL CHARACTERISTICS OF THE SYSTEM

The main characteristics of tramway systems were presented in Table 1.7. Table 4.1 provides additional data on the characteristics of tramway systems. In addition, the following should also be mentioned.

Table 4.1 Features and characteristic values related to tramway systems

Minimum horizontal alignment curvature radius	R_{cmin} = 20–25 m, preferred value $R_c \geq$ 30 m R_{cmin} = 15–18 m at shunting tracks (depots)
Types of track integration	• A single track per direction at the two opposite sides of the roadway • A double track on one side of the roadway • Central alignment (double track)
Types of stops integration	• Stop with a centre (island) platform • Stop with laterally staggered platforms • Stop with laterally opposed platforms
Types of power supply system (catenary overhead system integration)	• Central mast and opposite cantilevers • Lateral mast and double-track cantilevers • Catenary connected to laterally opposed masts • Catenary connected to building facades • Mixed catenary connection (on one side to lateral masts and on the other to building facades)
Vehicle length	Simple: 8–18 m, Articulated: 18–30 m, Multi-articulated: 25–45 m
Vehicle width	2.20–2.65 m (normal track gauge)
Commercial speed per tramway corridor category (without tram priority at level intersections)*	Class (E):V_c = 12–15 km/h Class (D):V_c = 16–18 km/h Class (C):V_c = 18–20 km/h Class (B):V_c = 20 km/h Class (A):V_c = 30 km/h
Impact of tram priority at signals on the commercial speed of the trams	Increase of commercial speeds by 15–25% for corridor classes (D) (V_{cmax} = 22.5 km/h) and (B) (V_{cmax} = 25 km/h)

Source: Adapted from Pyrgidis, C., 1997, Light rail transit: Operational, rolling stock and design characteristics, *Rail Engineering International*, No 1, 1997, Netherlands, pp. 4–7.

*Commercial speeds for each tramway corridor category result empirically considering an average distance between intermediate stops equal to 500 m, a halt time of 20 sec, and no priority at signals (Bieber, 1986).

4.3.1 Data related to track alignment and track superstructure

The cant of the outer rail at curved sections of the horizontal alignment is only deployed for the case of a 'fully exclusive tramway corridor' (class A). For all other tramway corridor categories, the cant is avoided.

Generally, in the vertical alignment, radii R_v that are smaller than 150–200 m should be avoided.

Based on the type of tramway corridor, the track may be laid in two distinct ways:

- For the case of a 'fully exclusive corridor' (class A) and in some cases, for 'exclusive protected corridor' (Class B), the rails used are similar to those of classic railway tracks (Vignoles/flat-bottom type), but of lighter construction (46 kg/m or 50 kg/m).
  For the construction of the superstructure, the excavation is usually 70 cm deep in order to ensure a unified sub-base. The track is ballasted or made on concrete (slab track). For intersections with road arteries, a special construction is foreseen.
- In the case of corridors of classes E, D, C and in many cases of B, in order not to prevent the movement of other vehicles and the crossing of pedestrians, special types of rails which are embedded in the pavement (grooved rails, Figure 4.14) are used. In both cases, in order to reduce the vibrations, resilient fastenings are used, elastic pads are placed under the sleepers, and antivibration layers may be employed under the ballast.

In special cases, high-damping, mass-spring systems (floating slab tracks) may be used. These systems (see Chapters 5 and 19) allow an even greater reduction of vibrations, however, they increase the implementation cost significantly.

4.3.2 Rolling stock data

Modern urban tramway vehicles are low floor, feature bidirectional movement, and their frame is constructed with rounded edges in order to protect pedestrians in the event of a collision.

To improve vehicle accessibility for people with disabilities, the rolling stock manufacturers and the network operators have adopted various measures such as:

- The construction of high platforms.
- The general reduction of the height of the vehicle floor.
- The reduction of the floor height only at the door ascending/descending spaces.
- The addition of an extra car with lower floor height.

The number of doors of a tram must necessarily be more than three (4–8), with a minimum width of 70 cm in the case of a single door and 1.20 m in the case of a double door. The height of the doors must necessarily be greater than 1.85 m. Hence, the passengers are better served at stops, and halt times are reduced (boarding–alighting time of about 20 sec).

The bogies of tramway vehicles must allow for the successful inscription of vehicles in curved track segments that have very small radii (up to 20–25 m), while at the same time they must be able to top speeds of 80–90 km/h in straight track segments (in which the track is of good ride quality). In this context, the bogies of modern tramway vehicles are different from the bogies used in other railway systems. More specifically, two alternative technologies of bogies are currently used, namely (Pyrgidis, 2004; Pyrgidis and Panagiotopoulos, 2012):

- Bogies with independently rotating wheels.
- Bogies with mixed behaviour (they operate as conventional bogies at straight paths of the track, and as bogies with independently rotating wheels in curves).

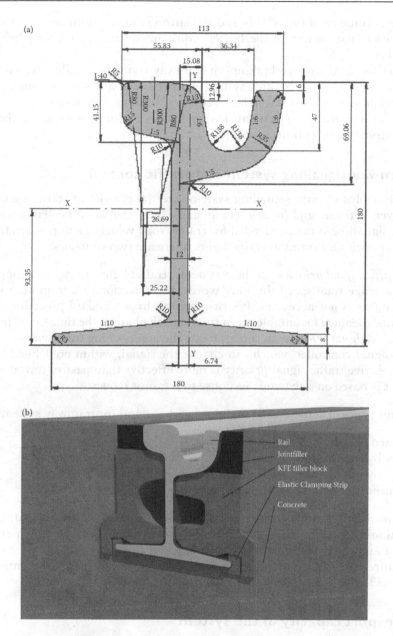

Figure 4.14 (a) Grooved rail Ri60N, cross section and (b) embedded grooved rail, sound damping system. (From edilon)(sedra, 2015.)

Regardless of the bogie's technology, in order to optimise their characteristics to meet the track alignment geometric data, and to achieve the desired performance, manufacturers may adopt (see Section 3.2.4):

- A small wheelbase (2a = 1.70–2.00 m instead of 2.50–3.00 m that is applied in conventional railway vehicles).
- A small wheel diameter ($2r_o$ = 0.60–0.70 m instead of 0.80–1.00 m that is applied in conventional railway vehicles).

The equivalent conicity of the wheels and the stiffness of the primary suspension springs constitute critical parameters of the bogies' construction for their lateral behaviour (see Section 3.2.4).

The gradual braking (service braking) of tramway trains is usually ensured by an electric braking system (rheostat braking system equipped with an energy regeneration system) which, in a second phase, is replaced by mechanical braking (brake discs, pads). The transmission is either pneumatic or electrical. The emergency braking is ensured by the additional effect of electromagnetic braking.

4.3.3 Tramway signalling system and traffic control

The basic principles of tram signalling systems are: (a) priority at traffic signal locations should be given to trams, and (b) at level crossings, there should be a collaboration between the different signalling systems intended for trams, road vehicles, and pedestrians.

Regarding priority for trams at traffic lights, there are two strategies:

Passive traffic signal priority: In these systems, traffic lights are set to turn green based on an average tram speed. In other words, the detection of a tram at crossings with traffic lights is not necessary. Priority is given by a standard procedure: favourable cycle time meaning favourable green time at each phase of the time-coordination cycle.

Active traffic signal priority: In this strategy, the approaching tram sends a signal to the traffic signal controller, which can change the signal, within predefined limits, in its favour. Active traffic signal priority is more effective than passive traffic signal priority, as it is based on a dynamic response to a transit request.

There are four types of active traffic signal priority systems for tramway systems:

- Dedicated priority by phasing changes.
- Priority by extended green time.
- Priority by phase and phase timing adjustment.
- Implementation of Intelligent Transport Systems (ITS) approaches.

In order to locate the tram's position, two different approaches are applied: the use of a Global Positioning Satellite (GPS) system or the use of special sensors, which are placed on the pavement and can detect the tram as it passes over them.

The literature states that tram priority at intersections can increase the commercial speed of trams by 35% (Pyrgidis and Chatziparaskeva, 2012; Fox et al., no date).

4.3.4 Transport capacity of the system

The majority of tramway systems transport 150–250 passengers per train (standing and seated) (Lesley, 2011). In any event, the capacity of the tram, for a given acceptable passenger density (e.g., 6 passengers/m^2, 4 passengers/m^2), may be calculated by taking into consideration the inner length and width of the vehicle and the number of passenger seats.

Figure 4.15 (Bieber, 1986) presents average transport system capacity values in relation to the trains' headway for four types of trams and two types of urban buses (considering average train/vehicle transport capacity values). The diagram is restricted to service frequencies between 6 trains/h and 60 trains/h. Frequencies greater than 60 trains/h cause problems to other traffic and reduce the level of service.

According to the literature, the headway of two successive trains ranges from 1.5 min up to 30 min, depending on the volume of passenger traffic, the day, and the hour of the day.

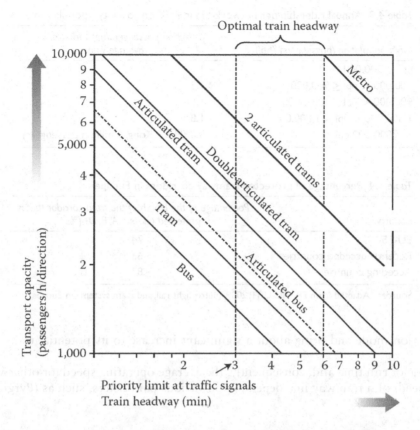

Figure 4.15 Train headway – transport capacity chart for urban public transport modes. (Adapted from Bieber, C. A., 1986, *Les choix techniques pour les transports collectifs*, Lecture Notes, Ecole Nationale des Ponts et Chaussées, Paris.)

The typical values range from 5 min (peak hours) to 15 min (off-peak hours). The lowest values of train headways are recorded for the Hong Kong tram system (1.5 min during peak hours).

Regarding the passenger traffic volume, Table 4.2 presents the annual passenger traffic per network-km for tramway systems of all European countries (ERRAC-UITP, 2009). Table 4.3 lists the results of statistical analysis on the annual ridership per network-km in relation to a city's population.

4.3.5 Run time and commercial speeds

Run time is one of the fundamental factors that determine the quality of service provided by a tramway network to its users. Low values of run time render the tram a more attractive

Table 4.2 Annual ridership per network-km for tramway systems in European countries

Country	Million passengers/network-km/year (average values)
EU-15	1.00
Recently acceded countries	1.97
Acceding countries	1.36

Source: Adapted from ERRAC-UITP, 2009, Metro, light rail, and tram systems in Europe.

Table 4.3 Annual ridership per network-km in relation to a city's population

City's population (passengers) (ppl)	Million passengers/network-km/year (average values)
ppl ≤ 300,000	1
300,000 < ppl ≤ 900,000	1.3
900,000 < ppl ≤ 1,5000,00	1.6
1,500,000 < ppl ≤ 1,800,000	1.8
> 5,000,000 ppl	3.8 (Hong Kong, 6 million passengers)

Table 4.4 Percentage of protected tramway corridors in Europe

Country	Percentage of the length of the tram corridor that is protected (classes A, B, D) (%)
EU-15	74
Recently acceded countries	63
Acceding countries	87

Source: Adapted from ERRAC-UITP, 2009, Metro, light rail, and tram systems in Europe.

transportation mode and bring about a significant increase to its potential transportation volume.

The value of run time and, consequently, the average operating speed (or otherwise commercial speed) of a tramway line depend upon various parameters, such as (Pyrgidis et al., 2013):

- The category of tramway corridors encountered along the route and the length per category.
- The signalling system applied at level crossings where the tram intersects with other transportation modes.
- The number of intermediate stops and the distance between them.
- The halt time.
- The citizens' awareness of the presence of trams and their attitude.
- The tram drivers' behaviour.
- The track alignment.
- The performances of the rolling stock.
- The number of level crossings.
- The time of the day.

The most important parameters are the category of the tramway corridor and the signalling system used.

The proportion of the length of the tramway corridor that is protected (classes A, B, D) to the total length of the tram route is usually high and ranges between 60% (e.g., in Zurich) and 90–100% (e.g., in Cologne, Karlsruhe). Table 4.4 lists the average protection percentages for all tramway networks in Europe (ERRAC-UITP, 2009).

With regard to commercial speeds, V_c values in the range of 13.5–35 km/h have been recorded in practice at various low-floor tram networks, while the most usual values are of the range 15–25 km/h.

Table 4.5 Average commercial speed of all tramway systems of European countries

Country	Commercial speed (average values) km/h
EU-15	22.6
Recently acceded countries	15.71
Acceding countries	21.10

Source: Adapted from ERRAC-UITP, 2009, Metro, light rail, and tram systems in Europe.

Table 4.6 Total implementation cost of tramway systems using mixed power supply technology (overhead catenary system + ground power supply system (APS) (2014 data))

City/France	Line length (km)	Percentage of line length with APS (%)	Cost per km (million €)
Tours	15	13.3	27.4
Orleans	12	17.5	30.3
Angers	12	12.5	32.9
Reims	12	16.7	31.2
Bordeaux	44	30.9	35.7

Table 4.5 lists the commercial speeds that are developed by average at all tramway networks in Europe (ERRAC-UITP, 2009).

4.3.6 Cost of implementing a tramway

For the assessment of the construction cost of tramway lines, 2012–2014 data are used. The power supply system, labour costs, and the extent of any underground or extensive surface civil engineering works/rehabilitation are key parameters that differentiate the total implementation cost.

The urban low-floor tramway systems cost between €14 M (Palermo) up to over €60 M per track-km.

The average total cost of construction per km of an urban tramway with overhead power supply system is in the range of €20–€23.5 M (€22.5 M for Europe, €20 M for Africa, and €23.5 M for North America). Any special civil engineering works may significantly increase these costs (e.g., €62.8 M for the Jerusalem tram and €84.5 M for the Seattle tram).

Table 4.6 gives the total implementation cost per track-km for five in-service tramway systems using mixed power supply technology (overhead catenary system + ground power supply system [APS]) (2014 data) (see also Table 20.5).

In Dubai, the APS system was applied to the total length of the network (10.6 km). The overall construction cost of the system reached €66.8 M per km (2014 data) due to the exclusive use of the APS, but mainly due to the high manufacturing cost of stations (indoor areas and air-conditioning).

The average cost per vehicle for an urban tram with an overhead catenary system amounts in the range of €2.5 M (width 2.40 m, length 32 m) (2014 data). This cost can reach up to €3.5 million or even more depending on the available equipment and the capabilities of the vehicle.

The average cost for an urban tram vehicle with a power supply system at ground level is approximately 15% bigger.

For tram-trains, the cost of a vehicle is significantly higher, and it approximates €4 M–€4.5 M.

4.4 INTEGRATION OF TRAMWAY CORRIDORS ACROSS THE ROAD ARTERIES

4.4.1 Types of integration of tramway corridors

The integration of at-grade tramway lines within the right-of-way can take place in different ways depending on the geometric and traffic characteristics of the road and the nature of the roadside land uses, as shown in greater detail in the following subsections.

4.4.1.1 A single track per direction at two opposite sides of the road

This option is more appropriate for roads that operate as one-way roads for all other traffic. Its main advantages are as follows (Figure 4.16):

- In the case of an overhead power supply system, masts can be installed on the side footpaths. As a result, a smaller right-of-way is required.
- At the stop areas, the existing sidewalks may well serve as part of the platforms.

Its main disadvantages are as follows:

- A noticeable difficulty in feeding the adjacent land uses, which necessarily takes place during specific hours of the day, especially during hours when the tram stops operating or operates with very low frequencies.
- In the case of small building blocks, the increased number of intersections can reduce the run time savings, which come as a result of the segregation of the tram from all other traffic.

4.4.1.2 Double track on one side of the road

In this case, in the vicinity of the tramway stops, it is required to build an islet in order to create a platform for the second vehicle. As a result, the road width is reduced. This problem can be solved by creating a recess, which, however, reduces the width of the sidewalk (Figure 4.17).

This is the simplest and least space-consuming integration; however, it has some negative impact on the local residents and their activities. In the case of small building blocks,

Figure 4.16 Placement of a single tramway track at the two opposite sides of the roadway. (From Thessaloniki Public Transport Authority, 2015.)

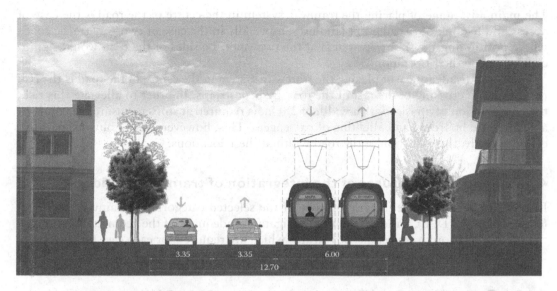

Figure 4.17 Placement of a double tramway track on one side of the roadway. (From Thessaloniki Public Transport Authority, 2015.)

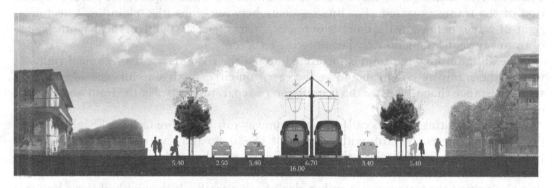

Figure 4.18 Placement of a double tramway track in the middle of the road using central electrification mast. (From Thessaloniki Public Transport Authority, 2015.)

the successive intersections with right-turning movements can reduce the run time savings, which come as a result of the segregation of the tram from all other traffic.

4.4.1.3 Central alignment

The tramway system is located in the centre of the right-of-way, usually in double-track superstructure (Figure 4.18). With this integration, there is no problem with turning road vehicle movements. In the case of overhead power supply, the integration of the tram in the centre of the road artery may be implemented in two ways:

- *Without electrification masts between the two tracks:* In this case, the power supply of tramway vehicles is achieved through wires, which are suspended on laterally opposed masts or connected to building facades. This integration offers a smaller right-of-way.
- *With electrification masts placed centrally between the two tracks* (Figure 4.18). This solution is preferable in terms of aesthetics, but it is more expensive and less favourable for the road traffic as it requires greater corridor width.

The main advantage of placing the tramway system in the centre of the road is the ease of access and feeding of the adjacent land uses, especially in the case of two-way traffic roads where positioning the tramway system at the two opposite sides of the road would significantly impede their operations.

The main disadvantage of placing the tramway system in the centre of the road is the risk regarding the crossing of the rest of the road by pedestrians. In order to alleviate this risk, the construction of an islet with a width of 2.0 m, is required at stops to ensure the comfortable and safe boarding and alighting of passengers. This, however, has an obvious negative effect, namely, the reduction of the road width at these locations.

4.4.2 Geometric features of the integration of tramway corridors

In order to decide on the integration type of the selected categories of tramway corridors (classes A, B, C, D, E), it is essential to investigate the adequacy of the geometric features of the road in relation to the geometric features which are required for each case for the operation of the tramway system.

4.4.2.1 Technical and Total Tramway infrastructure Right-Of-Way

The term *technical right-of-way* describes the minimum width required for the safe operation of the tramway system. For line segments between stops (the 'plain line'), technical right-of-way is defined by the number of tracks, their gauge, the width of the dynamic gauge of tramway vehicles, and the civil engineering structure gauge.

Depending on how the tram is integrated across the road artery, the technical right-of-way of the tramway system should be increased on either both sides or one side only, by such a distance that allows for the installation of separators between the tram and the rest of the traffic.

The final resulting width is called Total Tramway infrastructure technical Right-Of-Way (TTROW). Concrete bollards with a height of 15–20 cm and a width of 40 cm or greater (minimum width 30 cm) where plants are dibbled may be used as means of segregation. These bollards may also be used for posting signs or signals, which will serve the signalling not only of the tramway system but also of other traffic. If the width of these bollards is greater than 1.20 m, they can be used as intermediate stops for pedestrians during their movement on the level crossing. Apart from these segregation means, hatched lane (with a width of 40 cm), railings, wall separators, trees, and so on can also be used, depending on the type of integration of the tramway system in the road.

4.4.2.1.1 Total Tramway infrastructure technical Right-Of-Way at
Straight segments of the alignment (TTROWS)

In the case of double track, at straight paths, the structure gauge is equal to the dynamic gauge of vehicles increased by 100 mm at either side of the double track and by 200 mm between opposite moving tram vehicles.

The mathematical relationships that can be used in order to calculate the TTROWS for different ways of integrating tramway corridors across the road arteries, and depending on the category of tramway corridor, are shown hereunder:

1. Placement of a double tramway track at the centre of the roadway (central alignment):

$$TTROWS = 2 \times (b_{sw} + 0.1 + g_{dv}) + b_{em} + 0.2 \,(classes\ A, B, C, D) \tag{4.1}$$

$$\text{TTROWS} = 2 \times g_{dv} + b_{em} + 0.4 \,(\text{class E}) \tag{4.2}$$

where:
  b_{sw}: Width of separator.
  g_{dv}: Dynamic gauge width of tram vehicle.
  b_{em}: Width needed for the installation of electrification masts.

In case no electrification mast is foreseen, $b_{em} = 0$.
2. Placement of a double tramway track on one side of the roadway:

$$\text{TTROWS} = 2 \times (0.1 + g_{dv}) + b_{em} + b_{sw} + 0.2 \,(\text{classes A,B,C,D}) \tag{4.3}$$

$$\text{TTROWS} = 2 \times g_{dv} + b_{em} + 0.4 \,(\text{class E}) \tag{4.4}$$

If no electrification mast is foreseen, $b_{em} = 0$.
3. Placement of a single tramway track at the two opposite sides of the roadway:

$$\text{TTROWS} = 2 \times (b_{sw} + 0.1 + 0.1 + g_{dv}) \,(\text{classes A,B,C,D}) \tag{4.5}$$

$$\text{TTROWS} = 2 \times (0.2 + g_{dv}) \,(\text{class E}) \tag{4.6}$$

Table 4.7 provides the minimum values of the Total Tramway infrastructure Right-Of-Way which are required in order to integrate a tramway system across a road artery for a vehicle width of 2.30 m (dynamic gauge width $g_{dv} = 2.60$ m), for a separator with a width of $b_{sw} = 0.40$ m, for an electrification mast installation width of $b_{em} = 0.30$ m, and for corridor categories A, B, C, and D.

4.4.2.1.2 Total Tramway infrastructure Right-Of-Way in Curves (TTROWC)

A larger right-of-way is required in curves, and this is due to the following reasons:

• Primarily, it is due to the extra space required, because of the geometry, for the integration of the tracks across the road.
    This extra space depends on the angle between the two intersecting roads and on the type of integration of the tramway line across the road before and after the intersection. In this context, the required right-of-way of the tramway lines can be related and

Table 4.7 Total infrastructure right-of-way values for a tramway system at straight path

Types of integration of tramway corridors across the road	Straight path (TTROWS) (m)
Placement at one side of the roadway – single track	3.20
Placement at one side of the roadway – double track without an electrification mast	6.00
Placement at one side of the roadway – double track with a centrally placed electrification mast	6.30
Placement at the centre of the roadway – double track without a centrally placed electrification mast	6.40
Placement at the centre of the roadway – double track with a centrally placed electrification mast	6.70
Placement at the two opposite sides of the roadway – single track	6.40

be expressed via the minimum required width of intersecting roads (Chatziparaskeva and Pyrgidis, 2015).

• Secondly, it is due to the effects of vehicle end throw on the outside of a curve and centre throw on the inside of a curve (Figures 4.19 and 4.20) (Mundrey, 2000). These reach their greatest values in the middle of the circular arc of the curved segment. They directly depend on the length of the cars of the articulated tramway vehicle (they increase as the length increases).

Engineers can estimate the right-of-way at turns through appropriate design simulation.

4.4.2.2 Geometric integration of tramway corridors at curved sections of roads in the horizontal alignment

The geometric integration of a tramway corridor in curved sections of the right-of-way requires the horizontal alignment of the tram, including considerations for end and centre throw, to be able to match the horizontal alignment of the road artery and space constraints

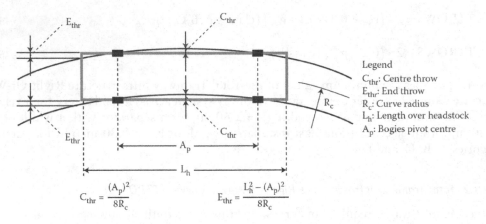

Legend

C_{thr}: Centre throw
E_{thr}: End throw
R_c: Curve radius
L_h: Length over headstock
A_p: Bogies pivot centre

$$C_{thr} = \frac{(A_p)^2}{8R_c} \qquad E_{thr} = \frac{L_h^2 - (A_p)^2}{8R_c}$$

Figure 4.19 Inscription of a railway vehicle in curves. Definition of end/centre throw. (Adapted from CRN CS 215, 2013, Engineering Standard Track, version 1.1, July 2013.)

Figure 4.20 Inscription of a double articulated tramway vehicle in curves. Effects of vehicle end and centre throw.

of the right-of-way in general; a minimum radius of horizontal alignment equal to $R_{cmin} = 25$ m is the usual minimum value considered.

A design simulation for all possible combinations regarding the integration type of the tramway track before and after the intersection, for intersection angles that vary between $\varphi_o = 90°$ and $\varphi_o = 170°$, has been developed in the literature (Chatziparaskeva and Pyrgidis, 2015). For this simulation, the following were considered:

- Static vehicle width equal to 2.30 m and dynamic vehicle width equal to 2.60 m.
- Curve radius in the horizontal alignment $R_c = 25$ m (minimum permitted).
- Integration without central electrification mast.
- Right turn and left turn.
- At the area of the turn, the tramway corridor is common (class E).

The design simulation provided the minimum required road widths b1 and b2 of the road arteries. Table 4.8 presents the indicative results for all possible combinations of integration of the tramway track, for intersection angle $\varphi_o = 120°$.

Regarding the symbols used for the integration type (column 1 of Table 4.8):

- The first letter indicates whether the corridor is 'exclusive' for trams or whether the use by other road vehicles is also permitted (classes B, D). More specifically, the following symbols are adopted:
  C: Corridor class C ('exclusive tramway corridor' without use by other road vehicles).
  F: Corridor classes B and D ('protected or separated tramway corridors').
- The second letter indicates the integration type of the double tramway track. More specifically the following symbols are adopted:
  A: Integration of the track at the left side of the road.
  Δ: Integration of the track at the right side of the road.
  K: Integration of the track in the middle of the road.
- The number 1 refers to single track, whereas the number 2 refers to double track.
- The symbol α indicates left turn movement, whereas the symbol δ indicates right turn movement.
- Finally, the symbol X indicates that integration is not possible.

As an example, the symbols F2Δ-F2K (δ) indicate transition by a right turn, from a tramway corridor category B or D, with a double track placed at the right side of the roadway, to a tramway corridor category B or D, with a double track placed at the centre of the roadway (Figure 4.21).

In order to enable the geometric integration of the tramway tracks at turns, even with the smallest allowable radius of 25 m, the two intersecting roads must have the available width that is calculated by following the procedure described above.

4.5 INTEGRATION OF STOPS

4.5.1 Types of stops integration

Terminals and stops constitute an important component of the tramway system. They are considered as structural elements of the tramway infrastructure and they fall under its operational facilities. Their presence in the system is necessary because they allow for the boarding and alighting of passengers to/from the trains.

Table 4.8 Integration of a double tramway track in curved sections of the roads in the horizontal alignment – required roadway width for road intersection angle $\varphi_0 = 120°$

Road width / Integration type	bl = 7 m	bl = 8 m	bl = 9 m	bl = 10 m	bl = 11 m	bl = 12 m	bl = 13 m	bl = 14 m	bl = 15 m	bl = 16 m	bl = 17 m	bl = 18 m	bl = 19 m	bl = 20 m
CA-CA (δ) CΔ-CΔ (α)	X	b1=8 b2=13.02	b1=9 b2=11.26	b1=10 b2=10.09	b1=11 b2=9.2	b1=12 b2=8.52	b1=13 b2=8.01	b1=14 b2=7.61	b1=15 b2=7.32	X	X	X	X	X
CA-CK (δ) CΔ-CK (α)	X	X	X	b1=10 b2=13.78	b1=11 b2=12	b1=12 b2=10.64	b1=13 b2=9.62	b1=14 b2=8.82	b1=15 b2=8.24	X	X	X	X	X
CA-CΔ (δ) CΔ-CA (α)	X	X	X	X	X	X	X	X	X	X	X	X	X	X
CK-CA (δ) CK-CΔ (α)	X	b1=8 b2=15.52	b1=9 b2=13.75	b1=10 b2=12.6	b1=11 b2=11.71	b1=12 b2=10.99	b1=13 b2=10.39	b1=14 b2=9.88	b1=15 b2=9.43	X	X	X	X	X
CK-CK (δ) CK-CK (α)	X	X	X	X	X	b1=12 b2=15.58	b1=13 b2=14.38	b1=14 b2=13.36	b1=15 b2=12.46	X	X	X	X	X
CK-CΔ (δ) CK-CA (α)	X	X	X	X	X	X	X	X	X	X	X	X	X	X
CΔ-CA (δ) CA-CΔ (α)	X	X	X	X	X	X	X	X	X	X	X	X	X	X
CΔ-CK (δ) CA-CK (α)	X	X	X	X	X	X	X	X	X	X	X	X	X	X
CΔ-CΔ (δ) CA-CA (α)	X	X	X	X	X	X	X	X	X	X	X	X	X	X
CA-FA (δ) CΔ-FΔ (α)	b1=7 b2=16.69	b1=8 b2=12.62	b1=9 b2=10.86	b1=10 b2=9.69	b1=11 b2=9	b1=12 b2=9	b1=13 b2=9	b1=14 b2=9	b1=15 b2=9	X	X	X	X	X
CA-FK (δ) CΔ-FK (α)	b1=7 b2=27.78	b1=8 b2=19.64	b1=9 b2=16.12	b1=10 b2=13.78	b1=11 b2=12.4	b1=12 b2=12.4	b1=13 b2=12.4	b1=14 b2=12.4	b1=15 b2=12.4	X	X	X	X	X

(Continued)

Table 4.8 (Continued) Integration of a double tramway track in curved sections of the roads in the horizontal alignment – required roadway width for road intersection angle φ_o = 120°

Road width/Integration type	bI = 7 m	bI = 8 m	bI = 9 m	bI = 10 m	bI = 11 m	bI = 12 m	bI = 13 m	bI = 14 m	bI = 15 m	bI = 16 m	bI = 17 m	bI = 18 m	bI = 19 m	bI = 20 m
CA-FΔ (δ) / CΔ-FA (α) / CA-FIF2 (δ) / CΔ-FIF2 (α)	X	X	X	X	X	X	X	X	X	X	X	X	X	X
CK-FA (δ) / CK-FΔ (α)	b1=7 b2=18.21	b1=8 b2=15.12	b1=9 b2=13.35	b1=10 b2=12.4	b1=11 b2=11.31	b1=12 b2=10.59	b1=13 b2=9.99	b1=14 b2=9.48	b1=15 b2=9.03	X	X	X	X	X
CK-FK (δ) / CK-FK (α)	b1=7 b2=30.82	b1=8 b2=24.64	b1=9 b2=21.1	b1=10 b2=19.2	b1=11 b2=17.02	b1=12 b2=15.58	b1=13 b2=14.38	b1=14 b2=13.36	b1=15 b2=12.46					
CK-FΔ (δ) / CK-FA (α) / CK-FIF2 (δ·α)	X	X	X	X	X	X	X	X	X	X	X	X	X	X
CΔ-FA (δ) / CΔ-FΔ (α)	b1=7 b2=18.11	b1=8 b2=18.11	b1=9 b2=18.11	b1=10 b2=18.11	b1=11 b2=18.11	b1=12 b2=18.11	b1=13 b2=18.11	b1=14 b2=18.11	b1=15 b2=18.11	X	X	X	X	X
CΔ-FK (δ) / CA-FK (α)	b1=7 b2=30.82	b1=8 b2=30.82	b1=9 b2=30.82	b1=10 b2=30.82	b1=11 b2=30.82	b1=12 b2=30.82	b1=13 b2=30.82	b1=14 b2=30.82	b1=15 b2=30.82	X	X	X	X	X
CΔ-FΔ (δ) / CΔ-FA (α) / CΔ-FIF2 (δ) / CΔ-FIF2 (α)	X	X	X	X	X	X	X	X	X	X	X	X	X	X
FA-CA (δ) / FΔ-CΔ (α)	X	X	b1=9 b2=10.74	b1=10 b2=9.69	b1=11 b2=8.91	b1=12 b2=8.3	b1=13 b2=7.84	b1=14 b2=7.48	b1=15 b2=7.23	b1=16 b2=7.05	b1=17 b2=6.94	b1=18 b2=6.9	b1=19 b2=6.4	b1=20 b2=6.4
FA-CK (δ) / FΔ-CK (α)	X	X	b1=9 b2=15.08	b1=10 b2=12.98	b1=11 b2=11.42	b1=12 b2=10.2	b1=13 b2=9.28	b1=14 b2=8.56	b1=15 b2=8.06	b1=16 b2=7.7	b1=17 b2=7.48	b1=18 b2=7.4	b1=19 b2=6.4	b1=20 b2=6.4

(Continued)

Table 4.8 (Continued) Integration of a double tramway track in curved sections of the roads in the horizontal alignment – required roadway width for road intersection angle $\varphi_o = 120°$

Road width / Integration type	bl = 7 m	bl = 8 m	bl = 9 m	bl = 10 m	bl = 11 m	bl = 12 m	bl = 13 m	bl = 14 m	bl = 15 m	bl = 16 m	bl = 17 m	bl = 18 m	bl = 19 m	bl = 20 m
FA-CΔ (δ) / FΔ-CA (α)	X	X	X	X	X	X	X	X	X	X	X	X	b2 = 6.4	b2 = 6.4
FK-CA (δ) / FK-CΔ (α)	X	X	X	X	X	X	b2 = 10.39	b2 = 9.88	b2 = 9.43	b2 = 9.05	b2 = 8.71	b2 = 8.41	b2 = 8.15	b2 = 7.92
FK-CK (δ) / FK-CK (α)	X	X	X	X	X	X	b2 = 14.38	b2 = 13.36	b2 = 12.46	b2 = 11.7	b2 = 11.02	b2 = 10.42	b2 = 9.9	b2 = 9.44
FK-CΔ (δ) / FK-CA (α)	X	X	X	X	X	X	X	X	X	X	X	X	X	X
FΔ-CA (δ) / FΔ-CΔ (α) / FIF2-CA (δ) / FIF2-CΔ (α)	X	X	X	X	X	X	X	X						
FΔ-CK (δ) / FK-CK (α) / FIF2-CK (δ·α)	X	X	X	X	X	X	X	X	X	X	X	X	X	X
FΔ-CΔ (δ) / FΔ-CA (α) / FIF2-CΔ (δ) / FIF2-CA (α)	X	X	X	X	X	X	X	X	X	X	X	X	X	X
FA-FA (δ) / FΔ-FΔ (α)	X	X	b2 = 10.34	b2 = 9.29	b2 = 9	b2 = 9	b2 = 9	b2 = 9	b2 = 9	b2 = 9	b2 = 9	b2 = 9	b2 = 9	b2 = 9.0
FA-FK (δ) / FΔ-FK (α)	X	X	b2 = 15.08	b2 = 12.98	b2 = 12.4	b2 = 12.4	b2 = 12.46	b2 = 12.4	b2 = 12.4	b2 = 12.4	b2 = 12.4	b2 = 12.4	b2 = 12.4	b2 = 12.4

(Continued)

Table 4.8 (Continued) Integration of a double tramway track in curved sections of the roads in the horizontal alignment – required roadway width for road intersection angle $\odot_0 = 120°$

Road width/ Integration type	bl = 7 m	bl = 8 m	bl = 9 m	bl = 10 m	bl = 11 m	bl = 12 m	bl = 13 m	bl = 14 m	bl = 15 m	bl = 16 m	bl = 17 m	bl = 18 m	bl = 19 m	bl = 20 m
FA-FΔ (δ) FΔ-FA (α) FA-FIF2 (δ) FΔ-FIF2 (α)	X	X	X	X	X	X	X	X	X	X	X	X	b1 = 19 b2 = 9 b2 = 9.4 (F1F2)	b1 = 20 b2 = 9.0 b2 = 9.4 (F1F2)
FK-FA (δ) FK-FΔ (α)	X	X	X	X	X	X	b1 = 13 b2 = 9.99	b1 = 14 b2 = 9.48	b1 = 15 b2 = 9.03	b1 = 16 b2 = 9	b1 = 17 b2 = 9	b1 = 18 b2 = 9	b1 = 19 b2 = 9	b1 = 20 b2 = 9
FK-FK (δ) FK-FK (α)	X	X	X	X	X	X	b1 = 13 b2 = 14.38	b1 = 14 b2 = 13.36	b1 = 15 b2 = 12.46	b1 = 16 b2 = 12.4	b1 = 17 b2 = 12.4	b1 = 18 b2 = 12.4	b1 = 19 b2 = 12.4	b1 = 20 b2 = 12.4
FK-FΔ (δ) FK-FA (α) FK-FIF2 (δ · α)	X	X	X	X	X	X	X	X	X	X	X	X	X	X
FΔ-FA (δ) FΔ-FΔ (α) FIF2-FA (δ) FIF2-FΔ (α)	X		b1 = 9 b2 = 18.11 X (F1F2)	b1 = 10 b2 = 18.11	b1 = 11 b2 = 18.11	b1 = 12 b2 = 18.11	b1 = 13 b2 = 18.11	b1 = 14 b2 = 18.11	b1 = 15 b2 = 18.11	b1 = 16 b2 = 18.11	b1 = 17 b2 = 18.11	b1 = 18 b2 = 18.11	b1 = 19 b2 = 18.11	b1 = 20 b2 = 18.11
FΔ-FK (δ) FA-FK (α) FIF2-FK (δ · α)	X		b1 = 9 b2 = 30.82 X (F1F2)	b1 = 10 b2 = 30.82	b1 = 11 b2 = 30.82	b1 = 12 b2 = 30.82	b1 = 13 b2 = 30.82	b1 = 14 b2 = 30.82	b1 = 15 b2 = 30.82	b1 = 16 b2 = 30.82	b1 = 17 b2 = 30.82	b1 = 18 b2 = 30.82	b1 = 19 b2 = 30.82	b1 = 20 b2 = 30.82
FΔ-FΔ (δ) FA-FA (α) FIF2-FIF2 (δ · α)	X	X	X	X	X	X	X	X	X	X	X	X	X	X

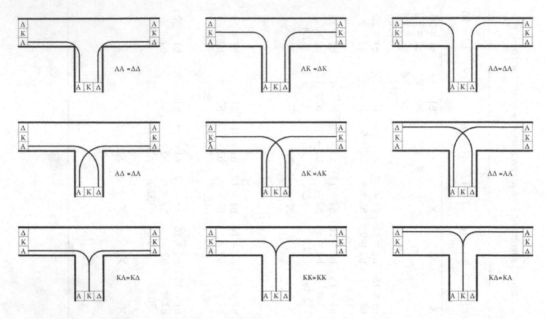

Figure 4.21 Types of integration of double-track tramway corridors across the road before and after the turn.

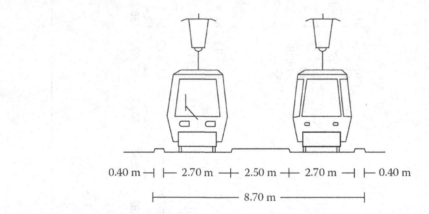

Figure 4.22 Stop with centre (island) platform. (Adapted from RATP D.D.E, 1994, Projet de rocade tramway en site propre entre Saint Denis et Bobigny: Schéma de principe, 1993, Paris, Février.)

Three categories of tramway stops may be considered (RATP D.D.E, 1994):

- Stop with centre (island) platform (Figure 4.22).
- Stop with laterally staggered platforms (Figure 4.23).
- Stop with laterally opposed platforms (Figure 4.24).

The level of service provided to the users of a tramway system at stops is determined by the degree to which certain parameters are satisfied. The key parameters (quality parameters) which reflect the user needs are:

- The acceptable distance between successive stops.
- The location of stops at areas where land uses constitute attractors of a large number of trips.

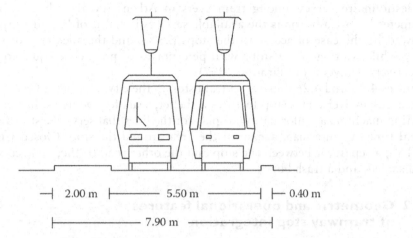

Figure 4.23 Stop with laterally staggered platforms. (Adapted from RATP.D.D.E, 1994, Projet de rocade tramway en site propre entre Saint Denis et Bobigny: Schéma de principe, 1993, Paris, Février.)

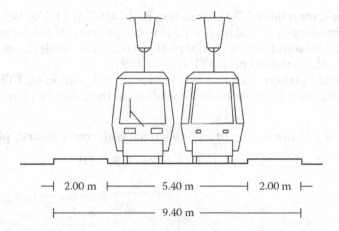

Figure 4.24 Stop with laterally opposed platforms. (Adapted from RATP.D.D.E, 1994, Projet de rocade tramway en site propre entre Saint Denis et Bobigny: Schéma de principe, 1993, Paris, Février.)

- The required halt time.
- The accessibility of the stop.
- The ability for quick transfer to other modes of public transport.
- The service of people with reduced mobility.
- The safety and comfort of passengers while waiting at the stops (seats).
- The information regarding the route and the next train arrival available for passengers.
- The easy supply of tickets.
- The interfaces between staff and users.
- The easy identification of the stop from afar.
- The protection of users against bad weather conditions (shelter).
- The aesthetics of the stop.
- The attractiveness of the stop (surface integration, location at areas with recreational activities, retail or medical facilities).

A questionnaire survey among tram users in Athens revealed that the most important parameter for a tram stop is the available services in terms of land uses (preference 43%), followed by the ease of access to the stops (28%), and the adequacy of information and safety while waiting at the stops with percentages of preference equal to 12% and 11%, respectively (University of Piraeus, 2007).

Figures 4.25 and 4.26 illustrate a tram stop in the city of Athens, Greece, and Grenoble, France, respectively. This stop features a shelter, seats for the users, lighting, an automatic ticketing machine, an information display for the lines that serve the stop and waiting times in real time, a route map, a map of the area around the stop, Closed Circuit TeleVision (CCTV), a separator between the stop and the other road traffic, an accessibility ramp for the disabled, and a trash bin.

4.5.2 Geometric and operational features of tramway stop integration

The location of stops and the choice of the type of platforms for a tramway network are based on the following geometric and operational criteria:

4.5.2.1 Geometric criteria

In the area of STops, the required Total Tramway Right-Of-Way (TTROWST) is larger than the respective right-of-way for a 'plain line', due to the presence of platforms.

The minimum allowed width for a central platform is 2.50 m, while the minimum allowed width for a side platform is 2.00 m (RATP D.D.E, 1994).

The mathematical equations that can be used for the calculation of TTROWST for different types of integration of tram stops, depending on the category of tramway corridor, are shown below:

1. Installation of a tramway stop of a double track with centre (island) platform:

$$TTROWST = 2 \times (b_{sw} + 0.1 + g_{dv}) + b_{cp} \text{ (classes A, B, C, D)} \tag{4.7}$$

Figure 4.25 Tramway stop at Athens, Greece. (Photo: A. Klonos.)

Figure 4.26 Tramway stop facilities, Lyon, France. (Adapted from Villetaneuse, C., 2008, available online at: https://fr.wikipedia.org/wiki/Ligne_3_du_tramway_de_Lyon (accessed 8 August 2015).)

$$\text{TTROWST} = 2 \times (0.1 + g_{dv}) + b_{cp} \text{ (class E)} \tag{4.8}$$

where b_{cp}: Width of the centre (island) platform.

2. Installation of a tramway stop of a double track with laterally opposed platforms:

$$\text{TTROWST} = 2 \times (b_{lp} + g_{dv}) + 0.2 \text{ (classes A, B, C, D, E)} \tag{4.9}$$

where b_{lp}: Width of the side platform.

3. Installation of a tramway stop of a double track with laterally staggered platforms:

$$\text{TTROWST} = b_{lp} + 2 \times g_{dv} + b_{sw} + 0.3 \text{ (classes A, B, C, D)} \tag{4.10}$$

$$TTROWST = b_{lp} + 2 \times g_{dv} + 0.3 \,(\text{class E}) \tag{4.11}$$

The length of the platforms must allow for the stopping of the tram of the greatest length. Normally, and as long as is allowed by the length of the building blocks, the length of the platform should be at least double in order to allow for the stopping of two coupled trams during the peak hours.

4.5.2.2 Operational criteria

At intersections where full priority is granted to trams, it is preferable to install the stop after the intersection, so that the arrival time of the tram can be accurately estimated, as the stopping time largely depends on the time required for the boarding and alighting of passengers. This allows for the delay of the intersecting traffic to be minimised.

In the case of an intersection where it is not desirable to give priority to trams, the most appropriate location of the stop is before the intersection, as the maximum delay time that can occur for the tram is equal to the length of the red phase of the light signal. Moreover, if the stopping time coincides with the red phase, there will be no delay.

For large uphill slopes, it is preferable to place the tram stop at the end of the slope.

In case of curves with small radii, stops should preferably be located after the curve.

The distance between two successive stops should generally be greater than 400–450 m and less than 750–800 m.

4.6 TRAMWAY DEPOT FACILITIES

4.6.1 General description and operational activities

The depot can be considered as the 'heart' of a tramway system. It is the starting point of all trams from which they commence their transport services for their passengers. In general, depots are spacious areas, which accommodate the trains when they are not in timetable service. Maintenance (light or heavy) also takes place in the same area. This includes small-scale repairs, sanding, and cleaning.

The establishment of a new tramway depot is a tough procedure since it takes place in an urban area with all the naturally ensuing problems. The main problem is finding a sufficiently large and available site; such large sites are rarely available inside the urban environment and are usually very expensive. Moreover, the selection of the location for the depot in an urban area almost always creates tensions and protests from neighbouring residents.

The location and design of the depot significantly affect the overall operational cost of the tramway system. The depot should ideally be located as close to the tramway network as possible, in order to minimise the 'dead' vehicle kilometres. Furthermore, all of the involved installations and facilities must be designed optimally, since any wrong estimation can increase the time and cost of activities performed, thereby increasing the total operational cost (Tramstore21, 2012a).

Table 4.9 presents the facilities of a tramway depot. Painting workshop can be characterised as an optional facility

The main design, constructional, and operational characteristics of the essential facilities are presented in the following (Tramstore21, 2012a,b,c,d,e,f; Verband Deutscher Verkehrsunternehmen 823, 2001).

Table 4.9 Facilities at a new tramway depot

Parking area/yard	Administration offices
Maintenance hall/workshop	Welfare facilities
Vehicle cleaning area	Waste storage
Warehouse (storage) area	Car-parking space for employees and visitors
Painting workshop	

4.6.1.1 Parking area/yard

During the design of this area, the primary objective is to achieve the maximum tram parking capacity and a smooth flow of trains. The length of parking tracks depends on the number of trams that will park at each track, as well as on the tram's length. The lateral distance between parking tracks should be sufficient in providing a corridor of about 1.50 m between the sides of two parallel-parked trams, in order to allow access by drivers and maintenance and cleaning staff. Thus, the size of the parking yard is the product of track length and track width, which depends on the number of tracks and their in-between distance.

Regarding the sheltering of the train parking area, there are three alternatives:

- Outdoor area.
- Sheltered area (which features a roof with or without side walls).
- Indoor area (features a roof, side walls, a front/end wall, and access doors).

4.6.1.2 Maintenance hall/workshop

This facility includes workshops for heavy maintenance, light maintenance, bogie maintenance, vehicle cleaning area, sanding plant, electronic systems unit, track maintenance (rails and catenaries), as well as facilities for the auxiliary equipment.

The length of each track depends on the length of the vehicles; normally, a maintenance track should be sufficiently long enough to service at least one tram.

The number of tracks depends on the number of trains that are served in the particular depot, on the multitude and type of activities performed within the light maintenance workshop and on the overall configuration, layout, and utilisation of the available space, maintenance-wise. In general, the maintenance facility area should be able to accept approximately 10% of the total number of trams normally served at the particular depot (Tramstore21, 2012a,b,e).

The lateral distance between two adjacent tram maintenance tracks should be sufficient in providing a corridor of about 3.5 m between the sides of two neighbouring trams. Within this space, the maintenance staff may move, place the required mechanical equipment, and perform all necessary activities.

4.6.1.3 Vehicle cleaning/washing area

For the washing of trains in most depots, either the 'drive-through system' or the 'gantry system' is used (Tramstore21, 2012c,f).

Regarding the positioning of these systems inside the depot area, it is considered preferable to locate them along the route section that the train follows from the moment it enters the depot till it reaches the parking area. With this configuration, 'dead' mileage can be avoided, and the vehicle may also enter the maintenance halls if required, in a state facilitating inspection by the maintenance personnel. Furthermore, many depots locate their sand silos between the entrance and the parking yard, namely, before the cleaning area (Tramstore21, 2012c,f).

4.6.2 Classification of tramway depots

Tramway depots are classified as follows:

- According to the means of transportation that they serve:
  - *Exclusively for tram use*: Only tramway vehicles are served.
  - *Mixed use*: Besides tram vehicles, other mass urban transit means, such as buses and trolleys, are also served.
- According to the activities performed within their area:
  - *Fully operating*: All required activities are performed in the depot (see Table 4.9).
  - *Limited operating*: A limited number of activities are performed. This may occur in two cases: in the first one, some activities are outsourced to third parties; in the latter case, the tramway system includes more than one depot and the required activities are shared among them.
- According to the depot's location within the network:
  - *Central*: When the depot is located in the centre of the network. This 'gravitational' location is preferred when the network follows a radial-shaped development.
  - *Terminal*: When the depot is located at either end of the network. This position is preferred when the network follows a linear-shaped development.
- According to the size of the ground plan area:
  - *Very small*: Serving up to 25 tramways.
  - *Small*: Serving 25–35 tramways.
  - *Medium*: Serving 35–65 tramways.
  - *Large*: Serving more than 65 tramways.
- According to its accessibility from the main track:
  - Through a junction.
  - At the end of the main track as an extension.

4.6.3 Main design principles and selection of a ground plan area

The designer of a tramway system must be aware of the required area of the site during the early stages of the study so as to make an initial estimate of the cost not only of the tramway depot but also of the whole project. On the other hand, the operator needs to know in advance the required area of the site in order to proceed with their search and the procedures that will be required for its acquisition as quickly as possible (e.g., expropriation).

Currently, there are no regulated specifications for the design of a tramway depot. The literature references (Tramstore21, 2012a) and (Verband Deutscher Verkehrsunternehmen 823, 2001), provide the basic design, construction, and operation principles without correlation with the train fleet (number of vehicles, train length). In this context, the design choices that are made and the final area of the tramway depot for which they are made are governed by the initiative of the designers and the recommendations of the system operators. The cost of implementing a tramway depot is very high and is around 20% of the infrastructure cost of a tramway system. The oversizing of the tramway depot increases the cost of the project significantly, while its undersizing results in problems in the system's operation.

The required area size of a tramway depot's ground plan depends on the following parameters:

- The fleet to be served (number of vehicles).
- The dimensions of the trains (length and width).
- The minimum allowed horizontal curve radius in the track alignment, and therefore, the geometric elements of switches and crossings.
- The layout of parking and maintenance tracks for the trains.
- The area size of main buildings and facilities.
- The maintenance policy applied by the operator.

In the initial fleet demand design, potential expansions should also be taken into account, regarding both the number of trams and their length, even mid-term.

It is recommended that the various facilities of the depot are interconnected. Additionally, the installation of a ring track (loop line), which circles around the area and is accessible from several points, is desirable. This allows trams to enter or exit the various facilities, such as the parking yard or the maintenance hall, without crossing through and occupying other areas (Tramstore21, 2012a; Verband Deutscher Verkehrsunternehmen 823, 2001). In general, the following routes should be possible without performing any manoeuvres:

- Entry → Parking Yard → Exit.
- Entry → Inspection and Cleaning → Parking Yard.
- Parking Yard → Inspection and Cleaning → Parking Yard.
- Entry → Maintenance hall → Parking Yard.
- Parking Yard → Maintenance hall → Parking Yard.

The warehouse (storage areas) should be located within or next to the maintenance workshop in order to reduce the transfer times of the various materials. Administration offices, welfare facilities, and the parking area for cars and motorcycles should be constructed at places where car and pedestrian movement does not conflict with operational tracks and tram movements.

Relevant literature (Chatziparaskeva et al., 2015) proposes a methodology which allows the estimation of the minimum required area of the ground plan of the various installations of a tramway depot and its total area, in relation to the fleet, the length of the trains, and the number and type of activities performed.

The whole approach is performed with the aid of two 'tools':

- Statistical data from existing tramway depots.
- Design simulation of the required facilities and integration into an overall layout/plan.

The paper concludes with:

- The formulation of simple mathematical expressions for calculating the useful area of individual facilities.
- The export of tables, which show the total area of the tramway depot.
- The configuration of the typical ground plans of the tramway depot sites which assign these findings for fleets between 15 and 80 vehicles and train lengths of 30, 35, and 40 m.

Table 4.10 provides example results of the design simulation of tramway depots for a fleet of 15–80 trains and for vehicle lengths of 30, 35, and 40 m.

Table 4.10 Example results from the design simulation – estimation of the required area E_d of the tramway depot's ground plan for a fleet of 15–80 vehicles and for vehicle lengths of 30, 35, and 40 m

Fleet	Total area E_d (m²) Ratio x,y Tram length 30 m	Total area E_d (m²) Ratio x,y Tram length 35 m	Total area E_d (m²) Ratio x,y Tram length 40 m
15	30,264 x = 1.4y	32,604 x = 1.4y	34,944 x = 1.4y
20	31,040 x = 1.1y	33,440 x = 1.1y	35,840 x = 1.40y
33	35,308 x = 1.7y	38,038 x = 1.5y	40,768 x = 1.3y
45	41,850 x = 1.1y	45,570 x = 1.2y	49,290 x = 1.2y
52	43,650 x = 1.6y	47,530 x = 1.6y	51,410 x = 1.7y
67	47,700 x = 1.6y	51,940 x = 1,6y	56,180 x = 1.5y
80	51,750 x = 0.9y	55,860 x = 1.7y	60,420 x = 1.6y

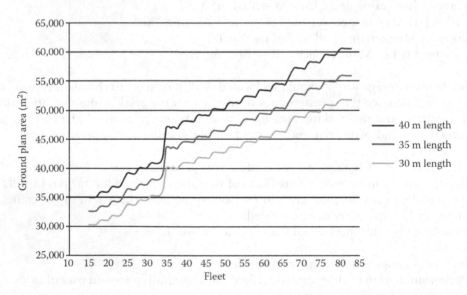

Figure 4.27 Variation in the total required ground plan area of the tramway depot in relation to the fleet, for different tram lengths. (Adapted from Chatziparaskeva, M., Christogiannis, E., Kidikoudis, C., and Pyrgidis, C. 2015, Estimation of required ground plan area for a tram depot, *Proc IMechE Part F: J Rail and Rapid Transit* 1–15, IMechE 2015 DOI: 10.1177/0954409715570714.)

Figure 4.27 illustrates the variation of the tramway depot's total ground plan area in relation to the fleet, for three different tram lengths.

Figure 4.28 illustrates the variation of the tramway depot's total ground plan area in relation to the vehicle's length, for different fleet size.

By studying Figures 4.27 and 4.28, as well as Table 4.10, the following conclusions can be reached:

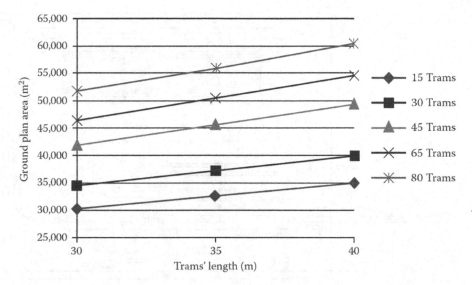

Figure 4.28 Variation in the total required ground plan area of the tramway depot in relation to the train length for different fleet sizes. (Adapted from Chatziparaskeva, M., Christogiannis, E., Kidikoudis, C., and Pyrgidis, C. 2015, Estimation of required ground plan area for a tram depot, *Proc IMechE Part F: J Rail and Rapid Transit* 1–15, IMechE 2015 DOI: 10.1177/0954409715570714.)

- The construction of depots for 15 vehicles (very small depots) requires an area of 30–35 acres.
- The construction of depots for 45 vehicles (medium depots) requires an area of 42–50 acres.
- The construction of depots for 80 vehicles (large depots) requires an area of 52–60 acres.

Figure 4.29 illustrates an example of a layout of the configuration of a tramway depot aiming to serve a fleet of 45 vehicles with a 35-vehicle length, as a result of a design simulation.

4.7 REQUIREMENTS FOR IMPLEMENTING THE SYSTEM

Generally, the tramway system is selected as a means of transport:

- When there is relatively low demand for travel ($\leq$ 10,000–15,000 passengers/h/direction).
- When it is sought to regenerate/upgrade an area and, generally, when it is desired to maintain the activities of an area.
- For cities facing a specific air pollution problem.
- When there is a high demand for travel (> 10,000 passengers/h/direction) and the subsoil or the lack of funding prevents the implementation of an underground solution.

The tram is the only urban public transport mode that results in the active removal of private cars from the areas through which it passes, with concurrent satisfaction of the demand. This is mainly achieved by integrating the tramway system across existing roads, thereby substantially reducing the street parking spaces for private cars.

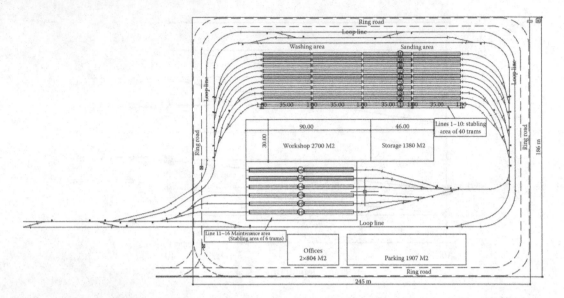

Figure 4.29 Example layout of the configuration of a tramway depot for 45 vehicles with a length of 35 m. (Adapted from Chatziparaskeva, M., Christogiannis, E., Kidikoudis, C. and Pyrgidis, C. 2015, Estimation of required ground plan area for a tram depot, *Proc IMechE Part F: J Rail and Rapid Transit* 1–15, IMechE 2015 DOI: 10.1177/0954409715570714.)

Implementing a tramway system is politically the easiest way to 'claim' a dedicated lane for public transport in the urban space. In addition, it is easier to make decisions regarding pedestrianisations in cities that feature a tramway system.

4.8 HISTORICAL OVERVIEW AND PRESENT SITUATION

4.8.1 Historical overview

The evolution of the trams consists of five distinct periods. The period of the horse-drawn tram, the transition period from horse power to electric power, the period of development of electric trams, the period of the dismantling of trams and, finally, the period of the reintegration of trams in the urban transportation systems.

4.8.1.1 The first horse-drawn tram

The first passenger tram in the world was the Swansea and Mumbles Railway in Wales which commenced operation in 1807 as a horse-drawn tram. From 1877 to 1929, this tram was powered by steam.

The first tramway lines were laid in the United States, specifically in Baltimore, in 1830, in New York in 1832 (New York–Harlem tram line), and in New Orleans in 1834 (the oldest tram network with continuous operation worldwide). In Europe, the first tramway line was laid in France, near St. Etienne, in 1838.

In 1853, the first tram with grooved rails commenced operation in the Broadway Avenue of New York. These new tracks were soon available also in Europe and were invented by Alphonse Loubat.

The new transport mode was disseminated relatively fast. By the end of the nineteenth century and the beginning of the twentieth century, several big cities worldwide featured horse-drawn tram transport.

4.8.1.2 The transition period from the horse-drawn tram to electrification

Mechanical systems developed rapidly, beginning with the steam-powered systems in 1873 and continuing with the electric trams after 1881, when Siemens presented the first electric-powered vehicle at the International Electricity Exhibition in Paris.

The steam-powered tram appeared in Paris in 1878. The prototype of an electric tram was developed by the Russian engineer Fyodor Pirotsky, who converted a horse-drawn tram into an electric tram. His invention was trialled in St. Petersburg, Russia, in 1880. In 1881, Werner von Siemens opened its first electric tram line in the world at Lichterfelde near Berlin.

In 1883, Magnus Volk constructed an Electric Railway (Volk's Electric Railway) along the east coast in Brighton, England. This 2-kilometre line remains in service until today and is the world's oldest electric tram, which is still functional.

The first major electrical system in Europe operated in Budapest since 1887.

Parallel advances took place during the same period in the United States – where Frank Sprague contributed to the invention of an electricity collection system using overhead wires. At the end of 1887, with the aid of this system, Sprague successfully installed the first large-scale electric train system in Richmond, Virginia (Richmond Union Passenger Railway).

Horse-drawn trams are still in operation in the Isle of Man, in the Bay Horse Tramway network, which was built in 1876. Similarly, Victor Harbor Horse Drawn Tram, which was constructed in 1894 (revived in 1986), is in operation in Adelaide, Australia. New horse-drawn tram systems were created at the Hokkaido Museum in Japan and at Disneyland.

4.8.1.3 The development of electric trams

It was not until 1914, that all the tramway networks in the world became electric. Electrification was technically perfected, and by the early 1930s, the electric tram has been the main means of urban transport worldwide.

4.8.1.4 The period of dismantling of tram networks

The emergence of the private car and improvements in the level of service provided by urban buses resulted in the rapid disappearance of the tram network in most Western and Asian countries by the end of 1950. In Paris, trams were abolished in 1938. In 1949 in the United States, only ten cities maintained tram lines.

The oldest system among all, namely the Swansea and Mumbles Railway, was bought by the South Wales Transport Company, which operated a bus fleet in the region, and it was eventually abolished in 1960.

The tram networks are no longer maintained or upgraded. As a result, the tram is discredited in the eyes of the passengers. Consequently, tram lines were slowly replaced by bus lines.

4.8.1.5 Restoration and reintegration of tramway systems

The situation began to change in favour of the tram around the mid-1980s. The 1990s marks the renaissance of trams worldwide. New modern vehicles were constructed. Their difference when compared with the old ones is such that it can be said that they constituted a brand new urban transport mode. The 'modern trams' are longer and more comfortable, they move almost noiselessly and are much faster, they have a modern design, and traction and braking are controlled electronically.

Nantes and Grenoble in France became the pioneer cities in the construction of modern tram systems. Their new systems were launched in 1985 and 1988, respectively.

The renaissance of the tram in North America began in 1978, when Edmonton, a city in Canada, acquired the German U2 system constructed by Siemens–Duewag. Three years later, the cities of Calgary, Alberta, and San Diego followed.

4.8.2 Present situation

All the data recorded and analysed in this section relate to the year 2019 (Pyrgidis et al., 2020). The raw data were obtained per country, per city, and per line, from various available sources and crosschecked. Afterwards, they were further manipulated for the needs of this chapter. The data relate to railway systems which meet the technical and operational characteristics that are attributed to trams and tram-trains as described in Section 4.3. They serve the city's urban space, and by majority, are referred to by one of the following terms: tram, tramway, streetcar, strassenbahn, stadtbahn, and tram-train. In some cases, they are referred to by the terms metro leger and light rail, but they have the characteristics of the tram and tram-train described in Section 4.3.

In total, 483 tram systems have been identified to still be in operation in a total of 458 cities. Of these systems, 29 (6%) can be characterised as tram-trains, 54 (11%) are heritage trams, while 400 (83%) are urban trams (Table 4.11). In total, 55 countries possess at least one tram system.

Particularly for trams, in several cities, multiple distinct systems have been identified with various dates of commencement of operations. Of the 458 cities 25 were identified to possess more than one system.

In order to make meaningful comparisons with the competitive urban mass railway systems (metros and monorails), in the statistical analysis that follows, the following assumptions were made:

- In the 25 cases of cities that possess more than one tram system, it was investigated whether these systems are connected seamlessly or not. If they are, they were counted as a single system and if not as multiple systems.
- Only systems that serve urban mass public transport were taken into account, meaning that heritage systems were not considered (except for those merged with current systems).
- Concerning network length, when multiple tramways were counted as a single system, their total length was added and associated with the date of commencement of operation of the chronologically first system.

This above process left 420 public transport (urban trams and tram-trains) systems to be included in the analysis, most of which are located in Europe (299), followed by Asia (57), North America (41), Africa (11), Oceania (7), and South America (5). Table 4.12 showcases

Table 4.11 Classification of tramway systems per continent and per type (2019 data)

Continent	Urban	Tram-train	Tourist	Total
Africa	11	0	1	12
Asia	57	2	6	65
Europe	283	22	12	317
North America	37	5	19	61
Oceania	7	0	7	14
South America	5	0	9	14
Total	**400**	**29**	**54**	**483**

Table 4.12 Countries in possession of the most tramway systems

Country	Number of public transport tramway systems in operation
Russia	62
Germany	56
USA	35
France	33
Ukraine	20

Table 4.13 Classification of tram-train systems per continent, country, and city

Continent	Country/City	Number
Asia	Japan/Toyama	2
	Japan/Fukui	
America	Canada/Calgary	5
	Canada/Edmonton	
	Mexico/Puebla	
	USA/Salt Lake City	
	USA/Seattle	
Europe	Austria/Gmunden	22
	Belgium/Oostende	
	Denmark/Aarhus	
	France/Esbly-Crécy	
	France/Lyon	
	France/Mulhouse	
	France/Nantes	
	France/Paris	
	France/Villejuif-Athis-Months	
	Germany/Chemnitz	
	Germany/Karlsruhe	
	Germany/Kassel	
	Germany/ Heidelberg	
	Germany/Nordhausen	
	Germany/Saarbrucken	
	Germany/Zwickau	
	Italy/Sassari	
	Netherlands/Rotterdam/Hauge	
	Spain/Alicante	
	Spain/Cadiz	
	Switzerland/Bex-Villars-Bretaye	
	UK/Sheffield	
Total		29

the five countries with the most tramway systems worldwide. Table 4.13 showcases the classification of the 29 tram-train systems per continent, country, and city. The longest tram system operates in Karlsruhe Germany. It is 262.4 km long and is classified as a tram-train.

As shown in Figure 4.30, most of the 420 public transport systems currently in operation are of normal gauge (220 systems; 52.3%), while a significant number (85 systems; 20.25%) has wide gauge (1,524 mm), followed by (72 systems; 17.1%) metric gauge (1,000 mm). Most systems utilise either low or low/high-floor rolling stock (191 and 112 tramway systems, respectively). However, as shown in Figure 4.31, high-floor vehicles are still in circulation for a significant number of systems (117 systems; 28%).

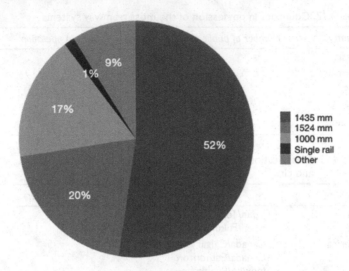

Figure 4.30 Distribution of the 420 public transport tramway systems based on track gauge.

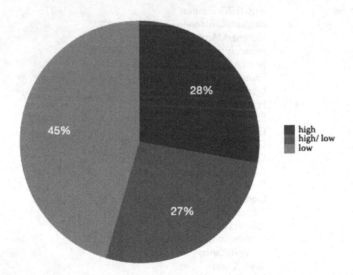

Figure 4.31 Distribution of the 420 public transport tramway systems based on floor height.

Figures 4.32 and 4.33 depict, respectively, the evolution of the 420 public transport tramway systems over the years both in terms of the number of systems that commenced their operation (and are up to today in revenue service) and the total length of their networks. Values were aggregated per decade.

Figure 4.32 indicates a peak of construction during the late 1800s followed by a period of decline. An increased rate of construction is once again observed in the past 4 decades.

Currently, 20 tram systems are in construction across 16 countries and are planned to span a total of 348 km (Table 4.14). All of them are to adopt low-floor vehicles, and 17 of them are to be of normal gauge. Of the 20 systems under construction, two of them refer to planned tram-train systems. It should be noted that entries in Table 4.14 refer only to entirely new systems and do not include extensions or new lines of existing systems

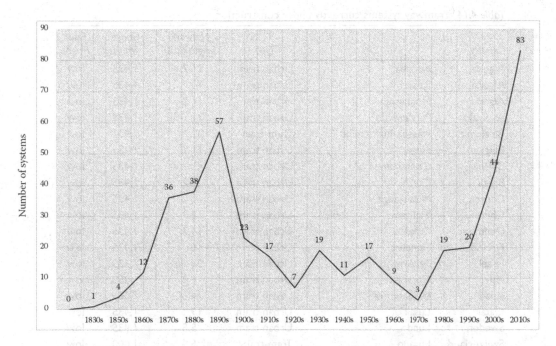

Figure 4.32 The number per decade of public transport tramway systems that were put into service and are still in revenue service as of 2019.

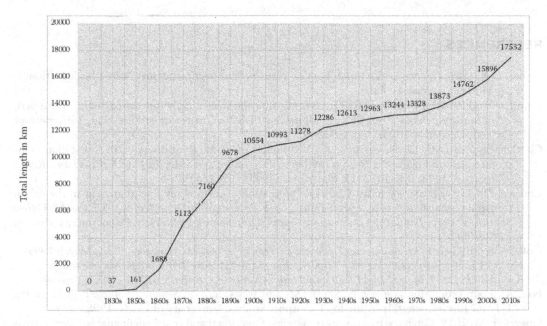

Figure 4.33 Evolution of tramways that are still in revenue service around the world in terms of the total network length.

Table 4.14 Tramway systems currently under construction

Country	City	Type	Network length (km)	Gauge (mm)	Floor height
Algeria	Annaba	Urban tram	21.7	1,435	low
Algeria	Batna	Urban tram	14	1,435	low
Algeria	Mostagnem	Urban tram	14.2	1,435	low
Australia	Parramatta	Urban tram	21	1,435	low
Belgium	Hasselt-Maastricht	Tram-train	32	1,435	low
Belgium	Liege	Urban tram	11.7	1,435	low
Bolivia	Cochabamba	Urban tram	42	1,435	low
Brazil	Cuiabá	Urban tram	22.2	1,435	low
Canada	Mississauga	Urban tram	18	1,435	low
China	Tianshui	Urban tram	12.9	1,435	low
Denmark	Odense	Urban tram	14.7	1,435	low
Finland	Tampere	Urban tram	16	1,435	low
Israel	Tel Aviv	Urban tram	23	1,435	low
Italy	Cosenza	Urban tram	9.4	950	low
Japan	Utsunomiya	Urban tram	14.6	1,065	low
Qatar	Lusail	Urban tram	19	1,435	low
Sweden	Lund	Urban tram	5.5	1,435	low
Switzerland	Lugano	Tram-train	5.9	1,000	low
USA	Bethesda-New Carrolton	Urban tram	26.1	1,435	low
USA	Tempe	Urban tram	4.8	1,435	low

REFERENCES

Bieber, C.A. 1986, *Les choix techniques pour les transports collectifs*, Lecture Notes, Ecole Nationale des Ponts et Chaussées, Paris.

Brand, C. and Preston, J. 2005, TEST (Tools for Evaluating Strategically Integrated Public Transport), *The Supply of Public Transport, A Manual of Advice*, Transport Studies Unit, University of Oxford, December 2003, Updated March 2005.

Chatziparaskeva, M., Christogiannis, E., Kidikoudis, C. and Pyrgidis, C. 2015, Estimation of required ground plan area for a tram depot, *Proceedings of IMechE Part F: J Rail and Rapid Transit*, pp. 1–15, IMechE 2015. doi: 10.1177/0954409715570714.

Chatziparaskeva, M. and Pyrgidis, C. 2015, Integration of a tramway line alignment in the urban transport system towards sustainability, *2nd International Conference 'Changing Cities: Spatial, Design, Landscape & Socio-economic Dimension'*, Porto Heli, Peloponnesus, Greece, 22–26/06/2015.

Collection CERTU. 1999, *Nouveaux Systèmes de Transports Guidés Urbains*, Paris, March 1999.

CRN CS 215. 2013, *Engineering Standard Track*, version 1.1, July 2013, CRN, Westborough.

ERRAC-UITP. 2009, *Metro, Light Rail and Tram Systems in Europe*. ERRAC –UITP, Brussels.

Fox, K., Halbo, C., Montgomery, F., Smith, M. and Jones, S. n.d., *Selected Vehicle Priority in the UTMG Environment*, Institute for Transport Studies, University of Leeds. 1998.

Guerrieri, M. 2019, Catenary-free tramway systems: Functional and cost-benefit analysis for a metropolitan area, *Urban Rail Transit*, Vol. 5(4), pp. 289–309, available online at: https://doi.org/10.1007/s40864-019-00118-y (accessed 8 November 2020).

Hass-Klau, C., et al. 2003, *Bus and Light rail: Making the right choice*, ETP, Brighton.

http://www.flickr.com/people/77501394@N00 kaffeeeinstein, 2008, available online at: http://en.wikipedia.org/wiki/CarGoTram (accessed 7 August 2015).

Lasart75, 2010, Available online at: http://en.wikipedia.org/wiki/Low-floor_tram) (accessed 7 August 2015).

Lesley, L. 2011, *Light Rail Developer's Handbook*, J. Ross Publishing, Miami.

Mundrey, J. 2000, *Railway Track Engineering*, 3rd edition, Tata McGraw-Hill Education, New York.

Pyrgidis, C. 1997, Light rail transit: Operational, rolling stock and design characteristics, *Rail Engineering International*, No. 1, 1997, Netherlands, pp. 4–7.

Pyrgidis, C. 2004, Il comportamento transversale dei carrelli per veicoli tranviari, *Ingegneria Ferroviaria*, October 2004, Rome, 10, pp. 837–847.

Pyrgidis, C. and Chatziparaskeva, M. 2012, The impact of the implementation of green wave in the traffic light system of a tramway line-The case of Athens tramway, *2nd International Conference on Road and Rail Infrastructure* (CETRA), 7-9/5/2012, Dubrovnic, Croatia, pp. 891–897.

Pyrgidis, C. and Panagiotopoulos, A. 2012, An optimization process of the wheel profile of tramway vehicles, *Elsevier Procedia Social and Behavioral Sciences*, Vol. 48, pp. 1130–1142.

Pyrgidis, C., Chatziparaskeva, M. and Politis, I. 2013, Investigation of parameters affecting the travel time reliability at tramway systems: The case of Athens, Greece, *3rd International Conference on Recent Advances in Railway Engineering – ICRARE 2013*, Iran University of Science and Technology, April 30–May 1, Tehran.

Pyrgidis, C., Tsipi, D., Dolianitis, A. and Barbagli, M. 2020, Urban mass railway transportation systems in revenue service at the end of 2019: Metro, tramway, monorail, *12th Annual Monorailex 2020, Virtual Workshop*, 19–22/9/2020.

RATP D.D.E. 1994, Projet de rocade tramway en site propre entre Saint Denis et Bobigny: Schéma de principe, February 1993, Paris.

Tramstore21. 2012a, Building sustainable and efficient tram depots for cities in the 21st century, Tramstore21 Publication, available online at: http://tramstore21.eu/ (accessed 28 March 2015).

Tramstore21. 2012b, Capacity of the tram stabling and maintenance area, Tramstore21 Publication, available online at: http://tramstore21.eu/ (accessed 28 March 2015).

Tramstore21. 2012c, Interior cleaning, Tramstore21 Publication, http://tramstore21.cu/ (accessed 28 March 2015).

Tramstore21. 2012d, Using the roof, Tramstore21 Publication, available online at: http://tramstore21.eu/ (accessed 28 March 2015).

Tramstore21. 2012e, Vehicle preventive maintenance and warranties, Tramstore21 Publication, available online at: http://tramstore21.eu/ (accessed 28 March 2015).

Tramstore21. 2012f, Washing policy, Tramstore21 Publication, available online at: http://tramstore21.eu/ (accessed 28 March 2015).

University of Piraeus. 2007, Evaluation of the level of service provided by the tramway of Athens and proposals for improvement interventions, Research Program, TRAM S.A., Athens, Greece.

Verband Deutscher Verkehrsunternehmen 823. 2001, Recommendations on the design of depots for LRVs and Tramcars, Verband, Cologne.

Villetaneuse, C. 2008, Available online at: https://fr.wikipedia.org/wiki/Ligne_3_du_tramway_de _Lyon (accessed 8 August 2015).

Chapter 5

Metro

5.1 DEFINITION AND DESCRIPTION OF THE SYSTEM

The metro, or metropolitan, or sometimes termed as 'underground railway' (Figure 5.1), is a system that exclusively uses electric traction and usually uses the traditional steel wheel on a rail guidance system (though sometimes rubber-tyred wheels are used, Figure 5.2), on an exclusive corridor, the largest part of which is underground and in any case is grade-separated from the rest of the urban road and pedestrian traffic.

In relation to other urban transport modes, the system is characterised by:

- High-frequency service (train headway up to 1 min).
- Large transport capacity (up to 45,000 passengers/h/direction) (Bieber, 1986).
- Movement, for a large percentage or for the entirety of the length on an underground exclusive corridor.
- High construction cost (€70–€150 M/track-km) or even higher in some cases.
- Long implementation period (in some cases even decades).

From an engineering point of view, it is a very complex and challenging project as it requires specialised knowledge regarding a variety of engineering disciplines (soil mechanics, structural mechanics, transportation engineering, architecture, power supply systems, low-voltage telecommunication systems, trackwork technologies, automated control systems, rolling stock technologies, computer systems, etc.).

5.2 CLASSIFICATION OF METRO SYSTEMS

5.2.1 Based on transport capacity

Based on the passenger volume they serve, metro systems are classified as follows:

- Heavy metro.
- Light metro.

The light metro is a hybrid solution between the heavy metro and tramway. Compared with the heavy metro, the light metro is characterised by lower transport capacity, lighter vehicles, and shorter distance between intermediate stops. It is commonly selected for the service of cities with a population between 500,000 and 1,000,000 inhabitants. On the other hand, the construction of a heavy metro is more appropriate for cities with a population greater than 1,000,000 inhabitants.

DOI: 10.1201/9781003046073-5

Figure 5.1 Athens metro system (steel wheels, driver). (Photo: A. Klonos.)

Figure 5.2 Lausanne metro system (rubber-tyred wheels – driverless). (Adapted from Amort, J. 2006, available online at: http://en.wikipedia.org/wiki/Rubber-tyred_metro#/media/File:Rame_m2_lausann e.JPG (accessed 7 August 2015).)

Table 5.1 compares some key constructional and functional characteristics of the two types of metros mentioned above.

5.2.2 Based on the Grade of Automation of their operation

Based on the Grade of Automation (GoA) of their operation, metro systems are classified into four categories. Figure 5.3 illustrates these four categories and presents the operational characteristics which determine the GoA for each category (Rumsey, 2009).

Table 5.1 Heavy metro/light metro: basic differences as regards their constructional and functional characteristics

	Light metro	Heavy metro
Distance between successive stops	400–800 m	500–1,000 m
Commercial speed	25–35 km/h	30–40 km/h
Integration in relation to the ground surface	At grade, underground and elevated	Mainly underground
Maximum transport capacity	35,000 passengers/h/direction	45,000 passengers/h/direction
Train formation	2–4 vehicles	4–10 vehicles
Train length	60–90 m	70–150 m
Vehicle width	2.10–2.65 m	2.60–3.20 m
Driving system	With driver or automated	With driver usually or automated

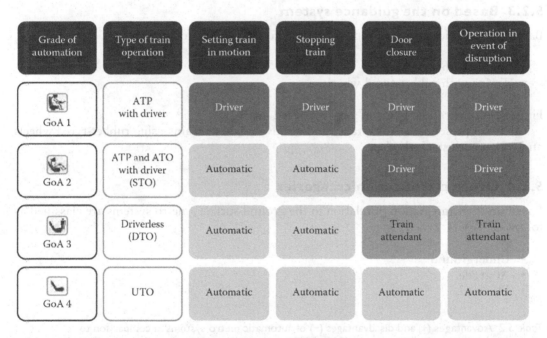

Figure 5.3 Classification of metro systems based on the Grade of Automation of their operation. (Adapted from UITP. 2013b, Press kit metro automation facts, figures and trends. A global bid for automation, UITP Observatory of Automated Metros, available at: http://www.uitp.org/sites/default/f iles/Metro%20automation%20-%20facts%20and%20figures.pdf (accessed 14 March 2015).)

More specifically:

GoA1: Operation with a driver – the driver of the train is actively involved throughout the driving activity. The train is only equipped with Automatic Train Protection (ATP) system.

GoA2: Semi-automatic Train Operation (STO) – there is a supervising driver who undertakes driving only in case of system failure and is responsible for opening and closing the doors. The train is equipped with ATP and Automatic Train Operation (ATO) systems.

GoA3: Driverless Train Operation (DTO) – the train moves without a driver. There is a train attendant who is responsible for the opening and closing of the doors and can intervene in case of system failure. The train is equipped with ATP and ATO systems.

GoA4: Unattended Train Operation (UTO) – the train moves automatically and all of the above operations are performed without the presence of a driver or an attendant. The train is equipped with ATP and ATO systems.

Generally, the train operation is considered to be automatic when the trains are driverless (GoA4 and GoA3). These two GoAs must be accompanied by the installation of automatic sliding gates along the platforms (Platform Screen Doors (PSD); see Section 5.4.4) in order to increase passenger safety.

Table 5.2 shows the advantages and disadvantages of automated metro systems compared with metro systems with driver.

5.2.3 Based on the guidance system

Based on the guidance system, metro trains are classified as follows:

- Trains with steel wheels.
- Trains with rubber-tyred wheels.

Figure 5.4 illustrates a bogie of a rubber-tyred metro.

Table 5.3 presents the advantages and disadvantages of trains using rubber-tyred wheels and trains with steel wheels.

5.2.4 Other classification categories

Based on their integration in relation to the ground surface, metro systems are classified as follows:

- Underground.
- At grade.
- Elevated.

Table 5.2 Advantages (+) and disadvantages (−) of automatic metro systems in comparison to conventional metro systems (with driver)

+ Driverless → Lower operation personnel cost
+ Operation independent of the availability of drivers → Regularity and flexibility of services
+ Human factor absence → Increased traffic safety
+ Automatic driving → More cost-efficient driving → Lower energy consumption → Reduced environmental impacts
+ Higher service frequency → Shorter trains for the same transport capacity → Smaller platform length
+ Unified speed, higher service frequency → Higher track capacity
+ Lower delays at the platforms, reduced time for manoeuvers at terminals → Reduced number of trains required for the accomplishment of all scheduled routes
− Driverless → Concerning feature for some of the system's potential passengers → Discouraging factor for using the transport mode
− Driverless → Fewer job positions
− Increased maintenance cost and additional personnel cost for system safety associated with the automation system itself

Figure 5.4 Mockup of a bogie of a M2 train. (Adapted from Rama, 2007, online image available from https://commons.wikimedia.org/wiki/File:Bogie-metro-Meteor-p1010692.jpg.)

Table 5.3 Advantages (+) and disadvantages (−) of metro trains with rubber-tyred wheels and trains with steel wheels

With rubber-tyred wheels	With steel wheels
+ Low rolling noise	− High rolling noise
− Increased noise when starting the train	
+ Greater accelerations	− Smaller accelerations
+ Ability to move along greater longitudinal gradients (up to 13%)	− Ability to move along smaller longitudinal gradients (up 5%)
− Increased energy consumption	+ Lower power consumption
− Greater maintenance cost (frequent tyre replacement)	+ Lower maintenance cost
− Much lower lateral stability of vehicles (lateral guiding wheels required)	+ Higher lateral stability of vehicles
− Lower axle loads	+ Greater axle loads
+ Reduced braking distance	− Increased braking distance
+ Increased passenger dynamic comfort	− Reduced passenger dynamic comfort

Finally, based on the network's layout, metro systems are classified into systems that adopt:

- Radial-shaped layout.
- Linear-shaped layout, with or without branches.
- Grid-shaped layout.

In most cities, the layout is mainly dictated, and thus explained by the gradual development of the metro system, and reflects the arrangement of the city functions itself (existing or planned). For new branches (extensions) or new networks, the deployment of a grid-shaped layout is preferable and is therefore sought for implementation. The reason for this is to avoid the risk of overloading the city centre. Unlike in other cities, a radial-shaped layout is selected aiming to boost the centre. Finally, in some cities, the network is rudimentary as it only includes a single line.

5.3 CONSTRUCTIONAL AND OPERATIONAL CHARACTERISTICS OF A METRO SYSTEM

The basic characteristics of metro systems were presented in Chapter 1, Table 1.7.
   In addition to those, the following are also mentioned.

5.3.1 Track layout

The alignment and, hence, the track layout characteristics of a metro line are largely determined by:

- The need to serve specific locations that are trip generators, which are located at a relatively short distance from each other.
- The need to deal with soil settlement when placed underground, which can be hazardous for the overlying structures.

All of the above impose an alignment that largely follows the road arteries above the ground surface. This leads to the adoption of a horizontal alignment which is characterised by a considerably large percentage of curved segments and radii ranging from R_c = 500 m to R_c = 150 m. As for the secondary lines (depot, sidings), radii can be reduced to R_c = 70–80 m.
   The longitudinal profile of the line is imposed by:

- The maximum longitudinal slope that a metro trainset can cope with.
- The need to limit the depth of excavations for the stations, and the various ventilation or other equipment shafts.
- The adoption of a line with a horizontal alignment profile and a longitudinal profile which allows for energy savings.

The maximum longitudinal gradient of a metro network ranges between i_{max} = 3% and i_{max} = 8%, though it is advisable that a gradient should not normally exceed 5%. At stations, depots, and generally at locations where trains are parked, the longitudinal gradient of the track should be less than i = 2%, in order to avoid a possible movement of trains in case the braking system is not activated.

5.3.2 Track superstructure

The track superstructure is usually made of a concrete slab (slab track). The introduction of this track bed system instead of the ballasted track is mainly due to the following reasons:

- A much lower annual maintenance cost – easier maintenance (in case of ballasted tracks, the limited width inside the tunnels complicates the maintenance work).
- A longer lifetime (50 years vs. 25 years).
- A lower height of track superstructure.
- The ability of road emergency vehicles to move on the track superstructure.
- A better behaviour under stress – greater lateral track resistance.

On the contrary, the implementation cost of slab track is greater than the cost of ballasted track, which is (€1,000–€1,200 per meter as compared to €500–€600 per meter, for the case of construction in a single-track tunnel) (2014 data).

Concerning metro systems, many techniques have been developed for slab track which differ with regard to the type and characteristics of their structural features as well as the construction and maintenance methods applied (Figures 5.5 and 5.6) (Quante, 2001; Ponnuswamy, 2004; Rhomberg, 2009). In parallel, for the same technique, differentiations are observed depending on whether the superstructure is laid in the 'plain' track, in areas of switches and crossings, in a depot, in twin-bore tunnels, or in single-bore double-track tunnels. Finally, special solutions are adopted for the areas where there is a need for protection against vibration and noise.

Among the first systems of slab track that were used in metro construction were the Rheda system (Figure 5.6) and the Zublin system.

The selection of a suitable system of slab track requires a multi-criteria approach. Table 5.4 presents a list of options for the main track superstructure components in the case of 'plain' track. It should be highlighted that these options are the most commonly used during recent years, based on international construction practice.

In the last decade, a tendency to use slab track systems with a direct fixing of the rails on the concrete slab is observed. This technique is gaining more and more ground in the market due to continuous improvement in the quality of the connection between the baseplate and the rail, as well as continuous development of materials used as elastic pads, it has significant advantages over the classical methods of slab track using sleepers. The only drawback in the use of these systems is their moderate ability to absorb noise and vibration.

Table 5.5 proposes some options for track superstructure components for areas that are sensitive to noise and vibrations. The system that can ensure a maximum reduction in the ground-borne noise is the floating slab; however, this is also the most expensive option.

Figure 5.5 Slab track, Stedef system. (Adapted from Jailbird, 2005, online image available at: https://en.wiki pedia.org/wiki/Railroad_tie.)

Figure 5.6 Slab track: The ties of the Rheda 2000 system before they are tightened on a concrete bed, Nuremberg-Ingolstadt high-speed railway line, Germany. (Adapted from Terfloth, S. 2004, available from https://commons.wikimedia.org/wiki/File:Schwellen_Rheda.jpg 2004.)

Table 5.4 Suggested track superstructure components of slab track for a 'plain' track

Track superstructure components	Option	Comment
Rails	UIC54, CWR	For $R_c < 600$ m, hardness 1,100 A For $R_c > 600$ m, hardness 900 A
Slab track system	Direct fastening systems without sleepers	Easy to construct and maintain
Fastenings	Spiral-shaped resilient fastenings	High elasticity and lateral resistance
Pads	Elastic rail and baseplate pads	• Plastic pads for the electrical insulation of the track • Elastomers for which the ratio of vertical dynamic to static stiffness is $K_{dyn}/K_{stat} < 1.5$ to reduce noise and vibration

Table 5.6 attempts a comparison of the different techniques that are used to address the ground-borne noise and vibrations in urban railway systems in terms of noise reduction and ease of maintenance.

Table 5.7 attempts a comparison of the implementation cost of the above techniques. The implementation cost of a floating slab is approximately €2.5 M per km (double track) (2014 data) (Figures 5.7–5.10).

The development of direct fixing systems that achieve a reduction of noise and vibrations similar to that achieved by the use of floating slabs (i.e., 30 dB) is in full swing. This may lead to the universal prevalence of direct rail-fixing systems over all other solutions.

A floating slab may be constructed with one of the following methods:

• With a continuous concrete slab, cast *in situ*.
• With a discontinuous slab that is made of prestressed concrete elements.

Table 5.5 Suggested structural slab track components for areas that are sensitive to noise and vibrations

Level of reduction of noise and vibrations	Option	Comments
Great	• Floating slab with discrete bearings (Figure 5.7) • Floating slab with elastomer strips (Figure 5.8) • Slab using springs • Floating slab with elastomer mat (Figure 5.9)	The use of floating slab with elastomer mat should be avoided due to the fact that it is very difficult to replace the elastomer in case of wear
Moderate	• Floating slab or slab with elastomer mat • Very flexible clip fastenings for the direct fixing of rails with preloading • Very flexible clip resilient fastenings for the direct fixation of rails	A techno-economic study is required for the selection of the optimal solution
Low	• Clip resilient fastenings • Rail web dampers (Figure 5.10)	A techno-economic study is required for the selection of the optimal solution

Table 5.6 Comparison of the techniques used as countermeasures for the ground-borne noise and vibrations in the case of urban railway systems

Noise countermeasure	Noise reduction (dB)	Ease of maintenance
Floating slab with elastomer mat	≈20	X
Floating slab with elastomer strips	≈25	√
Floating slab with discrete bearings	≈30	√
Floating slab with springs	≈20–25	√
Resilient fixing system	≈2–10	√√
Resilient fixing system with preloading (APT-ST)	≈10	√√
Very resilient fixing system with preloading (APT-BF)	≈20	√√
Rail web damper	≈2–5	√√

Note: X: Difficult maintenance, √: easy maintenance, √√: very easy maintenance.

Table 5.7 Cost factor for various noise reduction systems

Track superstructure type	Cost factor
Ballasted track and direct resilient fixing system of rails	I
Very resilient fixing system of rails	1.2–1.6
Floating slab	2.5–4.5

Table 5.8 provides a comparison between the two aforementioned techniques. According to the specifications of the project and the construction restrictions, the engineer should select the most advantageous solution.

5.3.3 Tunnels

Based on the number of bores (branches) and the number of tracks per bore, the underground sections of a metro network are classified into the two following categories:

• Single-bore double-track tunnels (one tunnel with two tracks).
• Twin-bore tunnels (two single-track tunnels).

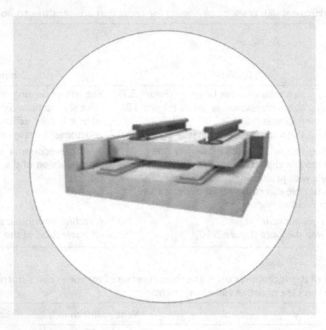

Figure 5.7 Floating slab with discrete bearings (point-like support). (Adapted from GETZNER. no date, Mass-Spring System, GETZNER company brochure, available at: http://www.getzner.com/en/downloads/brochures/ (accessed 14 March 2015).)

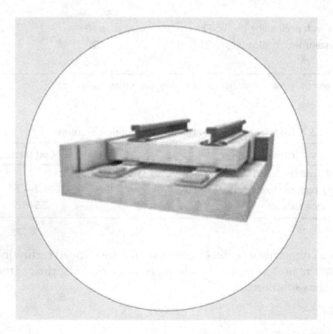

Figure 5.8 Floating slab with elastomer strips (linear support). (Adapted from GETZNER. no date, Mass-Spring System, GETZNER company brochure, available at: http://www.getzner.com/en/down loads/brochures/ (accessed 14 March 2015).)

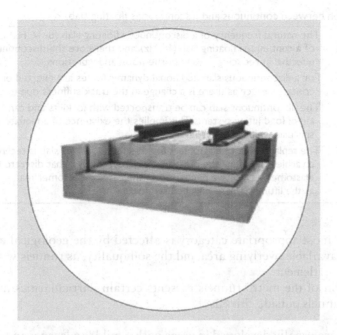

Figure 5.9 Floating slab with elastomer mat (full surface layer). (Adapted from GETZNER, no date, Mass-Spring System, GETZNER company brochure, available at: http://www.getzner.com/en/down loads/brochures/ (accessed 14 March 2015).)

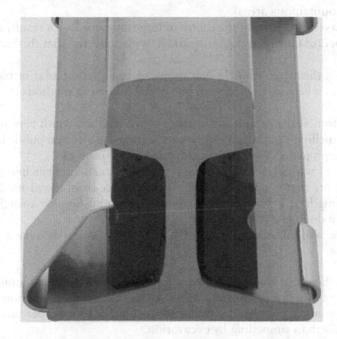

Figure 5.10 Placement of damping materials on the rail web. (From Vossloh, 2015.)

Table 5.8 Comparison between continuous and discontinuous floating slab

Ability to reduce ground-borne noise	The natural frequency of a discontinuous floating slab (8–16 Hz) is lower than that of a continuous floating slab (16 Hz), and therefore the discontinuous slab is more effective in reducing ground-borne noise and vibrations
Stress condition	For a discontinuous slab, additional dynamic forces are exerted on the wheel–rail contact surface as there is a change in the track stiffness due to the discontinuity
Ease of construction	The discontinuous slab can be transported with forklifts and can be fitted with the aid of load lifting systems. This implies the existence of adequate available open space as well as of appropriate access
Ease of maintenance	The replacement of elastomers for the discontinuous slab is technically much easier to achieve by lifting the prestressed slabs, provided that discrete bearings or elastomer strips have been used. In the case of elastomer mat, this is not feasible, as the lifting jack can only be used locally

The choice of the most appropriate category is affected by the geological and local conditions, that is, the available overlying area and the soil quality, as tunnels with a large diameter show greater settlement.

The construction of the metro tunnels presents certain particularities in relation to the construction of tunnels outside cities as:

- Large cities are usually developed in areas with a mild landscape, and soil materials at the influence depth are usually not rocky.
- The height of the soil overlying the tunnel is usually less, ranging between 10 m and 25 m (as compared to even hundreds of metres of overlying soil height in the case of large tunnels in mountainous areas).
- The metro passes underneath the centre of large cities and, as a result, any visible fault (or even suspected fault) is quickly identified (with anything that this may then trigger).

The cross section of the metro tunnel excavation can be either circular or rectangular.

A metro tunnel can be constructed with one of the following methods:

- With excavation, if the project is constructed at a shallow depth beneath the road surface. The tunnelling excavation involves relocating the affected public utility networks alongside the project as well as perpendicular to the project.
- With boring, by which open excavation is avoided. A general principle is that the upper limit of the bore that is opened must be at a distance from the ground surface that is at least the length of 1 diameter of the bore (e.g., about 6 m for a single-track tunnel, or 9.5 m for a double-track tunnel).
- With drilling and blasting (drill and blast). This option is very rare within city environments.

The cost of tunnelling and the comparison of costs for the different tunnelling methods depend on a large number of parameters. For shallow tunnels, construction by direct excavation is more economical compared to bored tunnelling, but for deeper tunnels, tunnelling by boring is cheaper than tunnelling by excavation.

Tunnelling by excavation is usually more economical compared to tunnelling by drilling. In tunnels with excavation, the following two techniques are applied:

- The technique of open trench (cut and cover) (Figure 5.11).
- The technique of excavating and backfilling (cover and cut).

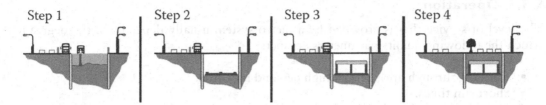

| Step 1 | Step 2 | Step 3 | Step 4 |

Figure 5.11 Stages of applying the cut and cover technique. (Adapted from FHWA. 2013, Technical Manual for Design and Construction of Road Tunnels – Civil Elements, Federal Highway Administration – U.S. Department of Transportation, Chapter 5 – Cut and Cover Tunnels, available at: http://www.fhwa.dot.gov/bridge/2013 (accessed 14 March 2015).)

Regarding tunnels without an open excavation, the following two techniques are mainly applied:

- The use of Tunnel Boring Machines (TBM).
- The New Austrian Tunnelling Method (NATM).

5.3.4 Rolling stock

A typical metro train commonly consists of 4–10 vehicles.

The construction materials that are commonly used for the vehicles are semi-stainless steel and aluminium.

The doors are divided into:

- Simple sliding.
- Double sliding.

Regarding the interior of the vehicles, the transport capacity of the train (passengers standing and seated) is usually 600–1,200 people. The percentage of seated to standing passengers varies from 25% to 45%.

Vehicles of a metro system are equipped with conventional bogies. The bogies must be characterised by:

- Small wheelbase ($2a$ = 1.80–2.20 m instead of 2.50–3.00 m – used for conventional railway vehicles).
- Small wheel diameter ($2r_o$ = 0.70–0.80 m instead of 0.90–1.00 m – used for conventional railway vehicles).

The equivalent conicity of the wheels and the stiffness of the springs of the primary suspension constitute the critical construction parameters of the rolling stock.

By using conventional bogies, and for horizontal alignment radii that are up to R_c = 70–80 m, there can be a combination of values of the equivalent conicity of the wheels and the stiffness of the springs of the primary suspension of the bogies that ensure the curve negotiation of wheelsets without wheel slip in curves, on one hand, and running speeds in excess of V_{max} > 80 km/h at a straight path, on the other hand; this is the desirable speed for straight segments. On the contrary, for smaller radii, curve negotiation of bogies is characterised by the exertion of extremely large values of the guidance forces (Joly and Pyrgidis, 1990; Pyrgidis, 2004).

5.3.5 Operation

The level of service that is provided by a metro system usually depends on the degree to which the following quality parameters are met:

- Running through areas with a high demand for travel.
- Short run times.
- Dense train headway/service during peak hours.
- Reliability of schedule.
- Appropriate pricing policy/ease in ticket supply.
- Passenger safety on the train and at stations.
- Air quality inside the vehicles/use of air conditioning.
- Passenger dynamic comfort during transport.
- Availability of passenger seats/satisfying train patronage.
- Clean trains.
- Accessibility/acceptable distance between successive stations.
- Service for the disabled.
- Integration with other transport modes/ability to provide park-and-ride services.
- Passenger information regarding the route onboard and at stations.
- Interfaces between staff and users.

5.3.5.1 Commercial speeds, service frequency, and service reliability

The commercial speed of a metro system (V_c = 30–40 km/h) is much greater than that of tramways (15–25 km/h) and greater than that of a monorail (15–40 km/h).

The usual practice which minimises the run time between two successive stops is that the vehicle develops the maximum speed by accelerating as much as possible and then decelerates at a slow, steady pace.

In modern metro systems, the headway between trains can reach even 1 min, while the operation of trains with a headway of more than 15 min discourages the use of the metro system. Usually, a different frequency is applied during the peak hours (higher frequency) than the frequency for the remaining operational hours of the system.

The delay time for a service is usually defined as any time interval that exceeds 3 min. The halt time is the sum of two distinct time values: the time taken for the opening and closing of the doors during stopping (3 s) and the time taken for passenger boarding and alighting. Theoretically, 2 persons/s can board or alight from a door that is 1.30 m wide. For example, an approximate value of around 14 s, is required for single-track platforms for vehicles with a transport capacity of 40 persons for each door with a width of 1.30 m.

The need for dense headways imposes specific functional requirements for the signalling system and renders the existence of an ATP system absolutely necessary.

Generally, the following systems can be distinguished:

- Conventional signalling systems.
- Systems utilising telecommunications for the data transmission (Communications-Based Train Control [CBTC]).

In conventional systems, the detection of trains is achieved with the aid of 'traditional' means such as track circuits (usually) or axle counters. Interlocking systems carry out route checks, control the moving components of the track (switches, flank protection), and give the appropriate signals to the driver for the route through either side signals or cab signalling. The protection of trains is achieved either at a specific point (by acting on the brake when the driver exceeds a restrictive value) or continuously (by continuously monitoring the speed of the train). The operation of individual interlocking systems takes place at a central post (centralised traffic control) where the traffic flow is monitored by a system usually termed as Automatic Train Supervision (ATS). Often the treatment of individual systems is performed automatically by the centre's computer equipment, while inspectors are assigned tasks such as traffic monitoring and intervention in case of emergencies (injuries, suicides, etc.).

In CBTC systems, the location of a train is monitored by the odometer of the train (regularly corrected via beacons that are placed on the track) which is usually sent to a single central interlocking system over a wireless telecommunication system. The central post issues all the commands and movement authorities for the route and the speed (thereby also integrating the protection function of the train over the same telecommunication system). The train separation principle is that of the 'moving block' which generally allows for denser headways.

The traffic control system (either conventional or CBTC-based) is integrated with various other functionalities such as passenger information systems and traffic management (computer-aided and sometimes fully automated train scheduling, with the dispatcher intervening only in cases of failures and degraded-mode operations).

5.3.5.2 Fare collection and ticket supply

An increasing number of metro systems prefer the installation of Automatic Fare Collection (AFC), in an effort to reduce ticket evasion.

An AFC system includes the following components:

- Ticket vending machines.
- Ticket offices.
- The ticket top-up (add credits) machines, via which the users can purchase their ticket with the aid of an electronic card (credit card) that they can top up with additional credits (money).
- The facilities that separate the area where passengers can enter only with a validated ticket from the rest of the station (paid/unpaid areas). These facilities are usually automated gates that open when a ticket is validated ('closed system') (Figure 5.12); however, in some cases, there is no physical barrier ('open' or 'honour' system) (Figure 5.13).
- Ticket sales, which are different from one system to another.
- An automated management system that consists of a central computer and is connected with all stations.

The components of the AFC system are connected electronically in order to record every single transaction.

The new fare collection systems use smart cards. These smart cards are cards that are scanned by a card reader placed near the platform entrance. The card reader records the passenger's entry point and exit point and reduces the corresponding amount for that particular trip from the card.

Smart cards were first used in Hong Kong in 1997 ('Octopus' card) and are now broadly used in many metro systems.

The latest developments in AFC technology include direct ticket payments through bank credit cards or mobile phones.

Figure 5.12 Automated gates which separate the area leading to the platforms from the rest of the station, West Kensington tube station. (Adapted from Mckenna, C. 2007, available online at: http://web.mit.edu/2.744/www/Results/studentSubmissions/humanUseAnalysis/jasminef/ (accessed 7 August 2015).)

Figure 5.13 Ticket validation systems without physical barriers, Akropoli Metro station, Athens, Greece. (Adapted from http://www.athenstransport.com/english/tickets/ (accessed 14 March 2015).)

5.3.5.3 *Revenues for the system operator*

Granting rights to third parties for managing areas for commercial use within metro stations may constitute an additional profitable source of income for the operating company, in addition to the fare revenues. There should, however, beset limits regarding the operation of these sites. Proper initial planning and proper management of the system are required, so as to not interfere with the primary objective of the stations, which is the unhindered and safe movement of passengers.

The selling of advertising rights at stations and metro trains can also bring significant income to the system operating company. There are two ways of advertising in a metro system:

1. Fixed advertising (advertising signs – billboards placed on the platforms and at locations across which passengers frequently pass when moving within the station; display of advertising messages on tickets; billboards inside and outside of trains).
2. Variable advertising – variable (mainly visual) advertising is done with the aid of modern electronic devices. Screens and high-definition televisions in various forms (LED, PDP) are used; these can be placed at the platforms, at strategic locations in the stations, and inside the trains.

5.3.6 Implementation cost

The implementation cost of a metro is very high (€70 M–€150 M per track-km, 2014 prices). The variation of the cost is high, as it is affected by a number of parameters such as (Davies, 2012; MacKechnie, no date; Clark, 2019):

- The percentage of network length that is underground, above ground (elevated), or at grade.
- The excavation method used (deep bore or cut and cover).
- The type of cross section of the tunnel (twin-bore tunnel or single-bore double-track tunnel).
- The quality of the subsoil.
- The depth of the line.
- The number of stations.
- The length of the platforms.
- The expropriations and land values.
- The labour cost and the material costs in each country.
- The technologies used for the various electromechanical and railway systems and the rolling stock.
- Various unforeseen but significant costs (e.g., the recovering of important archaeological findings and their processing).

The average cost of building a metro system, with a percentage of underground length of 75%, varies between €70 M and €100 M per km. For a 100% underground metro system, the cost varies between €100 M and €150 M per km (2014 data).

In recent years, there has been a trend of constructing metro systems with the smallest possible percentage of underground length in order for the cost of construction to be as economical as possible.

The cost of the rolling stock varies between €1.3 M and €2 M per car (2014 prices), so, for example, for a six-car train, the cost is approximately €8–€12 M, depending on the train length, width, passenger transport capacity, internal fittings, kinematic characteristics, performances, driverless/with driver, and so on.

The driverless metro systems require higher investments for the supply of the rolling stock, the control systems, and the protection systems both at the platforms and on the track (UITP, 2013a). On the contrary, the construction cost of the stations may become less, as the possibility of more frequent service allows reducing the length of platforms. However, in automated systems, the operational costs are almost half compared to the conventional ones (cost reduction for personnel, energy consumption reduction, but increased maintenance cost for protection systems).

5.4 METRO STATIONS

Metro stations are divided into three categories according to the functions they serve:

- Simple stations, whose only mission is to serve the area surrounding the station.
- Transfer stations, serving transfers between lines of the same metro network.
- Interchanges, where there is a connection with other transport modes (trams, buses, suburban rail services, etc.).

The stations constitute structural components of the system that are not usually constructed with the TBM. This is due to their usually rectangular cross sections, their different dimensions with respect to the tunnels, and, most importantly, due to their own different cross sections transversally to the alignment. The stations are usually constructed by excavation which requires the occupation of space on the ground surface, with all that this entails for the traffic in the city and for the activities of its inhabitants.

In this context, it is imperative that a complete study of the system's stations be performed before the beginning of the construction of the metro system. This is because, if certain construction and design options are not carefully and appropriately considered, there is an increased risk of failures and malfunctions either directly or in the mid-term; this will result in an actual construction cost that is significantly higher than the initially estimated cost. It should be noted that the construction of stations increases the total construction cost by 25–30%, while the construction of an underground station is 4–6 times more expensive than the construction of a surface station.

Three of the main design/construction issues of a metro system that are of concern to the engineers during the phase of the project construction are:

- The location/selection for stations.
- The depth of their construction.
- The method by which a station is constructed, as well as some of its structural elements.

The above three construction criteria are influenced by many parameters which influence one another, a fact that renders the selection a difficult task. The adoption of a suitable solution is a matter of knowledge, study, and research; however, the experience gained from similar projects remains an irreplaceable asset for both designers and constructors.

5.4.1 Location selection for metro stations

The location of metro stations is studied in accordance with the servicing of network users and, generally, the servicing of areas where there is a high travel demand. The unsuitable selection of the locations of stations can lead to failures and malfunctions such as increased walking time for pedestrians, lack of service for locations that constitute transport generators, unsuitable service for areas with increased travel demand (universities, stadiums, hospitals, etc.), and inability to service 'park-and-ride' facilities. Finally, some external factors arising from the location of the stations (such as the expropriation of areas where the stations will be built). If not properly addressed, they may result in delays in the stations' construction as well as in an increase in the construction cost.

The location selection for metro stations depends on:

The trip characteristics of potential users: One of the issues that need to be addressed initially is to determine the number of persons who want to travel, where they want to go, when, and how often. To collect this information, appropriate transport studies are necessary. The selected location of the stations must serve the travel demand.

The accessibility of stations: The stations should be placed at intersections of major roads, close to squares, at locations of mass entertainment (stadiums, shopping centres), hospitals, universities, and public services.

The availability of space for the construction of metro stations: The stations should be located in areas of the city where there is an available surface area for the installation of the construction site and the performance of the excavation, even temporarily.

The distance between stops: Maintaining an average and acceptable distance between two successive metro stations is necessary. This distance should be shorter in areas where the population density is high, and it should be longer in areas where the density is low.

The land uses: Identification studies of land uses along the alignment of the metro system are necessary. The stations should be located in areas where land uses justify their presence and require the presence of a high-capacity transport mode such as the metro.

The design of the alignment of the line: The optimum solution should be obtained with regard to the geometry of the alignment (both in horizontal and in vertical profile), while at the same time it should be designed so as to achieve the maximum commercial train speed.

The connection to other public transport modes: Metro stations should be placed in locations that allow the transfer to other transport modes and, generally, locations that ensure complementarity among the available transport modes.

The terminals: Finally, with regard to terminals, these should be installed in places that enable easy connection of the metro with other modes of transport, such as railway stations, airports, and bus stations, while ensuring that there is enough space for 'park-and-ride' services which are essential for a high level of service. The track layout of the metro terminals must allow the performance of the necessary train manoeuvring and the connection with areas of depots where trains will be parked, maintained, repaired, and cleaned.

The chance to recover important archaeological artefacts/sites during excavations: The processing of such findings may end up delaying the project (Venizelou Station, Thessaloniki Metro, Greece).

5.4.2 Construction depth of metro stations

The construction depth of metro stations should be selected carefully, taking into account all related parameters such as the tunnel's depth for safe tunnelling, the ground conditions, and the area-specific characteristics (land availability, public utility networks, archaeology, neighbouring and overlaying buildings, etc.), while an unsuitable construction depth can have adverse consequences such as minor or major impacts or even damages to neighbouring and overlying buildings, structures or networks, increased effects from the presence and pressure of the groundwater (resulting in continuous pumping of water during construction, or to the overdimensioning of the system's components), the possible crossing with public utility networks, and archaeological findings (resulting in a large increase in the cost of implementation and duration of construction). The worst possible impact that may occur is the uncontrolled subsidence of the overlying soil and ground during the excavation or the construction of the stations, leading to possible damage to any overlying buildings.

More specifically, the construction depth of metro stations depends on the following:

Characteristics of the soil: The necessary geotechnical/geological studies and tests (both *in situ* and laboratory) must be performed in order to determine the characteristics of the subsoil and the presence of groundwater zones in the area where it is planned to construct the station. These studies largely determine the technical feasibility of the project in relation to the construction depth.

Archaeological findings: The presence and relative risk of archaeological zones should be estimated by competent archaeological services, so as to avoid excavations at those areas or perform them at a greater depth, if possible.

Public utility networks: Public utility networks, which include water supply, sewage sanitation, gas supply and liquid fuels, electricity grid, and telecommunications network, are present in all major cities. The design and construction of a metro system must respect their presence, the complexity of their structure, the identity of their construction, and their behaviour during the construction and operation of the project.

Overlying buildings: The passage of a metro line under sensitive, old, or even listed buildings has significant drawbacks. If such a design cannot be avoided, it is required to perform studies to assess the anticipated displacements and subsidence of the ground surface and the impact of these displacements on the static behaviour of all the structures within the zone of influence.

Seismicity: In earthquake-prone areas, the construction of a metro network is of a special nature. It is needed to thoroughly examine the influence of the presence of underground structures (tunnels, stations, etc.) on the surface ground acceleration. The maximum and minimum acceptable values of certain parameters, such as the depth of the tunnels and the horizontal distance between an overlying structure and the tunnel axis, beyond which the difference in the surface ground acceleration cannot be neglected, should also be taken into account in all the structural risk analyses undertaken for the buildings and structures within the metro works zone of influence.

5.4.3 Construction methods

The construction methods that are applied for metro stations concern the following structural elements of a station:

a. The construction of the station's 'shell'.
b. The surface constructions and particularly:
   • The station's entrances.
   • The protrusion from the ground of elements used for ventilation or other electro-mechanical structures.
   • The lifts to the street level.
c. The number of levels of the station.
d. The stations' architecture related to the functional and operational design of stations.

5.4.3.1 Construction of the station's shell

Initially, the working site is installed, the preparatory work is performed in the site area, and the public utility networks are relocated in order to release the necessary space for construction activities. In continuation, archaeological excavations are carried out down to the level where no more archaeological findings exist. Then, the construction of the temporary civil works can proceed depending on the construction methodology (piles, sheet piles, diaphragm walls, etc.), although in some cases if the archaeological excavation is deep, for example, 5–10 m, these temporary civil work structures are necessary in order to enable the archaeological investigation itself to be carried out.

If the construction methodology is of the 'cut and cover' principle, and once all the station excavations are completed down to the lowest level, then waterproofing and concreting of the permanent civil works structures begin with a 'bottom-up' sequence, and this is carried out up to the roof slab of the station, after which the ground surface is finally reinstated.

If instead, a 'top-down' construction methodology is followed, the sequence starts by constructing the peripheral piles or diaphragm walls, followed by the archaeological excavations. The construction activities are then followed by the roof slab construction by *in situ* concreting, then the excavations down to the first underground level are performed, followed by the first-level slab construction, then the excavations down to the second level are performed, followed by the second-level slab construction, and this sequence is followed until the concretion is done down to the lowest level. With this methodology, when reaching the lowest level, the civil works construction is fully completed as well.

Alternatively, in an NATM-type of station construction method, a shaft is constructed, occupying a relatively small surface area at street level, and access is made possible through that vertical shaft in order to construct the station platforms in a fully underground manner. As a consequence, NATM-type stations are usually deep stations. The original shaft, constructed by cut and cover from the street level, is typically used to house the concourse and electromechanical equipment areas of the station.

5.4.3.2 Surface construction

Entrances: The design and location selection of all surface structures must be coordinated with the existing features of the surface conditions and with any future interventions that are planned by any public or local authorities. In fact, the study of surface structures must be coordinated with existing and future structures relating to:

- Provisions of roads, sidewalks, and pavements.
- Transfer facilities to and from other modes of transport, such as bus stops, taxi ranks, car parks, and so on.
- Buildings.
- Public utility networks.
- Arrangements of public spaces, parks, and gardens, as well as the reinstatement of the areas near the station.

The requirements for the location of the stations' entrances are:

The provision of direct access: The exact location of entrances should ensure an easy layout for vertical communication and the elimination of underpasses, or at least the minimisation of their length to the greatest possible extent.

The penetration of natural light: The greatest possible direct vertical communication between the street level and the public reception area allows the penetration of natural daylight (Figure 5.14), which is desirable, as it contributes to improving the quality of space at the public reception level and ensures the smoothest possible transition from the natural lighting of the exterior spaces to the dark underground areas of the metro.

Figure 5.14 Metro station entrance made of Plexiglas, Canary Wharf tube station, London, UK. (Adapted from Chmee2, 2013, available online at: https://commons.wikimedia.org/wiki/File:Canary_Wharf_tube_station_in_London,_spring_2013_(3).JPG (accessed 7 August 2015).)

Figure 5.15 Entrance to the Angel station, London Underground, UK. (Adapted from Sunil060902, 2008, available online at: http://en.wikipedia.org/wiki/Angel_tube_station (accessed 7 August 2015).)

The integration with the existing conditions of the surface: The entrances shall be located so as to protect and enhance the natural and built environments, while at the same time they must not disturb the road traffic and the movement of pedestrians. The siting of station entrances is usually performed:

- In existing open spaces (public squares or small parks).
- On sidewalks, if the required space is available.
- In existing buildings (Figure 5.15).

The entrances can be designed appropriately in order to be integrated with existing buildings. In such cases, the cost of land acquisition and the difficulties and construction cost, as opposed to the expected benefits, should be taken into account.

The integration in the city's historic sites: Particular attention should be paid to the location and the design of entrances in areas of archaeological interest and in areas where archaeological findings may be revealed during the construction of the metro. In such cases, in addition to the necessary coordination with the competent authorities, the study of the entrances must also be coordinated and integrated harmoniously in the layout of the archaeological sites. Also, a special study is required for the location and configuration of the stations' entrances in places that are of particular importance within the historical centre of the city. In such cases, the entrances should be designed so as to blend with the exterior appearance of traditional buildings.

5.4.3.3 Number of station levels

The number of levels of metro stations depends on the needs and specific functional and operational characteristics of each station separately. Stations may have from one to several (e.g., five) underground levels depending on the station design. Starting from the top, these levels usually contain:

- Street level, where various covered or noncovered accesses to the station are located, together with emergency exits and ventilation openings/grilles.
- Public reception area level (usually called 'concourse level') – with passenger movements and ticket purchasing/validation. Staff rooms are usually located on this level too.
- Electromechanical and railway systems – technical areas equipment level.
- Platform level.
- Under-platform levels including cable network and piping network corridors, pumping rooms/sumps, etc.

Between the concourse level and the platforms, there may be a 'transfer level' for transfer to another metro line, if the station serves more than one line, or simply as an intermediate level in deep stations. Also, stations can be designed so that electromechanical equipment areas may be isolated at one or more separate levels to avoid housing together with the passenger movement areas or could be blended within the station levels together with the public areas.

The construction of an underground project, such as the metro, requires not only the construction of conventional stairs but also the installation of escalators and lifts for facilitating the vertical movements of passengers.

According to statistics, the level of service to be provided by the escalators is typically:

- Sixty passengers/min/m of width of the escalator moving to an upper level.
- Seventy-five passengers/min/m of width of the escalator moving to a lower level.
- The capacity limit of escalators is 135 passengers/min/m of width.

Besides escalators, moving walkways (travelators) are also used and are particularly useful for faster and more comfortable movement of the public within the same level of a metro station. The desired level of passenger service is:

- Hundred passengers/min/m width of moving walkway.

Finally, lifts are used in stations with one or more levels, with one or more lifts leading from the street level to the 'unpaid' part of the concourse area, and more lifts leading from the concourse area to the platforms (typically two for the case of side platform stations, that is, with one lift per platform and one lift only in the case of centre platform stations).

5.4.3.4 Station architecture

Each station should function and operate providing comfortable and safe stay and movement to its users, bearing in mind the requirements for adaptability to the particular environment of integration for each case. The promotion of a specific and special identity for the station, in conjunction with the standardisation and the originality of styles and colours, contributes decisively to the attractiveness of the system. Regarding the standardisation of the architecture of metro stations, three alternatives are adopted:

- Standard principles of construction and architectural expression, which can then be adapted to each different type of station. In this case, the key structural and architectural features of each station are uniform. Appropriate adjustments to the needs of each station provide each station with a relative specificity.
- Separate architectural design for each station based on its functional, structural, and morphological specifications. In this case, the system as a whole acquires the necessary uniformity through the use of the same materials and the use of standard signposting.
- Design of a typical station. This solution is used in case there are no requirements for a special architecture. However, the 'typical' station only refers to the architectural approach as the different station depths, geotechnical conditions, location of entrances at street level, functional requirements of electromechanical systems, and so on never lead to 'typical' station concepts.

In all cases, the use of finishing materials that are easy to maintain and replace is required. Figures 5.16 and 5.17 illustrate the architecture that was chosen for metro stations of a European metropolis.

5.4.4 Platforms

Platforms are the areas where boarding and alighting of passengers to and from the trains take place. At the same time, platforms also serve as waiting areas for the passengers.

5.4.4.1 Layout of platforms

The platforms can be placed either between the two main tracks (central platform) or at both sides of each track (side platforms) (Figures 5.18 and 5.19), while side platforms can also be placed on entirely different levels within a station.

The layout of the central platform is the most economical solution; however, it often causes jams in passenger flows while requiring very careful marking to guide and orientate passengers.

In some stations, the central platform is accompanied by two side platforms. Although this solution requires more space, it is functionally ideal because it allows the boarding of passengers from the central platform and the alighting of passengers on the side ones.

Figure 5.16 Port Dauphine station entrance, Paris, France. (Adapted from Clericuzio, P. 2004, available online at: http://en.wikipedia.org/wiki/Paris_M%C3%A9tro) (accessed 7 August 2015).)

Figure 5.17 Solna Centrum station, Stockholm, Sweden. (Adapted from Halun, J. 2013, available online at: https://commons.wikimedia.org/wiki/File:20130601_Stockholm_Solna_centrum_Metro_statio n_6879.jpg (accessed 7 August 2015).)

5.4.4.2 *Platform dimensions*

The width of the platform is determined by the anticipated traffic during peak hours. The minimum width of the platform is 2.50 m while the usual width is between 3.50 m and 4.00 m. Greater widths are foreseen for busy stations.

Figure 5.18 Metro station – Central platform, Athens, Greece. (Photo: A. Klonos.)

Figure 5.19 Metro station – side platforms. (Photo: A. Klonos.)

In the area of the platforms, no columns should be present as they obstruct passenger movements (Figures 5.20 and 5.21) and reduce visibility.

Depending on the platform's use, its surface can be divided into the following zones:

- The safety zone, with a width of 0.50 m measured from the edge of the platform which should not be used.
- The concentration zone, which is used by passengers waiting to board the trains. The density of passengers in this zone is estimated at 2 persons/m^2.

Figure 5.20 Central metro platform with columns, Athens, Greece. (Photo: A. Klonos.)

Figure 5.21 Side platform with columns, Sofia, Bulgaria. (Adapted from Krussev, E. 2012, *A Walk through Sofia's New Metro Tunnel (Part I)*, online image, available at: http://www.publics.bg/en/publicatio ns/112/A_Walk_through_Sofia%E2%80%99s_New_Metro_Tunnel_(Part_I).html (accessed 25 January 2012).)

- The traffic zone located behind the concentration zone, with a width of about 1.50 m for the movement of passengers alighting from the trains.
- The equipment area, which is actually the remaining width of the platform. Cash desks, electronic ticketing distributors, and so on are placed in this zone.

Figure 5.22 Automated gates at metro stations (half height), Ōokayama station, Tokyo, Japan. (Adapted from Tennen-Gas, 2008, available online at: https://commons.wikimedia.org/wiki/File:Platform _screen_doors_003.JPG (accessed 7 August 2015).)

The total width of the platform depends to a large extent on the importance that is given to the concentration zone in relation to the variation in the number of passengers boarding the trains.

Regarding the height of the platform, it should be such that when the train is stopped, the vehicle's floor is on no occasion lower than the level of the platform's floor.

The gap between the floor of a stopped vehicle and the platform must be minimised. Usually, this gap should be a few centimetres wide and no more than 5 cm.

Finally, as mentioned in Section 5.2.2, for driverless metro systems, automated sliding gates are usually installed along the platform called Platform Screen Doors (PSDs) or platform edge doors. Based on this technology, the platform is separated from the tracks by these transparent doors. The doors of the train and the platform open simultaneously, only when the train is stopped at a prespecified position. These partition doors, depending on their height from the platform floor, maybe 'half height' (Figure 5.22; PSDs height is approximately 1.5 m) or 'full height' (Figure 5.23; PSDs height is approximately 2.2 m). Depending on the platform design, the space above the PSDs maybe left open (Figure 5.23) or maybe closed with suitable architectural panels up to the platform ceiling, thus completely isolating the platform and track areas (Figure 5.24).

5.5 DEPOT FACILITIES

Metro depots, depending on the activities performed, are distinguished as fully operational and partly operational.

In a metro network, as in the case of trams, the depot is of particular importance for the efficiency of the system.

The metro system is characterised by continuous extensions and dynamic adaptation to the passengers' transport requirements, resulting in changes in the fleet size and,

Figure 5.23 Automated gates at metro stations (full height), Copenhagen, Denmark. (Photo: A. Klonos.)

Figure 5.24 The platform screen door of Ecological District Station of Kaohsiung MRT (full height – isolated platform from tracks). (Adapted from Shack, 2008, online image available online at: https://commons.wikimedia.org/wiki/File:Platform_screen_door_of_Ecological_District_Station.jpg.)

consequently, changes in its maintenance requirements. Therefore, during the construction of a new metro system, the construction of a fully operational depot constitutes common practice; this is not the case with the depots that are constructed at a later stage to serve the network expansions.

When there is more than one depot in a network, an allocation of the performed activities among the available depots is very often the selected option by the system operator.

Depending on their number and their location in relation to the main line of the network, some depots perform only specific activities (e.g., only parking, light maintenance, or inspection), while others offer full-scale maintenance facilities.

The first stage of the design of a new metro depot is the selection of the area where it will be built. The geographic position of the depot is selected based on the following criteria:

- Sufficient ground plan area.
- Acceptable length of 'dead' vehicle-kilometres.
- Slight landscape (small height variations across its area).
- Availability of suitable land.
- Ability to integrate the depot into the existing land use – ability to access the road.

The planning and dimensioning process is difficult due to the great disparity between different metro systems and the lack of standards.

Metro depots exhibit major differences in comparison with tram depots. More specifically:

- Different horizontal alignment radii R_c are adopted. For tramway depots, a minimum horizontal alignment radius of $R_c = 17$–18 m is used, preferably $R_c = 20$ m, while for metro depots, a radius of $R_c = 70$–80 m is used.
- The length of parking and maintenance tracks in the respective areas is different between the two depot types, since the metro trains are longer (60–150 m) than those of the tramway (30–45 m). However, it should be clarified that metro vehicles can be detached from the train and can be led individually to the maintenance tracks.
- In the case of the metro, the depot is usually located close, yet outside the urban area, as opposed to the case of the tramway, where it is normally required to search for an area within or at the boundaries of the urban area.
- The metro network usually has a radial shape. This allows for more options in searching for an area for the construction of the depot. The tramway network usually has a 'linear' shape which reduces the options for the depot site considerably.
- The two depots require different facilities for the maintenance of the bogies (due to the different floor heights of the vehicles). The larger size of the engineering equipment that is used at the metro depot and the higher number of electronic and other systems, in comparison with the tramway, result in a requirement for larger maintenance areas and more staff.
- The comparatively large horizontal alignment radii that are adopted at metro depots render the construction of the ring track (loop line), which interconnects the various facilities, more difficult, compared to tramway depots. This fact mainly imposes the entry to and exit from the parking and maintenance area from the same area (i.e., through bidirectional train movement, Figures 5.25 and 5.26). Also, having an access redundancy, that is, with two entry/exit points for the trains, is considered as a significant operational advantage.

As in the case of tramway systems, the estimate of the depot's ground plan area is an important tool in the selection process.

Pyrgidis et al. (2015) attempt an estimate of the required ground plan area of the premises and facilities of a metro depot with the aid of data collected from metro depots that are either existing or under construction. More specifically, after statistical analyses, the average values of the surface area of the individual installations and of the total ground plan area of the depot per train of the design fleet were calculated.

Table 5.9 provides the results obtained from this process.

Figure 5.25 Metro depot – entry and exit in parking tracks from the same side. Bidirectional train movement by necessity.

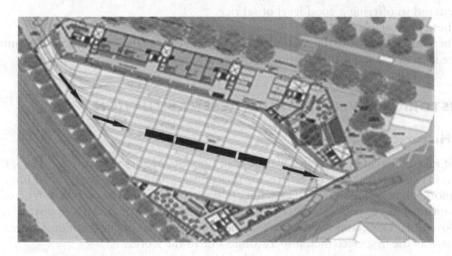

Figure 5.26 Tramway depot – entry and exit in the parking tracks from the opposite side.

Table 5.9 Estimated ground plan area per trainset for the various facilities and for the depot in total

Facility	Estimated required ground plan area per trainset (m²)
Parking area	432.34
Maintenance area	196.55
Warehouse	84.23
Management – staff buildings	38.22
Private car parking area	126.57
Total ground plan area	3,546.63

Source: Adapted from Pyrgidis, C., Chatziparaskeva, M., and Siokis, P. 2015, Design and operation of tramway and metro depots – Similarities and differences, Approved for presentation at the International Conference Railway Engineering 2015, 30/06–1/07, 2015, Edinburgh.

5.6 REQUIREMENTS FOR IMPLEMENTING THE SYSTEM

The metro system is a mass transit system that features many advantages that can be instrumental in improving a city's level of transport service.

However, due to the particularities of its construction (underground work) and the high implementation cost, in order to determine the feasibility of the construction, an extensive feasibility study must be carried out in advance.

As it is proposed and described in Chapter 21, before the financial and socioeconomic evaluation of the construction of a metro project, a thorough study of all the issues that pertain to the technical and operational feasibility of the project has to be undertaken. This particular approach is called 'technical and operational applicability verification' (Pyrgidis, 2019; Chatziparaskeva, 2018, Chatziparaskeva and Pyrgidis, 2018). It may be integrated within the feasibility study and constitute its initial stage, or it may be conducted independently (see Section 21.1)

The metro is usually selected as a transport mode for cities with populations greater than 1 million inhabitants and in cases where:

- There is a high demand for travel (> 10,000 people/h/direction).
- There is a persistent specific air pollution problem in a city or a region.
- The other public transport systems (e.g., buses) operate in saturation conditions, and despite any improvements (routing, use of longer vehicles), they fail to meet the current demand in offering a good level of service.
- There is available funding.
- There are spaces available in the system's periphery for the installation of terminals of the system itself and the creation of parking areas and bus terminal facilities.

5.7 HISTORICAL OVERVIEW AND PRESENT SITUATION

5.7.1 Historical overview

The first metro system in the world was built in London. More specifically, in 1863, the first tunnel was excavated in the city centre, for the railway connection between Paddington and Farringdon. The first 'dedicated' metro line, however, was a line that was constructed at the City (South London), between Stockwell and King William Street. This line was launched on 4 November 1890, and is now part of the Northern line of the London Underground. This line was the first electrified underground line in the world.

Metro systems exist and operate worldwide, while numerous new metro systems and expansions of existing ones are under construction.

5.7.2 Present situation

The data recorded and analysed in this section relate to the year 2019. The raw data were obtained per country, per city, and per line, from various available sources, and cross-checked. Afterwards, they were further manipulated for the needs of this section.

It should be noted that in the case of cities where there are metro systems that are managed by different operators, these systems were registered as one system. The data refer to urban rail systems that commenced operation before 1 January 2020, and met the technical and functional characteristics attributed to metro systems as described in Section 5.3. These systems:

- Mainly serve the urban centre of a city.
- Move underground for the largest part of their route.

- Develop a maximum running speed of $V_{max} \leq 110$ km/h, while their commercial speed varies between $V_c = 25$ and 50 km/h.

The majority of the above systems are referred to in the city in which they operate by the terms 'Metro', 'Subway', 'Underground', 'Metropolitan Railway', 'Metrorail', and 'U-Bahn'. In some cases they are referred to by various terms or brands such as 'Rapid Transit', 'Rail Transit', 'Sky Train', 'Light Rail Transit System', and 'Tren Urbano', but they all feature the characteristics of the metro system described above.

Additional information regarding track gauge, wheel material, GoA, type (light or heavy), and the number of lines were also recorded. Regarding the guidance system, if a single line of the whole metro network utilises rubber-tyred wheels, the system was classified as rubber-tyred. Regarding automation, metros were classified based on their highest level of automation amongst their lines. Systems were classified as either driverless or with driver. It should be noted that in the case of cities where there are more than one metro systems that are managed by different operators, these systems were registered as one system.

In total, 181 metro systems have been identified to be in operation as of 2019, most of which (166) are heavy metro systems. Most systems are located in Asia (89), followed by Europe (54), North America (18), South America (17), Africa (2), and Oceania (1) as indicated in Figure 5.27. Table 5.10 highlights the five countries in possession of the most metro systems.

The vast majority of systems (143 systems, 78.5%) are of normal gauge. Most systems (153 systems, 84.5%) utilise only steel wheels, and only 35 systems are in possession of at least one line that is driverless.

Figures 5.28 and 5.29 depict, respectively, the evolution of metros over the year, both in terms of the number of systems that commenced their operation and the total length of their networks.

As indicated by Figure 5.28, metro systems appear to rise significantly in popularity after the 1950s.

Table 5.11 presents the ten cities with the largest metro network length worldwide. It should be noted that the city with the smallest metro network length is the city of Catania in Italy; the length of its metro network is only 3.8 km.

Table 5.12 presents the metro systems that operate either exclusively or partially at a GoA4.

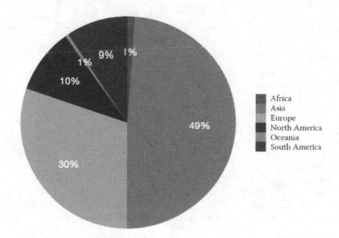

Figure 5.27 Distribution of metro systems based on location.

Table 5.10 Countries in possession of the most metro systems

Country	Number of metros
China	36
India	12
USA	12
Japan	10
Brazil	9

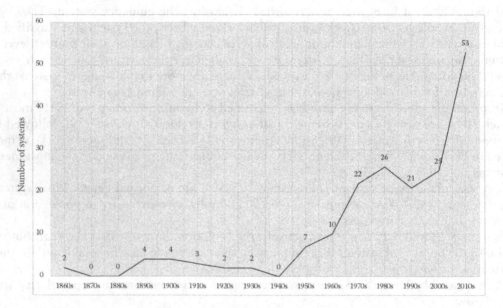

Figure 5.28 The number per decade of metro systems that were put into service.

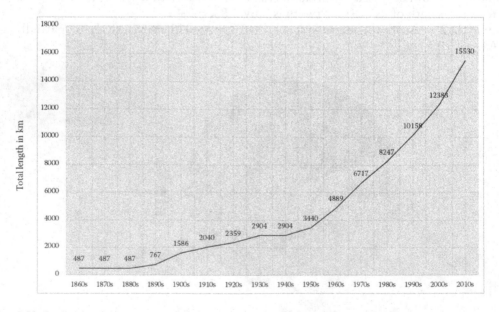

Figure 5.29 Evolution of metros around the world in terms of the total network length.

Metro 205

Table 5.11 Ten cities with the largest metro network length in the world

No.	Country	City	Total length (km)
1	China	Beijing	699.5
2	China	Shanghai	676
3	China	Guangzhou	514.8
4	Russia	Moscow	408.1
5	UK	London	402
6	USA	New York	394
7	China	Nanjing	377
8	India	Delhi	368
9	South Korea	Seoul	353.2
10	China	Wuhan	339

Table 5.12 Metro systems with a Grade of Automation of 3 or 4 (2019 data)

No.	Country	City
1	Australia	Sydney
2	Brazil	São Paulo
3	Canada	Vancouver
4	China	Guangzhou
5	China	Hangzhou
6	China	Hong Kong
7	Denmark	Copenhagen
8	France	Lille
9	France	Lyon
10	France	Paris
11	France	Rennes
12	France	Toulouse
13	Germany	Nurnberg
14	Hungary	Budapest
15	Italy	Brescia
16	Italy	Milan
17	Italy	Torino
18	Japan	Nagoya
19	Japan	Osaka
20	Japan	Tokyo
21	Japan	Yokohama
22	Malaysia	Kuala Lumpur
23	Portugal	Lisboa
24	Qatar	Doha
25	Singapore	Singapore
26	South Korea	Seoul
27	Spain	Barcelona
28	Switzerland	Lausanne
29	Taiwan	Taipei
30	Thailand	Bangkok
31	Turkey	Ankara
32	United Arab Emirates	Dubai
33	UK	Glasgow
34	UK	London
35	USA	Miami

Table 5.13 Metro systems under construction (2019 data)

Country	City	Network length (km)	Gauge (mm)	Type
Argentina	Cordoba	32.9	1,000	Heavy
Canada	Montréal	80	1,435	Light
Cote D'Ivoire	Abidjan	37.5	1,435	Heavy
Ecuador	Quito	22	1,435	Heavy
Greece	Thessaloniki	9.6	1,435	Light
Pakistan	Lahore	27.1	1,435	Heavy
India	Pune	54.6	1,435	Heavy
Iraq	Baghdad	22	1,435	Heavy
Kazakstan	Astana	22.6	1,524	Light
Russia	Chelyabinsk	5.7	1,524	Heavy
Russia	Krasnoyarsk	23	1,524	n.d.
Saudi Arabia	Riyadh	176	1,435	Heavy
Taiwan	Taichung	31.8	1,435	Heavy
Turkey	Gebze	15.6	1,435	Heavy
Vietnam	Hanoi	13.1	1,435	Heavy
Vietnam	Ho Chi Minh City	19.7	1,435	Heavy

Currently, as shown in Table 5.13, 16 metro systems are in construction across 13 countries and are planned to span over 590 km. Most of them are to be of normal gauge and heavy. It should be noted that entries in Table 5.13 refer only to entirely new systems and do not include extensions or new lines of existing systems.

REFERENCES

Amort, J. 2006, Available online at: http://en.wikipedia.org/wiki/Rubber-tyred_metro#/media/File: Rame_m2_lausanne.JPG (accessed 7 August 2015).

Bieber, C.A. 1986, *Les choix techniques pour les transports collectifs*, Ecole Nationale des Ponts et Chaussées, Paris.

Chatziparaskeva, M. and Pyrgidis, C. 2018, A decision making tool for the applicability verification of tramway projects, *8th International Symposium on Speed up and Sustainable Technology for Railway and Maglev Systems*, Barcelona, Sidges, Spain, September.

Chatziparaskeva, M. 2018, *Methodology and Mathematical Tools for the Applicability Verification of the Implementation of a Tramway Line*, PhD Dissertation, School of Civil Engineering, Aristotle University of Thessaloniki, Thessaloniki.

Chmee2, 2013, Available online at: https://commons.wikimedia.org/wiki/File:Canary_Wharf_tube _station_in_London,_spring_2013_(3).JPG (accessed 7 August 2015).

Clark, J. 2019, Comparison of cost and construction times of first metro lines in Asia, on line available online at: https://livinginasia.co/comparison-of-first-metro-lines-in-asia/ (accessed 6 April 2021).

Clericuzio, P. 2004, Available online at: http://en.wikipedia.org/wiki/Paris_M%C3%A9tro) (accessed 7 August 2015).

Davies, A. 2012, Why do subways cost so much more here than elsewhere?, available online at: http:// blogs.crikey.com.au/theurbanist/2012/02/15/why-do-subways-cost-so-much-to-build-here-tha n-elsewhere/ (accessed 14 March 2015).

FHWA. 2013, *Technical Manual for Design and Construction of Road Tunnels – Civil Elements*, Federal Highway Administration, U.S. Department of Transportation, Chapter 5 – Cut and Cover Tunnels, available online at: http://www.fhwa.dot.gov/bridge/2013 (accessed 14 March 2015).

GETZNER. n.d., *Mass-Spring System*, GETZNER company brochure, available online at: http://www.getzner.com/en/downloads/brochures/ (accessed 14 March 2015).

Halun, J. 2013, Available online at: https://commons.wikimedia.org/wiki/File:20130601_Stockholm _Solna_centrum_Metro_station_6879.jpg (accessed 7 August 2015).

Jailbird, 2005, Online image, available online at: https://en.wikipedia.org/wiki/Railroad_tie (accessed 6 April 2021).

Joly, R. and Pyrgidis, C. 1990, Circulation d'un véhicule ferroviaire en courbe – Efforts de guidage, *Rail International*, No. 12, December 1990, pp. 11–28.

Krussev, E. 2012, *A Walk through Sofia's New Metro Tunnel (Part 1)*, online image, available online at: http://www.publics.bg/en/publications/112/A_Walk_through_Sofia%E2%80%99s_New :Metro_Tunnel_(Part_1).html (accessed 25 January 2012).

MacKechnie, C. n.d., How much do rail transit projects cost to build and operate?, available online at: http://publictransport.about.com/od/Transit_Projects/a/How-Much-Do-Rail-Transit-Proje cts-Cost-To-Build-And-Operate.htm (accessed 14 March 2015).

Mckenna, C. 2007, Available online at: http://web.mit.edu/2.744/www/Results/studentSubmissions/ humanUseAnalysis/jasminef/ (accessed 7 August 2015).

Online image, available online at: http://www.athenstransport.com/english/tickets/ (accessed 14 March 2015).

Ponnuswamy, S. 2004, Ballastless track for urban transit lines, *IE(I) Journal-CV*, Vol. 85, pp. 149–158.

Pyrgidis, C. 2004, *Il comportamento transversale dei carrelli per veicoli tranviari*, Ingegneria Ferroviaria, Ottobre 2004, No. 10, Roma, pp. 837–847.

Pyrgidis, C., Chatziparaskeva, M. and Siokis, P. 2015, Design and operation of tramway and metro depots – Similarities and differences, *13th International Conference Railway Engineering–2015, 30/06–1/07 2015*, Edinburgh.

Pyrgidis, C. 2019, Applicability verification – A supporting tool for undertaking feasibility studies on urban railway systems, *9th International Congress ICTR 2019, 24–25 October*, Athens.

Quante, F. 2001, Innovative track systems, Technical construction, available online at: http://www .promain.org/images/counsil/08_Selection_of_Tack_Systems.pdf (accessed 14 March 2015).

Rama. 2007, Online image, available online at: https://commons.wikimedia.org/wiki/File:Bogie -metro-Meteor-p1010692.jpg (accessed 06 April 2021).

Rhomberg, H. 2009, Modern superstructures for railway high speed lines, types, characteristics and installation technology, available online at: http://conference.europoint.eu/highspeed2009/ hubertrhomberg (accessed 14 March 2015).

Rumsey, A. 2009, IRSE seminar on communications-based train control, International Technical Committee of the IRSE, available online at: http://www.irse.org/knowledge/publicdocuments /2009_10_01_IRSE_Seminar_on_Communications_Based_Train_Control.pdf (accessed 14 March 2015).

Shack. 2008, Online image, available online at: https://commons.wikimedia.org/wiki/File:Platform _screen_door_of_Ecological_District_Station.jpg(accessed 06 April 2021).

Sunil060902. 2008, Available online at: http://en.wikipedia.org/wiki/Angel_tube_station (accessed 7 August 2015).

Tennen-Gas. 2008, Available online at: https://commons.wikimedia.org/wiki/File:Platform_screen _doors_003.JPG (accessed 7 August 2015).

Terfloth, S. 2004, Available online at: https://commons.wikimedia.org/wiki/File:Schwellen_Rheda .jpg 2004 (accessed 14 March 2015).

UITP. 2013a, Observatory of automated metros, available online at: http://metroautomation.org/ (accessed 14 March 2015).

UITP. 2013b, Press kit metro automation facts, figures and trends – A global bid for automation, UITP Observatory of Automated Metros, available online at: http://www.uitp.org/sites/default /files/Metro%20automation%20-%20facts%20and%20figures.pdf (accessed 14 March 2015).

Chapter 6

Monorail

6.1 DEFINITION AND DESCRIPTION OF THE SYSTEM

The monorail is an electrified light-rail passenger transport system. This transport mode (in a typical manner, an articulated train) is formed with a small number of vehicles (2–6 and rarely 8) and, in most cases, moves via rubber-tyred wheels, on an elevated permanent way (guideway). The guideway is essentially a beam, which takes over the traffic loads and guides and supports the vehicles (guidebeam).

The system often covers short distances in the range of S = 1.5–12 km. When used for public transport, its length may be significantly longer. It develops maximum running speeds of V_{max} = 60–100 km/h and commercial speeds of V_c = 15–40 km/h.

Traditionally, the monorail was selected as a means of transport:

- When there is a need for a transport mode that will serve movement within amusement parks, zoos, etc.
- For the transportation of passengers along short distances and through areas that are of particular interest in terms of view.

In recent years, monorails are increasingly used for purposes exceeding recreational use and are also lately being introduced for air-rail links, as a means to circumvent land scarcity issues in congested cities (e.g., China, Indonesia, and South Korea), for movement within shopping malls, etc.

6.2 CLASSIFICATION OF THE MONORAILS AND TECHNIQUES OF THE SYSTEM

Figure 6.1 provides a classification of monorails based on a series of criteria.

6.2.1 Based on train placement in relation to the guidebeam

Depending on the way the trains are placed in relation to the guidebeam, monorail systems fall into three categories:

- *Straddled systems*: In straddled systems, the train sits above a beam, surrounding it (Figures 6.2 and 6.3). The guidebeam's profile is of either orthogonal form or type I. The vehicle 'embraces' the beam, thereby providing safety against derailment.

DOI: 10.1201/9781003046073-6

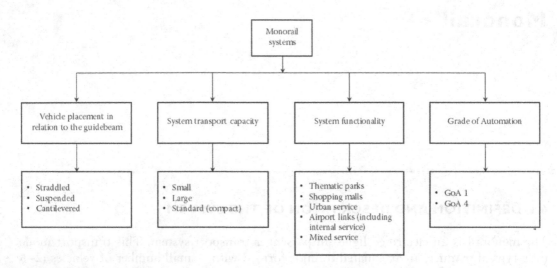

Figure 6.1 Classification of monorails.

Figure 6.2 Straddled monorails: (a) in Seattle, USA. (Adapted from Fitzgerald, G., 2005, available online at: http://en.wikipedia.org/wiki/Seattle_Center_Monorail) (accessed 7 August 2015).) (b) In Dubai, UAE.

The train moves with the aid of DC electric motors that are placed between the vehicles. Those motors trigger a system of either perpendicular or transverse wheels in rolling relatively to the guidebeam.

The wheels are connected with the car body through an electronically controlled active suspension. This levels the vehicle's floor to the station's platform, thus allowing the safe boarding and alighting of passengers.

• *Suspended systems*: In the case of suspended monorails (Figure 6.4), the vehicle is placed (suspended) below the guidebeam. The suspended systems produce greater potential visual intrusion since the total height of the vehicle and the guidebeam is approximately 6 m.

• *Cantilevered systems*: In the case of cantilevered monorails, the opposing moving trains share the same large-width beam. The trains balance with the aid of rubber-tyred wheels on surfaces that lie on the side ends of the beam (Figure 6.5). This technique has not yet been applied in practice.

Figure 6.3 Guideway of straddled monorail in Russia, Moscow. (Adapted from Lutex, 2009, available online at: en.wikipedia.org/wiki/Moscow_Monorail (accessed 7 August 2015).)

6.2.2 Based on transport capacity

According to their transport capacity, straddled monorails can be classified into three categories (Kato et al., 2004):

- Small monorails.
- Large monorails.
- Standard (compact) monorails.

In small straddled monorail systems, the width of the guidebeam (660–700 mm), the width of the vehicles (2.30–2.64 m), the length of the train (20–30 m), and the axle load (8 t) are relatively small. The train is composed of two carriages, each of which has a capacity of 60–80 passengers and can carry, considering a headway of 5 min, approximately 2,000 passengers/h/direction (the dimensions and the transport capacity of the vehicles differ according to the manufacturer).

In large straddled monorail systems, the width of the guidebeam (850–900 mm), the width of the vehicles (3.0 m), the length of the train (50–90 m), and the axle load (10–11 t) are relatively larger. The train is formed with 5–6 and rarely 8 carriages, each of which has a transport capacity of 100–170 passengers and can carry, considering a headway of 5 min, approximately 12,500 passengers/h/direction.

In compact straddled monorail systems, the train is composed of 3–4 carriages, each of which has a transport capacity of 60–100 passengers and can carry, considering a headway of 5 min, approximately 4,800 passengers/h/direction.

(a)

(b)

Figure 6.4 Suspended monorails (a) in Wuppertal, Germany (Photo: A. Klonos), and (b) in the Ueno Zoo, Japan (Ueno Zoo Monorail Taito-ku, 2005) (out of service since 2019).

6.2.3 Based on system techniques

During the application and the development of monorails, various guidance techniques were used: Lartigue, Alweg, steel box beam, inverted T systems, Urbanaut, Safege, I-beam, and double-flanged (suspended systems).

Figures 6.6 and 6.7 illustrate the techniques that are used for monorail systems.

6.2.4 Based on functionality/services provided

Based on functional/operational aspects and, more specifically, based on the nature and extent of the services they provide, monorail systems may be divided into the following categories:

Figure 6.5 Cantilevered monorail system. (Adapted from Online image. Available at: http://www.monorails .org/tMspages/TPindex.html (accessed 11 March 2015).)

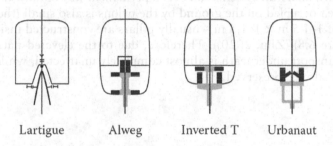

Lartigue   Alweg   Inverted T   Urbanaut

Figure 6.6 Straddled monorail systems – guidance techniques. (Adapted from Online image. Available at: www.urbanaut.com (accessed 11 March 2015).)

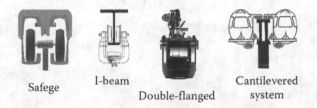

Safege   I-beam   Cantilevered
                     Double-flanged   system

Figure 6.7 Suspended and cantilevered monorail systems – guidance techniques. (Adapted from Online image. Available at: http://www.monorails.org/tMspages/TPdoub.html (accessed 11 March 2015); Online image. Available at: http://www.monorails.org/tMspages/TPmbeam.html (accessed 11 March 2015); and Online image. Available at http://fr.wikipedia.org/wiki/Monorail (accessed 11 March 2015).)

- Monorails for recreational use or servicing thematic parks.
- Monorails for urban public transport.
- Monorails servicing airports (airport links with the city centre or airport internal service).
- Monorails servicing malls, university campuses, etc.
- Mixed service monorails.

6.3 CLASSIFICATION OF THE MONORAILS AND TECHNIQUES OF THE SYSTEM

6.3.1 Permanent way

The permanent way is almost always elevated and usually consists of two guidebeams (one beam per traffic direction 'double track'). Hence, the superstructure of monorail systems comprises two beams with sufficient distance between them, so that the two opposite moving trains can cross paths (Figures 6.2 through 6.4).

A few monorail systems dispose of only one guidebeam (single track) operating either in a loop (amusement park monorails) or by employing bypasses at stations so that a single beam can be used for bi-directional operation (Shonan suspended monorail) (Kennedy, n.d.).

The beam is usually made of concrete, while in some cases (e.g., Aerobus system) it is made of steel. Its width is smaller than the vehicle's width. It ranges from 0.6 m to 1.0 m while the height is approximately twice as long (1.5 m). The beam runs continuously along the entire line, thus creating a unified element, which is constantly visible by observers and which creates visual nuisance (Pyrgidis and Barbagli, 2018).

The guidance beams are supported structurally by pillars made of reinforced concrete. These pillars are located along the route at regular intervals and their cross section may be rectangular, circular, or 'Y-shaped' in various variations.

Small or standard monorail systems run over a permanent way of only 4.5 m (two beams). Moreover, the area occupied on the ground by the pylons is also small (their base typically has dimensions of 1–1.5 m × 1–1.5 m – usually pillars are constructed inside a 2–3 m wide green zone (Figure 6.8) (Zhu, 2020)). Therefore, due to the elevated nature of the structure, any road transport underneath is almost completely unaffected, while pedestrian and bicycle paths may also be preserved.

Figure 6.8 Zone occupied on the ground by the pillars of a monorail system, Daegu Urban Railway Line 3 Monorail System (The Monorail Society).

When there is road traffic underneath the track, the minimum recommended height clearance is 5 m. However, for aesthetical reasons, there is a tendency to use much higher pillars, usually around 10–12 m high. In order to reduce the visual intrusion and allow for larger spans, a very effective technique is to use arc-shaped hunched girders for guidance. The span of the steel or concrete pillars ranges from 15 m to 24 m.

In cases where the traffic volume of road vehicles under the monorail infrastructure is high, the pillars can be replaced by a special frame structure (straddle bents, Figure 6.9) (Monorail Society/Pedersen, n.d.).

Different architectural solutions for pillars may, to an extent, reduce the visual nuisance related to them. An opportunity lies with covering pillars with aesthetically pleasing elements, which may include various forms of vegetation, coloured elements, or even reflective panels. In that way, their perceived visual nuisance is reduced to a large extent (see Chapter 19).

As regards the track geometry alignment design, the minimum radius of the horizontal alignment is R_c = 70 m for large monorails and R_c = 45–40 m for small monorails (R_c = 40 m, at the depot), whereas the minimum radius of the vertical alignment is R_v = 500 m. The longitudinal gradients are up to 10% (the Lotte World system in South Korea has an i_{max} = 20% gradient).

In order to reduce the effect of the centrifugal force at horizontal curves, the beam bends not only at the horizontal level but also at the vertical level (development of cant).

The line length of monorail systems (S) worldwide usually ranges between 1.5 km and 10 km, while the distance between successive stops is usually longer than the respective distance for other urban means of transport (800–1,500 m). In cases of urban service monorail lines, the length (S) usually ranges between 10 km and 30 km. The longest straddled monorail network in the world is the Chongqing monorail (China, 98.5 km). One of its two lines has a length of 55.5 km and is the longest monorail corridor worldwide. In Cairo, Egypt, one of the lines under construction has a length of 54 km and the total length of the network is 96 km. The longest suspended monorail line worldwide is in Japan (Chiba City, 15.2 km).

Figure 6.9 Special frame structure (straddle bents) used for the Okinawa monorail system in Japan. (Adapted from Monorail Society/Pedersen, K., no date. Okinawa Monorail, another Monorail Society Exclusive available online at: http://www.monorails.org/tMspages/Okinawa1.html (accessed 13 July 2015).)

The switching is achieved by one of the following four techniques (Pedersen, n.d.):

1. Segmented switch: The segmented track allows the beam to go from a straight position to a curved one.
2. Suspended monorail switch: It is applied to the suspended systems. Safege has pivoting horizontal plates inside the box beam that act as running surfaces in either direction of the switch (Figure 6.10). Siemens (SIPEM) uses a technology in which a vertical plate pivots inside the switch.
3. Beam replacement switch: A straight section of beam pivots to the side, while a curved section moves into place.
4. Rotary switch: Common to people-mover class, steel-beam monorails and rotating switches replace a straight section of track with a curved one.

Owing to the weight and size of the guidebeams, the process of switching is much slower in relation to the respective process for the conventional railway (12–20 s against 0.6 s).

6.3.2 Rolling stock

The trains are articulated. The maximum running speed is V_{max} = 60–100 km/h, and the acceleration/deceleration values range between 1.0 m/s^2 and 1.2 m/s^2.

The vehicles run on either 4-axle or 2-axle bogies using air suspension. The maximum axle load is Q_{max} = 10–11 t.

The vehicles' width ranges between 2.30 m and 3.00 m according to the type of the mono-rail (small, large, and standard). The length of utilised rolling stock varies, though typically train sets of 50–90 m are used. The height of the floor of the vehicle over the guidance beam ranges from 45 cm to 1.13 m.

Figure 6.10 Switching diverters – Shonan monorail, Japan. (Adapted from online image available at: https://en.wikipedia.org/wiki/Shonan_Monorail, 2005.)

The motors are fed by two electrical lines (bus bars), placed laterally to the guidebeam. The triple-phase current of 500 V converts to 1,500 DC or 750 DC, and with the aid of an electronic voltage regulator, the smooth acceleration and deceleration of the train may be achieved. An emergency generator, installed in the central station, takes over the system's supply, in the case of a power failure, so that the vehicles can be driven to the nearest stations. Owing to the elevated permanent way, it is possible to use solar panels along the guidebeam in order to gain energy for the power supply (Marconi Express Bologna).

Owing to their rubber-tyred wheels, the vehicles are sensitive to ice and snow. As a result, heating of the beam could be required, resulting in an increase in energy consumption. On the other hand, the vehicles move quietly without causing 'electromagnetic pollution'.

The carrying capacity of tyres is reduced as speed increases, due to the heat that is generated at the contact area (for instance, tyres are replaced every 130,000 km for the systems in Haneda and Okura).

The internal space of monorail vehicles is generally comfortable and aesthetically pleasing. The larger width of vehicles, when compared to tram rolling stock, allows for the formulation of more free space and the construction of larger windows.

The transport capacity of a standard monorail train with four vehicles (4 persons/m^2) often ranges between 200 and 400 passengers (seated and standing).

6.3.3 Operation

The commercial speed is usually V_c = 15–40 km/h. In order for the system to be attractive, the following must be true: (a) for lines that serve thematic parks: $V_c \geq 15$ km/h, (b) for lines that serve urban transport: $V_c \geq 25$–30 km/h, and (c) for lines that serve any other use: $V_c \geq 20$ km/h.

The headway ranges between 3 min and 15 min (minimum headway: 1.0 min), whereas the transport volume capacity of the monorail system depends on the size and the headway. In theory, maximum transport volume capacities of 20,000–25,000 passengers/h/direction can be achieved (trains of 600 seats scheduled every 1.5 min). It should be noted that in line 15 of the Sao Paolo Monorail in Brazil, a train of 8 vehicles with a trainset capacity of 1,000 passengers is utilised. A system transport capacity of 40,000 passengers/h/direction is achieved.

The track requires very low maintenance, and the rubber-tyred wheels are replaced approximately every 160,000 km (Kennedy, n.d.).

The elevated guideway imposes difficulties in evacuating the trains in case of emergency. Suspended monorails often have their vehicles' doors at the floor level, connected with stairs or slides, as in the case of aircraft.

In straddled systems, the evacuation can be performed.

- From the front or the rear vehicle, in a standing train, provided that the trains are articulated and the free movement of passengers along the train is feasible.
- Laterally with the aid of a standing train approaching on the second guidebeam (side evacuation).
- With the aid of a stair, accessed from the ground.
- With an emergency passenger platform (Las Vegas system, Figure 6.11).

Monorail systems move either with a driver or automatically (driverless systems). The use of DTO (Driverless Train Operation)/ATO (Automatic Train Operation) techniques at monorail systems is increasing in order to reduce operational cost. The first fully automatic unmanned monorail was constructed in Japan and commenced operation during the Osaka Expo in 1970.

Figure 6.11 Emergency walkway on Las Vegas Monorail, USA. (Adapted from Mikerussell at en.wikipedia., 2007, online image available at: https://en.wikipedia.org/wiki/Bombardier_Innovia_Monorail#/m edia/File:Monorail_incoming.jpg (accessed 8 August 2015).)

The total implementation cost of a double-track monorail system (totally elevated permanent way) ranges between €30 M and €90 M per track-km (2014 data), depending mainly on the transport capacity (www.lightrailnow.org/myths/m_monorail001.htm, https://pe destrianobservations.wordpress.com/2013/08/24/monorail-construction-costs/, and www .monorails.org/tMspages/HowMuch.html).

The cost of large-type monorails is almost double that of the small type. The cost of suspended systems is higher than that of the straddled systems (by 20%).

6.4 ADVANTAGES AND DISADVANTAGES OF MONORAIL SYSTEMS

6.4.1 Advantages

- It occupies a small footprint for the pillars; thus, the expropriation costs are low (Figure 6.8). The guideway does not interfere with the existing road transport infrastructure. Especially when it is used at zoos, animals are protected from run overs.
- It does not pollute the atmosphere and moves almost noiselessly. (Usually, the measured noise levels generated by monorails during operation are about 75 dBA. These are lower than the average of 90 dBA generated by light-rail systems.)
- It offers a panoramic view for its passengers.

- The construction time is relatively low (construction rate is approximately 8 km/year).
- The rubber adhesion allows higher acceleration, sharper curves, and steeper grades than the metro.
- The implementation cost is almost three times lower than the implementation cost of an underground system (metro) and two times lower than that of an elevated metro.

6.4.2 Disadvantages

- Its transport volume capacity is relatively low, although flexible modular transport system capacities are possible.
- It does not allow for a direct connection with another railway system (system incompatibility with other systems).
- The evacuation of the trains and the removal of passengers in case of an accident or immobilisation of the train on the track are difficult (especially in the case of suspended systems).
- The infrastructure and the rolling stock are often constructed by different manufacturers, whose techniques may be incompatible with each other. Often, rolling stock manufacturers conclude agreements with infrastructure providers.
- The implementation cost is almost three times higher than the implementation cost of the tram.
- Access to the boarding platforms is difficult due to the elevated stations.
- The system causes visual intrusion due to its elevated permanent way. In the case of a monorail, visual nuisance is of higher impact when compared with solutions primarily situated on the surface of the ground, since the structural elements of its infrastructure have generally large dimensions, while at the same time several structural elements are present throughout (guidance beams, escape ways) or for a large section (pillars) of the route. Most such elements are elevated, thus creating a second level of construction on top of the existing urban infrastructure and hindering further visibility towards the sky and generally the upwards visual range of any observer and at any point in time. Concurrently, the mass of the structural elements of a monorail hides large portions of the urban landscape thus hindering visibility towards the city itself (Pyrgidis and Barbagli, 2019). On the other hand, in the case of a monorail system, visual nuisance is less impacting than that caused by an elevated metro operating on heavier infrastructure.

6.5 REQUIREMENTS FOR IMPLEMENTING THE SYSTEM

Generally, the monorail is selected as a means of transport under the following circumstances:

- When there is a need for a transport mode that will serve movement within amusement parks, zoos, and so on.
- For the transportation of passengers over small distances, and in areas that are particularly interesting in terms of their view.
- For the connection of urban areas of the same altitude, where there is a natural barrier hindering their connection (e.g., water).
- Finally, in environmentally sensitive areas where there is a great demand for transportation, and it is not feasible to integrate a surface or an underground railway system.

In recent years, monorails are increasingly used not only for recreational purposes but also for urban public transport, for serving connection with airports, movement within shopping malls, and so on.

6.6 HISTORICAL OVERVIEW AND PRESENT SITUATION

6.6.1 Historical overview

The first monorail system was constructed in Russia in 1820 by Ivan Elmanov. Efforts for the development of single-beam railways, as an alternative to conventional railways, started at the beginning of the nineteenth century.

One of the first monorail systems that were developed was that of the French engineer Charles Lartigue, who constructed a line between Ballybunion and Listowel in Ireland in 1888. This line closed down in 1924 as it was ruined during the Irish Civil War. The Lartigue system uses a central single rail for support and movement, and two rails, placed at a lower level on both sides of the central rail, for guidance.

During the period 1900–1950, several systems were investigated and were either abandoned at the design phase or remained as prototypes, without ever being developed. In 1901, the Behr system was proposed for development between Liverpool and Manchester. In 1910, the Brennan system was also proposed for use in a coal mine in Alaska.

The first suspended monorail system was constructed in Wuppertal in 1901 (Figure 6.4a). The technique that was applied is that of the 'double-flange' (double-flanged, Figure 6.7). This system is the oldest monorail system in operation worldwide.

From 1950 to 1980, monorails were constructed in Japan (1957, Ueno Zoo, Figure 6.4b), in Disneyland of California (1959), in Seattle (1962), in Walt Disney World in Florida (1971), in Hawaii (1976), and in other areas. However, the use and application of monorails were very limited, as it was for all public transport systems due to their competitiveness with the private car. The guidance techniques that were used during this period were the Alweg technique for straddled systems and the French technique Safege for the suspended ones (Figures 6.6 and 6.7).

Since the 1980s and up until today, the interest in the use of monorail systems has revived due to traffic congestion and urbanisation.

The most recently launched monorail system is Marconi Express in Bologna, Italy (November 2020), and then the monorail system in the Flower Expo Park of Yinchuan, China (2018). In 2016, five monorail systems were launched. This number includes the first monorail to launch in Africa, namely, the Calabar Monorail in Nigeria. The same year also saw the commencement of operation of two monorail systems in China (the Chengdu Airtrain and the Shenzhen BYD Skyrail). In the same year, the Ashgabat Monorail was launched in Turkmenistan and the Da Nang Monorail in Vietnam. Finally, two systems were launched in 2015. Specifically, the Daegu Metro Line 3 in South Korea and the Xi'an Qujiang Monorail in China. The most recently discontinued monorail systems are the Ueno Zoo monorail in Japan (2019), the Chiang Mai Zoo monorail in China (2019), the Chester Zoo monorail in Great Britain (2019), the monorail of Magdeburg in Germany (2014), and the monorail of Sydney in Australia (2013).

6.6.2 Present situation

All the data analysed in the following section relate to the year 2019. The raw data were obtained both per country and per line, from various available sources, and cross-checked. Afterwards, they were further manipulated for the needs of this chapter.

There are 49 monorail systems in operation worldwide (Figure 6.12), of which 24 are in Asia, 11 in Europe, 10 in North America, and 2 in South America. Oceania and Africa each have 1 system. At country level, most operational systems are located in the United States and in Japan both of which have 9 systems. (The Marconi Express monorail system in Bologna, Italy, is not included in this list.)

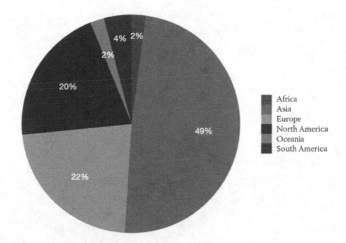

Figure 6.12 Distribution of the 49 monorail systems based on location.

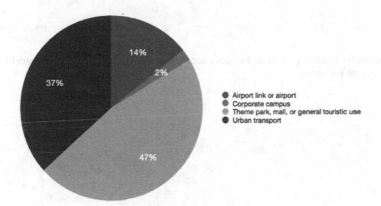

Figure 6.13 Distribution of the 49 monorail systems based on provided services.

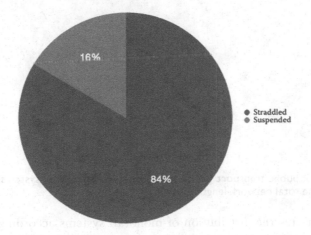

Figure 6.14 The number per decade of public transport (urban and airport use) monorail systems that were put into service.

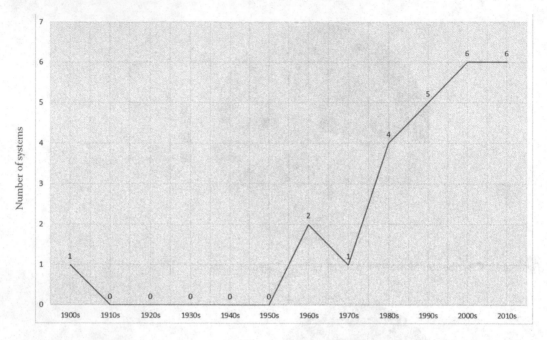

Figure 6.15 Evaluation of public (urban and airport use) monorail systems around the world in terms of the number of systems.

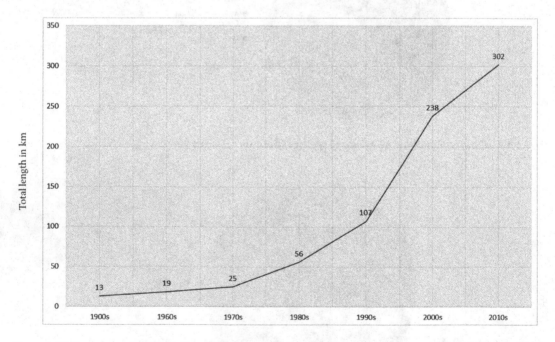

Figure 6.16 Evolution of public transport (urban and airport use) monorail systems around the world in terms of the total network length.

Figure 6.13 illustrates the distribution of monorail systems according to their provided services. As shown in this figure, the main use for monorail systems is the service of thematic parks, recreational areas, malls, and university campuses (24 systems in total, 49%),

Table 6.1 Monorail systems under construction (2019 data)

Name	Continent	Country	City	Expected date	Length	Placement	Type	GoA
Cairo Monorail	Africa	Egypt	Cairo	2023	96	Straddled	Urban Service	Driverless
Kai Tak monorail	Asia	China	Hong Kong	2023	9.0	Straddled	Urban service	n.d.
Wuhu Metro	Asia	China	Wuhu	2020	46.2	Straddled	Urban service	With Driver
Zunyi Rapid Transit System	Asia	China	Zunyi	n.d.	50.0	n.d.	Urban service	n.d.
QOM Monorail – Line M	Asia	Iran	Qom	n.d.	7.0	Straddled	Urban service	With Driver
Yellow Line	Asia	Thailand	Bangkok	2022	30.4	Straddled	Urban service	With Driver
MRTA Pink Line '2020'	Asia	Thailand	Bangkok	2021	34.5	Straddled	Urban service	Driverless
Marconi Express	Europe	Italy	Bologna	In operation from November 2020	5.0	Straddled	Airport service	Driverless
Krasnogorsk Monorail	Europe	Russian Federation	Krasnogorsk	2020	13.0	Suspended	Urban service	Driverless

whereas their use as an urban transport mode is increasing in recent years, with 18 systems (36.7%) being of pure urban transportation use and 7 systems (14.3%) serve airports. Forty-one out of the 49 systems (i.e., approximately 84%) are straddled (Figure 6.14). The total length of the 49 monorail networks is over 410 km.

Of the 25 systems that serve urban areas and airports (public transport monorails):

- Seven (28%) are of Grade of Automation 4 (driverless), while the remaining 18 (72%) have a driver.
- Eighteen (72%) are straddled and 7 (28%) are suspended.
- Japan has the most systems (7), followed by the United States (3).

Figures 6.15 and 6.16 depict, respectively, the evolution of the 25 public transport monorails over the years, both in terms of the number of systems that commenced their operation and the total length of their networks.

The total length of the 25 public transport monorails is just over 300 km.

According to 2019 data, 8 monorail systems are in construction across 5 countries and are planned to span a total of 291.1 km (Table 6.1). (The Marconi Express, as it is mentioned in Section 6.6, has started operation in November 2020.) All meant for public transport use. Entries in Table 6.1 concern entirely new systems and do not include extensions or new lines of existing systems.

REFERENCES

Fitzgerald, G. 2005, Available online at: http://en.wikipedia.org/wiki/Seattle_Center_Monorail) (accessed 7 August 2015).

https://pedestrianobservations.wordpress.com/2013/08/24/monorail-construction-costs/ (accessed 11 March 2015).

Kato, M., Yamazaki, K., Amazawa, T. and Tamotsu T. 2004, Straddle-type monorail systems with driverless train operation system, *Hitachi Review*, Vol. 53(1), pp. 25–29.

Kennedy, R. n.d., *Considering Monorail Rapid Transit for North American Cities*, report available online at: http://www.monorails.org/webpix%202/ryanrkennedy.pdf (accessed 11 March 2015).

Lutex. 2009, Available online at: en.wikipedia.org/wiki/Moscow:Monorail (accessed 7 August 2015).

Mikerussell at en.wikipedia. 2007, Online image, available online at: https://en.wikipedia.org/wiki/Bombardier_Innovia_Monorail#/media/File:Monorail_incoming.jpg (accessed 8 August 2015).

Monorail Society/Pedersen, K. n.d., Okinawa monorail, another monorail society exclusive!, available online at: http://www.monorails.org/tMspages/Okinawa1.html (accessed 13 July 2015).

Online image. Available online at: en.wikipedia.org/wiki/Shonan_Monorail (accessed 8 August 2015).

Online image. Available online at: http://fr.wikipedia.org/wiki/Monorail (accessed 11 March 2015).

Online image. Available online at: http://www.monorails.org/tMspages/TPdoub.html (accessed 11 March 2015).

Online image. Available online at: http://www.monorails.org/tMspages/TPindex.html (accessed 11 March 2015).

Online image. Available online at: http://www.monorails.org/tMspages/TPmbeam.html (accessed 11 March 2015).

Online image. Available online at: www.urbanaut.com (accessed 11 March 2015).

Pedersen, K. n.d., *The Switch Myth*, report and online images available online at: http://www.monorails.org/tmspages/switch.html (accessed 11 March 2015).

Pyrgidis, C. and Barbagli, M. 2018, Methodology for the applicability verification of a monorail system, *10th International Monorail Association Annual Conference (Monorailex 2018)*, Berlin, Germany, 13–15 September 2018.

Pyrgidis, C. and Barbagli, M. 2019, Evaluation of the aesthetic impact of urban monorail systems, *11th International Monorail Association Annual Conference (Monorailex 2019)*, Chiba, Japan, 24–26 November 2019.

Ueno Zoo Monorail Taito-ku. 2005, Available online at: https://commons.wikimedia.org/wiki/File: UenoZooMonorail1280.jpg (accessed 7 August 2015).

www.lightrailnow.org/myths/m_monorail001.htm (accessed 11 March 2015).

www.monorails.org/tMspages/HowMuch.html (accessed 11 March 2015).

Zhu, E. 2020, Technical futures of multi-standard monorail systems, *International Monorail Association Virtual Workshop*, 19–22 September 2020.

Chapter 7

Automatic passenger transport railway systems of low- and medium-transport capacity

7.1 DEFINITION

The means of transport classified as automatic passenger railway systems of low and medium transport capacity operate in an exclusive corridor by using either individual vehicles with a capacity of 3–25 persons or low- and medium-transport capacity trains (50–250 persons, standing and seated). They are automated (driverless systems) and their rolling system includes at least one iron element (steel wheels on rails or rubber-tyred wheels on steel guideway). They belong to the general category of automated transport modes on 'fixed right-of-way' (Automated Guideway Transit, AGT), which also includes higher capacity systems (e.g., driverless metro, monorail) (ACRP Report 37, 2010; Wikipedia, 2015c).

Depending on the traction system, they are classified into two categories (RPA, 2012):

- Cable-propelled systems.
- Self-propelled electric systems.

7.2 CABLE-PROPELLED RAILWAY SYSTEMS

7.2.1 General description and classification

The cable-propelled railway transportation systems use vehicles propelled or hauled by cables. They move using either rubber-tyred wheels or steel wheels, moving on conventional rails, on steel beams of type I profile, on guideway made of reinforced concrete, or finally, on steel truss elements. They are classified into two main categories:

I. Systems aimed to connect areas of very high altitude difference. They are characterised by high longitudinal slopes (usually $i > 10$–15%). The funicular, the cable railway, and the inclined elevator belong to this category. These systems are investigated in detail in Chapter 10.
II. Systems aimed to serve connections of slighter longitudinal slope ($i < 10$–15%) (Andréasson, 2001).

In this chapter, the second category (II) is investigated exclusively. These systems fall under the category of Automated People Movers (APMs). They serve trips within small or medium distances (from $S = 300$ m up to $S = 12{,}000$ m) and their transport capacity can reach up to 8,000 passengers/h/direction (for distances up to 4 km). The running speed is $V_{max} = 35$–50 km/h.

DOI: 10.1201/9781003046073-7

Depending on whether the vehicles can be detached from the pulling cable or not, they are distinguished into detachable (continuously circulating configuration) and non-detachable systems (shuttle-based configuration) (Figure 7.1) (Dale, 2014a– d; Wikipedia, 2015a).

Based on their transport capacity, they are classified into two categories: cable trains (trainsets) and cable cars (individual vehicles) (Figures 7.2 and 7.3).

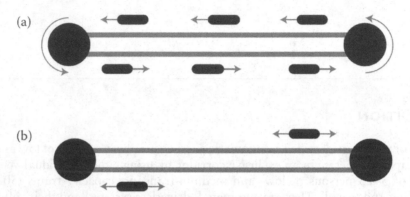

Figure 7.1 Cable-propelled railway transportation systems. (a) Detachable systems (continuously circulating configuration). (b) Non-detachable systems (shuttle-based configuration). (Adapted from Dale, S. 2014a, *Cable Cars, Lesson* I: *Introduction*, available online: http://gondolaproject.com/2010/07/09 /cable-cars-lesson-I-introduction/2014 (accessed 7 April 2015).)

Figure 7.2 Cable train. (Adapted from Wikipedia, 2015b, *Cable Liner*, available online: http://en.wikipedia.org /wiki/Cable_Liner,2014 (accessed 7 April 2015).)

Figure 7.3 Cable car. (From Doppelmayr, 2011.)

The cable train is composed of 2–6 vehicles. The vehicles are attached to a steel pulling cable and are interconnected. The train accelerates and decelerates through the pulling cable. Depending on the configuration of the superstructure, the trains do not change pulling cables at the terminal or intermediate stations (non-detachable systems), or they have the ability to change cables (detachable systems).

The cable cars (many small individual vehicles) are pulled through a constantly moving cable. The vehicles do not accelerate or decelerate via the pulling cable as in the previous case. Upon arrival at the station, they are detached from the pulling cable and the deceleration system of the station is activated (decelerator–conveyor–accelerator system). After the completion of passenger boarding and alighting, the acceleration system of the station is activated, guiding the vehicles out of the station. Through the acceleration system of the station, the speed of vehicles is equalised with the speed of the pulling cable, allowing thus, without being noticed by the passengers, the smooth reattachment of vehicles with the pulling cable. This system has low dwell times ($t_s < 90$ s) and is suitable when there is a constant flow of passengers.

7.2.2 Constructional and operational features of the systems

7.2.2.1 System 'principles' and superstructure configurations

The operation of cable-propelled railway systems may be performed by three different 'principles': shuttle, loop, and continuous movement (ACRP Report 37, 2010; Wikipedia, 2015a).

7.2.2.1.1 Shuttle 'principle'

The shuttle 'principle' does not allow the detachment of the train from the cable; it is used for small route lengths (up to 3 km) and is characterised by relatively high dwell time. The transport capacity of the system depends on the length of the system and the number of intermediate stations. It may reach up to 6,000 passengers/h/direction.

Figure 7.4 illustrates the various possible superstructure configurations in case of operation by applying the shuttle 'principle'.

In case (a) (single-lane shuttle), the system operates with only one train, which runs between the two terminals, along the single track without any bypass area, by using just one pulling cable. The total number of stations amounts to two (the terminals), but it is also possible to have up to three intermediate stations. This simple superstructure configuration is preferred in cases of low-demand transport work.

In case (b) (single-lane shuttle with bypass), the system operates with two trains, running among the terminal stations, along a single track with a bypass area, by using either one or two pulling cables (Dale, 2014d).

The first subcase (one pulling cable) allows for very steep longitudinal slopes and is used in funiculars (see Chapter 10). The bypass area is set necessarily in the middle of the route and the intermediate stations (0–3) are located at symmetrical distances.

The second subcase (two pulling cables) allows for the bypass area to be set on a different point, other than the middle of the route. This configuration enables the operation of one of the two trains, by providing 24-h service (Figure 7.5).

The configuration of case (c) (dual-lane shuttle) operates with two trains, running among the terminal stations, along a double track without any bypass area, by using two separate pulling cables. It is possible to operate the system with up to three intermediate stations. This configuration ensures a high level of service, while outside rush hours only one train can operate as in the case of the single-lane shuttle.

Finally, the configuration of case (d) (dual-lane shuttle with bypasses) operates with four trains. It consists of two parallel single-lane shuttles with bypass, doubling the system performance and providing a high level of service, while outside rush hours it can operate only with one train as in the case of the single-lane shuttle with bypass.

7.2.2.1.2 Loop 'principle'

The loop 'principle' allows for train detachment from the cable, and at the same time serves a large number of intermediate stations, as well as the simultaneous use of many trains.

In Figure 7.6, the three possible superstructure configurations (single loop, double loop, and pinched loop) are displayed in the case of operation of cable-propelled railway systems with the loop 'principle'.

The loop can be single (Figure 7.6a) or double (Figure 7.6b).

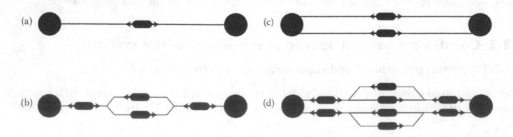

Figure 7.4 Various possible superstructure configurations in the case of operation of cable-propelled railway systems with the shuttle 'principle'. (a) Single-lane shuttle, (b) single-lane shuttle with bypass, (c) double-lane shuttle, and (d) double-lane shuttle with bypass. (From Lea+Elliott Inc., 2015.)

Figure 7.5 Station of a cable-propelled railway system that uses superstructure configuration of single-lane shuttle with bypass with two cables – Venice, Italy. (Adapted from Luca, F., 2010, Online image available at: https://commons.wikimedia.org/wiki/File:Venezia_-_Fermata_Marittima_people_mover.jpg (accessed 8 August 2015).)

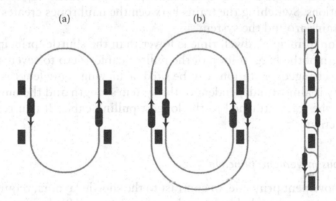

(a)   (b)   (c)

Figure 7.6 Possible superstructure configurations in the case of operation of cable-propelled railway systems with the loop 'principle'. (a) Single loop, (b) double loop, and (c) pinched loop. (From Lea+Elliott Inc., 2015.)

In a single-loop case, due to the movement of trains toward one direction only, the level of service is low, since passengers who wish to move to the previous station are forced to run throughout the system, in order to reach their destinations.

In the double-loop case, this issue does not exist, as the movement of trains is possible in both directions (Figure 7.7). This possibility allows for maintenance procedures and the continuation of system operation in case of a temporary problem related to one of the two loops. The double-loop configuration holds double transport capacity compared with the single-loop configuration.

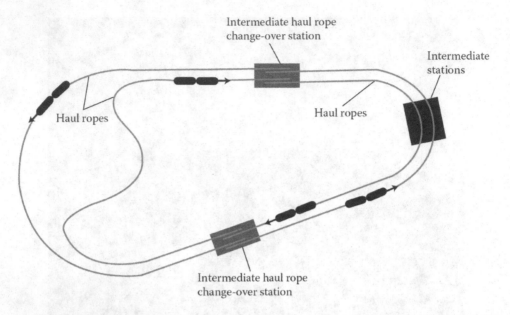

Figure 7.7 Superstructure configuration in the case of double loop 'principle'. (From Doppelmayr, 2011.)

The configuration of the pinched loop is similar to that of the dual shuttle, with the difference that the track change equipment is required at terminal stations (turntable). This layout minimises dwell times, as it allows for the simultaneous movement of more than three trains along the track. This design uses several haul rope loops, which adjoin and overlap one another at the stations. Switching the trains between the haul ropes creates a synchronised, circular flow of trains around the system.

By using the loop 'principle', dwell time is lower than the shuttle 'principle' and is determined by the length of the biggest loop of the pulling cable. Zero to seven stations and two to three haul rope changeover stations can be sited, achieving route lengths of up to 12 km. Transport capacity is almost independent of the system's length and the number of stations, and is defined by the train attached to the longest pulling cable. It can reach up to 6,000 passengers/h/direction.

7.2.2.1.3 Continuous movement 'principle'

The continuous movement principle, in contrast to the shuttle layouts, requires two separate tracks, since vehicles are detachable from the cable (Figure 7.8). It operates on individual vehicles (cable cars). It is characterised by exceptionally low dwell time, less than 1 min, regardless of the system's length. It allows for the setting up of up to 13 intermediate stations, and 2–3 haul rope changeover stations, achieving long system lengths (approximately up to 9 km). Possible superstructure configurations are of linear shape (straight layout), with circular single and double loops. The transport capacity of the system is independent of its length and the number of stations and can reach up to 4,500 passengers/h/direction.

7.2.2.2 Guideway

The guideway of cable-propelled railway systems can be integrated into the surrounding area either at ground level, under the ground, or on an elevated structure.

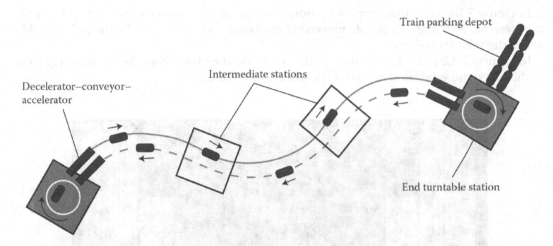

Figure 7.8 Continuous movement 'principle' – linear shaped-superstructure configuration. (From Doppelmayr, 2011.)

Figure 7.9 Steel guideway of cable-propelled railway systems supported by pillars made of reinforced concrete. (From Doppelmayr, 2011.)

In elevated systems, the guideway is normally implemented on steel beams (coupled) in truss form, which are placed on supports made of reinforced concrete (Figure 7.9). Tracks do not require heating in case of low temperatures; moreover, the use of variable spacers between the coupling and the supports allows for easy counterbalancing of any differential settlement.

They can also use a guideway made entirely of reinforced concrete.

Platforms are designed in order to be placed on the same level as the floor of the vehicles, facilitating in this way, passenger boarding and alighting. Platform length is defined based on the number of vehicles of each train. For the safety of passengers, the station platform is separated by sliding doors (PSD) from the traffic corridor.

In Figure 7.10, a schematic representation of the rolling system of the cable-propelled railway systems (cable-driven people movers) is displayed and in Figure 7.11 the pulling cable under the vehicles is shown.

In Figures 7.12 and 7.13, the sheaves that carry on the cable along the straight segments of the track and guide it in curved track sections are illustrated.

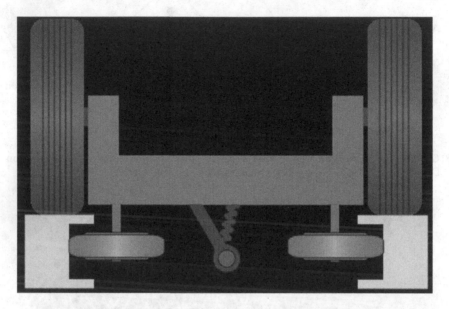

Figure 7.10 Schematic representation of the rolling system of cable-propelled railway systems. (Adapted from Wikipedia, 2015d, *Venice People Mover*, available online: en.wikipedia.org/wiki/Venice_People_Mover (accessed 7 April 2015).)

Figure 7.11 Pulling cable of cable-propelled railway systems. (Adapted from Doppelmayr, 2011, Online image available at: http://en.wikipedia.org/wiki/Cable_Line (accessed 8 August 2015).)

Figure 7.12 Guidance sheaves of cable-propelled railway systems. (From Doppelmayr, 2011.)

Figure 7.13 Guidance sheaves of cable-propelled railway systems. (From Doppelmayr, 2011.)

7.2.3 Advantages and disadvantages

7.2.3.1 Advantages

- Environmentally friendly, since they have low energy consumption and zero gas emissions.
- Highly reliable, to a percentage of more than 99.5%, as their operation is not affected by extreme weather conditions (strong winds, snow, and very high temperatures).
- Systems running on an elevated structure are adjusted to soil settlement, due to the low total weight (vehicles + superstructure).
- Maintenance works are comparatively smaller and fewer, and no facilities are required, to be set outside the line. Maintenance works are normally performed in a special area below the platform level.
- All the advantages of an automated transport system (reliability, safety, and low operational cost).

7.2.3.2 Disadvantages

- Continuous rolling noise due to the movement of the cable, which causes annoyance when the system is integrated into densely populated areas.
- Non-expandable systems.
- Limited length and transport capacity.
- The presence of many curved segments increases energy consumption due to friction in guidance sheaves.

7.2.4 Requirements for implementing the system

Cable-propelled railway systems are chosen to usually serve movements within airport areas, at big hotel complexes, casinos, conference and health centres, and in areas of educational or corporate campuses.

In the urban environment, they may feed heavier railway transport systems, as extensions or by connecting their lines.

Cable-propelled railway systems of continuous movement are suitable for any application shorter than 8 km. Usually, they serve public parking areas, hospitals, universities, or shopping malls. They connect central railway stations or metro stations with suburbs.

In some cases, such as in the city of Perugia, Italy, they serve purely urban movements (Figures 7.14 and 7.15). Known as the MiniMetro, the Perugia system can operate so

Figure 7.14 Cable-propelled railway system of Perugia, in Italy (MiniMetro). (From LEITNER AG/SPA, 2015.)

Figure 7.15 Guideway of the cable-propelled railway system of Perugia (MiniMetro). (From LEITNER AG/ SPA, 2015.)

frequently that the dwell time is almost non-existent. Extending to 3.2 km in length, the system currently has seven intermediate stations. Vehicles can be inscribed into horizontal alignment radii of up to 30 m and move in longitudinal slopes up to 15%. Vehicles bear rubber-tyred wheels and they are 5 m long with a maximum transport capacity of 50 passengers. In stations, vehicles are automatically detached from the pulling cable and conveyed through the station by an independent conveyor system.

Table 7.1 displays the constructional and operational features of various cable-propelled railway systems of low- and medium-transport capacity that move on relatively slight longitudinal slopes.

Table 7.2 displays indicatively the total implementation cost (infrastructure + rolling stock) per length-km for some systems (MetroTram, 2014).

7.3 SELF-PROPELLED ELECTRIC SYSTEMS

7.3.1 General description and classification

These self-propelled railway systems, depending on their power supply system, are distinguished into two categories based on those using:

- Batteries.
- Outside power feeding.

Systems that use lower-capacity vehicles (3–25 persons) belong to the first category. These vehicles are power supplied by lithium-ion or lead-acid batteries that provide autonomy in the region of about 60–75 km. Battery charging is performed either in the platforms or in a specially formed charging area, within the maintenance facilities of the system.

They serve short-distance movements of individuals or small groups without making intermediate stops. The vehicles launch from the starting point either upon travel request or based on scheduled itineraries (Bly and Teychenne, 2005).

Table 7.1 Technical and operational features of various cable-propelled railway systems of low- and medium-transport capacity that move in relatively slight longitudinal slopes

Name	Country/city	Track superstructure configuration	Starting year of operation	Line length (m)	Running speed (km/h)	Train headway (s)	Train transport capacity (persons)	Transport system capacity (passengers/h/direction)	Number of vehicles per train/number of trains	Number of terminal stations (intermediate stations)
Mandalay Bay Tram System I	USA	Dual-lane shuttle	1999	838	36	300	160	1,300	5/2	4(2)
System II						220		1,900		2
Air–Rail Link	United Kingdom/Birmingham	Dual-lane shuttle	2003	585	36	120	160	1,600	2/2	2
International Airport Link	Canada/Toronto	Dual-lane shuttle	2006	1,473	43.2	250	196	2,150	6/2	3(1)
International Airport Shuttle	Mexico/Mexico	Single-lane shuttle	2007	3,025	45	650	104	600–800	4–6/1	2
Tronchetto–Piazzale Roma Shuttle	Italy/Rome	Single-lane shuttle with bypass	2010	870	29	190	200	3,000	4/2	3(1)
MGM City Center Shuttle	USA/Las Vegas	Dual-lane shuttle	2009	650	37.8	150	132	3,000	4/2	3(1)
NDIA Shuttle	Qatar/Doha	Dual-lane shuttle	2013	500	45	110	190	6,000	5/2	2
Cabletren Bolivariano	Venezuela/Caracas	Pinched loop	2012	2,100	47	270	212	3,000	4/4	5
Oakland Airport Connector	USA	Pinched loop	2014	5,100	50.4	280	148	1,900	4/4	3
Perugia Minimetro	Italy	Continuous movement	2008	3,200	36–43.2	90	25	<3,000	1/25	7
Cincinnati Concourse Train	USA	Dual-lane shuttle (underground)	1994	400		132		5,700	3/2	3(1)
Detroit Express Tram	USA	Single-lane shuttle with bypass	2002	1,100		192		4,000	2/2	3(1)
Minneapolis St. Paul Airside	USA	Pinched loop	2004	800		186		1,700	2/2	4
Minneapolis St. Paul Landside	USA	Dual-lane shuttle (underground)	2001	400		84		5,200	3/2	2
Tokyo Narita	Japan	Dual-lane shuttle with bypass	1992	300		108		9,800	1/4	2
Zurich Skymetro	Switzerland	Pinched loop (underground)	2003	1,100		150		4,500	2/3	2

Source: Adapted from ACRP Report 37., 2010, *Guidebook for Planning and Implementing Automated People Mover Systems at Airports*, Research sponsored by the Federal Aviation Administration, TRB Washington, DC.

Table 7.2 Indicative total implementation cost (infrastructure + rolling stock) of cable-propelled systems of low- and medium- transport capacity (2014 data)

Cable-propelled system	Cost (€ M/length-km)
Birmingham Air–Rail, UK	12.7
Oeiras Municipality (Lisbon West) SATUO, Portugal	19.5
Perugia Minimetro, Italy	32.0
Venice, Italy	26.5

Source: Adapted from MetroTram., 2014, Information portal about guided public transportation in Europe since 1980. *MetroTram*, available online: http://www.metrotram.it/index.php?vmcity =VENEZIA&ind=0&num=1&lang=eng&vmsys=fun (accessed 7 April 2015).

They are classified into Personal Rapid Transit (PRT) and Group Rapid Transit (GRT) systems.

In PRTs (or podcars), the vehicles have a transport capacity of 3–6 persons and they move exclusively within a unique traffic corridor (Gilbert and Perl, 2007).

In GRTs, the vehicles have a higher transport capacity (20 persons) and they can be operated either in a unique traffic corridor or in a network.

Systems of medium-transport capacity, which are normally used for service in airport areas, belong to the second category. They fall under APM systems.

7.3.2 Battery-powered systems

The concept and development of PRT/GRT systems started in the 1950s. Don Fichter and Edward Haltom (Monocab) were the first to start researching for such systems as well as for alternative means of transport. The research and development of the PRT/GRT systems in Europe started in 1967 with the Aramis system that was, however, abandoned as a project in 1987.

The first systems were set in operation in Japan during 1975–1976 (the CVS system was only in operation for 6 months after 8 years of research) and in the United States in 1975 (the Morgantown system that is still in operation). During the 2000–2010 decade, a more systematic development, essentially in PRT/GRT systems, took place and several systems were set in operation (Masdar PRT, Abu Dhabi, UAE, 2010; Heathrow pod, Heathrow Airport, Great Britain, 2011; and Business Park Rivium GRT, Capelle aan den IJssel, the Netherlands, 2009).

Most systems of this category run on rubber-tyred wheels in a guideway made of asphalt or concrete. On the basis of the criterion set in Section 7.1 concerning the rolling system, these systems are not included in the railway transportation systems and are therefore not investigated further.

A number of systems that met the criterion of Section 7.1 and can be characterised as railway systems are currently under research, in the stage of prototype construction, or in the stage of testing (Janic, 2014; www.advancedtransit.org, nd). Two systems that can be classified in this category are the Vectus/POSCO ICT (POSCO ICT Co., Ltd, 2015) and the Modutram system (Vega, 2018). The former is already in operation as of 2014 in the city of Suncheon, South Korea. The latter is being tested since 2015 on a test track in Guadalajara, Mexico.

The Vectus/POSCO ICT can operate both with batteries (autonomy for distances up to 4–5 km) and power collecting shoe (3rd and 4th rail power feeding system). The system is characterised by short connection length, high running speeds (V_{max} = 60 km/h), and very small vehicle headways (4–10 s). Actually, this system operates exclusively with outside power feeding (3rd and 4th rail) (see Table 7.3 and Figure 7.16).

Table 7.3 PRT automated passenger railway system currently in operation (electric power feeding)

Name of system/ manufacturer	Category of system	Country/ city-starting date of operation	Features
Vectus/POSCO ICT	PRT	S. Korea/ Suncheon 4/2014	Route length: 4.64 km of elevated, bi-directional guideway Fleet: 40 vehicles Vehicle transport capacity: 6–9 passengers Max. speed: 35–50 km/h System transport capacity: 1,500 passengers/h/direction Power supply system: exclusively 3rd and 4th Rail Power Supply System DC 500 V Minimum vehicle headway: 5–6 s Rolling system: iron wheels made of polymer material running on rails

Source: Adapted from POSCO ICT Co., Ltd, 2015. SunCheon SkyCube PRT Project Overview v1.1.

Figure 7.16 Vectus/POSCO ICT PRT automated passenger railway system, Suncheon, South Korea. (From POSCO ICT Co., Ltd., 2015.)

The vehicles are equipped with an automatic obstacle tracking and detection system, which allows acceleration or immobilisation of vehicles in case of obstacle detection.

Each vehicle is equipped with a two-way communication system between the passengers and the system control centre. Moreover, the vehicles are air-conditioned and they have a closed-circuit monitoring system with cameras. There are Liquid Crystal Display (LCD) screens in passenger seats, which provide useful information during the trip.

The floor of the vehicles is usually located on the same level as the boarding and alighting platform.

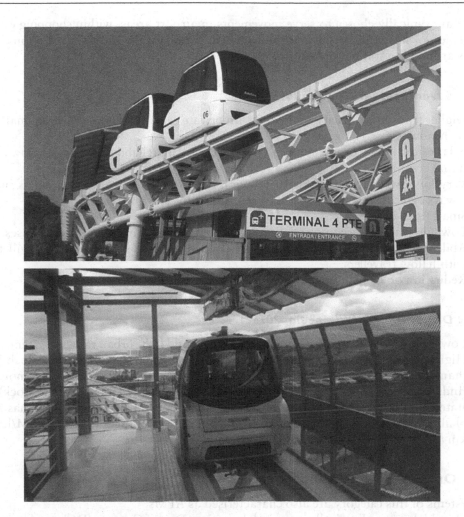

Figure 7.17 The Modutram system (Online images available at: https://solarimpulse.com/companies/modut
ram) (accessed 25 February 2021).

Vehicles maintain a minimum running headway time of 5–10 s to prevent collision. This is ensured through the closed-telecommunication system and automatic detectors.

The Modutram (or Autotren) system was constructed in 2014 in Guadalajara, Mexico. It consists of a fleet of ultra-light, electric and driverless trainsets, which are formed from 1 to 4 wagons, for up to 24 passengers. The vehicles have rubber tyres running on steel rails (Figure 7.17). Its permanent way is exclusive, elevated, or underground with lanes, whose width is half the width of conventional bus or train lanes. The passenger stations are always located in parallel lanes. Vehicles circulate in the main lanes at 4 s intervals, stopping only at those stations where they will go up or down, operating in a similar way to an elevator where all routes are express and scheduled on demand (https://www.tuv.com/landingpage /en/global-rail/lam-references/modutram.html, 2020). The cost of an Autotren corridor is 20% higher than Bus Rapid Transit (BRT) systems (Vega, 2018).

PRT systems operate essentially as feeder modes of heavier railway transport systems, as an alternative to other slower transport modes, such as buses, or moving on foot.

They are typically chosen to serve movements in airport areas, within shopping centres, universities, or hospitals, for local trips in new cities, and for the supply/distribution of passengers around the stations of railway systems.

7.3.2.1 Advantages (Wikipedia, 2015c)

- Higher transport capacity compared with private cars, due to the ability of small vehicle headways.
- All the advantages of an automated transport system (reliability, safety, and low operational cost).
- Very low dwell time. Outside rush hours, the level of service is increased as there is always typically an available vehicle waiting at the starting point.
- Small right-of-way.
- Low run time, since commercial speeds of PRTs, are double than those of buses.
- Ability of exclusively private transport. Passengers choose whether they will travel with fellow travellers.
- Reduction of air pollution.

7.3.2.2 Disadvantages (Wikipedia, 2015c)

- Lower transport capacity compared with buses and light urban railway systems.
- High total implementation cost per length-km compared with buses, but much lower than the cost required for the construction of a light railway system (tram, monorail) (Indicative cost €8–11 M/length-km, double track, infrastructure + rolling stock). It is stated that the system at Heathrow (it uses rubber-tyred wheels on asphalt) has a cost of about €10 M/length-km (ATRA, 2014a) and the system at Suncheon €10 M/length-km (POSCO ICT Co., Ltd, 2014) (2014 data).

7.3.3 Outside power feeding systems

The systems of this category are also characterised as APMs.

APM vehicles are electrically powered by either Direct Current (DC) or Alternating Current (AC) provided by a power distribution subsystem along the guideway. Vehicle propulsion may be provided by either 750 or 1,500 V DC rotary motors, 480 or 600 V AC rotary motors, or by AC Linear Induction Motors (LIM).

Self-propelled vehicles or trains run in an exclusive double-track corridor through rubber-tyred wheels or steel wheels on rails. Their maximum running speed fluctuates between V_{max} = 50 and 75 km/h.

They were initially developed to serve urban areas and then they were applied mostly in big airports. According to 2010 data, APMs were in operation in 44 airports worldwide (electric and cable-propelled in total) (ACRP Report 37, 2010). Their permanent way is integrated into the surrounding space, on the surface, underground, and in elevated cross sections.

The latest systems have been constructed in order to facilitate the movement of passengers and airport employees within the airport's security zone [movement between check-in (terminal) areas and aeroplane gates]. These systems are characterised by high transport capacity [8,500–9,000 passengers/h/direction (passengers with carry-on baggage only)], high speed, low train headway (2 min) and longer connection lengths.

More recently, they have been used to connect terminal stations with areas outside the airport's security zone, such as parking areas, car rental services, hotels and other related employment and activity centres. These systems are characterised by lower transport

Table 7.4 Technical and operational features of various electric automated railway systems of low- and medium-transport capacity – systems in operation in airports

Name	Country	Track superstructure configuration	Starting year of operation	Line length (km)	Train headway (min)	Transport system capacity (passenger/h/ direction)	Type of provided service	Number of vehicles per train/number of trains
Gatwick Airport Transit	United Kingdom/ London	Dual-lane shuttle, elevated	1987	1.2	2.6	4,200	Landside conveyance	3/2
Madrid Barajas Airport	Spain	Pinched loop, underground	2006	2.2	2	6,500	Airside conveyance	3/6
Miami International Airport	USA	Dual-lane shuttle, elevated	1980	0.3	2	6,750	Airside conveyance	3/2
New York–John F. Kennedy International Airport	USA	Pinched loop, primarily elevated	2003	13.0	2–4	3,780	Landside conveyance	1–2/32
Kansai International Airport	Japan/Osaka	Four single-lane shuttles with bypasses, elevated	1994	2.2	2–2.5	14,400	Airside conveyance	3/9
Paris Roissy Charles de Gaulle Airport	France	Shuttle, underground	2007	0.6	2	4,500	Airside conveyance	2/3
Washington Dulles International Airport	USA	Pinched loop, underground	2010	2.3	2	6,755	Airside conveyance	3/29
Tampa International Airport	USA		2018	3.05			Landside conveyance	
Soekarano–Hatta Airport	Indonesia		2017				Airside conveyance	2/3
Munich Airport	Germany	Dual-lane shuttle, underground	2016	1.5		10,900	Landside	
Dubai International Airport (Terminal 1)	UAE		2016				Landside	5/18

Source: Adapted from ACRP Report 37., 2010 Guidebook for Planning and Implementing Automated People Mover Systems at Airports, Research sponsored by the Federal Aviation Administration, TRB Washington, DC; airmundo.com/en/blog, 2018.

capacity [3,000 passengers/h/direction, 50 passengers/vehicle (passenger all baggage)] and higher train headway (3 min).

Self-propelled APM systems have high requirements as far as the infrastructure is concerned demanding heavy concrete structures for the guideway and heating of the guideway running surfaces in winter. Thanks to guideway switches they can be operated in the pinched-loop mode. Self-propelled APMs are extremely reliable and reach availability rates of more than 99.5%.

Table 7.4 displays the constructional and operational features of various self-propelled electric railway systems in operation in airports.

REFERENCES

ACRP Report 37. 2010, *Guidebook for Planning and Implementing Automated People Mover Systems at Airports*, Research Sponsored by the Federal Aviation Administration, TRB, Washington, DC, available online at: airmundo.com/en/blog, 2018 (accessed 8 November 2020).

Andréasson, I. 2001, Innovative transit systems survey of current developments, *Logistik Centrum VINNOVA Report VR 2001*: 3, The Swedish Agency for Innovation Systems, January 2001.

ATRA. 2014a, *Heathrow*, available online at: http://www.advancedtransit.org/advanced-transit/applications/heathrow/2014 (accessed 7 April 2015).

Bly, P. and Teychenne, P. 2005, Three financial and socio-economic assessments of a personal rapid transit system, *American Society of Civil Engineers*, pp. 1–16. doi: 10.1061/40766(174)39.

Dale, S. 2014a, *Cable Cars, Lesson 1: Introduction*, available online at: http://gondolaproject.com/2010/07/09/cable-cars-lesson-1-introduction/2014 (accessed 7 April 2015).

Dale, S. 2014b, *Cable Cars, Lesson 2: Single Loop Cable Shuttles*, Online image, available online at: http://gondolaproject.com/2011/03/23/cable-cars-lesson-2-single-loop-cable-shuttles/2014 (accessed 7 April 2015).

Dale, S. 2014c, *Cable Cars, Lesson 3: Dual Loop Cable Shuttles*, Online image, available online at: http://gondolaproject.com/2011/05/17/cable-cars-lesson-3-dual-loop-cable-shuttles/2014 (accessed 7 April 2015).

Dale, S. 2014d, *Cable Cars, Lesson 4: Dual By-Pass Shuttles*, available online at: http://gondolaproject.com/2011/05/24/cable-cars-lesson-4-dual-by-pass-shuttles/2014 (accessed 7 April 2015).

Doppelmayr. 2011, Online image, available online at: http://en.wikipedia.org/wiki/Cable_Line (accessed 8 August 2015).

Gilbert, R. and Perl, A. 2007, Grid-connected vehicles as the core of future land-based transport systems, *Energy Policy*, Vol. 35(5), pp. 3053–3060.

Janic, M. 2014, *Advanced Transport Systems - Analysis, Modeling and Evaluation of Performances*, Springer, London.

Luca, F. 2010, Online image available online at: https://commons.wikimedia.org/wiki/File:Venezia_-_Fermata_Marittima_people_mover.jpg (accessed 8 August 2015).

MetroTram. 2014, Information portal about guided public transportation in Europe since 1980, *MetroTram*, available online at: http://www.metrotram.it/index.php?vmcity=VENEZIA&ind=0&num=1&lang= eng&vmsys=fun (accessed 7 April 2015).

Online image, available online at: http://en.minimetro.com/Application-area/For-Traffic-Planners (accessed 7 April 2015).

Online images, available online at: https://solarimpulse.com/companies/modutram) (accessed 25 February 2021).

POSCO ICT Co., Ltd. 2014, Korea's first Personal Rapid Transit (PRT), SkyCube, available online at: http://globalblog.posco.com/koreas-first-personal-rapid-transit-prt-skycube/.

POSCO ICT Co., Ltd. 2015, SunCheon SkyCube PRT project overview v1.1.

RPA. 2012, Impact assessment study concerning the revision of Directive 2000/9/EC relating to cableway installations designed to carry persons, *Final Report Prepared for DG Enterprise and Industry RPA*, October 2012.

Vega, I.P. 2018, The autotren a solution for Latin American cities developed in Guadalajara, available online at: https://udgtv.com/noticias/jalisco/autotren-solucion-ciudades/

www.advancedtransit.org, nd (accessed 8 November 2020).

www.tuv.com/landingpage/en/global-rail/lam-references/modutram.html,   2020.   (accessed   8 November 2020).

Wikipedia. 2015a, *Cable Car*, available online at: en.wikipedia.org/wiki/Cable_car, 2014 (accessed 7 April 2015).

Wikipedia. 2015b, *Cable Liner*, available online at: http://en.wikipedia.org/wiki/Cable_Liner, 2014 (accessed 7 April 2015).

Wikipedia. 2015c, *Personal Rapid Transit*, available online at: http://en.wikipedia.org/wiki/Personal rapid transit (accessed 7 April 2015).

Wikipedia. 2015d, *Venice People Mover*, available online at: https://en.wikipedia.org/wiki/Venice _People_Mover (accessed 7 April 2015).

Chapter 8

Suburban railway

8.1 DEFINITION AND CLASSIFICATION OF SUBURBAN RAILWAY SYSTEMS

The length of the route and the frequency of the service are usually used to distinguish three types of passenger railway transport systems: (a) the suburban railway; (b) the commuter or peri-urban or urban rail, and (c) the regional railway.

The term 'suburban railway', according to international practice, usually refers to an electrical passenger railway transport system, whose features are adapted for commuting transportation service within the geographical boundaries of large urban agglomerations (suburbs and satellite regional centres) (Figure 8.1). Suburban railway systems cover distances of 10–40 km. They are defined by very high frequency services (usually trains run every 5–30 min), commercial speeds of 45–65 km/h, and transport capacity of up to around 60,000 passengers/h/direction. Suburban services started in Berlin, Germany, at the end of the nineteenth century. From then on, increasingly more cities have integrated the suburban railway into their transport system.

'Commuter' or 'periurban' or 'urban rail' covers distances of 30–50 km. The track is usually electrified and the operation is defined by relatively high frequency services (usually trains run every 20–60 min) and commercial speeds of 50–70 km/h.

Finally, the 'regional railway' covers distances of 50–150 km. The track can be electrified or not and one of the two terminal stations is usually located in a small or medium urban centre. The operation is defined by relatively lower frequency services (usually 1–3 h) and higher commercial speeds (70–100 km/h).

8.2 CONSTRUCTIONAL AND OPERATIONAL CHARACTERISTICS OF THE SUBURBAN RAILWAY

The main characteristics of the suburban railway are given in Table 1.7. In addition to those characteristics, the following should also be noted:

- The track design speed of lines servicing exclusively suburban trains varies between V_d = 120 and 160 km/h.
- The track must be double so as to achieve sufficient track capacity and to reduce run time.
- A suburban train typically runs at grade for most of the journey; however, it can also run underground (city centres) or on elevated infrastructure. The total length of the route requires fencing on both sides of the track.

DOI: 10.1201/9781003046073-8

Figure 8.1 (a) Electric suburban railway, double-deck, Switzerland. (b) Electric suburban railway, EMU, Greece. (Photo: A. Klonos.)

- To provide suburban services, Electrical Multiple Units (EMU) are usually used. Apart from the railcars, loco-hauled trains and push-pull trains are also used. The total number of vehicles usually varies from 2 to 8 and may fluctuate depending on the transport demand.
- As the route is relatively long in comparison with urban transit (metro, tramway, monorail), a higher level of comfort during the trip is required and seated accommodation is increased (Figure 8.2). More specifically, the percentage of seated passengers should reach 50% of the total number of available seats.
- The vehicles must have a large number of external doors that ought to be wide enough to facilitate rapid boarding and alighting and reduce halt times at stops. When the commuting service is very busy, since the length of platforms cannot be easily increased, double-deck carriages may be used (Figure 8.1a).

Figure 8.2 Internal arrangement in vehicles of suburban trains, Russia. (Photo: A. Klonos.)

- The usual frequency for the service varies from 15 min to 30 min. The lowest value of train headways is 1.5 min. The lowest acceptable frequency for a suburban train is one train per hour.
- The use of a clock-face timetable (e.g., every xx.03 and xx.33 and so on) is advisable since it helps the passengers to easily memorise the departure times, while at major stations it is suggested to fix departure times at round minutes (e.g., 9:10 instead of 9:12).
- The traction characteristics of the powered vehicles must be able to provide swift acceleration, and the train must be equipped with particularly efficient braking systems.
- The transport capacity of the trains can vary between 250 and 1,500 passengers with carriages of moderate-high-floor or high-floor. The transport capacity of the suburban railway can reach 60,000 passengers/h/direction (e.g., trains at 1.5 min intervals with an individual train transport capacity of 1,500 passengers) with values varying usually between 5,000 and 40,000 passengers/h/direction.
- The signalling and train protection system must allow short headways. In the case of suburban services being mixed with other types of services (e.g., regional services, long-distance passenger, freight), the requirements increase. In such cases, some stations should allow overtaking, and if the traffic volume is high, quadruple tracks may be used.

In order to serve demand during peak hours, there are various solutions. Their application presumes analytical data for the origin and destination of the passengers. These solutions are:

- *Skipping of stations*: Trains do not stop at every station. They are distinguished under categories (e.g., A, B, etc.), and each category serves particular stops. This method results in an increase of the commercial speeds of the trains. On the other hand, it creates some problems for the users, such as:
  - The obligation to harmonise services for transfers from stops of one category to stops of another category.
  - The complexity of the service system.

- *The zonal timetabling method*: The difference from the previous method is that in this method the trains serve all of the stations of one or more designated areas, and then they are directed to the main station of the line (the terminal) without any stop. This method is suitable for cases where the demand is greater at a limited number of stations, instead of being evenly distributed throughout the line. Problems from the application of this method arise when there is a demand for service among the zones that are not connected.
- *Return without passengers*: It is the method that is applied in cases where the demand is limited to one direction. In such cases, trains moving in the other direction are running empty, and with an accelerated nonstop timetable in order to assume earlier duties along the more demanding direction.
- *Increase in the number of vehicles of the train formation – Exploitation with Multiple Units (MU)*: In this case, longer platforms are necessary.

8.3 ADVANTAGES AND DISADVANTAGES OF THE SUBURBAN RAILWAY

8.3.1 Advantages

The great development of the suburban railway comes as a result of its advantages, namely:

- Flexibility to adapt to the demand.
- Dynamic comfort of passengers.
- Relatively high commercial speed.
- Great passenger transport capacity.
- Reliable services.
- Frequent services.
- Increased transport safety.
- Low burden on the environment.

8.3.2 Disadvantages

The main disadvantages of the suburban railway are:

- The disturbance caused during its construction.
- The high construction cost of the infrastructure (€10–20 M per track-km according to 2019 data), and that it is reduced significantly when the existing infrastructure is used.

8.4 REQUIREMENTS FOR IMPLEMENTING THE SYSTEM

The suburban railway can operate by using either existing or new infrastructure.

In the case where suburban services are provided on the existing infrastructure, the following issues must be addressed (Pyrgidis, 2008):

1. Can the length of the connection justify suburban services?
2. Are there any attractive advertisements and interest generating items/signages placed at intervals along the rail route?

3. Is the existing passenger transport volume or the potentially developed one, significant? Is there seasonal fluctuation in the transport volume? Is there any demand for transport between intermediate stops?

4. Does the railway infrastructure allow the development of the desired speeds and, subsequently, the desired run times?

5. Is there sufficient available rolling stock suitable for suburban services (frequent services, adequate 'dynamic comfort' of passengers, etc.)?

6. Do the stations provide a high level of service to the passengers? Is there the possibility of transfer to and, more importantly, of integration with other means of transport (park and ride service facilities, buses, etc.)? Are the railway stations located in close proximity to the city centres that need to be served? Are the stations and the stops safe?

7. Does the capacity of the line allow the operation of additional trains and, specifically, the number that is required in order to achieve frequent services? Do trains of any other category run on the line (freight, regional trains, high-speed trains), and if so, are the timetables well harmonised?

8. Is the line electrified, in order for the desired traction characteristics to be assured and, most importantly, in order to avoid any environmental consequences along the transit area as a result of the increased number of trains?

9. Is the line equipped with an electric signalling system? Additionally, is there an Automatic Train Protection System installed?

10. Are there many level crossings (if any)? Are they equipped with automatic barriers? Is there any fencing along the line?

11. Is it possible to ensure that there is adequate space for the construction and the effective operation of areas of maintenance, repair, and parking of rolling stock for the specific network of suburban service?

12. Does the operation of trains create problems of noise pollution and vibrations that can be successfully addressed?

13. Can the issues 4–12 be dealt with using relatively low-cost interventions?

14. Finally, and most importantly, are the users of the other competing means of transport in favour of the creation of a suburban railway in that area?

In the case of completely new infrastructure, the issues that must be examined are numbers 1, 2, 3, 11, 12, 13, and 14, whereas all the rest must conform to the specifications for which the construction is tendered.

8.5 AIRPORT RAILWAY LINKS

8.5.1 Railway services between urban centres and neighbouring airports

The modes of ground transportation to/from airports can be distinguished into:

- Roadway systems including private cars, taxis, buses, and coaches.
- Railway systems including metro, tramway, monorail, suburban/regional, and intercity trains (high or conventional speed).

Transport mode selection is affected by a number of factors, such as service regularity, travel cost, punctuality, safety, security, overall travel time (door-to-door), frequency, comfort level (mainly ease of transporting luggage), availability of facilities for persons with special needs, etc. In this context, it may be worth noting that, for a large number of people, getting to the airport can be the most stressful part of the overall journey, especially when travelling to the airport during peak hours (Odoni and Neufville, 2013).

Today, there is a large number of airport rail links in place that excel in many of the aforementioned factors. More specifically, they have the advantage of not being overly affected by extreme weather conditions and traffic congestion. In fact, railway systems offer a high level of regularity and punctuality, something that is vital for people using airports, whose main goal, especially when on outbound journeys, is not to miss their flights. At the same time, rail provides fast services (station-to-station) at a relatively low cost. However, issues may arise when passengers have to transport a number of large and/or heavy pieces of luggage.

Based on 2017 data, a study sample of 193 airport rail links in operation was identified, serving 152 airports and 138 urban centres in 44 countries, mainly in Europe and Asia (Pyrgidis et al., 2018). It is noteworthy that out of the 152 airports, 116 were being served by a single type of railway system, 31 by two types, and 5 by three types. In addition, 12 urban centres were being served by at least two airports, each with its own airport–urban centre rail link.

The data analysis for the aforementioned 193 airport rail links has allowed a statistical overview to be made as regards the following characteristics:

- The development in the number of airport–urban centre rail links in relation to that of airports.
- The average airport–urban centre distance served by the different types of railway systems.
- The railway system type of the airport link in relation to urban centre population size.

In addition, 33 airport rail links also found that were either under construction or the construction of which had been given the go ahead at the time of compilation of this study (end of 2017), which indicates that the trend to link urban centres to airports by rail will continue in the future.

8.5.2 Development in the number of airport–urban centre rail links in relation to that of airports

Whilst airport–urban centre rail links have been in place since the 1920s to transport people to/from airports, it was not until the 1970s that their introduction saw a significant increase, with a prominent rise in the 1990s and onward. This is well illustrated in Figure 8.3, which

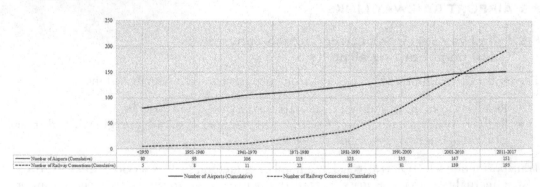

	<1950	1951-1960	1961-1970	1971-1980	1981-1990	1991-2000	2001-2010	2011-2017
Number of Airports (Cumulative)	80	93	106	113	123	135	147	151
Number of Railway Connections (Cumulative)	5	8	11	22	35	81	139	193

——— Number of Airports (Cumulative)   ----- Number of Railway Connections (Cumulative)

Figure 8.3 Number of new airports and new airport–urban centre rail links that went into operation during the period pre-1950 through 2017.

shows the introduction of the 193 identified airport–urban centre rail links in relation to that of the 152 airports they serve, per 10-year intervals till the end of 2017 (the years before 1950 are grouped into a single interval). As can be observed from Figure 8.3, over the past 30 years, there has been a relatively stable increase in the number of airports that went into operation whilst, in contrast, there has been a significant rise in the number of airport rail links. Especially since the 1990s, there has been a stark rise in the number of airport rail links. It is worth noting that a quarter of the aforementioned 193 rail links became operational after 2010.

It was found that the 152 airports under study are mainly served by metro, suburban/regional and intercity railway systems.

8.5.3 Average airport–urban centre distance served by the different types of railway systems

In Figure 8.4, for the identified 193 airport rail links under study, the average airport–urban centre distance served by the different types of railway systems is shown. As can be observed, airports that are served by suburban/regional railway systems are, on average, located at a distance of 27 km from the urban centre. Those served by intercity railway systems follow with an average distance of 20.4 km, and then monorail systems (18.2 km), metro systems (16.8 km) and, finally, tramway systems (16.6 km).

Metro systems generally serve airports that are located at a distance of up to 30 km from the urban centre and are frequented by a large annual number of air travel passengers. Tram systems mainly serve distances of 5–20 km, whilst beyond that their presence is almost non-existent. Monorail systems serve airports that are located at distances shorter than 25 km. It should be noted that the relatively low number of monorail systems (six) included in the study sample were located in Asia where population numbers are significantly high. Suburban/regional and intercity railway systems serve all distances, independently from the annual number of air travel passengers, since they serve longer routes that go beyond the airport–urban centre section.

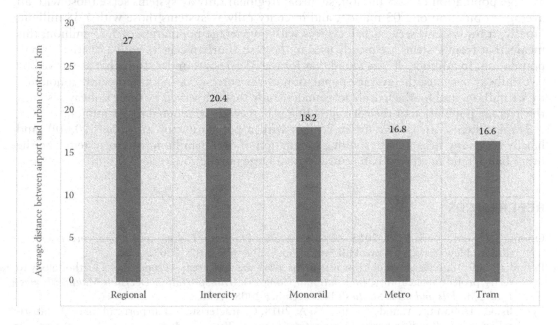

Figure 8.4 Average airport–urban centre distance per type of railway system.

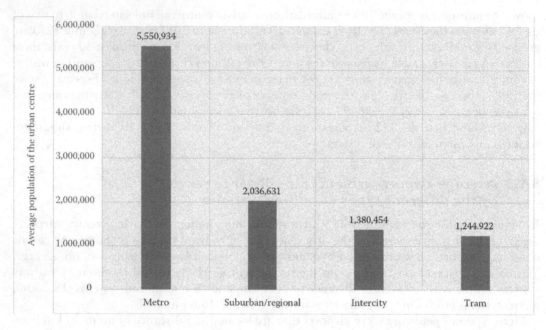

Figure 8.5 Average population of urban centres served by a sole airport and a single railway system, per type of railway system (excluding monorail systems as only three met sample selection criteria).

8.5.4 Number and type of railway systems serving as airport links in relation to urban centre population size

In Figure 8.5 the average urban centre population that is served by the 116 out of the 152 airports under study, which are served by a single railway system, per type of system, is shown. As can be observed from Figure 8.5, metro systems serve urban centres that have an average population of 5.55 million, suburban/regional railway systems serve those with an average population of 2.03 million, and intercity railway systems those with 1.38 million. Finally, tram systems serve urban centres with an average population of 1.25 million; this means that tram systems are mostly used in the case of urban centres with a relatively small population. In addition, it was found that for the 31 airports under study that are served by two railway systems, the average population of the corresponding urban centre amounted to 3.1 million, and for the five airports under study that are served by three railway systems, the average population of the corresponding urban centre amounted to 5.1 million.

Tram systems often serve urban centres with a population of less than 500,000, and hardly serve any urban centres with a population of over 1 million. Metro systems, on the other hand, tend to serve urban centres with a large population (over 500,000).

REFERENCES

Odoni, A.R. and Neufville, R. 2013, *Airport Systems: Planning, Design, and Management*, Second edition, New York: McGraw Hill Education, ISBN: 978-0-07-177058-3.

Pyrgidis, C. 2008, Integration of railway systems into suburban areas – Proposal of a methodology to verify project applicability, *International Congress 'Transportation Decision Making: Issues, Tools, Models and Case Studies'*, 14 November 2008, Venice.

Pyrgidis, C., Dolianitis, A. and Papagiannis, A. 2018, Characteristics of airport rail links: A statistical overview, *Rail Engineering International*, No. 1, 2018, pp. 4–6.

Chapter 9

Rack railway

9.1 DEFINITION AND DESCRIPTION OF THE SYSTEM

The operational limits and the economic exploitation of a rail transport system are directly dependent on the track's geometrical design. More specifically, the greatest problems are caused by large longitudinal gradients. In these cases, the movement is a function of the available traction power, the total weight of the train and the wheel–rail adhesion.

The coefficient of friction between steel and steel cannot exceed a certain value. Therefore, the longitudinal gradient of the railway track which allows for the economic exploitation of the network is predefined (see Table 1.3).

In the case of gradients that are greater than 50–70‰, an additional force is required on inclines in order to overcome the force of gravity. This force is added to the movement resistance. The required braking force, which depends on the gradient and, consequently, the reliability of the braking system also plays an important role.

To ensure the required additional traction and braking force, two techniques are used:

- The cog railway.
- The traction by cables (cable-propelled systems, see Chapter 10).

The track superstructure of a rack railway consists of two conventional rails plus a toothed rack rail in-between (Figures 9.1 and 9.2). The wheelsets of the powered vehicles are equipped with one or more cog driving wheels that are arranged either horizontally or vertically (Figure 9.3). The vehicles (powered and trailers) are usually equipped with cog braking wheels, which along with the driving wheels mesh with the rack rail in-between the two main rails, and thus ensure the necessary supplementary traction and braking efforts.

This transport system is mostly used in mountainous areas and tourist resorts when the track alignment includes longitudinal gradients greater than 50–70‰.

9.2 CLASSIFICATION OF RACK RAILWAY SYSTEMS

9.2.1 Type of cog rail

During the development of rack railways, there were three cog rail systems that were mainly used. The most commonly used cog system nowadays is the Abt system. The other two well-known systems are the Riggenbach system and the Strub system.

The Abt system was invented by Roman Abt in 1882, a Swiss locomotive engineer (Wikipedia, 2015b). This system limits the discontinuities that arise when the traction effort is applied (Figures 9.4 and 9.5).

DOI: 10.1201/9781003046073-9

Figure 9.1 Track superstructure of a rack railway, Drachenfells, Germany (Photo: N. Pyrgidis.)

Figure 9.2 Track superstructure of a rack railway, OSE, Kalavryta, Greece. (Photo: A. Klonos.)

The Riggenbach system was invented by Niklaus Riggenbach in 1863. The system is more complex and expensive to build than the other systems (Wikipedia, 2015b). However, it can sustain greater forces while effectively keeping the wear at a minimum level.

Figure 9.6 shows a cog wheel using the Riggenbach system, while Figure 9.7 illustrates a track equipped with the Riggenbach system.

The Strub system was invented by Emil Strub in 1896. It is the simplest of all rack systems in use and has become increasingly popular (Wikipedia, 2015b).

Figure 9.8 shows a cog wheel using the Strub system, while Figure 9.9 illustrates a track equipped by the Strub system.

Figure 9.3 Driving axle of a powered rack railway vehicle, OSE, Greece. (Photo: A. Klonos.)

Figure 9.4 Cog wheel used on Abt system. (Adapted from Softeis in Deutschen Museum, 2004. Online image available at: en.wikipedia.org/wiki/Rack_railway (accessed 8 August 2015).)

Figure 9.5 The Abt cog system, OSE, Kalavryta, Greece. (Photo: A. Klonos.)

Figure 9.6 Cog wheel used on Riggenbach system, Switzerland. (Photo: A. Klonos.)

Among other less used rack systems are Locher (also known as Punch system) (Figure 9.10) (Wikipedia, 2015a), Lamella (also known as the Von Roll system; Figure 9.11), Riggenbach Klose, Marsh, and Morgan (Figure 9.12).

The Locher 'herringbone' system is implemented with a pair of opposed horizontal pinions engaging teeth cut into the sides of the rack, so that it does not depend on gravity for the engagement of the teeth, and can be used on extreme longitudinal gradients. The sole example is the Pilatus cog railway, Switzerland, 1885, with a maximum gradient of 480‰. A similar system is that of the Krasnoyarsk ship lifting line in Russia. Finally in Panama, a special system that resembles the Riggenbach system is used (Treidellokstrecken).

Overall, out of a total of 59 rack railways in service (2019 data), the most frequently used system is the Abt system (37.3%), followed by Riggenbach system (27.11%), and the Strub system (22%).

Figure 9.7 The Riggenbach cog system, Arth Goldau, Switzerland. (Photo: A. Klonos.)

Figure 9.8 Cog wheel used on Strub system, Switzerland. (Photo: A. Klonos.)

The Lamella system is found in five lines, while the Locher and Marsh systems are found in one line (Pilatus Bahn and Mount Washington Cog Railway, respectively). In specific lines, a combination of different cog rail systems, which is only achievable through the use of Riggenbach, Strub, and Lamella systems, is observed (Jehan, 2003).

All the above data and the data recorded and analysed in the following relate to the year 2019. The raw data were obtained both per country and per line, from various available sources, and were cross-checked (Syofiardi Bachyul, 2009; Wikipedia, 2015a- c; Bruse's Funiculars Net, n.d.). Afterwards, they were further manipulated for the needs of this chapter.

Figure 9.9 The Strub cog system, Chamonix, France. (Photo: A. Klonos.)

Figure 9.10 Cog wheel used on Locher system, Switzerland. (Photo: A. Klonos.)

9.2.2 Type of adhesion along the line

The rail systems using the rack/cog rail are split into two categories based on the type of adhesion:

- Purely toothed (racked) tracks.
- Mixed adhesion operation (dual rack and adhesion system).

The first category consists of all the tracks in which the rolling of vehicles throughout the entire line is ensured by artificial adhesion (rack rod/rail). These lines typically include comparatively very high longitudinal gradients, serve much shorter distances, and usually connect mountainous tourist resorts with the flat areas below them.

Figure 9.11 The Lamella cog system, Capolago, Switzerland. (Photo: A. Klonos.)

Figure 9.12 The Morgan cog system. (Adapted from Goodman Manufacturing Company, 1919, *Goodman Mining Handbook for Coal and Metal Mine Operators, Managers, Etc.* Online image available at: en .wikipedia.org/wiki/Rack_railway (accessed 8 August 2015).)

The second category comprises the lines that include track sections of very steep gradients, in which the traction of vehicles is ensured additionally by the use of a rack rail, but also sections where rolling is ensured through conventional adhesion only (Machefer Tassin, 1971; Dunn, 1980; Avenas, 1984). This category covers a wider range of lines (touristic, urban, etc.).

For these networks, special systems have been designed for the rolling stock, which allow an easy transition from two to three rails while the train is in motion as inclines increase (Carter and Burgess, Inc., 2006).

From a total of 59 rack railway systems in service worldwide, 27 (2019 data) are dual rack and adhesion systems, while 32 are purely racked (Table 9.1).

Table 9.1 Number of cog railway systems in service per continent and per country (2019 data)

Continent	Country	Number of cog railway systems with purely rack tracks	Number of dual rack and adhesion systems	Total
Europe	Austria	2	1	3
	Czech Republic	0	1	1
	France	3	2	5
	Germany	2	2	4
	Greece	0	1	1
	Hungary	1	0	1
	Italy	2	1	3
	Slovakia	1	1	2
	Spain	0	2	2
	Switzerland	13	11	24
	Russia	1	0	1
	UK	1	0	1
America	Brazil	1	1	2
	Panamas	0	1	1
	USA	3	0	3
Asia	India	0	1	1
	Indonesia	1	0	1
	Japan	0	1	1
Oceania	Australia	1	1	2
Total		32	27	59

9.3 EVOLUTION OF THE SYSTEM AND APPLICATION EXAMPLES

The very first cog railway was opened in 1869 on Mount Washington in New Hampshire, USA, and still operates with its original steam locomotives. The idea to build a cog railway up to the top of the mountain was attributed to Sylvester Marsh in 1857.

Rigi-Bahnen (Rigi railways) in Switzerland opened in 1873 as the first cog railway to operate in Europe (Jungfrautours, n.d.).

By the end of the nineteenth century, rack railways had spread rapidly all over Europe and America. The main reason for this rapid expansion was the fact that at that time, conventional steam locomotives were not able to pull heavy loads on gradients higher than 45‰. This was to change in the next century with the development of advanced electric locomotives, which could cope with inclinations of up to 100‰ without the need for rack wheels. Thus, the majority of rack railway lines were consequently replaced, closed down, or continued operation for touristic reasons; mostly those of relatively low longitudinal gradients.

During the last 30 years, only three new rack railway lines have been built and put into operation:

- The Quincy and Torch Lake Cog Railway (Hancock–Quincy Mine's), in the United States, 701 m long (1997).
- The Skitube Alpine Railway (Bullocks Flat–Perisher Valley–Blue Cow) in Australia, 8,500 m long (1988).
- The Panoramique des Domes line in France, 5,200 m long (May 2012).

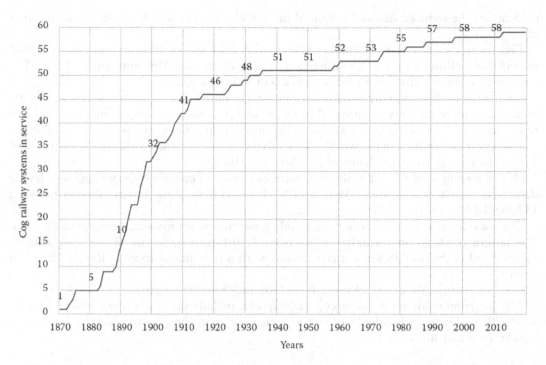

Figure 9.13 The number per decade of rack railway systems that were put into service (end 2019).

It has to be noted that:

- Some lines were replaced by a conventional adhesion track. The most recent examples are the Paola–Cosenza line in Italy (1987) and the St. Gallen–Appenzell line in Switzerland (2018).
- Some lines that had seized operation were refurbished and commenced operation once again (for example, the Montserrat line in Barcelona closed in 1957 and recommenced operation in 2003).
- The line Pikes Peak Cog Railway in the United States will remain closed until 2021.
- No new system is under construction.
- Twenty-four more rack railway systems, which were previously constructed but are not currently in operation, were recorded.

Table 9.1 presents the number of cog railway systems in service in the world, per continent, and per country. The vast majority of these systems are located in Switzerland.

Figure 9.13 depicts the evolution of the 59 rack railway systems in terms of the number of systems per year that commenced their operation (and are still in operation as of 2019 end).

9.4 CONSTRUCTIONAL AND OPERATIONAL FEATURES OF RACK RAILWAY SYSTEMS

9.4.1 Track alignment

The cog rail technique can be applied to all track gauges that are characterised as 'small' track gauges. It performs better in tracks of metric and narrow gauges (1.00, 0.80, and

0.75 m), as the reduced size and weight of the vehicles require smaller traction and braking efforts.

The most commonly used track gauge in the field of cog railway systems is 1,000 mm (51%), followed by 1,435 mm gauge (17%) and, thirdly, 800 mm gauge (12%). The Krasnoyarsk ship lifting line in Russia has a 9,000 mm gauge.

Rack railway lines usually involve a single track and very steep longitudinal gradients, which determine the maximum speed in ascent and descent. The longitudinal gradient can range from 50‰ to as much as 480‰, while usual values lie between 200‰ and 250‰. The steepest cog railway in the world is Pilatusbahn in Switzerland, with maximum gradient 480‰. In station areas, gradients of 20–70‰ are allowed.

Out of a sample of 57 rack railway systems currently in operation for which the relevant data were available (out of the 59 in total), 34 (60%) have a maximum gradient between 15% and 25%.

The total length of the 32 in-service purely rack lines is approximately equal to 202 km. The length of these lines usually ranges from 4,500 m to 6,000 m. Wengernalpbahn in Switzerland is the world's longest cog railway with a continuous toothed line of 19,110 m length.

Dual rack and adhesion tracks usually exhibit much longer lengths.

The horizontal alignment of a cog railway can include curve radii not lower than R_c = 85–90 m. The cant is calculated in the same way as in the case of secondary, local, or regional railway lines.

9.4.2 Track superstructure

Rack railway tracks are usually placed at grade. However, there are also some examples of underground tracks (e.g., the Skitube Alpine line in Australia, with a length of 8.5 km that features two tunnels that are 3.3 km long (Bilson Tunnel) and 2.6 km long (Blue Cow Tunnel), respectively, and the Bavarian Zugspitze Railway).

Given that most tracks are integrated into difficult mountainous terrain, a specific track bed design is needed. The minimum depth of the ballast – measured from the bottom to the underside of the sleepers – is 20 cm. The rails are common wide-foot rails and need to have higher mechanical resistance than conventional rails. All types of sleepers are used. The distance between sleepers follows the same principles as in the case of conventional adhesion tracks. The flange clearance 'σ' must not exceed 5 mm.

The use of toothed tracks necessitates a more sophisticated implementation of switches and crossings (Figures 9.14–9.16).

The construction cost of a purely toothed cog rail single-track infrastructure normally varies between €10 M and €15 M/km (Carter and Burgess, Inc., 2006; http://www.revolvy .com/main/index.php?s=Panoramique).

As previously mentioned, in France in 2012, the 'Panoramique des Domes' line was put into service. The construction of this line commenced in 2010 and completed in 2012. The line has a length of 5,200 m and a gauge of 1,000 mm. It utilises the Strub system that is present throughout its length. For the construction of this line a total of €86 M were spent, which translates to €16.54 M/km (Colona d' Istria, 2012).

9.4.3 Rolling stock

Most purely toothed cog railway trains comprise one or two vehicles, with a total transport capacity of 100–200 passengers (seated and standing) (Figure 9.17).

Figure 9.14 The Strub cog system switch, Wengernalpbahn, Switzerland. (Photo: A. Panagiotopoulos.)

Figure 9.15 The Abt cog system switch. (Adapted from Hapestof, 2009, online image available at: http://cs. wikipedia.org/wiki/Ozubnicov%C3%A1_dr%C3%A1ha (accessed 8 August 2015).)

When designing rolling stock intended to operate on steep longitudinal gradients, particular effort is put into minimising the weight of the vehicles.

The power vehicles are equipped with toothed wheels to enforce traction and braking. The trailer vehicles have at least one toothed wheel per axle to ensure proper braking and to avoid slipping when moving downhill. Moreover, the power vehicles are placed, for safety reasons, either in the rear when ascending (pushing the other vehicles), or in the front when descending (pulling the other vehicles).

Figure 9.16 The Riggenbach cog system switch. (Adapted from Hapestof, 2009, online image available at: http://cs.wikipedia.org/wiki/Ozubnicov%C3%A1_dr%C3%A1ha (accessed 8 August 2015).)

The safety operation of the system necessitates a reliable braking system, capable of functioning properly under any climatic or operational conditions. Therefore, rolling stock intended for high longitudinal gradients is equipped with:

- A 'main' brake, which enforces constant speed when descending.
- Two independent brakes that apply on the driving toothed wheels or on the braking wheels and which can effectively immobilise the vehicle in the most adverse section of the track.
- A braking system that is automatically activated once the train speed exceeds the allowed speed limit. This system is indispensable in lines with gradients i > 125‰.
- A system that allows the automatic braking of the powered vehicles, while traction is still applied.
- A supplementary safety brake which, in case of steep acclivities, prevents the gravity-triggered backward movement of the vehicles.

In certain cases of very steep gradients, the passenger seat level – for obvious reasons – is inclined with regard to the track level (Figure 9.18).

The touristic nature of most cog railways requires increased visibility from inside the vehicles over the picturesque landscapes and highlands that they cross. This need for panoramic views translates into much larger window surfaces than in conventional vehicles.

All modes of traction have been used in rack railways: steam, diesel, and electric.

In case of electric traction, the overhead catenary would require special attention during inclement weather that could increase the cost of this alternative. Diesel units do not feature this problem, but may have additional problems, such as additional noise and air pollution (Carter and Burgess, Inc., 2006).

However, electric traction features a major advantage in comparison with diesel traction. In diesel traction, the braking equipment is rapidly discarded because of the sheer amount

Figure 9.17 Electric cog trainset with two vehicles, Drachenfells, Germany (Photo: N. Pyrgidis.)

Figure 9.18 Seat arrangement in cog railway vehicles in case of very steep gradients, Pilatusbahn, Switzerland. (Photo: A. Klonos.)

of energy consumed in order to retain constant speed when descending. In electric traction though, this problem is eliminated, since it is possible to recuperate part of the energy that is wasted during braking.

Historically, the first electric cog railway was constructed in Salève (1891–1893) by the engineer, Thurry. The oldest steam cog railway in Europe is Achenseebahn, Austria (opened in 1889).

Out of a total of 59 rack railway systems in service (2019 data), the majority are exclusively electrically powered (67.2%). Out of these systems, four use three-phase current. Finally, six systems are exclusively steam loco-hauled.

Biodiesel locomotives operate in Mount Washington and Nilgiri Mountain Railway (India). The cost of cog railway rolling stock is very high (€3–€5 M per train) (2014 data).

9.4.4 Operation

Cog railway systems can be divided into two categories from the operation point of view. The first is related to providing year-round operation for daily passenger transport. The second category refers mainly to tourist traffic, where in some cases the cog railway itself is the attraction.

According to 2019 data, 90% of cog rail systems operate throughout the whole year, and the remaining operate seasonally (usually from March to October).

Purely toothed cog railways usually operate at speeds of $V = 15$–25 km/h (maximum speed can go up to $V_{max} = 40$ km/h) when ascending. When moving downhill, however, safety restrictions impose lower speeds. The commercial speed of a cog railway system varies from $V_c = 7.5$–20 km/h.

In dual rack and adhesion systems, trains run at $V = 20$–30 km/h on the toothed sections (speeds up to 40 km/h have been recorded). Maximum speeds of $V_{max} = 75$ km/h have been recorded on the conventional track sections.

Figure 9.19 illustrates the allowed descending speed as a function of the longitudinal gradient (‰). Curve 2 corresponds to modern-type vehicles, curve 1 with old-type vehicles, and curve 3 relates to 2-loco traction (dual-power vehicles) (Loosli, 1984).

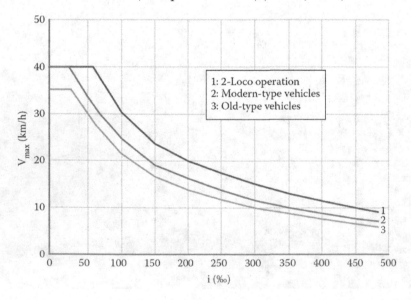

Figure 9.19 Maximum allowed speed V_{max} for cog railway in descent, as a function of longitudinal inclination i of the track (‰). (Adapted from Loosli, H., 1984, *Le chemin de fer à crémaillère – ses particularités et domaines d'application*, Revue Technique Sulzer, No 2, pp. 17–20.)

The frequency of the service depends mainly on the demand and on the kind of services offered.

Rack railway provides the possibility of intermediate stops; the transport system capacity is a function of:

- The connection length S.
- The commercial speed V_c.
- The vehicle transport capacity C_v.
- The dwell time t_{ts} at the two terminal stations.

Considering that S = 6,000 m, V_c = 15 km/h, C_v = 200 passengers, and t_{ts} = 6 min in each terminal, the system's transport capacity is calculated at 400 passengers/h/direction.

Table 9.2 provides the main characteristics of some purely racked lines indicatively.

9.5 ADVANTAGES AND DISADVANTAGES OF RACK RAILWAY SYSTEMS

9.5.1 Advantages

- Cog railway operation is not affected by climatic conditions and/or by the weather, and it is also considerably environmentally friendly.

Table 9.2 Main characteristics of purely racked lines (indicatively)

Line/country	Cog system	Length (m)	Traction system	Max gradient (%)	Track gauge (mm)
Petit train de la Rhune (France)	Strub	4,200	Electric (3-phase)	25	1,000
Stuttgart Rack Railway (Germany)	Riggenbach	2,200	Electric	17.8	1,000
Schafberg Railway (Austria)	Abt	5,850	Steam + diesel	26	1,000
Jungfraubahn (Switzerland) (the rack railway that operates at the greatest height in Europe)	Strub	9,300	Electric (3-phase)	25	1,000
Pilatus Bahn (Switzerland) (the steepest rack railway in the world)	Locher	4,800	Electric	48	800
Wengernalpbahn (Switzerland) (the longest continuous rack railway in the world)	Strub + Lamella	19,110	Electric	25	800
Snowdon Mountain Railway (UK)	Abt	7,600	Steam + diesel	18.2	800
Fogaskerekű Vasút (Hungary)	Strub (specific profile)	3,700	Electric	11	1,435
Štrbské Pleso–Štrba (Slovakia)	Lamella	4,757	Electric	15	1,000
Principe-Granarolo (Italy)	Riggenbach	1,130	Electric	21.4	1,200
Corcovado Train (Brazil)	Riggenbach	3,800	Electric (3-phase)	33	1,000
Mount Washington Cog Railway (USA)	Marsh	4,800	Steam + biodiesel	37.4	1,422
Skitube Alpine Railway (underground 5.9 km) (Australia)	Lamella	8,500	Electric	12.5	1,435

Especially for the dual rack and adhesion system:

- It is suitable for lines with varying longitudinal gradients, as it offers the possibility of operating both in conventional and artificial adhesion. This significantly reduces the amount of earthworks required and allows for the design of a line that offers a good cost/utility ratio.
- It allows for connection with a pre-existent network without excluding the possibility of future extension.

9.5.2 Disadvantages

- The cost of rolling stock is high due to the specialised equipment.
- The running speeds and the transport capacity of cog railway are significantly smaller when compared with conventional railways.
- It requires specialised staff for operation and maintenance.
- The braking system is complicated.
- Switches and crossings display technical difficulties in realisation.

9.6 REQUIREMENTS FOR IMPLEMENTING THE SYSTEM

In general, purely toothed cog railway is preferred when considering distances of S = 4–20 km, with constant longitudinal gradients of i = 50‰–250‰ (max 480‰) and relatively high transport demand. It is mostly used for passenger transport and, in very few cases, also for freight. Its contribution is mainly directed towards leisure activities (tourism, excursions).

Alternatively, the problem of steep longitudinal gradients can be solved by cutting through the cliff mass (tunnel), or by helical or spiral track alignment design (e.g., St. Gotthard Pass in Switzerland), or by use of successive course-reversing stubs (switchback-design type, 'z'), as in the case of the rail pass through the Andes in Peru.

Dual rack and adhesion systems are preferred for connections of S = 10–50 km, which include sections of high longitudinal gradients that normally limit conventional adhesion. In this case, a mixed adhesion operation very often provides a more efficient solution economy-wise, than changing the track alignment or introducing more powerful locomotives.

REFERENCES

Avenas, J. 1984, *Le métro de Lyon*, La vie du rail, No. 1971, Décembre 1984.
Bruse's Funiculars Net. n.d., *Funicolare Principe – Granarolo*, available online at: http://www.funi culars.net/line.php?id=342 (accessed 17 May 2012).
Carter and Burgess Inc. 2006, *Little Cottonwood Canyon Transit Analysis*, available online at: http: //wfrc.org/Previous_Studies/2006%20Little%20Cottonwood%20Canyon%20SR210%20 Transp%20Study%20Aug06/9%20Appendices.pdf. (accessed 7 August 2015).
Colona d' Istria, G. 2012, Le Panoramique des Dômes boit l'eau !, *Le point*, available online at: https ://www.lepoint.fr/societe/le-panoramique-des-domes-boit-l-eau-27-05-2012-1465761_23.php.
Dunn, J. 1980, *Modern Trains*, New English Library, London.
Goodman Manufacturing Company. 1919, *Goodman Mining Handbook for Coal and Metal Mine Operators, Managers, Etc.* Online image, available online at: en.wikipedia.org/wiki/Rack_r ailway (accessed 8 August 2015).
Hapestof. 2009, Online image, available online at: http://cs.wikipedia.org/wiki/Ozubnicov%C3%A1 _dr%C3%A1ha (accessed 8 August 2015).

Jehan, D. 2003, *Rack Railways of Australia*, 2nd edition, Illawarra Light Railway Museum Society, ISBN 0-9750452-0-2, New South Wales.

Jungfrautours. n.d., *A Summary of Exceptional Mountain Railway and Cable Car Facts*, available online at: http://www.jungfrautours.ch/index.php?option=com_content&view=article&id=73&I temid=325&lang=en (accessed 26 July 2012).

Loosli, H. 1984, *Le chemin de fer à crémaillère – ses particularités et domaines d'application*, Revue Technique Sulzer, No. 2, pp. 17–20.

Machefer Tassin, Y. 1971, *A propos de crémaillère … une solution Nippo – Brésilienne originale*, La vie du rail, No. 1971, Décembre 1984, pp. 8–11.

Online image, available online at: http://www.revolvy.com/main/index.php?s=Panoramique%20des%20 D%C3%B4mes%40wiki%3aPanoramique_des_D%C3%B4mes (accessed 20 May 2015).

Softeis in Deutschen Museum. 2004, Online image, available online at: en.wikipedia.org/wiki/Rack_railway (accessed 8 August 2015).

Syofiardi Bachyul Jb. 2009, PT KA set to revive Padang-Sawahlunto railway line, *The Jakarta Post*, available online at: http://www.thejakartapost.com/news/2009/01/16/pt-ka-set-revive-padangs awahlunto-railway-line.html (accessed 17 May 2012).

Wikipedia. 2015a, *List of Funicular Railways*, available online at: http://en.wikipedia.org/wiki/List_of_funicular_railways (accessed 11 April 2015).

Wikipedia. 2015b, *Rack Railway*, available online at: en.wikipedia.org/wiki/Rack_railway (accessed 11 April 2015).

Wikipedia. 2015c, *Steep Grade Railway*, available online at: en.wikipedia.org/wiki/Steep_grade_r ailway (accessed 11 April 2015).

Chapter 10

Cable railway systems for steep gradients

10.1 DEFINITION AND DESCRIPTION OF THE SYSTEM

For movement on longitudinal gradients that are greater than gradients on which a conventional railway may climb, apart from the rack rail technique, cable hauling can also be used.

On the basis of the technique that is applied for vehicles' traction, cable railway systems for steep gradients are classified into three categories:

- Funicular (non-detachable cable-propelled vehicles for steep gradients).
- Cable railway (detachable cable-propelled vehicles for steep gradients).
- Inclined elevator.

The funicular (or inclined plane, or inclined railway or cliff railway) (Figure 1.37) operates using two vehicles which move on rails with the aid of a cable; one of the vehicles is ascending while the other one is descending. The cable rolls over pulleys mounted on the track (Figures 10.1 and 10.2). The smooth movement of the cable and of the whole system is ensured by means of an electric motor, placed at the highest point of the network. The vehicles are permanently connected to both ends of the cable and they start and stop simultaneously. The ascending vehicle uses the gravitational force of the descending one (counterbalance system). This system connects distances $S < 5$ km, which have continuous gradients that usually vary between 300‰ and 500‰ (maximum recorded gradient: $i_{max} = 1,100‰$).

The cable railway also uses vehicles which run on conventional rails with the aid of a cable which is moving constantly and at a constant speed. The difference between the two systems is that for the cable railway, the vehicles can be detached from the traction cable as they are not permanently connected to it. The vehicles can stop independently by detaching from the cable, and they may start again by reattaching to the cable. This process can be done automatically and manually (San Francisco cable-propelled system, USA).

The inclined elevator (or inclined lift or inclinator) is a variant of the funicular. It operates using a single vehicle. The vehicle is either winched up at the station at the top of the inclined segment where the cable is rolled around a winch drum, or the weight of the vehicle is balanced by a counterweight, hence the system operates as a funicular. It usually connects distances $S < 1.5$ km (maximum recorded value $S_{max} = 1.36$ km) that have large continuous gradients (typically 400–700‰) which may reach significantly high values (systems Eiffel Tower, Paris, France: $i_{max} = 4,761‰$; Olympic Stadium, Montreal, Canada: $i_{max} = 1,880‰$; Katoomba Scenic, Australia: $i_{max} = 1,288‰$).

The oldest cable railway system for large longitudinal gradients in the world is the inclined elevator of Reisszug (Wikipedia, 2015a). This is a private single track with a length of $S = 190$ m and a gradient of $i_{max} = 670‰$ which serves the transport of people (three passengers) and goods (2,500 kg) to the Hohensalzburg Castle in Salzburg, Austria. It was first

DOI: 10.1201/9781003046073-10

Figure 10.1 Cable-guiding pulleys, Barcelona, Spain.

Figure 10.2 Cable rolling through pulleys, Saint Moritz, Switzerland.

documented in 1515. The track originally used wooden rails and a hemp hauling rope, while the mechanism itself was being operated by horses or people. Today steel rails, steel cables, and an electric motor have replaced the old technology, but the track still follows the same route.

The newest cable railway system in the world is the funicular Stoosbahn in the Swiss canton of Schwyz (Wikipedia, 2015b, c; Bruce's Funiculars Net, n.d.). It connects the Hintere Schlattli with the village and mountain resort of Stoos, above Morschach. It opened in December 2017 and replaces the older Drahtseilbahn Schwyz–Stoos funicular, operating since 1933. The new line has a length of 1.74 km and a maximum gradient of 1,100‰ (47.7°). It is the steepest funicular railway in the world. It can serve 1,500 passengers/h/direction and can develop maximum speeds of 26 km/h (run time 4–7 min) (Standseilbahnen, 2015).

10.2 THE FUNICULAR

10.2.1 Evolution of funiculars and application examples

Modern funicular railways operating in urban areas date from the 1860s. The first line of the funiculars of Lyon opened in 1862, followed by other lines in 1878, 1891, and 1900. The Giessbachbahn railway constructed in 1879 in Switzerland was Europe's first funicular railway used for transporting passengers.

Based on 2019 data, a total of 248 funicular systems are recorded in the world. Table 10.1 presents the number of 'funiculars' per continent and per country. The majority of 'funiculars' in operation are in Switzerland (50), followed by Japan (22) and Italy (21).

All the above data and the data recorded and analysed in the following sections relate to the year 2019. The raw data were obtained both per country and per line, from various available sources, and cross-checked. Thereafter, they were further manipulated for the needs of this study.

Figure 10.3 depicts the evolution of the 248 funiculars in terms of the number of systems per year that commenced their operation (and are still in operation as of the end of 2019).

10.2.2 Constructional and operational features of funiculars

10.2.2.1 Infrastructure

The operation of all funicular railway systems for steep gradients is performed using the shuttle 'principle' (see Section 7.2.2.1).

Regarding the track superstructure configuration, it is either a single-lane shuttle with bypass type or a dual-lane shuttle type.

In the first case, the system operates using two vehicles which intersect while moving between terminals, along a single track, using a pull cable. The passing loop is located halfway.

In the second case, the system operates using two vehicles which move between terminals, along a double track without passing loop, using a pull cable.

In both cases, the counterbalance system is used, and the intermediate stations (0–3) are located at symmetrical distances.

With regard to the number of rails used to form the track superstructure, funiculars can be classified into three categories:

Two-rail superstructure configuration with passing loop (Figures 10.4c, 10.5, and 10.6): This type of superstructure does not require switches and crossings. The vehicles feature specially arranged wheels (see Figure 10.4c). More specifically, only the wheels of one side

Table 10.1 Number of 'funiculars' per continent and per country (2019 data)

Continent	Country	Number of 'funiculars'
Europe (175)	Austria	11
	Azerbaijan	1
	Czech Republic	3
	Croatia	1
	France	16
	Georgia	1
	Germany	14
	Greece	1
	Hungary	1
	Italy	21
	Lithuania	2
	Luxembourg	1
	Norway	2
	Poland	3
	Portugal	8
	Romania	1
	Russia	2
	Slovakia	1
	Spain	11
	Sweden	2
	Switzerland	50
	Turkey	3
	Ukraine	1
	United Kingdom	13
	Luxembourg	1
	Wales	5
North America (14)	Canada	2
	United States	12
South and Central America (19)	Argentina	1
	Brazil	6
	Chile	10
	Colombia	1
	Mexico	1
Asia (38)	China	4
	Hong Kong	2
	India	2
	Israel	1
	Japan	22
	Lebanon	1
	Malaysia	1
	Thailand	3
	Vietnam	2
Africa (1)	South Africa	1
Oceania (1)	New Zealand	1
Total		248

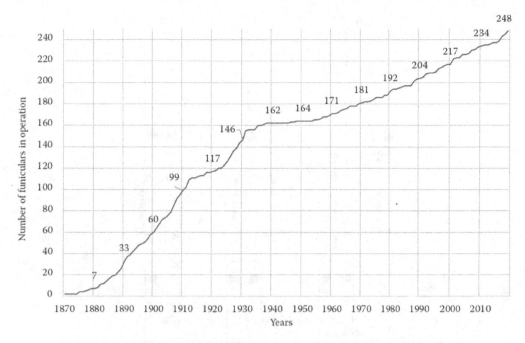

Figure 10.3 The number per decade of funiculars that were put into service (end of 2019).

of the wheelset feature a flange which is in fact double. The intersecting vehicles have their wheels with a flange at opposite sides, and this allows them to follow different tracks. This configuration significantly reduces costs.

Three-rail superstructure configuration with passing loop (Figures 10.4b and 10.7): In superstructure layouts using three rails, the middle rail is shared by both cars. The superstructure is wider than the two-rail layout, but the passing loop section is simpler to build.

Four-rail superstructure configuration (Figures 10.4a and 10.8): In some four-rail funiculars (Figure 10.8), the upper and lower sections are interlaced while having a single platform at each station.

Table 10.2 presents the basic constructional and operational characteristics of *funiculars*.

Funiculars operate on tracks with various gauge values. The most common gauge values are 2e = 1,000 mm and 1,067 mm.

In the Kintetsu Ikoma funicular line in Nara, Japan, both two-rail and four-rail configurations are combined in the track superstructure.

As regards the horizontal alignment, it may either be straight or have curved sections. The curve radii of the horizontal alignment that have been recorded are no less than $R_c = 85$–90 m.

Funiculars, according to a sample of 238 systems, typically (over 68%) have a length shorter than 1 km. According to the same sample, the mean length is 814 m. Only four funiculars were found to have a length greater than 3 km and to be still in service (end of 2019 data).

Figure 10.9 depicts the evolution of the funiculars (a sample of 238 systems) over the years in terms of the total length of their lines.

The total length of the funicular lines in service, is around 200 km (end of 2019).

The three larger (in terms of length) lines are:

- Flattach Molltaler Gletscher, in Austria (S = 4,827 m).
- Sierre-Montana-Cranz, in Switzerland (S = 4,192 m).
- Pitzaler Gletscherbahn, in Switzerland (S = 3,693 m).

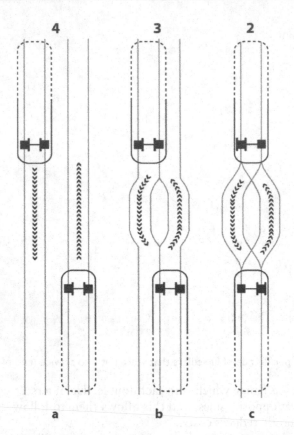

Figure 10.4 Track superstructure configurations of 'funiculars'. (a) Four-rail superstructure configuration. (b) Three-rail superstructure configuration with passing loop. (c) Two-rail superstructure configuration with passing loop. (Adapted from Cmglee at English Wikipedia., 2010, Online figure available at: http://en.wikipedia.org/wiki/Funicular.)

Regarding the maximum gradient, the values range from 80‰ and up to 1,100‰. Most systems (34%) have a gradient between 300‰ and 500‰. The mean value of the maximum gradient is 454‰.

The three steepest lines are:

- Stoosbahn, in Switzerland, (i_{max} = 1,100‰).
- Hungerburgbahn, in Austria, (i_{max} = 1,036‰).
- Reina Victoria, in Chile (i_{max} = 1,010‰).

The vast majority of funicular lines are constructed at grade. However, underground systems also exist (e.g., Pitzaler Gletscherbahn in Tirol, Austria; Val Gardena Ronda Express in Italy; Carmelite in Israel; and Metro Alpin Saas-Fee in Switzerland). One line, the Schwebebahn, in Dresden, Germany, is suspended.

The track superstructure can be ballasted, made on concrete, or made on steel. The rails are mounted on the track superstructure with special rail fastenings at intervals of about 80 cm.

The total implementation cost of a funicular system varies between €20 M and €30 M/track-km (2017 data) (Audit Scotland, 2009; *The Herald*, 2014; *The Guardian*, 2017).

Figure 10.5 Two-rail superstructure configuration with passing loop (single-lane shuttle with Bypass), Salzburg, Austria. (Photo: A. Klonos.)

Figure 10.6 Two-rail superstructure configuration with passing loop (single-lane shuttle with Bypass), Cape Town, South Africa.

Figure 10.7 Three-rail configuration funicular, Angel flight in Los Angeles, USA. (Adapted from Sullivan, J., 2005, Online image available at: http://www.10best.com/interests/16-fantastic-funiculars/ (accessed 8 August 2015).)

Figure 10.8 Four-rail configuration funicular (dual-lane shuttle), Valparaíso, Chile. (Adapted from Bahamondez, M., 2006, Online image available at: https://commons.wikimedia.org/wiki/File: Ascensores_de_Valparaiso.jpg (accessed 8 August 2015).)

Table 10.2 Basic constructional and operational characteristics of 'funiculars'

Connection length	Usually $S < 1,000$ m, $S_{min} = 39$ m, $S_{max} = 4,827$ m
Running speed	Usually $V < 20$ km/h, $V_{min} = 3.6$ km/h, $V_{max} = 50.4$ km/h
Longitudinal gradient	Usually $i = 300–500‰$ $i_{min} = 80‰$, $i_{max} = 1,100‰$
Track gauge	Various (usually $2e = 1,000$ mm and $1,067$ mm)
Track superstructure configuration	• Single track – 2 rails – passing loop • Double track – 3 rails – passing loop • Double track – 4 rails – without passing loop
Vehicle transport capacity	10–420 passengers, usually 50–80
Traction cable gripping	Permanently attached
Number of cars	2 cars – one ascending, one descending

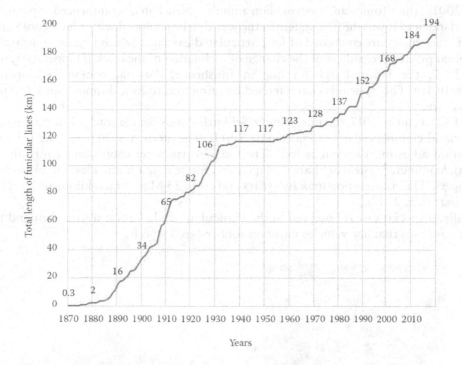

Figure 10.9 Evolution funiculars around the world in terms of the total network length.

10.2.2.2 Rolling stock

The main principle of 'funicular' operation is that the two cars are permanently attached to each other by a cable which runs through a pulley at the top of the incline (Figure 10.4). Counterbalancing of the two cars, with one ascending and one descending, minimises the energy needed to lift the ascending car. The cars can be attached to a second cable running through a pulley at the bottom of the incline in case the applied gravity force is insufficient to activate the mechanism and operate the vehicles on the slope.

'Funiculars' rank amongst the safest means of transport. This is due to the many safety devices used (Doppelmayr, n.d.).

Braking is ensured through:

- A 'service' brake, which can bring the entire system to a standstill. It is applied in some specific cases: power failure, insufficient voltage, and so on.

- A 'security' brake, which can immobilise the entire system at any point along the route. This brake is activated in case of cable sliding or slipping, violation of the vehicle's speed restriction, opening of the door while moving, and so on.
- Two 'auxiliary' brakes, which are automatically applied if the hauling cable brakes, or if the vehicle's speed exceeds a certain threshold.

The vast majority of 'funiculars' use electricity as an energy source, but there still exist funiculars powered by water (e.g., Elevador do Bom Jesus, Braga, Portugal; Nerobergbahn-Wiesbaden, Germany) or operated mechanically (e.g., Siclau, Romania, Greenwood Forest Park, Wales).

In 2000, the world's first 'funicular' powered by solar energy was put into operation, in Livorno, Italy (Montenero line) (Museo Galilleo, n.d.).

In 2001, the 'funicular' system Fun'ambule, Neuchatel, commenced operation in Switzerland, and it was the first system in the world to be equipped with a level compensation system. Its vehicles are composed of four articulated carriages which remain constantly in a horizontal position regardless of the change of inclination of the track (Doppelmayr, n.d.).

In 2007, the Hungerburgbahn line in Innsbruck, Austria, commenced operations (Figure 10.10). This system is characterised by numerous technical innovations and unique architecture.

On 15 December 2017, the line Schwyz/Schlattli–Stoos (single track of normal gauge) in Switzerland commenced operation (Figure 10.11). The Stoos funicular is designed with an inclination adjustment system as well. The four 34-passenger cabins on each train of this touristic/commuter system are barrel-shaped and rotate to maintain a level floor surface for passengers. The construction took five years and cost €25 M per track-kilometer (2017 data) (*The Guardian*, 2017).

Finally, a special case is observed in the Funiculaire de la Grotte d'Aven Armand line in France, since on that line vehicles move on rubber-tyred wheels.

Figure 10.10 Hungerburgbahn, Innsbruck, Austria. ((a) Adapted from Innsbrucker Nordkettenbahnen Betriebs GmbH, 2008, *The Hungerburg funicular*, online, available at: http://www.nordkette .com/en/cable-railways/history/the-hungerburg-funicular.html (accessed 26 July 2012); (b) www.guentheregger.at 2013.)

Figure 10.11 The Stoos funicular system in Switzerland (Adapted from Pakeha, 2017, On line image available https://en.wikipedia.org/wiki/Stoosbahn#/media/File:Betriebsaufnahme_der_neuen_Stoosbahn,_Dezember_2017_@(2).jpg (accessed 10 April 2020).)

10.2.2.3 Operation

Cable railway systems operate on constant speed, which, in the case of 'funiculars', ranges from 3.6 km/h to 50.4 km/h (Wurzeralmbahn, Austria, i = 300‰). In most cases, V < 20 km/h. Concerning the vehicle's passenger transport capacity, funiculars can accommodate up to 420 persons (Vomero–Centrale, Naples, Italy); however, the usual figures for most systems range between 50 and 80 passengers.

The maximum transport system capacity is recorded in the Naples Chiaia Funicular in Italy with a value of 9,500 passengers per hour and direction, while for a sample of 90 systems, for which data were available, the average transport system capacity is 1,445 passengers per hour and direction.

Transport capacity depends on the connection length S, the vehicle's running speed V, the vehicle's passenger transport capacity C_v and the dwell times at the terminal stations t_{ts}.

For a service that has no intermediate stops, and for S = 1,000 m, V = 15 km/h, C_v = 100 passengers and t_{ts} = 2 min in each terminal, the system's transport capacity results in 1,000 passengers/h/direction.

10.3 THE INCLINED ELEVATOR

The operation of inclined elevators is performed using the shuttle 'principle' (see Section 7.2.2.1).

Regarding the track superstructure configuration, it is either a single-lane shuttle or a dual-lane shuttle type.

In the first case, the system operates using one single vehicle which moves between terminals, along a single track, without any passing loop, using a single pull cable. The total number of stations is two, that is, the two terminals, but it is also possible to have up to three intermediate stations.

In the second case, the system operates using two vehicles which move independently from one another on a single track each, without intersection, using a single pull cable.

It is, in essence, a double single-lane shuttle. This configuration ensures a high level of service, while off-peak, only one train may operate as a single-lane shuttle. There are only a few systems in the world that operate using this system. Some classic examples are the system of Montmartre, Paris, France (Figure 10.12), the Universeum system in Gothenburg, Sweden, the Odessa Funicular system in Ukraine, and the system in Spa, Belgium.

During its 1991 renovation, the Montmartre funicular railway was converted from a traditional four-rail funicular railway where the two cabins counterweigh each other, to two totally independently operated inclined lifts. This allows one cabin to remain in service if the other must be taken out of service for maintenance. The two cabins are completely automatic in operation, using the weight of the passengers as a determining factor for when it is time to depart. Therefore, especially at busy times, it can happen that both cabins will be travelling in the same direction at the same time (Wikipedia, 2015d).

Table 10.3 summarises the main constructional and operational characteristics of inclined elevators.

The longest inclined elevator system is recorded in Switzerland. Specifically, the Piotta–Ritom Funicular has a length of 1,369 m. Typically, such systems have lengths shorter than 300 m.

Like with funiculars, metric gauges (1,000 mm and 1,067 mm) are more common.

The Lärchwand–Schrägaufzug line in Austria features the largest width among all 'inclined elevator' lines in the world (width equal to 8,200 mm).

Usually gradients vary between 400‰ and 700‰.

The recorded speeds of inclined elevators range from 1.8 km/h to 39.6 km/h (Dorfbahn, Tirol, gradient 53.5‰). Usual speed values do not exceed 8 km/h. Vehicle passenger transport capacity usually ranges between 40 and 60 people, but there are cases, such as the Dorfbahn in Tirol, which can carry up to 270 passengers.

For a service without any intermediate stops, and for $S = 300$ m, $V = 8$ km/h, $C_v = 40$ passengers and $t_{ts} = 2$ min in each terminal, the system's transport capacity results in 564 passengers/h/direction.

Figure 10.12 The Montmartre funicular system in Paris, France (double inclined elevator). (Adapted from Breuer, R., 2004, Online image available at: http://en.wikipedia.org/wiki/Montmartre_Funic ular,2014 (accessed 8 August 2015).)

Table 10.3 Main constructional and operational characteristics of inclined elevators

Connection length	Usually $S < 300$ m, $S_{min} = 21$ m and $S_{max} = 1,369$ m
Running speeds	Usually $V < 10$ km/h, $V_{min} = 1.8$ km/h and $V_{max} = 39.6$ km/h
Longitudinal gradient	Significantly high values are recorded. Usually $i = 400\%o–700\%o$
Track gauge	Various
Track superstructure configuration	Single track − 2 rails − without passing loop
Vehicle transport capacity	3–270 passengers (usually 40–60)
Traction cable gripping	Permanently attached
Number of cars	1 car

10.4 ADVANTAGES AND DISADVANTAGES OF CABLE RAILWAY SYSTEMS FOR STEEP GRADIENTS

10.4.1 Advantages

- They can move on very steep gradients.
- They are environment-friendly, regarding both air and noise pollution.
- 'Funiculars' are used as a feeder transport service for urban railway networks at areas with large relief.

10.4.2 Disadvantages

- They provide service for short distances only.
- Transport system capacity is rather restricted and fixed.
- Seamless transit and connection with other systems is not possible.
- System operation requires specialised facilities and rolling stock.

10.5 REQUIREMENTS FOR IMPLEMENTING THE SYSTEM

'Funiculars' are typically implemented when the distances that are to be connected do not exceed 5 km, when there are continuous and steep gradients and a relatively high traffic demand (1,000–2,000 passengers/h/direction).

'Inclined elevators' are designed for even shorter distances (< 1.5 km), with continuously present and very steep gradients and relatively low transport volume (200–700 passengers/h/direction).

In hilly cities, cable-propelled railway systems are used:

- For passenger transport (in some cases, they are considered as public transport systems – e.g., the Petrin funicular in Prague, Czech Republic; the St Jean–Fourvière/St Just line in Lyon, France).
- For the transport of skiers to the ski resorts.
- As 'inclined lifts' to serve hotels' guests.
- For industrial use – power station funiculars are the most commonly used ones (UK, Switzerland).
- Some of the 'funiculars/inclined elevators' constitute attractions at thematic parks themselves.

According to 2019 data, there are three cable railway systems that are currently under construction and are expected to be put into service in the coming years. These are the following:

- Provincetown, Pilgrim Monument Inclined Elevator (Massachusetts)
- Springfield, Pynchon Plaza Elevator (Massachusetts)
- Qiddiya Project Funicular (Saudi Arabia)

The fact that during the last 10 years, 15 new funicular systems have been put in service while many existing ones were renovated/upgraded proves that this transport mode remains in consideration as an alternative for cases of a complex landscape.

REFERENCES

Audit Scotland. 2009, *Review of Cairngorm Funicular Railway, Prepared for the Auditor General for Scotland*, available online at: http://www.audit-scotland.gov.uk/docs/central/2009/nr_0 91008_cairngorm_funicular_railway.pdf (accessed 24 April 2015).

Bahamondez, M. 2006, Online image, available online at: https://commons.wikimedia.org/wiki/File: Ascensores_de_Valparaiso.jpg (accessed 8 August 2015).

Breuer, R. 2004, Online image, available online at: http://en.wikipedia.org/wiki/Montmartre_Funic ular, 2014, (accessed 8 August 2015).

Bruce's Funiculars Net. n.d., *Funiculars Database*, available online at: http://www.funiculars.net/ database.php (accessed 23 January 2021).

Cmglee at English Wikipedia. 2010. Online figure, available online at: http://en.wikipedia.org/wiki/ Funicular (accessed 07 August 2015)

Doppelmayr. n.d., *Funicular Railways*, available online at: http://www.doppelmayr.com/en/products /funicular-railway/brochure/browse-the-brochure-58 (accessed 24 April 2015).

Gunther, E., 2013, *Photography*. Available online at: www.guentheregger (accessed 07 August 2015).

Innsbrucker Nordkettenbahnen Betriebs GmbH. 2008, *The Hungerburg Funicular*, available online at: http://www.nordkette.com/en/cable-railways/history/the-hungerburg-funicular.html (accessed 26 July 2012).

Museo Galilleo. n.d., *Funicolare di Montenero [Funicular of Montenero]*, available online at: http:// brunelleschi.imss.fi.it/itineraries/place/FunicolareMontenero.html (accessed 26 July 2012).

Paheka. 2017, Online image, available online at: https://en.wikipedia.org/wiki/Stoosbahn#/media/Fil e:Betriebsaufnahme_der_neuen_Stoosbahn,_Dezember_2017_(2).jpg (assessed 10 April 2020)

Standseilbahnen. 2015, *6430.01 Schwyz Schlattli – Stoos*, available online at: http://standseilbahnen .ch/schwyz-schlattli-stoos.html (accessed 24 April 2015).

Sullivan, J. 2005, Online image, available online at: http://www.10best.com/interests/16-fantastic-funiculars/ (accessed 8 August 2015).

The Herald. 2014, *Cost of Constructing Funicular Railway*, available online at: http://www.hera ldscotland.com/sport/spl/aberdeen/cost-of-constructing-funicular-railway-1.93349   (accessed 24 April 2015).

Wikipedia. 2015a, *Reisszug*, available online at: http://en.wikipedia.org/wiki/Reisszug (accessed 5 September 2014).

Wikipedia. 2015b, *List of Funicular Railways*, available online at: http://en.wikipedia.org/wiki/L ist_of_funicular_railways (accessed 23 January 2021).

Wikipedia. 2015c, *Liste_der_Standseilbahnen*, available online at: http://de.wikipedia.org/wiki/L iste_der_Standseilbahnen (accessed 23 January 2021).

Wikipedia. 2015d, *Montmartre Funicular*, available online at: http://en.wikipedia.org/wiki/Montma rtre_Funicular (accessed 24 April 2015).

Chapter 11

Organisation and management of passenger intercity railway transport

11.1 SERVICES AND BASIC DESIGN PRINCIPLES OF PASSENGER RAILWAY TRANSPORT

The process of the design of railway services is generally aimed at the definition of the services that will be provided to its customers (Lambropoulos, 2004). It takes place at strategic, tactical, and operational levels:

- At the strategic level (long term), alternatives are examined, and decisions are taken regarding the necessary investment in infrastructure and rolling stock.
- At the tactical level (mid-term), the infrastructure and the rolling stock are considered to be decided upon and the timetables that will be offered to the customers are designed. In addition, the rolling stock providers are determined and issues concerning the staff are addressed.
- At the operational level (short term), timetables are already determined. At this level, emergencies and special incidents (e.g., long delays, equipment failures, track works, accidents, and events) are treated and personnel shifts are determined.

The following services are offered in passenger railway transport (Samuel, 1963; Alias, 1985; Lambropoulos, 2004):

- Urban services (tramway, metro, monorail, low and medium transport capacity railway systems).
- Suburban services.
- Regional services.
- Long-distance services of conventional speed (conventional or tilting trains).
- Long-distance services of high or very high-speed (conventional or tilting trains).
- International passenger services (conventional-speed, high-speed, or very high-speed trains).
- Local services.
- Miscellaneous services (airport links, touristic).

In most of the above cases, the railway infrastructure facilitates more than one of the aforementioned services. A typical example of this is the fact that the same track can accommodate suburban, regional, and intercity trains. The infrastructure manager of such a mixed traffic composition passenger network must organise a combination of services in the

DOI: 10.1201/9781003046073-11

provided infrastructure in a way that all services are supplementing each other while at the same time avoiding any potential conflict.

As regards the design of passenger railway transportation, the following parameters are mainly considered in its definition (Lambropoulos, 2004):

- The type of services offered by the railway system to its potential customers in terms of scheduled services (timetables) and added services.
- The origin and destination stations of itineraries as well as the intermediate stations.
- The departure/arrival times of trains at terminals and intermediate stations.
- The service frequency.
- The type of rolling stock to be used and the train formation.
- Commercial issues such as fares and their structure.

In particular, for timetable construction, it is essential that run times of trains between stations are known in advance. These run times are either calculated (using simulation models) or measured using a chronometer. The hierarchy of trains, namely, the classification of trains based on their priority, also plays an important role. Trains of higher priority are given priority at a track block section over trains of lower priority.

The timetabling procedure for passenger trains may regard:

- A system that is already in operation. This case requires either adjusting to any probable demand changes or offering new 'products' to its clientele.
- A brand new railway link under construction. In this case the target is to ensure the best choices for infrastructure and rolling stock so that the system will:
  - Offer a high level of service to its users.
  - Be economically viable for the operator.
  - Respect the environment.

The following sections provide an analysis of the long-distance passenger services only, as all the other services either have been already analysed or will be analysed in other chapters.

11.2 SERVICE LEVEL OF INTERCITY PASSENGER RAILWAY TRANSPORT: QUALITY PARAMETERS

The assessment of the quality of service for an intercity passenger train service is based on the degree of satisfying the following quality parameters:

- Short run time and, most importantly, a run time that is competitive with that of other means of transport.
- Service frequency.
- Reliability and regularity of scheduled services.
- Relatively low and competitive fares.
- 'Dynamic comfort' for passengers during transportation.
- Comfort while travelling (ample legroom and space, cleanliness and aesthetics of trains, air conditioning).
- Security during transportation.
- Security at railway stations.
- Ease in issuing and purchasing tickets.

- Special services provision, for example, services for accompanying private cars, night trains, restaurant, cafeteria, and other similar services.

For international passenger services, two additional parameters have to be considered:

- Minimise delays at border stations.
- Ensuring interoperability.

The basic criterion for assessing the above parameters is their comparison with the relative parameters of other competitive means of transport.

The conventional-speed intercity services involve connections of large urban centres separated by more than 150 km and potentially for typical services up to 1,000 km (when no train change is involved). The maximum running speeds are V_{max} = 160–200 km/h, whereas commercial speeds are V_c < 150 km/h. Intercity trains are usually scheduled to run every 1–3 h. In these cases, trains have a rather limited number of stops (the average distance between stops ranges from 50 to 150 km), but much emphasis is put on speed and mostly on passengers' comfort (dynamic, space) (Jorgensen and Sorenson, 1997). The halt time of a passenger train at a station is contingent on a series of factors such as the construction characteristics of platforms and trains and the number of passengers boarding and alighting at each station. This time span is usually around 2 min.

The high-speed and very high-speed intercity services involve connections of large urban centres and mainly connections between both edges of high-volume 'bi-pole trips'. They usually serve distances bigger than 300–400 km, the average being around 500 km (e.g., Paris–Lyon, Tokyo–Osaka, and Frankfurt–Hamburg) reaching more than 1,500 km (maximum distance is recorded in China: 2,439 km, Hong Kong–Beijing Xi). In these journeys, the number of stops is even more limited (the average distance varies between 150 and 250 km); however, the choice of stopping points depends on the characteristics of the connection. When choosing the stopping points, the train operator must balance the cost of the stop (lost time, increased energy consumption for acceleration) and the potential gain in passenger volume. A particularly strong emphasis is placed on the reliability and punctuality of scheduled services. The maximum running speeds that trains can achieve are V_{max} = 200–350 km/h, while commercial speeds are V_c > 150 km/h (see Chapters 1 and 12). High-speed trains are scheduled to run usually at intervals of 1–2 h. The rolling stock that is scheduled for a connection is specially designed and manufactured, as is the railway infrastructure and the signalling system.

According to 2020 data, and based on the definition that was provided in Section 1.4.2, seven countries are currently providing very high-speed services (S $\geq$ 400 km, V_c $\geq$ 180–200 km/h, V_{max} $\geq$ 250 km/h). These countries are China, France, Italy, Japan, South Korea, Russia, and Spain. Very high-speed services are characterised by frequent train departures (e.g., every hour) and fewer intermediate stops (e.g., every 250 km). China offers very high-speed services along several routes within its network.

In Table 11.1 information is presented on the longest routes per each of the seven aforementioned countries in which very high-speed services are provided.

International – high and conventional speed – trains fall into a special category of intercity trains. These trains must abide by interoperability standards. An emphasis is particularly placed on facilitating border crossing and on minimising delays at borders.

Regardless of the category of services offered in terms of speed, it is essential that the running speed remains constant during the biggest part of the route in order to secure a wisely managed and correctly operating intercity railway network.

Table 11.1 Information on the longest route per country over which very high-speed services are provided (2020 data)

Country	Route (Origin–destination)	V_{max} (km/h)	Longest route (km)	V_c (km/h)
China	Hong Kong (West Kowloon)–Beijing Xi	350	2,439	273
Spain	Madrid Atocha–Barcelona Sants	310	621	248.4
	Sevilla Santa Justa–Barcelona		1,093	197.5
France	Paris–Lyon–Marseille St. Charles	320	750	238.1
Italy	Rome Tiburtina–Milano Rogoredo	300	560.3	212.8
Japan	Tokyo–Aomori	320	675	203.5
	Tokyo–Hakodate		824	185.7
South Korea	Seoul Main–Busan	305	418	185.8
Russia	St. Petersburg–Moskow	250	650	185.7

Table 11.2 Classification of trailer passenger vehicles for intercity trains based on various criteria

Classification criteria	Trailer passenger vehicle type
Internal arrangement	• Vehicles with compartments • Vehicles with individual passenger seats
Number of decks	• Single deck • Double deck
Maximum running speed	• Conventional-speed vehicles ($V_{max} < 200$ km/h) • High-speed vehicles ($V_{max} \geq 200$ km/h)
Floor height	• High floor • Floor at platform level • Combined solutions
Level of service	• Exclusively A Class vehicles • Exclusively B Class vehicles • A and B Class vehicles (C Class vehicles have typically been discontinued)
Specific use (indicatively)	• Passenger coach • Sleeping car • Dining car • Lounge car • Luggage car • Observation car • Specialised types (hospital car, driving trailer, dome car, private car, etc.)
Technology of equipped bogies/axles	• Vehicles with conventional bogies • Vehicles with Jakobs-type bogies • Vehicles with bogies with independently rotating wheels • Vehicles with self-steering axles
Tilting capability	• Active tilting vehicles • Passive tilting vehicles • Non-tilting vehicles

11.3 ROLLING STOCK FOR PASSENGER INTERCITY RAILWAY TRANSPORT

Classification of trailer passenger vehicles for intercity trains may be based on: (a) their internal arrangement; (b) the number of decks; (c) the maximum allowed running speed (V_{max}); (d) the floor height; (e) the level of service provided; (f) their specific use; (g) the technology of equipped bogies/axles; and (h) the capability or not for tilting. Table 11.2 provides a list of the various types of trailer passenger vehicles based on each of the aforementioned classification criteria.

11.4 SCHEDULING OF PASSENGER TRAIN SERVICES

In most railway corridors, the system of clockface timetable has been in use for many years. Trains run at regular and quite dense time intervals (e.g., 6:12, 7:12, and 8:12). This system ensures that timetables are easily memorised by the users and that the railway establishes itself as a reliable system of round-the-clock services available to the user. However, sometimes services are less frequent outside the peak hours.

Nowadays there is a tendency for organising an interconnecting network of services that run on a clockface timetable (Lambropoulos, 2004). In this context:

- The main corridors where fast trains move at rhythmic headway of 1 h ('mainstream' services) are determined.
- The corridors where feeding trains move rhythmically toward the main axis corridors are determined.
- At junction stations, it is ensured that the trains of one branch correspond to those of the other.
- It is also ensured that trains arrive in such a way that connections with and transfers to all other lines and destinations become feasible.

This system has been coded as an integrated clockface timetable and is widely used today in Switzerland, Austria, the Netherlands and, partly, in Germany ('Inntegraler Taktfahrplan').

It should be noted here that the various categories of railway services as well as the differentiation between them is sometimes quite difficult to discern (there are, for instance, systems and services that are found both in metros and suburban trains, in regional and intercity trains, etc.).

The passenger train services are provided on a constant basis according to itineraries which are announced in advance. This is something that is taken for granted with regard to passenger services as opposed to freight ones.

11.5 CASE STUDY: SELECTION AND PURCHASE OF ROLLING STOCK

Rolling stock procurement must be based on a thorough study of the needs and, generally, of the railway system's 'environment'.

In particular, the purchase of new rolling stock must allow the system operator to:

- Renew the rolling stock.
- Increase the quality of the provided passenger and freight services and be prepared to respond to any increase in the transportation demand.
- Cover any shortage in rolling stock or replace technically defective rolling stock, the reparation of which is not economically viable.
- Take advantage, in the best possible way, of the new possibilities offered by infrastructure that is already completed or is expected to be completed in the near future.
- Increase the competitiveness of the railway services over other means of transportation (land, air) and claim a larger share in the passenger transportation market.
- Respond to interoperability needs.
- Increase the safety level of train traffic.

Significant factors that influence the design process of the procurement are:

- The useful lifespan of rolling stock.
- The development of the railway infrastructure.
- The type of services provided.

As regards the first factor, the viability of all railway vehicles, according to international practice, is considered to range between 30 and 35 years, given that maintenance and repair work of vehicles is carried out according to the approved specifications (Baumgartner, 2001). A reconstructed vehicle is considered to have a useful lifespan of 15 years on completion of its reconstruction.

The railway infrastructure comprises the track as well as all the civil engineering structures, systems, facilities, and premises ensuring the train traffic. The constructional characteristics of the aforementioned components must allow the running of the trains at the track design speed (V_d), whereas the track maintenance and other predictable or non-predictable parameters (level crossings, etc.) define the permissible track speed (V_{maxtr}). The objective of the infrastructure manager is to ensure that the permissible track speed is close to the track design speed. As regards the design speed of the rolling stock V_{rs} moving on the railway infrastructure, if major upgrades of infrastructure are not foreseen during its useful lifespan, it should be equal to or slightly higher than the track design speed.

The methodology that is used to calculate the necessary rolling stock that needs to be acquired for a specific network is described in the following. The steps that ought to be followed are briefly illustrated in the flowchart presented in Figure 11.1 (Pyrgidis, 2010). To better understand this methodological approach, an example concerning the case of dedicated passenger traffic operation with different offered services (intercity, suburban, and regional) is also provided in Figure 11.2.

11.5.1 Step 1: assessment of the existing situation

This step includes the gathering of specific data related to the railway infrastructure to the available rolling stock and the train timetable schedule.

Concerning the railway infrastructure, this step involves recording the number of tracks (double and single), the traffic mix (whether passenger and freight trains run on the network), the traction system, the track design speed, the civil engineering structures (tunnels, bridges, overpasses, etc.), the lineside systems (level crossings, electrification, signalling, and telecommunication systems), and the facilities and premises (stations, depots, etc.) for each line of the network (see Table 11.7, column 2).

Concerning the existing available railway rolling stock, this step involves recording the elements featured in Table 11.3, as per the category of rolling stock.

The assessment of the existing available railway rolling stock is followed by the recording of the passenger and freight service connections of the network.

The elements that are necessary for a full recording of passenger service connections are indicated in Table 11.4. In this table, a very small radial-shaped railway network of three corridors (A-B, A-C, and A-D) is considered as an example (Figure 11.2). This network is exclusively used by passenger trains.

The processing of the timetable provides the elements related to the service frequency to each connection (Table 11.5).

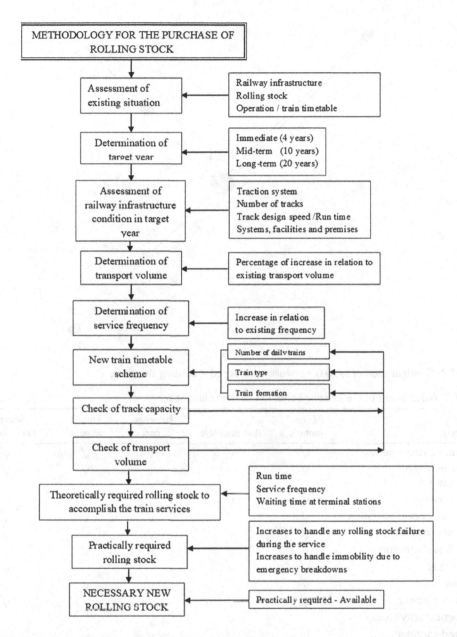

Figure 11.1 Methodological approach – steps for the calculation of the purchase of rolling stock. (Adapted from Pyrgidis, C., 2010, Estimation process of future rolling stock needs for a railway network, *5th International Congress on 'Transportation Research in Greece'*, 27–28 September 2010, Volos, Congress Proceedings, CD.)

Finally, Table 11.6 indicates a hypothetical scenario of an existing situation of the rolling stock that satisfies the service conditions of passenger trains considered in Table 11.4.

In the case of freight traffic, additional data regarding the train type in relation to its service frequency (regular, periodic, and optional), the train weight, the train type in relation to the load processing (e.g., unit train) the wagon type per train, etc. are collected.

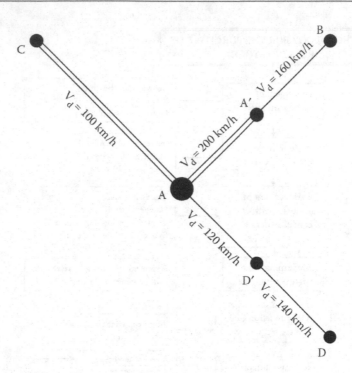

Figure 11.2 Configuration of railway network considered – existing situation.

Table 11.3 Assessment of the existing condition of the rolling stock

Elements	Loco motives	Railcars/MUs	Passenger cars	Freight wagons	Shunting locomotives
Model/manufacture series	√	√	√	√	√
Country of origin/ manufacturer	√	√	√	√	√
Traction system	√	√			√
Number of units in operation	√	√	√	√	√
Nature of procurement	√	√	√	√	√
Year of first circulation	√	√	√	√	√
Design speed	√	√	√	√	√
Axle load	√	√	√	√	√
Total motor power	√	√			√
Transport capacity (seats)		√	√		
Dedicated corridor	√	√	√	√	
Description/formation		√			
Interior configuration			√		
Characterisation in terms of functionality of the vehicle			√	√	

11.5.2 Step 2: determination of the target year

Once the data of the existing available rolling stock are recorded, the target year must be determined in order to evaluate the rolling stock needs; this target can be immediate (4 years), mid-term (10 years), or long term (20 years).

Table 11.4 Data recorded for each service – hypothetical indicative example

Origin–destination (connection)	Train category (service provided)	Type of train	Traction system	Run time	Daily trains per direction	Seats per train	V_{maxtr} (km/h)	V_{rs} (km/h)
A-B/B-A	Intercity express	Loco-hauled	Diesel	4 h 15 min	3/3	252	160	200 (LC) 160 (TC)
A-B/B-A	Intercity	Railcar	Diesel	5 h	5/5	220	160	160
A-C/C-A	Suburban	Railcar	Electric	1 h	16/16	180 (190 standing)	100	120
A-D/D-A	Regional	Railcar	Diesel	1 h 30 min	5/5	144	140	160
A-D/D-A	Regional	Loco-hauled	Diesel	1 h 50 min	3/3	504	140	160 (LC) 160 (TC)

LC: Locomotive.
PR: Single railcar.
TC: Trailer vehicle of a loco-hauled train.
TC': Trailer vehicle of a railcar.

Table 11.5 Service frequency as a result of Table 11.4 – hypothetical indicative example

Connection	Daily trains per direction	Service frequency (line operating 16 h daily)
A-B/B-A	8/8	Every 2 h
A-C/C-A	16/16	Every 1 h
A-D/D-A	8/8	Every 2 h

Table 11.6 Current situation of the rolling stock – hypothetical indicative example – hypothetical scenario

Category/type/formation of trains	Number of units in operation	Age (years)	Design speed V_{rs} (km/h)	Train transport capacity (seats)
Diesel railcars (1PR + 3TC' + 1PR)	6	10	160	220
Diesel railcars (1PR + 1PR)	4	5	160	144
Electric railcars (1PR + 2TC' + 1PR)	5	10	120	180 + 190 standing
Diesel locomotives (LC)	4	2	200	–
Diesel locomotives (LC)	4	20	160	–
Passenger cars (TC)	40	15	160	63

11.5.3 Step 3: assessment of the situation in the target year

The determination of the target year is accompanied by the adoption of a scenario related to the completion of the ongoing or planned works, which aim at the improvement of the railway infrastructure. This procedure is perhaps the most important one in terms of a successful evaluation of the needs in rolling stock. It should also be highlighted that on most occasions the new works are those that guide and set the requirements for the purchase of the new rolling stock. The whole network is divided into tracks and into track sections, and the works implementation scenario includes, for each separate track (or track section) of the network, the following elements:

- The traction system (diesel traction or electrification).
- The number of tracks (single or double track).
- The track design speed and the anticipated connection run time.

The hypothetical scenario may also include the configuration of new elements for facilities and premises expected to have a significant impact on the availability of the rolling stock and on the provided freight transport services, and thus on the setting up of the investment programme for the procurement of the rolling stock. Examples of such facilities and premises are the stations and the depots.

In the hypothetical example presented in this section, mid-term planning is being considered (after 10 years, see Table 11.7).

11.5.4 Step 4: determination of the transport volume target

At this stage, an assumption of an increase of the passenger transport volume is made (based on the results of an appropriate feasibility study) and the percentage of variation of such volume is defined. In the hypothetical example presented in this section, a 60% increase of

Table 11.7 Rail infrastructure conditions in the target year – hypothetical indicative example – hypothetical scenario

(1)	Existing situation (year 2014) (2)	Year 2024 (target year, medium-term approach) (3)
Corridor A-B Section A-A'	Double track Diesel traction V_d: 200 km/h V_{maxtr}: 160 km/h (due to the absence of electrification) Minimum run time: t = 2 h 15 min	Double track Electrification V_d: 200 km/h V_{maxtr}: 200 km/h Minimum run time: t = 2 h
Corridor A-B Section A'-B	Single track Diesel traction V_d: 160 km/h V_{maxtr}: 160 km/h Minimum run time: t = 2 h	Double track Electrification V_d: 200 km/h V_{maxtr}: 200 km/h Minimum run time: t = 1 h 30 min
Corridor A-C	Double track Electrification V_d: 100 km/h V_{maxtr}: 100 km/h Minimum run time: t = 1 h	Double track Electrification V_d: 140 km/h V_{maxtr}: 140 km/h Minimum run time: t = 45 min
Corridor A-D Section A-D'	Single track Diesel traction V_d: 120 km/h V_{maxtr}: 120 km/h Minimum run time: t = 1 h	Double track Diesel traction V_d: 160 km/h V_{maxtr}: 120 km/h Minimum run time: t = 45 min
Corridor A-D Section D'-D	Single track Diesel traction V_d: 140 km/h V_{maxtr}: 140 km/h Minimum run time: t = 30 min	Double track Electrification V_d: 160 km/h V_{maxtr}: 160 km/h Minimum run time: t = 25 min

all connections is considered (see Table 11.8). Thus, this creates a need for a redefinition of the train timetables and for the elaboration of a new train timetable scheme.

11.5.5 Step 5: determination of the service frequency target

The next step consists of the determination of the service frequency target and, in particular, of the increase in service frequency in relation to the existing frequency. The frequency is related to the level of service that the railway company wishes to offer to its customers (see the example in Table 11.9). The increase of speeds, which entails a reduction of the run time, has a positive impact on such a change.

11.5.6 Step 6: new train timetable scheme

Having taken into consideration all the previous steps, a new train timetable scheme is set for the target year. In this plan, an indicative example of which is given in Table 11.10, the following data are recorded for each connection:

- The train category (services provided).
- The type of train in terms of traction and formation.
- The train transport capacity (seated passengers for intercity and regional services, seated and standing passengers for suburban services).

Table 11.8 Hypothetical indicative example – estimation of the number of passengers that must be served by the new routing design for the target year

Existing situation	Maximum number of users *that can be served* daily, per connection per direction
	A-B: 252 × 3 = 756
	A-B: 220 × 5 = 1,100
	A-C: 370 × 16 = 5,920
	A-D: 144 × 5 = 720
	A-D: 504 × 3 = 1,512
	10,008 passengers daily in total (one direction)
	Number of users *that are served* daily per connection per direction (hypothetic)
	A-B: 1,600 (< 1,856)
	A-C: 5,300 (< 5,920)
	A-D: 2,100 (< 2,232)
	9,000 passengers in total (one direction) < 10,008
Target year	Maximum number of users *that have to be served*, daily per connection per direction
	14,400 per day in total (one) direction (increase 60%)
	A-B: 2,560
	A-C: 8,480
	A-D: 3,360

Table 11.9 Hypothetical indicative example – target year – desired service frequency

Connection	Daily trains per direction	Service frequency (track operating 16 h daily)
A-B/B-A	16/16	Every 1 h
A-C/C-A	24/24	Every 40 min
A-D/D-A	16/16	Every 1 h

Table 11.10 Hypothetical indicative example – scenario: passenger trains scheduled for the target year

Connection (origin–destination)	Train category (services provided)	Type of train in terms of traction and formation	Seats per train	Maximum running speed (km/h)	Run time	Daily trains per direction
A-B/B-A	Intercity Express	Electric loco-hauled trains (1LC + 5TC)	315 = 5 × 63	200	3 h 30 min	8/8
	Intercity	Electric railcars (1PR + 2TC' + 1PR)	180		4 h	8/8
A-C/C-A	Suburban	Electric railcars (1PR + 4TC' + 1PR)	320 (320 standing)	140	45 min	24/24
A-D/D-A	Regional	Diesel loco-hauled trains (1LC + 5TC)	315 = 5 × 63	160	1 h 10 min	16/16

- The maximum running speed V_{max} (between stations) ($V_{maxtr} = V_{max}$).
- The run time.
- The number of accomplished services per day.

Some of the basic principles that must be followed during the design of the new timetable scheme are:

- Efficient utilisation of the railway infrastructure.
- Exclusive routing of electric trains on tracks electrified for the total of their length.
- Routing of diesel trains at connections where only certain track sections are electrified and no stops at major hub stations are included.

- Use of interchange only when the connection to the final destination is direct and of suburban nature.
- The design speed V_{rs} of the new trailer vehicles must be at least equal to the maximum track design speed V_d applied in the network (in our case $V_{dmax} = 200$ km/h).

11.5.7 Step 7: checks on corridor track capacity and transport volume

After the elaboration of the new timetable scheme, checks are conducted with regard to track capacity and transport volume.

The percentage of availability of each corridor in relation to its track capacity should be greater than or equal to 25%. Should the new service schedule not satisfy the above criterion, appropriate interventions need to be made. These relate to an adjustment of the number of trains, of the type of trains, as well as of their formation, aiming at a correspondence between the existent corridor track capacity and the new timetable scheme.

Checking the transport volume includes checking the accomplishment of the transport volume that was previously defined as the target. Having calculated the total availability of passenger seats with an assumption of an average occupancy of passenger trains of 70%, one must examine if the timetable scheme adopted satisfies the achievement of the transport volume target at the forecast period.[1] Should the result of the check be negative, appropriate changes are made related, as in the case of the track capacity, to interventions in the number of trains, to the type of trains, and to their formation.

In our case study, the transport volume check is positive.

Transport volume check:
A-B: $([8 \times 5 \times 63] + [8 \times 180]) \times 0.70 = 2{,}772 > 2{,}560$
A-C: $(24 \times 620) \times 0.70 = 10{,}752 > 8{,}480$
A-D: $(16 \times 5 \times 63) \times 0.70 = 3{,}528 > 3{,}360$

11.5.8 Step 8: in theory – required rolling stock for the performance of scheduled services

To calculate the theoretically required rolling stock for the performance of the train services, the following factors are taken into consideration:

- Run time.
- Service frequency.
- Waiting time of trains at the terminal stations.

Required number of locomotives

1. *Connection A-B/B-A, electric loco-hauled trains:* The run time of electric loco-hauled trains is equal to 3 h and 30 min (one-way trip). Hence, for a complete itinerary (round trip), a total of 7 h are required. The waiting time at each terminal is considered to be equal to 1 h. These values are practically implying the realisation of intercity services every hour (at hourly intervals).

   Therefore, the total time that is required by a train in order to run a full service is $3.5 + 3.5 + 1 + 1 = 9$ h. On the basis of the new timetable scheme, the service frequency

---

[1] In the case of suburban trains, the number of standing passengers is also recorded.

operated by the electric loco-hauled trains is 2 h. The number of electric locomotives theoretically required for the performance of services is calculated: 9 h / 2 h = 4.5. This number is rounded to the next highest integer (i.e., 5).

2. *Connection A-D/D-A, regional services, and diesel loco-hauled trains:* The run time of diesel loco-hauled trains is equal to 70 min. The waiting time is considered to be equal to 50 min. Therefore, the total time that is required by a train in order to run a full service is 1.166 + 1.166 + 0.833 + 0.833 = 3.998 h. On the basis of the new timetable scheme, the service frequency operated by the diesel loco-hauled trains is 1 h. Thus, as in the previous case, 3.998 h / 1 h = 3.998 and consequently four diesel locomotives are required.

Required number of railcars

1. *Connection A-B/B-A, long-distance (intercity) services – electric railcars:* The total time that is needed by a train in order to run a full itinerary is 4 + 4 + 1 + 1 = 10 h. On the basis of the new timetable scheme, the service frequency operated by the electric railcars is 2 h. Thus, as in the previous case, 10 h / 2 h = 5 and consequently five electric railcars are required.

2. *Connection A-C/C-A, suburban services – electric railcars:* Given that the provided services are suburban, the waiting time at each terminal is considered to be equal to 30 min. The total time that is needed by a train in order to run a full service is 0.75 + 0.75 + 0.5 + 0.5 = 2.5 h. On the basis of the new timetable scheme, the service frequency operated by the electric railcars is 0.666 h. Thus, as in the previous case, 2.5 h / 0.666 h = 3.75 and consequently four electric railcars are required (these calculations do not take into account any additional trains needed to serve peak hours).

Required number of passenger cars:

The total number of passenger cars is 5 × 5 + 4 × 5 = 45.

11.5.9 Step 9: practically required rolling stock

First of all, it must be assumed that, in the target year, all railway vehicles in service shall have an age of 30 or less (useful lifespan). Moreover, the number of calculated locomotives and railcars should be increased (Baumgartner, 2001):

- At first by 10%, in order to take into account the existence of spare locomotives/railcars at nodal points of the network so as to be ready to handle any rolling stock failure during the service.
- Second by 20%, in order to take into account the immobility resulting from the need to keep a train in the depot for repair due to emergency breakdowns.

Apart from the number of locomotives and railcars, an increase of the passenger cars by 20% is also required, in order to take into account their immobility in the depot for emergency breakdowns.

Concerning freight transport, a scenario usually based on the variation of the freight transport volume is adopted to estimate the freight wagons. This variation may relate to:

- Increase of the transport volume.
- Change of the transported cargo type and the offered services.

- Change of the freight transport connections.
- Combinations of the above.

The increase of the transport volume is particularly boosted by the improvement of the level of services provided and, more specifically, of the following quality parameters:

- Reduction of route run time.
- Reliability and regularity of train services.
- Frequency of train services.
- Safety against theft.
- Competitive fares in comparison with the fares of other transport means.
- Offering of special services.

The suggested percentages for additions to the number of calculated locomotives are the same as for the case of passenger transport. Additional reserves are necessary due to the fact that freight wagons are subject to periodic inspection, which results in an important part of the fleet permanently undergoing inspection/repair. Moreover, in order to take into account traffic peak periods, the theoretically required freight wagons are increased by a small percentage.

11.5.10 Step 10: required rolling stock

Having estimated the practically required rolling stock (Table 11.11, column 3, i.e., theoretically required plus increases), the necessary rolling stock is estimated for the target year (Table 11.11, column 5).

Table 11.11 Hypothetical indicative example – calculation of the required rolling stock for the target year

Category/type/ formation of the required rolling stock (1)	Theoretically required (2)	Practically required (3)	Available (4)	Necessary (5)	Comments (6)
Electric locomotives V_{rs} = 200 km/h	5	8	0	8	
Diesel locomotives $V_{rs} \geq$ 160 km/h	4	6	4 V_{rs} = 200 km/h	2	The other four locomotives in operation will be 30 years old The traction characteristics of the available rolling stock also need to be checked
Electric railcars (1PR + 2TC' + 1PR) V_{rs} = 200 km/h	5	8	0	8	
Electric railcars (1PR + 4TC' + 1PR) V_{rs} = 140 km/h	4	6	0	6	The existing rolling stock can also be used but its design speed (120 km/h) does not satisfy the performance of the upgraded track (140 km/h)
Passenger cars V_{rs} = 200 km/h	45	60	0	60	The existing rolling stock will be old. Due to its low design speed (160 km/h), it can be used only in the A-D connection

It is estimated by subtracting the available rolling stock in the target year (Table 11.11, column 4) from the practically required rolling stock.

Therefore,

Necessary rolling stock = practically required rolling stock − available rolling stock.

REFERENCES

Alias, J. 1985, *La voie ferrée – Exploitation technique et commerciale*, Lecture Notes, Vol. 3, ENPC, Paris.

Baumgartner, J.P. 2001, *Prices and Costs in the Railway Sector*, Laboratoire de l'intermodalité des transports et de planification, EPFL, Lausanne, Switzerland.

Jorgensen, M. and Sorenson, S. 1997, *Estimating Emissions for Air Railway Traffic, Report for the Project MEET: Methodologies for Estimating Air Pollutant Emissions from Transport*, Department of Energy Engineering, Technical University of Denmark, Lyngby, Denmark, available online at: www.inrets.fr/infis/cost319/MEETDeliverable17.pdf.

Lambropoulos, A. 2004, *Design of Rail Freight Transport*, Lecture Notes, MSc Programme 'Design, Organisation and Management of Transportation Systems', AUTh, Faculty of Engineering, Thessaloniki.

Pyrgidis, C. 2010, Estimation process of future rolling stock needs for a railway network, *5th International Congress on 'Transportation Research in Greece'*, 27–28 September 2010, Volos.

Samuel, H. 1963, *Railway Operating Practice*, Odham's Press Limited, London, 1963.

Chapter 12

High-speed networks and trains

12.1 DISTINCTION BETWEEN HIGH SPEEDS AND CONVENTIONAL SPEEDS

As already mentioned in Section 1.4.2, based on speed, a railway system may be distinguished into one of the following three categories:

- Conventional-speed rail.
- High-speed rail.
- Very high-speed rail or super-fast rail.

Regarding conventional-speed rail, a single distinguishing criterion is generally accepted and is related to the track design speed ($V_d < 200$ km/h). In Section 1.4.2, a specific approach for distinguishing railway systems into high and very high speed is proposed. Based on that approach the definition differs for the railway infrastructure manager and for the railway operator.

In Chapter 11 (Section 11.2), data was provided regarding high-speed and very high-speed services provided by various railway systems around the world. In the following sections, data are provided regarding the infrastructure of these systems as well as the rolling stock circulating on it. Specifically, for these two constituents of a railway system, the following criteria are adopted:

- *Regarding railway infrastructure:*
  - High-speed tracks: 250 km/h $> V_{maxtr}$ $(V_d) \geq 200$ km/h.
  - Very high-speed tracks or super-fast tracks: V_{maxtr} $(V_d) \geq 250$ km/h.
    Assuming that the track is maintained to optimal conditions, the maximum permissible track speed (V_{maxtr}) will coincide with the track design speed (V_d).
    Based on 2019 data, several countries possess high-speed lines ($V_d \geq 200$ km/h).
    Based on the same data (2019), 19 countries possess very high-speed lines ($V_d \geq 250$ km/h) (see Table 12.5).
- *Regarding rolling stock:*
  - High-speed trains: 250 km/h $\geq V_{max} \geq 200$ km/h.
  - Very high-speed trains or super-fast trains: $V_{max} > 250$ km/h.

where V_{max}: Maximum running speed.

The data that are presented in the following sections regard both high-speed and very high-speed networks. For the sake of simplification, the term 'high speed' is used and refers to both.

DOI: 10.1201/9781003046073-12

12.2 HIGH-SPEED TRAIN ISSUES

The increase of speed beyond a specific value creates a series of issues and possible problems, the handling of which requires special interventions regarding both the rolling stock and the track.

The systematic operation of high-speed trains in the last 30 years allowed clear identification of these problems and in some cases the setting of crucial speed limits, beyond which they arise. The basic problems caused by the development of high speeds are the following (La vie du rail, 1989; Pyrgidis, 1993, 1994; Profillidis, 2014):

- Increase of the train's aerodynamic resistance.
- Problems arising in tunnels.
- Dysfunction of the lineside signalling.
- Increase of the braking distance of the train.
- Requirements for high tractive power.
- Lateral instability of vehicles in straight paths.
- Special requirements in the track geometry, as well as the horizontal and vertical alignment of the design.
- Noise pollution of the surrounding environment.
- Severity of damage in case of an accident (collision or derailment).
- Increase of vertical dynamic loads.
- Decrease in the dynamic comfort of passengers.
- Troublesome passage over switches and crossings.
- Intensity of the aerodynamic effects and their impacts during the movement of trains in the 'open' track, and their passage along the station platforms.

It is characteristic that the above problems increase proportionally and non-linearly to speed, resulting, beyond a specific limit, in the development of prohibitive conditions for a conventional railway.

The causes that will determine the maximum speed in the future, which may not be outperformed by the wheel–rail system, should be sought in these problems (and mainly in the increase of the aerodynamic resistances and the braking distance).

The above problems are described and discussed in the following:

- *Increase in the aerodynamic resistances of the train*: The resistance W_m of a train moving at a constant speed V on a straight path without longitudinal slopes is expressed by the following equation (the Davis equation) (Metzler, 1981):

$$W_m = A_w + B_w \cdot V + C_w \cdot V^2 \tag{12.1}$$

   where:
   A_w, B_w, C_w: Parameters depending on the characteristics of the rolling stock.

   The first term, A_w, is independent of the speed of the train and represents the rolling resistances. The second term, $B_w \cdot V$, is proportional to the speed and represents the various mechanical resistances (rotation of the axles, transmission of movement, etc.), as well as the air friction resistances along the train's lateral surface. The third term, $C_w \cdot V^2$, changes in proportion to the square of the speed and represents the aerodynamic resistance (aerodynamic drag).

   According to Figure 12.1, a speed change from 200 km/h to 300 km/h results in a change of the aerodynamic resistance of the train by 100%, while the mechanical

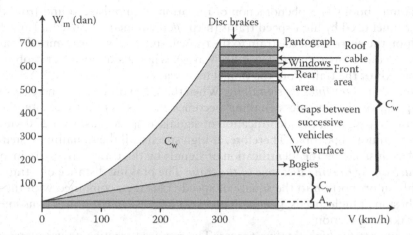

Figure 12.1 Change in the resistances of movement with respect to speed V (TGV-A, formation: 1 power vehicle + 10 trailer vehicles + 1 power vehicle). (Adapted from Profillidis, V. A., 2014, *Railway Management and Engineering*, Ashgate; La vie du rail, 1989, L'Atlantique à 300 km/h, Octobre.)

resistances remain literally unchanged. At high speeds, the aerodynamic resistances determine, therefore, the total movement resistance of the train and, hence, the required motor power of the power vehicles.

- *Problems arising in tunnels*: During the passage of a high-speed train ($V_p \geq 200$ km/h) through a tunnel, the following aerodynamic problems arise (Maeda, 1996; Profillidis, 2014):
  - *Sudden change of pressure*: The passage of a train through a tunnel is always accompanied by a fluctuation in the pressure exerted frontally and laterally to the train. This fluctuation is more annoying to the passengers, as the time, within which it takes place, gets less, and therefore, as the speed of the train's movement gets higher. The great differences in the pressure inside the tunnel, and also at its entry and exit, may cause earache and headache to the passengers.

    Experimental tests have shown that, for speeds of more than 200 km/h, a notable reduction of the acoustic comfort of passengers is observed. It is mentioned indicatively that the TSI that relate to the construction and operation of the railway tunnels focus mainly on the maximum permissible change in pressure (ΔP_{max}) generated inside the tunnels. In this context, it is required that the maximum change (ΔP_{max}) in the pressure along an interoperable train should not exceed 1,000 Pa during the crossing of the tunnel, for the maximum speed permitted by the specific civil engineering structure.
  - *High aerodynamic resistances*: During the passage of a train through a tunnel, the aerodynamic resistances, for the same speed and the same train formation, are much higher than those generated at the surface sections of the track. The major impacts resulting from the increase in the aerodynamic resistances inside a tunnel are the increase of the exerted forces on the train, which results in greater energy consumption.
  - *Interaction of trains running in opposite directions*: In case of a double-track tunnel, the crossing of trains running at speeds of more than 220 km/h may cause damage to the rolling stock (particularly breaking of window panes), due to the increased pressure waves that are generated.

- 'Tunnel boom' is a phenomenon of radiation of impulsive sound from the exit of a tunnel used by high-speed trains (https://en.wikipedia.org/wiki/Tunnel, 2015).
- Upon the entrance of the train to the tunnel, shock waves are emitted from the inlet to the outer environment (entrance waves), while a similar effect is observed upon the exit of the train from the tunnel (exit waves).
- *Dysfunction of the lineside signalling*: When the speed of the train increases, the visual perception of the lineside signalling becomes increasingly difficult. In bad weather conditions (e.g., fog), the identification of signals at speeds just over 220 km/h is troublesome, if not impossible. Therefore, at high speeds, all the signalling systems that are based exclusively on the identification of signals by the train drivers are incompatible.
- *Increase of the braking distance of the train*: The braking distance of a train increases roughly in proportion to the square of speed. This fact, combined with the reduction of adhesion at high speeds, generates an extra increase in the power consumed, during the braking operation.
- *Requirements for high tractive power*: The required tractive power increases in proportion to the third power of speed. The great nominal motor power required at high speeds, combined with the need for instant supply of great tractive powers to the network, makes the electrification of trains a necessary condition for the development of high speeds. At this point, it should be mentioned that, at speeds $V > 160$ km/h, a discontinuity may be observed in the contact between pantograph/overhead power wires, resulting in problems concerning the electrification of trains.
- *Instability in straight paths*: The speed parameter involved in the expression and value of creep forces determines the lateral stability of the vehicles to a great extent. At lower speeds, the movement is stable. Over a specific speed, the movement becomes unstable causing oscillations of high amplitude, contact of the wheel flanges with the rails, and lateral forces that may cause lateral displacement of the track (Pyrgidis, 1990).
- *Special track geometry alignment requirements*: The following three individual problems are identified:
  - *High centrifugal forces in the horizontal alignment curves*: During the movement of a railway vehicle in curves, centrifugal forces develop in curves, the value of which increases in proportion to the square of speed. To reduce these forces, it is required to adopt greater curvature radii on the horizontal alignment and apply higher cant in high-speed lines. An overly high cant value causes problems to the coexistence of fast and slow trains in the same network.
  - *High vertical accelerations in the curved segments of the vertical alignment*: During the passage of trains from the curved segments of the vertical alignment, the vertical accelerations increase in proportion to the square of speed.
  - *High impact of geometric track defects*: Numerous measurements taken in various networks have shown that the impact of dynamic loads on the track superstructure increases in proportion to the speed and is directly proportional to the ride quality of the track, that is, the geometric track defects.
    The geometric track defects comprise the main cause of additional dynamic stresses that are generated by the interaction track-rolling stock (Alias, 1977; Esveld, 2001).
- *Noise pollution of the surrounding environment*: For speeds of up to 300 km/h, the noise level increase is a function of the third power of speed, while for higher speeds, the acoustic annoyance increases in proportion to the sixth power of speed (La vie du rail, 1989).
- *Severity of accidents*: In case of collision between trains or collision of trains with obstacles on the track, the material damage is more severe and the likelihood of injuries is higher. The same also applies to the case of derailment.

- *Increase of vertical dynamic loads*: The increase of speed does not affect significantly the load change caused by the suspended masses of the vehicle (car body), since the vertical accelerations of the car body increase less quickly than the speed, and they may be restricted by reducing the natural frequency of the car body or by ensuring a relatively good track quality (Alias, 1977).

    In contrast, the semi-suspended masses of the vehicle (bogies), and particularly, the unsuspended masses (wheelsets) change significantly with the increase of speed and increase the total value of the vertical dynamic load (Pyrgidis, 1990).

- *Passage over switches and crossings*: Proper operation of high-speed network requires the passage of trains over switches and crossings with speeds higher than those applied in conventional networks. This requirement automatically generates new requirements in terms of the design and construction of switches and crossings.

- *Reduction of the dynamic comfort of passengers*: The increase of speeds automatically implies the increase of vertical and lateral accelerations of the car body, which have a direct impact on the dynamic comfort of passengers.

- *Intensity of the aerodynamic effects and their impact on the 'open' track and platforms*: In 'open' track sections, the following may occur:
  - During the passage of a train at high speed, the pressure along the lateral surface of the train as well as in the area adjacent to it changes and vibrations may affect the residential windowpanes located near the railway track.
  - During the crossing of trains running in opposite directions at high speed, the pressure distribution along the trains affects their dynamic behaviour.

On platforms, the air flow field that is generated intensifies with speed and may cause the following:
  - Loss of balance, difficulty in walking, and passengers or staff on the platforms near the tracks to be pushed violently.
  - Ballast turbulence with the risk of injuring people on the platforms and causing damage to the rolling stock.

12.3 SPECIFICATIONS AND TECHNICAL SOLUTIONS FOR THE ACHIEVEMENT OF HIGH SPEEDS

12.3.1 Track geometry alignment characteristics

The application of high speeds automatically implies reviewing the geometry of the track alignment and necessarily setting new values concerning the cant of track, the longitudinal slopes, the radii of curvature of the horizontal and vertical alignment, the distance between track centres, and the tolerance limits for the geometric track defects.

12.3.1.1 Selection of horizontal alignment radii – case study

Given that conventional bogies are used, it is not possible to ensure simultaneously high speeds in straight paths, and good inscription of wheelsets in curves; the horizontal alignment of the track layout should be characterised by the highest possible rate of straight paths, and by curved sections with the highest possible radii of curvature.

The selection of the horizontal alignment radii should be made based on two criteria (Pyrgidis, 2003):

1. The 'physical' behaviour of the vehicle, by applying the mathematical equations:

$$R_{cmin} = \frac{11.8 \cdot V_{max}^2}{U_{max} + I_{max}} \tag{12.2}$$

$$R_{cmin} = \frac{11.8 \cdot \left(V_{max}^2 - V_{min}^2\right)}{E_{cmax} + I_{max}} \tag{12.3}$$

where:

V_{max}, V_{min}: Maximum speeds, respectively, of the fastest and slowest trains that operate in the track (km/h).

U_{max}: Maximum permissible track cant (mm).

I_{max}: Maximum permissible track deficiency (mm).

$$I_{max} = \frac{2e_o}{g} \cdot \gamma_{ncmax} \tag{12.4}$$

E_{cmax}: Maximum permissible track excess (mm).

R_{cmin}: Minimum curvature radius in horizontal alignment (m).

γ_{ncmax}: Maximum permissible residual lateral acceleration (m/s^2 or g).

$2e_o$: Theoretical distance between the running surfaces of the right and the left wheel when centred $\approx$ distance between the vertical axis of symmetry of the two rails = 1,500 mm (normal gauge track).

From the results of Equations 12.2 and 12.3, the higher value of R_{cmin} is retained.

In case that, given that all the trains in the network run at the same speed, only Equation 12.2 applies (it is assumed that $160 \le U_{max} \le 200$ mm).

The above equations take into account:

a. The passage speed of the fastest and slowest train operating on the track.
b. The categories of the operating trains (passenger, freight, etc.) and their rate of proportion.
c. The behaviour of the human body in lateral accelerations and ensure, for the desired speeds, the selection of values for the horizontal alignment radii that meet the lateral dynamic comfort of passengers (physical behaviour of the vehicle).

By contrast, they ignore the constructional characteristics of the rolling stock that will operate on track to a great extent.

2. The 'geometric' behaviour of the vehicle:

Regardless of the results arising from the analytical calculation, it should be examined, mainly for speeds $V \ge 250$ km/h, whether the constructional characteristics of the bogies of the rolling stock which are going to circulate on the track allow, for the specific radii of curvature, the proper inscription of the vehicle bogies (Pyrgidis, 2003).

Specifically, if the bogies are characterised as too stiff on the level of the primary suspension (e.g., longitudinal and transversal stiffness K_x, K_y > 10^7 N/m, see Chapter 3), then the value of the horizontal alignment radius, which was calculated analytically, should increase by a rate.

Table 12.1 displays indicatively for:

a. V_{max} = 300 km/h, U_{max} = 180 mm, γ_{ncmax} = 0.2/0.5/0.7 m/s^2, K_y = 10^7 N/m.
b. And for various values of the longitudinal stiffness K_x of the primary suspension of vehicles:

Table 12.1 Selection of horizontal alignment radii on high-speed tracks

V_{max} = 300 km/h U_{max} = 180 mm	Horizontal alignment radius R_{co} (m) as it derives from simulation models			Horizontal alignment radius R_{cmin} as it derives from analytical Equation 12.2 and the rate of its required increment		
	γ_{ncmax} = 0.2 m/s²	γ_{ncmax} = 0.5 m/s²	γ_{ncmax} = 0.7 m/s²	γ_{ncmax} = 0.2 m/s² R_c = 5,042 m	γ_{ncmax} = 0.5 m/s² R_c = 4,140 m	γ_{ncmax} = 0.7 m/s² R_c = 3,699 m
K_x = 3.5 × 10⁷ N/m	6,310	6,850	7,120	25.1%	65.4%	92.5%
K_x = 1.5 × 10⁷ N/m	5,770	6,130	6,310	14.4%	48.0%	70.6%
K_x = 8 × 10⁶ N/m	4,870	5,140	5,320	0%	24.1%	43.8%
K_x = 5 × 10⁶ N/m	3,970	4,150	4,330	0%	0%	17.0%
K_x = 2 × 10⁶ N/m	1,990	2,080	2,170	0%	0%	0%

i. The theoretically minimum horizontal alignment radius R_{co}, which ensures the inscription of bogies without wheel–flange contact and without wheel slippage, that arises from the application of simulation models describing the lateral behaviour of railway vehicles (see assumptions in Chapter 3) (Joly and Pyrgidis, 1990; Pyrgidis, 1990; Pyrgidis, 2003).

ii. The minimum horizontal alignment radius R_{cmin} that arises from the application of Equation 12.2 and that does not take into account the stiffness of the primary suspension.

iii. The percentage by which the value of the above radius R_{cmin} should be increased, for the value R_{co} to arise, emerging from the application of simulation models.

At this point, it should be highlighted that the final rate of increment depends on the landscape and on any impact of the track layout on the surrounding built environment.

CASE STUDY

Intercity trains with a maximum running speed of V_{max} = 250 km/h, intercity trains with a maximum running speed of V_{max} = 200 km/h, and suburban trains with a maximum running speed of V_{max} = 140 km/h are planned to be put into service on normal gauge double-track line. The maximum permissible residual lateral acceleration is γ_{ncmax} = 0.07 g, the maximum permissible track cant is U_{max} = 180 mm, and the maximum permissible track excess is E_{cmax} = 100 mm. The rolling stock is to utilise conventional technology bogies.

You are tasked to select the minimum horizontal alignment curve radius.

Solution:

By applying Equation 12.4, for $2e_o$ = 1,500 mm and γ_{ncmax} = 0.07 g, it is derived that I_{max} = 105 mm.
By applying Equation 12.2, for V_{max} = 250 km/h, U_{max} = 180 mm, and I_{max} = 105 mm, it is derived that R_{cmin} = 2,892 m.
By applying Equation 12.3, for V_{max} = 250 km/h, V_{min} = 140 km/h, I_{max} = 105 mm, and E_{cmax} = 100 mm, it is derived that R_{cmin} = 2,469 m.

Of the aforementioned two values of R_{cmin}, the highest value is chosen and rounded up to the nearest 500 m. Therefore, a final value of R_{cmin} = 3,500 m is selected.

At this stage, a check for the transversal and lateral stiffness of the springs of the primary suspension of the bogies is required. There are three possible scenarios:

- The rolling stock of the very high-speed trains (V_{max} = 250 km/h) has already been selected and the primary suspensions are deemed to not be too stiff (e.g., $K_x = K_y = 8 \times 10^6$ N/m). In such a case, the choice of R_{cmin} = 3,000 m is adequate.
- The rolling stock of the very high-speed trains (V_{max} = 250 km/h) has already been selected and the primary suspensions are deemed to be too stiff (e.g., $K_x = K_y = 4 \times 10^7$ N/m). In such a case, the initially chosen radius of R_{cmin} = 3,000 m should maybe be increased. (In the relevant track superstructure regulations of a railway network, it should typically be stated by what percentage the horizontal alignment curve radii should be increased based on the characteristics of the bogies of the rolling stock).

- The rolling stock of the very high-speed trains (V_{max} = 250 km/h) has not yet been selected. In such a case, the future manufacturer of the rolling stock should be made aware of the elements of the horizontal alignment of the line on which the trains will circulate. Thus, they may choose the appropriate characteristics for the bogies.

12.3.1.2 Distance between track centres

In high-speed networks, regardless of the traffic load, a double track is exclusively used for safety reasons. The track gauge is normal or broad.

In networks of track design speeds at V_d = 200 km/h, the minimum distance between track centres is set at Δ = 4.20 m.

In networks of track design speeds at V_d = 300 km/h, the minimum distance between track centres is set at Δ = 5.00 m.

The increase of the distance between track centres at these speeds is imposed due to the intensification of the aerodynamic effects in the event of crossing of trains running in opposite directions.

12.3.1.3 Longitudinal slopes

In the case of operation of exclusively passenger trains, due to lower axle load, lighter trains and electrification and high longitudinal slopes, in the region of i = 3–4%, may be applied. The adoption of high longitudinal slopes limits the extent of the civil engineering structures.

12.3.2 Track superstructure components

The better ride quality of the track superstructure allows the reduction of the vertical dynamic stresses, ensuring a smoother rolling of the trains and better track robustness. The systematic operation of high-speed trains, as well as the numerous tests, confirmed the following:

- Use, throughout the network, of exclusively Continuous Welded Rails (CWR).
- Fixing of sleepers/rails, assisted by exclusively double elastic fastenings and the use of heavy rails of UIC 60 type (60 kg/m).
- Use of concrete sleepers (monoblock or twin-block sleepers).
- Adoption of a minimum ballast layer thickness of 35 cm filled with high-hardness materials or, alternatively, the adoption of slab track.
- Homogenisation of track stiffness characteristics.

With regard to crossings, the use of swing nose crossings with a movable point frog (Figure 12.2) is required.

12.3.3 Civil engineering structures

12.3.3.1 Tunnel traffic

The reduction of the aerodynamic impacts that emerge in a tunnel is achieved by reducing the ratio:

$$\frac{S_l}{S_u}$$

Figure 12.2 Movable point frog in a crossing area of a turnover. (From Voestalpine, 2015.)

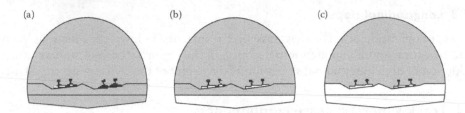

Figure 12.3 Cross sections of railway tunnels: (a) excavated, (b) effective, and (c) useful.

where:
S_1: The area of the cross section of the front surface of the train.
S_u: The area of the 'useful' cross section of the tunnel (Figure 12.3).

This ratio is called 'blockage ratio drag coefficient' of the train in the tunnel, and (Profillidis, 2014):

- For single-track tunnels, it is considered equal to 0.30–0.50.
- For double-track tunnels, it is considered equal to 0.14.

Table 12.2 displays the required area of the useful cross section S_u of the tunnel in relation to passage speed V_p in a double-track tunnel, for a cross section of the frontal surface of the vehicle, of approximately $S_1 = 10$ m^2.

The increase of the useful cross section of the tunnel reduces all the aerodynamic impacts mentioned in Section 12.2.

- *Operation exclusively of trains with airtight car-body structure*: The use of special trains reduces the annoyance of passengers from the fluctuation in pressures exerted frontally and laterally on the train.
- *Adopting a twin-bore tunnel instead of a single-bore double-track tunnel*: With twin tunnels, the crossing of trains running in opposite directions is prevented, resulting in

Table 12.2 Required area of the useful cross section S_u of a double-track
tunnel for high speeds

V_{pmax} (km/h)	160	200	240	300
S_u (m²)	40	55	71	100

Figure 12.4 High-speed tunnel entrance, Reisberg Tunnel, Germany. (Photo: A. Klonos.)

the reduction of waves from outer pressures that may cause problems to the passengers, the rolling stock, and the freight.
• *Specially shaping the tunnel inlets* (Figure 12.4): With this shaping, the 'tunnel boom' effect is reduced. The area of the lateral cross section of an entrance is usually 1.4 times greater than that of the lateral cross section of the tunnel.

12.3.3.2 Passage under bridges

The operation of trains at very high speeds requires special handling, with regard to the selection of the height clearance (h) of the rolling surface of the rails from the lower surface of the bridge carrier (Figure 12.5).

This distance, in an electrified railway line, also depends, among other things, on the train-passage speed. The lifting of the contact wire, for speeds in the region of 230–300 km/h, may take values in the region of 12 cm at the catenary support points and slightly greater in the middle of the opening (SNCF, 1984; UIC, 1986; Pyrgidis, 2004).

In the case that the civil engineering structure is of small width ($L_T \leq 20$ m), the installation of the catenary is performed by free passage, and the civil engineering structure is centred in the middle of the catenary opening for the best possible utilisation of the available height (Figure 12.6).

Table 12.3 provides the values of height clearance h, for passage speeds $V_p = 250$ km/h, for various widths of civil engineering structures L_T, and for various heights of the catenary contact wire h_{fc}, with $h_{fc} = 5.20–5.75$ m.

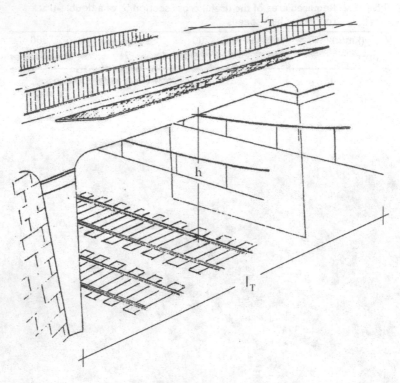

Figure 12.5 Passage of an electrified railway track under a road bridge – height clearance h, civil engineering structure width L_T, and civil engineering structure length l_T. (Figure based on SNCF, 1984, *Ligne aérienne de traction électrique en courant alternatif monophasé 25 kV–50 Hz*, internal document, Paris.)

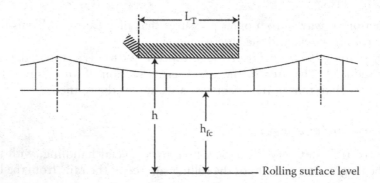

Figure 12.6 Civil engineering structures of small width ($L_T \leq 20$ m) – free passage of catenary. (Figure based on SNCF, 1984, *Ligne aérienne de traction électrique en courant alternatif monophasé 25 kV–50 Hz*, internal document, Paris.)

For the drafting of Table 12.3, it was assumed that:

$$h_o = 0.10 \text{ m}, C_{hmin} = 0.555 \text{m}, I_1 = 0.48 \text{m}$$

where:

    h_o: Track uplifting following maintenance works.
    C_{hmin}: Constructional height in the middle of the catenary opening.
    I_1: Isolation distance of wire-grounded structures.

Table 12.3 Height clearances of an electrified railway line – civil engineering structure deck for various values of L_T and h_{fc} – passage speed V_p = 250 km/h – free catenary passage

L_T (m) h_{fc} (m)	Height clearance h (m)							
	15	20	25	30	35	40	50	60
5.20	6.38	6.42	6.47	6.53	6.59	6.68	6.87	7.10
5.35	6.53	6.57	6.62	6.68	6.74	6.83	7.02	7.25
5.50	6.68	6.72	6.77	6.82	6.89	6.98	7.17	7.40
5.75	6.93	6.97	7.02	7.08	7.14	7.23	7.42	7.65

Source: Adapted from Pyrgidis, C., 2004. *Technika Chronika*, 24 (1–3), pp. 75–80.

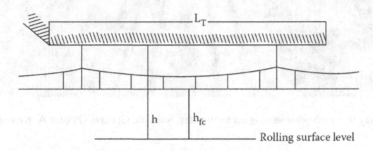

Figure 12.7 Civil engineering structures of large width – catenary fixing exclusively under the civil engineering structures. (Figure based on SNCF, 1984, *Ligne aérienne de traction électrique en courant alternatif monophasé 25 kV–50 Hz*, internal document, Paris.)

As can be seen from Table 12.3, in cases of railway track passage under civil engineering structures with width L_T > 35 m, the required height clearance h increases steeply. Under these conditions, for reasons of cost of the construction, in cases of civil engineering structures of large width, another method may be adopted for their installation (Figure 12.7).

12.3.3.3 Track fencing

In high-speed networks, fencing on each side of the track is compulsory throughout the track. This aims (Figure 12.8):

- To reduce the likelihood of accidents which involve railway vehicles and pedestrians or animals.
- To protect fauna and particularly larger mammals.

12.3.3.4 Noise barriers

On the sides of the line, mainly in inhabited areas, noise-protection walls are constructed, specially designed in order to prevent noise transmission effects (Figure 12.9) (Rechtsanwalt et al., 2002).

The basic difference, compared with the noise barriers placed in highways, is due to the impact – upon train passage – of pressure-sub-pressure waves on the noise barriers, which should be taken seriously into account in dimensioning (Rechtsanwalt et al., 2002; Schweizer Norm. SN 671 250a, 2002).

Figure 12.8 Railway line with wire fencing on both sides, Rapsani, Greece. (Photo: A. Klonos.)

Figure 12.9 Concrete noise barriers, Aula Bridge, Germany. (Photo: A. Klonos.)

The size of this impact depends mainly on:

- The square of speed of the passing train.
- The aerodynamic shape of the train.
- The shape-figure of the noise barrier.
- The distance of the noise barrier from the track centre.

A uniform load q is considered as a characteristic value of the impact of the pressure-sub-pressure wave, the value of which is given in Figure 12.10 in relation to the distance of the

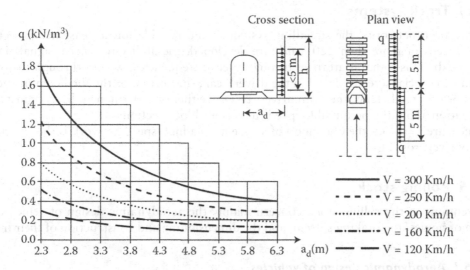

Figure 12.10 Change of the uniform load q in relation to the lateral distance a_d of the noise barrier from the track centre for different speeds. (Adapted from Schweizer Norm. SN 671 250a., 2002, *Schweizerischer Verband der Strassen – und Verkehrsfachleute* (VSS), May.)

Table 12.4 Values of the coefficients K and K' for decrease and increase, respectively, of the aerodynamic load that apply to the noise barriers

Type of rolling stock	K
Trains with conventional shape	0.85
Trains with aerodynamic shape	0.60
Noise barrier dimensions	*K'*
Noise barrier height ≤ 1 m	1.30
Noise barrier length ≤ 2.50 m	1.30
Typical value	1.00

Source: Adapted from Schweizer Norm. SN 671 250a., 2002, *Schweizerischer Verband der Strassen – und Verkehrsfachleute (VSS)*, May 2002.

noise barrier from the track centre and for different speeds. For speed V = 300 km/h and lateral distance a_d = 2.30 m from the track centre, the maximum value of q is observed (q = 1.8 kN/m²) (Schweizer Norm. SN 671 250a, 2002).

Then the values of q that arise from the diagram of Figure 12.10 increase or decrease by a coefficient K or K' depending on the type of the rolling stock and the dimensions of the noise barrier (Table 12.4).

12.3.3.5 Handling aerodynamic effects in an 'open' track and on platforms

With respect to the 'open' track, the problems are handled through the aerodynamic design of trains, the increase of the distance between track centres, and the placement of special protective walls and longitudinal noise barriers along the track.

Regarding the platforms, the problems are handled with the reduction of the train-passage speed in the station areas, the observance by the passengers and the staff on platforms, of a minimum permissible distance from the tracks, and the installation of ventilation ducts on platforms (Baker, 2003; Baker and Sterling, 2003).

12.3.4 Track systems

In high-speed networks, the signalling system should provide for, at least in the sections where speeds of more than 220 km/h are developed, the ability to receive signals in the driver's cab. This signal is attained through special frequencies, which, through the track, transmit to special screens placed in the driver's cab, the current or the forthcoming indication to be observed. Thus, the train drivers receive, either on a permanent basis or instantly, information about the permissible speeds in various block sections.

A measure to reduce the likelihood of accidents in a high-speed network is the absence of railway level crossings.

12.3.5 Rolling stock

Interventions to the rolling stock concern mainly the aerodynamic design of vehicles, the design of bogies, the braking system, and the dimensioning and construction of their frame.

12.3.5.1 Aerodynamic design of vehicles

12.3.5.1.1 Reduction of the coefficient K_1

The aerodynamic resistances W_α of the trains are given by the mathematical equation:

$$W_\pm = C_w \cdot V^2$$

where:

$$C_w = K_1 \cdot S_1 + K_2 \cdot L_{tr} \cdot p \tag{12.5}$$

K_1: Frontal wind force coefficient (a parameter that depends on the shape of the 'nose' and the 'tail' of the train).
S_1: Area of the cross section of the front surface of the train.
L_{tr}: Length of the train.
p: Perimeter that encloses the rolling stock laterally, up to rail level (rolling stock outline).
K_2: Side-wind force coefficient (parameter depending on the lateral external surface of a vehicle/train).

The reduction of coefficient K_1 allows the reduction of coefficient C_w, and as a result, the reduction of the aerodynamic resistances.

The finite element method is the most suitable for the design of the aerodynamics of vehicles. It enables the application of airflow forces to every part of the vehicle, resulting in the shape of the vehicle that favours the development of high speeds, and minimises energy consumption.

12.3.5.1.2 Reduction of coefficient K_2

The diagram of Figure 12.1 lists all the constructional parameters of a vehicle that affect the aerodynamic resistances of a train, as well as the rate of contribution of each of them to the value of coefficient C_w. The shapes of the 'nose' and the 'tail' of the train are directly related to the value of coefficient K_1, while all the other parameters concern coefficient K_2.

The use of trains with shared bogie vehicles is increasingly adopted. These fixed formation trains are equipped with special-type bogies (Jakob-type bogies). In these trains, two vehicles are successively straddled on each 2-axle bogie, thereby reducing the number of bogies to $(n_b/2) + 1$ (where n_b is the original number of bogies), and hence, the gap between the coupling of the vehicles, as well as the total weight of the train (Figure 12.11).

All the above conditions allow a significant reduction of coefficient K_2 and, hence, of the aerodynamic resistances.

12.3.5.2 Design of bogies

Vehicles intended to move at very high speeds are equipped, in their vast majority, with conventional bogies (see Section 3.2.1). Given that, with this type of bogies, it is not possible to attain simultaneously high speeds in straight paths and good inscription of wheelsets in curves, a large percentage of the network's length is constructed in a straight path, and the constructional parameters of the vehicle are optimised, in order to ensure high speeds in straight paths (see Section 3.3).

At the same time:

- The profile of the wheel treads is fixed in frequent intervals (every 300,000 km), so that their initial conicity is maintained.
- The static axle load is reduced to 16–17 t.
- The unsuspended and semi-suspended masses (wheelsets, bogies) are reduced.

12.3.5.3 Braking system

Safe braking at high speeds is attained by combining various braking systems and devices, such as disc brakes made of specially processed steel, rheostatic braking, electromagnetic shoes, and anti-lock braking system (wheel slide protection).

Figure 12.11 Articulated train formation with shared bogie vehicles, TGV Atlantique, bogie Y-237. (Adapted from Fabbro, SNCF Médiathèque, 1989.)

12.3.5.4 Vehicle design: construction

From the conception of the idea of high-speed trains, the engineers' choices were guided by caution. The dimensioning of the rolling stock and the track is performed with high safety considerations. The durability of the framework of the vehicles is increased, and the bumpers of the leading vehicle are strengthened. At the same time, a series of auxiliary devices and automations ensure the smooth movement of the train in case of failure of specific functions and warns the train drivers of potential problems.

12.3.5.5 Implementation cost

The infrastructure cost of a high-speed, double-track line varies from €10 to €40 M per track-km (2014 data). It depends on the percentage of line length constructed with slab track, the length of civil engineering structures (tunnels, bridges), and the difficulty of the topographic relief.

12.4 HISTORICAL REVIEW AND CURRENT SITUATION OF VERY HIGH-SPEED NETWORKS AND TRAINS

The application of high-speed networks began in Japan in 1964 with the operation of the Shinkansen train in the Tokyo–Osaka line (maximum running speeds V_{max} = 210 km/h, connection length S = 515 km).

In Europe, the operation of high-speed trains began in the early 1980s. The French railways were the first to operate high-speed trains in their network. Specifically, in the autumn of 1981, the TGV Paris Sud-Est train was routed in the Paris–Lyon new line, originally with a maximum running speed of V_{max} = 260 km/h which was increased (from 1983) to V_{max} = 270 km/h. It was followed, in the autumn of 1989, by the TGV Atlantique train, which was routed with V_{max} = 300 km/h on the homonymous corridor, serving areas in the western and south-western France.

Today, the technical developments in the field of the rolling stock, as well as in the field of the track, allow a railway train to move in complete safety, on a track of good ride quality, at speeds of 350 km/h.

The operation of high-speed trains today occurs either in new lines with high-speed specifications or in upgraded existing tracks. In both cases, the trains use conventional or tilting technology.

An identification of all the countries that possess at least one track that may be characterised as very high-speed (meaning that in the entirety or the majority of its length $V_d \geq 250$ km/h or $V_{max} \geq 250$ km/h) was attempted (Pyrgidis et al., 2020).

According to data last updated at the end of 2019, 19 countries belong in this group. Specifically, the countries that possess at least one very high-speed track are Austria, Belgium, China, France, Germany, Italy, Japan, Morocco, South Korea, the Netherlands, Poland, Russia, Saudi Arabia, Spain, Switzerland, Taiwan, Turkey, the UK, and Uzbekistan.

Table 12.5 depicts for each country (column 2) the total length of very high-speed tracks (column 3), the percentage of worldwide very high-speed tracks associated with that country (column 4), the maximum design speed of a least one track for that country (column 5), the maximum running speed (column 6), and finally the subtotals of lengths S of track per specific track design speed V_d (column 7).

Table 12.5 Countries possessing at least one track that may be characterised as very high speed – length of very high-speed tracks per country

(1)	Country (2)	Length (km) (3)	Percentage over total length (%) (4)	V_{dmax} (km/h) (5)	V_{max} (km/h) (6)	V_d (S) (7)
1	China	27,016	63.92	350 (380)	350	380 km/h (142 km), 350 km/h (11,712 km), 300 km/h (1,858 km), 250 km/h (13,304 km)
2	Spain	2,789	6.60	350	320	350 km/h (1,750 km), 300 km/h (924 km), 250 km/h (115 km)
3	France	2,556	6.05	320	320	320 km/h (1,273 km), 300 km/h (1,283 km)
4	Japan	2,495	5.90	320	320	320 km/h (675 km), 300 km/h (554 km), 285 km/h (515 km), 260 km/h (751 km)
5	Italy	1,017	2.41	300	300	300 km/h (734 km), 250 km/h (283 km)
6	Germany	1,288*	3.05	300	300	300 km/h (518 km), 280 km/h (426 km), 250 km/h (344 km)
7	Turkey	745	1.76	250	250	250 km/h (745 km)
8	Russia	1,110**	2.63	250	250	250 km/h (1,110 km)
9	South Korea	880	2.08	350	305	350 km/h (412 km), 300 km/h (245 km), 250 km/h (223 km)
10	Taiwan	354	0.84	300	300	300 km/h (354 km)
11	Belgium	214	0.51	300 (320)	320	300 km/h (172 km), 260 km/h (42 km)
12	Netherlands	125	0.3	300	300	300 km/h (125 km)
13	UK	114	0.27	300***	300	300 km/h (114 km)
14	Uzbekistan	600	1.42	250	200	250 km/h (600 km)
15	Austria	43	0.1	250	230	250 km/h (43 km)
16	Saudi Arabia	453	1.07	300	200	300 km/h (453 km)
17	Morocco	183	0.43	350	300	350 km/h (183 km)
18	Poland	224	0.53	250****	200	250 km/h (224 km)
19	Switzerland	57	0.3	250	200	250 km/h (57 km)
	Total	**42,263**	**100%**			**42,263**

*The length of very high-speed tracks in Germany is shorter since for the Berlin–Hanover line (258 km), the running speed fluctuates between 160 km/h and 250 km/h.
**The length of very high-speed tracks in Russia is shorter since for the line Moscow–Nizhny Novgorod (460 km), running speeds of 250 km/h are allowed in short segments.
***Eurostar trains run at up to 300 km/h in the UK but do not carry domestic passengers.
****Poland currently has no high-speed lines operating at speeds above 200 km/h. The Central Rail Line, which links Warsaw to Katowice and Kraków, was designed with an alignment that permits 250 km/h.

The data recorded in Table 12.5 relate to the end of the year 2019. The raw data were obtained both per country and per line, from various available sources, and cross-checked[1]. Afterward, they were further manipulated for the needs of this chapter.

The total length of lines where the developed running speeds V_{max} are greater than 250 km/h amounts to 42,263 km worldwide. China has the longest lines (S = 27,016 km) holding

[1] Data were collected from various internet sources as well as a number of issues of the *Railway Gazette International* and *International Railway Journal* from 2015 to 2019.

a percentage of 63.92 %, followed by Spain (S = 2,789 km, 6.60%), France (S = 2.556 km, 6.05%), and Japan (S = 2,495 km, 5.90%).

Table 12.6 provides, for all countries (19 in total) that possess at least one line that may be characterised as very high speed, data that concern both their high and very high-speed lines. Specifically, it includes their length, the year in which they commenced operation, and the maximum running speed (design speed) that can be developed.

The total length of the high and very high-speed lines in these 19 countries amounts to 52,072 km with China possessing 63.42% (33,029 km).

Figure 12.12 showcases the distribution of very high-speed lines around the world based on their design speed and based on the data of column (7) of Table 12.5. It can be observed that lines with speed $V_d = 250$ km/h and 250 km/h $< V_d \leq 300$ km/h amount to closely 40% of the worldwide length each, while lines with speed 300 km/h $< V_d \leq 380$ km/h amount to about 20%.

In Figure 12.13, an evolution of the construction of very high-speed tracks over the past years is presented. Data are presented for 5-year segments. The period of 1964–2004 was taken as a single segment.

Table 12.7 displays the basic constructional and operational characteristics of very high-speed trains (without a tilting car body) that operate currently in new and upgraded very high-speed tracks worldwide.

12.5 INTEROPERABILITY ISSUES

The term 'railway interoperability' implies the capacity of the trans-European railway system to allow safe and continuous circulation of trains among its various segments, achieving the required performance in specific lines. This capacity is ensured by a set of regulatory, technical, and operational requirements that must be satisfied. These requirements were set by Directive 96/48, which also established the above definition.

The railway interoperability concerns two different cases of railway operation:

- *High-speed networks* and specifically the operation of trains in the categories of tracks I, II, and III, as they were defined in Section 1.4.2.
- *Conventional-speed networks* that include new tracks which are designed for speeds less than 200 km/h or for existing tracks which are upgraded. However, the track design speeds remain less than 200 km/h.

In order for the trans-European network to be implemented, the railway interoperability required the high-speed railway system to be analysed in subsystems. These subsystems are described in detail in Annex II of Directive 96/48, and they are the following:

- Infrastructure.
- Rolling stock.
- Energy.
- Control-command and signalling.
- Operation and traffic management.
- Maintenance.
- Telematic applications for passenger and freight services.

The issuance of Directive 96/48 marked the beginning of the development and drafting of the TSI for each of the above subsystems, which comprise the essential elements for the achievement of interoperability.

Table 12.6 High-speed railway lines per country

Line	Length (km)	Starting year of operation	V_{max} (V_d) (km/h)
Japan			
Tokyo–Shin Osaka (Tokaido line)	515	1964	285
Shin Osaka–Hakata (Sanyo line)	554	1972/1975 in stages)	300
Tokyo–Shin Aomori (Tohoku line)	675	1982/2010 (in stages)	320–260
Omiya–Niigata (Joetsu line)	270	1982	240
Takasaki–Nagano (Hokoriku line)	117	1997	260
Nagano–Kanazawa (Hokoriku line)	228	2015	260
Hakata–Kagoshima–Chuo (Kyushu line)	257	2004/2011	260
Shin Aomori–Shin Hakodate (Tohoku line)	149	2016	260
Total	2,765		
France			
Paris–Lyon (TGV Paris-Sud-Est)	409	1981/1983 (in stages)	300
Paris–Le Mans/Tours (TGV Atlantique)	284	1989/1990 (in stages)	300
Lyon–Valence and Lyon detour (TGV Rhône-Alpes)	115	1992/1994 (in stages)	300
Interconnection TGV Nord-TGV Sud-Est	57	1994/1996 (in stages)	300
Paris–Lille–Calais (TGV Nord Europe)	333	1993	300
Valence–Marseille/Nimes (TGV Mediterranée)	243	2001	320
Vaires-sur-Marne–Baudecourt (TGV Est)	300	2007	320
Figueres–Perpignan (French section)	25	2011	300
Villers les Pots–Petit Croix (TGV Rhin-Rhone-part of the eastern branch)	140	2011	320
	106	2016	320
Baudecourt–Strasbourg (LGV East Europe (second phase))	182	2017	320
	302	2017	320
Le Mans–Rennes (LGV Bretagne Pays de la Loire)	60	2018	300
Tours–Bordeaux (LGV Sud Europe Atlantique)	2,556		
Contournement de Nimes–Montpellier			
Total			
Spain			
Madrid–Seville	472	1992	300
Madrid–Barcelona	621	2003/2008 (in stages)	310 (350)
Cordoba–Malaga	155	2006/2007 (in stages)	300 (350)
Madrid–Valladolid	179	2007	300
Madrid detour	5	2009	200
(Madrid)–Valencia	363	2010	300 (350)
Motilla del Palancar–Albacete	63	2010	300 (350)
Ourense–A Coruna (1,668-mm track gauge to be converted to normal gauge)	152	2011	300 (350)
Albacete–Alicante	171	2013	300 (350)
Madrid–Toledo	21	2005	250
Barcelona–Figueres	131	2013	300
Figueres–Borders	20	2010	300

(Continued)

Table 12.6 (Continued) High-speed railway lines per country

Line	Length (km)	Starting year of operation	V_{max} (V_d) (km/h)
Saragossa–Tardienta (Huesca)	79	2003	200
Santiago–Vigo	94	2015	250
Valladolid–Leon	163	2015	350
Olmedo–Zamora	99	2015	200
Anteguera–Granada	122	2019	300
Valencia–Castelllo	62	2019	350
Total	**2,972**		
Germany			
Augsburg–Munich–Olching (Munich–Augsburg)	43	1977–2011	200–230
Hamm–Bielefeld (Railway Hamm–Minden)	67	1980	200
Augsburg–Donauwörth (Nuremberg–Augsburg)	36	1981	200
Hanover–Würzburg	327	1988/1991(in stages)	280 (250 in tunnels)
Cologne–Duisburg	64	1991	200
Mannheim–Stuttgart	99	1991	280 (250)
Dinkelscherben–Augsburg	20	1992	200
Hanau–Gelnhausen (Kinzigtal Bahn)	16	1993	200
Berlin–Hanover	258	1998	160–250
Koln–Frankfurt (Cologne–Rhine/Main)	177	2002/2004 (in stages)	300
Munster–Bremen–Hamburg (Wanne–Eickel–Hamburg Railway)	288	1982/1991	200
Mannheim–Frankfurt	78	1991	200
Leipzig–Riesa (Leipzig–Dresden)	66	2002	200
Nuremberg–Ingolstadt (Nuremberg–Munich)	89	2006	300
Munich–Petershausen (Nuremberg–Munich)	29	2006	300
Berlin–Halle/Leipzig	187	2006	200
Erfurt–Leipzig/Halle	123	2015	300
Koln–Duren (Koln–Aachen)	42	2003	250
Rastatt South Offenburg (Karlsruhe–Basel)–	44	2004	250
Hanover–Hamburg	170	1987	200
Hamburg–Berlin	286	2004	230
Ebensfeld–Erfurt	100	2017	300
Total	**2,609**		
Italy			
Rome–Florence	254	1978/1992 (in stages)	250
Rome–Naples	205	2006/2009 (in stages)	300
Turin–Milan	125	2006/2009 (in stages)	300
Padova–Venice (Mestre)	25	2007	300
Bologna–Florence	79	2009	300
Milan–Bologna	215	2008/2009 (in stages)	300
Milan–Treviglio	27	2007	300
Naples–Salerno	29	2008/2009	250
Bologna–Verona	114	2009	200
Milan (Treviglio)–Brescia	58	2016	300
Total	**1,131**		

(Continued)

Table 12.6 (Continued) High-speed railway lines per country

Line	Length (km)	Starting year of operation	V_{max} (V_d) (km/h)
South Korea			
Gyeongbu HSR corridor (phase 1 and 2)	412	2004/2014 (in stages)	305 (350)
Osong–Gwangju	184	2015	300
Iksan–Yeosu Expo (Jeolla line)	180	2011	230
Suseon–Pyoengataek	61	2016	300
Seoul–Gangneung	223	2017	250
Total	1,060		
United Kingdom			
Channel Tunnel Rail Link (sections 1 + 2)	114	2003/2007	300
London–Newcastle–Edinburgh (East Coast Main Line)	632	2000	200
London Euston–Rugby–Edinburgh/Glasgow (West Coast Main line)	645	2004	200
Total	1,391		
Taiwan			
Taipei–Kaohsiung	345	2007	300
Taipei–Nangang	9	2016	300
Total	354		
China			
Beijing–Shanghai (via Xuzhou–Nanjing)	1,318	2008–2011	350
Urumqi–Lanzhou	1,776	2014	250
Baoli–Xian	148	2013	300
Xian–Zhengzhou	455	2010	350
	2,379		
Guiyang–Liuzhou–Guangzhou	856	2014	300
Jiangyou–Chengdu–Leshan	314	2014	200
	1,170		
Shanghai–Hangzhou	150	2010	350
Hangzhou–Changsha	933	2014	350
Changsha–Huaihua	416	2014	300
	1,499		
Beijing–Wuhan	1,119	2012	350
Wuhan–Guangzhou	968	2009	350
Guangzhou–Shenzhen–Hong Kong	142	2018	380
	2,229		
Changchun–Jilin	111	2010	250
Ha'erbin–Dalian	904	2012	350
Panjin–Yingkou	89	2013	350
	1,104		
Tianjin–Qinhuangdao	261	2013	350

(Continued)

Table 12.6 (Continued) High-speed railway lines per country

Line	Length (km)	Starting year of operation	V_{max} (V_d) (km/h)
Qinhuangdao–Shenyang	404	2003	250
	665		
Nanjing–Hefei	166	2008	250
Hefei–Wuhan	351	2009	250
Wuhan–Yichang	293	2012	250
Yichang–Chengdu	921	2009/2013	200
	1,731		
Nanjing–Hangzhou	251	2013	350
Hangzhou–Ningbo	152	2013	350
Ningbo–Wenzhou–Fuzhou–Xiamen	841	2009/2010	250
Xiamen–Shenzhen	502	2013	250
	1,746		
Hengyang–Liuzhou	498	2013	200
Liuzhou–Nanning	223	2013	250
	721		
Longyan–Xiamen	171	2012	200
Ganzhou–Longyan	114	2012	200
	285		
Jinan–Qingdao	364	2008	250
Shijiazhuang–Taiyuan	190	2009	250
Qingdao–Rongcheng	299	2014	250
	853		
Qinzhou–Beihai	100	2013	250
Nanning–Qinzhou–Fangchenggang	162	2013	250
	262		
Nanchang–Jiujiang	131	2010	250
Nanchang (Xiangtang)–Putian	635	2013	200
	766		
Guangzhou–Zhuhai (main line)	117	2012	200
Guangzhou–Zhuhai (Xinhui branch)	27	2011	200
	142		
Xian–Taiyuan	570	2014	250
Hefei–Bengbu	131	2012	300
Nanning–Guangzhou PDL	577	2014	250
Chengdu–Dujiangyan	86	2010/2014	200–220
Haikou–Sanya (Hainan Eastern Ring Railway)	308	2010	250
Wuhan Metropolitan	253	2013/2014	250
Zhengzhou–Kaifeng (Central Plain Metropolitan Intercity Rail)	50	2014	200

(Continued)

Table 12.6 (Continued) High-speed railway lines per country

Line	Length (km)	Starting year of operation	V_{max} (V_d) (km/h)
Maoming–Zhanjiang	103	2013	250
Xinhuang West–Guiyang North	286	2015	350
Hefeibeicheng–Funzhou	858	2015	350
Harbin North–Qiqihar South	269	2015	250
Shenyang South–Dandong	208	2015	250
Tianjin–Yujiapu	43	2015	350
Jilin–Huichun	360	2015	250
Nanjing East–Anqing	258	2015	250
Nanning–Baise	223	2015	250
Chengdu–Chongqing	307	2015	300
Zhengzhou–Xuzhou	362	2016	350
Chongqing North–Wenzhou North	247	2016	250
Guiyang North–Kunming	482	2016	350
Kunming–Baise	487	2016	250
Shijiazhuang–Jinan	319	2017	250
Zhangjiakou–Hohhot	287	2017	250
Wuhan–Jiujiang	224	2017	250
Baoji South–Lanzhou	401	2017	250
Xian–Chengdu	658	2017	250
Huaibei–Xiaoxianbei	27	2017	250
Guzhou–Jiujiang	334	2017	250
Chongqing–Guiyang	346	2018	200
Guangtong–Dali	173	2018	200
Jiangmen–Maoming	265	2018	200
Shenzhen–West	26	2018	350
Yuanping–Taiyuan	113	2018	250
Harbin–Jiamusi	343	2018	200
Hangzhou–Houangshan	287	2018	250
Harbin–Mudanjiang	293	2018	250
Qingdao–Yancheng	429	2018	200
Huaihua–Hengyang	318	2018	200
Jinan–Qingdao	308	2018	350
Tongren–Yuping	47	2018	200
Chengdu–Yaan	140	2018	200
Nanping–Longyan	248	2018	200
Shenyang–Chengde	505	2018	350
Tongliao–Xinmin	197	2018	250
Shantou–Meizhou	122	2018	200
Leshan–Jianwei–Pingshan–Yibin	145	2019	250
Meizhou–Chaoshan	122	2019	250
Rizhao–Dawangshuan–Qufu	235	2019	350
Wuhan (via Xiaogan)–Yummeng–Shiyan	399	2019	350
Shangqiu–Hefei	378	2019	350
Zhengzhou–Xiangyang	389	2019	350
Zhengzhou–Fuyang	276	2019	350
Xuzhou–Yancheng	313	2019	250
Lianyunggang–Huaian	105	2019	250
Qianjiang–Changde	335	2019	200
Beijing–Zhangjiakou	147	2019	350
Nanchang–Ganzhou	418	2019	350
Total	33,029		

(Continued)

Table 12.6 (Continued) High-speed railway lines per country

Line	Length (km)	Starting year of operation	V_{max} (V_d) (km/h)
Turkey			
Ankara–Istanbul	533	2009/2014	250
(Ankara)–Polatli–Konya	212	2011	250
Total	745		
Netherlands			
HSL–Zuid Schiphol–Dutch border	125	2009	300
Total	125		
Belgium			
Brussels–French border (HSL-1)	71	1997	300
Leuven (Brussels)–Liege (HSL-2)	61	2002	300
Liege–German border (HSL-3)	42	2009	260
Antwerp (Brussels)–Dutch border (HSL-4)	40	2009	300
Total	214		
Russia			
Moscow–St Petersburg	650	2009	250
Finnish border–St Petersburg	160	2010	200
Moscow–Nizhny Novgorod	460	2010	250 (at small parts)
Total	1,270		
Uzbekistan			
Tashkent–Samarkand	344	2011	200 (250)
Samarkand–Bukhara	256	2016	200 (250)
Total	600		
Austria			
Vienna–St Polten	43	2012	250–230
St Polten–Linz	133	2001/2014	200–230
Linz–Wels	25	1993	200
Wels–Punchheim	51	2012	200–230
New Unterinntalbahn (Kundl–Baumkirchen)	40	2012	220
Total	292		
Poland			
Grodzisk Mazowiecki–Zawiercie railway			
Total	224	2014	200 (250)
	224		
Switzerland			
Mattstetten–Rothrist			
Erstfeld–Biasca (Gothard base tunnel)	42	2007	200
Total	57	2018	200 (250)
	99		
Morocco			
Tanger–Kenitra			
Total	183	2018	300 (350)
	183		
Saudi Arabia			
Medina–Jeddah–Mecca			
Total	453	2018	200 (300)
	453		

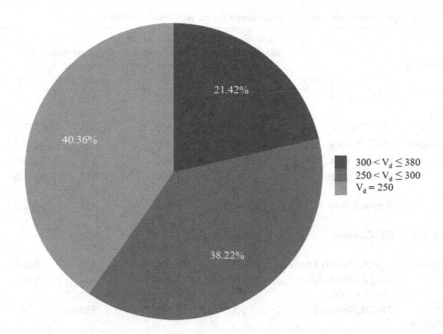

Figure 12.12 Distribution based on track design speed of very high-speed lines worldwide.

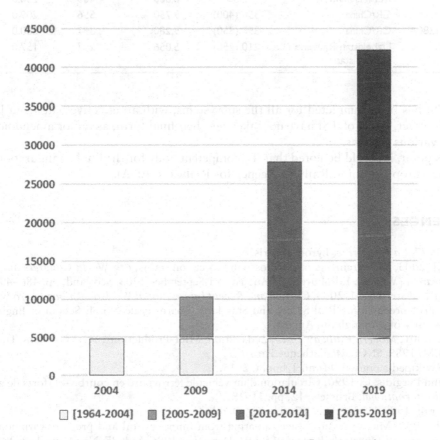

Figure 12.13 Evolution of worldwide high-speed tracks construction in 5-year periods.

Table 12.7 Very high-speed conventional technology trains worldwide (indicative table)

Name and type of train	Operator/Country	Maximum running speed (rolling stock design speed) (km/h)	Total nominal motor's power (kW)	Transport train capacity (passengers)	Train length (m)	First operation
Series W7 Shinkansen	JR West/Japan	275	12,000	934	302	2014
TGV Atlantique	SNCF/France	300 (300)	8,800	485	237.5	1989
TGV Euroduplex	SNCF/France	320 (320)	9,280	556	200.0	2017
TGV POS	SNCF/France	320 (320)	9,600	579	203.0	2006
ETR 575 AGV	NTV. Italo/Italy	300 (360)	7,500	450	201.0	2012
ETR1000 Zefiro	Trenitalia/Italy	300 (400)	9,800	457	202.0	2017
ICE 3 Class 407 Velaro	DB/Germany	320(320)		406		2011
KTX-Wongang	KORAIL/South Korea	305 (330)	8,800	410	201.0	2017
Talgo 350	SRO, ADIFF, RENFE/ Saudi Arabia	300 (350)	8,000	404	215.0	2018
700T THSR SHINKANSEN	THSRC/Taiwan	300	10,260	989	304	2007
HT80100	TCDD/Turkey	300	8,000	483	200.7	2016
S103	RENFE/Spain	350	8,800	403	200.0	2007
CR400BF	CR/China	350 (400)	9,750	556	209.0	2017
MTR CRH380A	CR/China	350 (350)	9,280	357	200.0	2018
Afrosiyob	Uzbekistan Railways / Uzbekistan	210 (250)	5,056	257	157.0	2011

The TSI has been completed for all the subsystems, with an effective date from January 2003. However, issues of TSI have not taken yet their final form, as minor amendments are made in various sections.

At this point, it should be noted that the competent body for the final configuration of the TSI is the European Union Railway Agency for Railways (ERA).

REFERENCES

Alias, J. 1977, *La voie ferrée*, Eyrolles, Paris.
Baker, C.J. 2003, Measurements of the crosswind forces on trains, *6th World Congress on Railway Research, (WCRR)*, Edinburgh, 28 August to 1 September 2003, Scotland, pp. 486–491.
Baker, C.J. and Sterling, M. 2003, *Current and Recent International Work on Railway Aerodynamics*, A report prepared for Rail Safety and Standards Board, Issue No. 4, School of Engineering, University of Birmingham, August.
Esveld, C. 2001, *Modern Railway Track*, 2nd edition, MRT-Productions, West Germany, Duisburg..
Fabbro, J.M. 1989, SNCF Médiathèque.Paris.
https://en.wikipedia.org/wiki/Tunnel_boom, 2015.
Joly, R. and Pyrgidis, C. 1990, Circulation d'un véhicule ferroviaire en courbe – Efforts de guidage, *Rail International*, Brussels, 12, pp. 11–28.
La vie du rail. 1989, L'Atlantique à 300 km/h, October.
Maeda, T. 1996, Micro-pressure wave radiating from tunnel portal and pressure variation due to train passage, *Quarterly Report of RTRI*, December 1996, Vol. 37, No. 4, pp. 199–203.

Metzler, J.M. 1981, *Géneralités sur la traction*, Lecture Notes, ENPC, Paris.

Profillidis, V.A. 2014, *Railway Management and Engineering*, Ashgate, Farnham.

Pyrgidis, C. 1990. *Etude de la stabilité transversale d'un véhicule ferroviaire en alignement et en courbe – Nouvelles technologies des bogies – Etude comparative*, Thèse de Doctorat de l' ENPC, Paris.

Pyrgidis, C. 1993, High-speed trains and the environment, *Rail Engineering International*, Netherlands, 4, pp. 13–17.

Pyrgidis, C. 1994, High-speed rail: Meeting the technical challenges, *Rail Engineering International*, 3, pp. 23–28.

Pyrgidis, C. 2003, High speed railway networks – Selection of minimum horizontal curve radii, *Proceedings of the 6th World Congress in Railway Research (WCRR)*, 28 August to 01 September 2003, Edinburgh, pp. 902–908.

Pyrgidis, C. 2004. Passage through of high-speed track under civil engineering structures – Calculation of minimum overhead clearance, *Technika Chronika*, Vol. 24(1–3), pp. 75–80.

Pyrgidis, C., Savvas, S. and Dolianitis, A. 2020, Classification of intercity and regional passenger railway systems based on speed – A worldwide overview of very high-speed infrastructure, rolling stock and services, *Ingegneria Ferroviaria*, No. 9, September 2020, pp. 635.

Rechtsanwalt, M.G., Popp, C. and Stoyke, B. 2002, *Deutscher Heilbäderverband e.V Hinweise zum Schutz gegen Schienenlarm*, Bund fur Umwelt und Naturschutz Deutschland, February 2002.

SNCF. 1984, *Ligne aérienne de traction électrique en courant alternatif monophasé 25kV–50Hz*, Internal document, Paris.

Schweizer Norm. SN 671 250a. 2002, *Schweizerischer Verband der Strassen – und Verkehrsfachleute (VSS)*, May 2002.

UIC. 1986, FICHE 606-2, *Installation of 25 kV and 50 or 60 Hz Overhead Contact Lines*, 4th edition, 1 January 1986, Paris, France.

Chapter 13

Tilting trains

13.1 DEFINITION AND OPERATING PRINCIPLE OF TILTING TECHNOLOGY

Tilting trains (Pendolino, trains à caisse inclinable, Neigezug) are conventional railway trains equipped with special technology that allows the car body of all vehicles, when it negotiates a horizontal curve, to rotate with respect to its longitudinal axis by an angle φ_t (tilting angle). As a result, the total rotation angle of the car body with respect to the horizontal level is equal to the sum of the track's cant angle δ_p plus the tilting angle φ_t (Figures 13.1 through 13.3).

As analysed in Figure 13.1, tilting technology further increases the lateral component of the vehicle's weight B_{ty} resulting in a decrease of the effect of the residual centrifugal force F_{nc} on the passengers, thereby improving their lateral dynamic comfort.

a. Conventional railway vehicle ($F_{nc} = F_{cfy} - B_{ty} = F_{cf} \cos \delta_p - B_t \sin \delta_p$)
b. Tilting railway vehicle $\left(F_{nc}^{'} = F_{cfy}^{'} - B_{ty}^{'} = F_{cf} \cos\left(\delta_p + \varphi_t\right) - B_t \sin\left(\delta_p + \varphi_t\right) \right)$, $F_{nc}^{'} < F_{nc}$

where:

F_{nc}, $F_{nc}^{'}$: Residual centrifugal force
F_{cf}: Centrifugal force
B_t: Vehicle's weight
B_{ty}, $B_{ty}^{'}$: Lateral component of the vehicle's weight
B_{tz}, $B_{tz}^{'}$: Vertical component of the vehicle's weight

Based on the above, tilting technology may potentially result in the following two alternative improvements regarding the level of service provided to passengers:

- It can ensure a lower residual centrifugal force $F_{nc}^{'} \left(F_{nc}^{'} < F_{nc} \right)$ on the car-body's level in curved segments of the track for the same passage speed V_p compared with conventional trains. As a result, it can also ensure a smaller lateral residual acceleration $\gamma_{nc}^{'} \left(\gamma_{nc}^{'} < \gamma_{nc} \right)$ and thus a higher level of lateral ride comfort for passengers. In this case, the residual centrifugal acceleration γ_{nc} at wheelsets' level and the forces exerted on the track (guidance forces F_{ij} in the case of flange contact, creep forces X_{ij}, T_{ij}) remain the same for both technologies (Table 13.1).

It can enable an increase of the passage speed in curves, keeping on the car-body level, the same value as that of the residual lateral acceleration γ_{nc} adopted for conventional trains

DOI: 10.1201/9781003046073-13

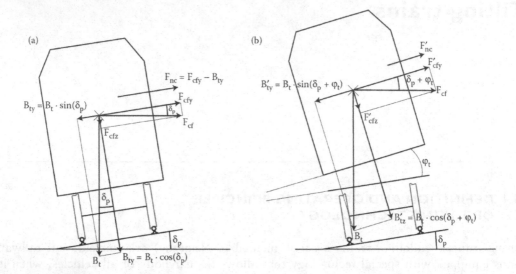

Figure 13.1 Railway vehicle motion in a curved section of the track. Track with cant. (a) Conventional railway vehicle (Adapted from Pyrgidis, C., and Demiridis, N., 2006, The effects of tilting trains on the track superstructure, *1st International Congress, 'Railway Conditioning and Monitoring' 2006*, IET, 29–30 November 2006, Birmingham, UK, Conference Proceedings, pp. 38–43; Pyrgidis, C., 2009, Tilting trains/conventional trains – Comparison of the lateral forces acting on the track, *Ingegneria Ferroviaria*, April 2009, Rome, No 4, pp. 361–371.)

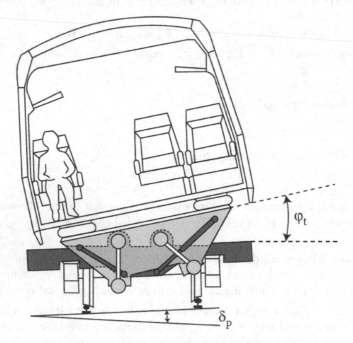

Figure 13.2 Tilting car-body railway vehicle negotiating a curved track section – Track with cant. (Adapted from ABB (n.d.), The fast-train technology for faster passenger services on existing railway lines, Leaflet, Manheim, Germany.)

Figure 13.3 Tilting car-body railway vehicle negotiating a curved track section, Silenen, Switzerland. (Photo: A. Klonos.)

Table 13.1 Tilting and conventional car-body trains running with the same speed – effects on horizontal curves

| | | Lateral residual acceleration | | |
Car-body technology	Passage speed	Car-body level	Bogies and wheelset level	Guidance forces/creep forces
Conventional car body	V_p	γ_{nc}	γ_{nc}	$F_{ij}/X_{ij}, T_{ij}$
Tilting car body	V_p	$\gamma'_{nc} < \gamma_{nc}$	γ_{nc}	$F_{ij}/X_{ij}, T_{ij}$

Source: Adapted from Pyrgidis, C., 2009, Tilting trains/conventional trains – Comparison of the lateral forces acting on the track, *Ingegneria Ferroviaria*, April 2009, Rome, No 4, pp. 361–371.

Table 13.2 Tilting trains running with higher speed than conventional car-body trains – effects on horizontal curves

| | | Lateral residual acceleration | | |
Car-body technology	Passage speed	Car-body level	Bogies and wheelset level	Guidance forces/creep forces
Conventional car body	V_p	γ_{nc}	γ_{nc}	$F_{ij}/X_{ij}, T_{ij}$
Tilting car body	$V'_p > V_p$	γ_{nc}	$\gamma'_{nc} > \gamma_{nc}$	$F'_{ij} > F_{ij},\ X'_{ij}, T'_{ij} > X_{ij}, T_{ij}$

Source: Adapted from Pyrgidis, C., 2009, Tilting trains/conventional trains – Comparison of the lateral forces acting on the track, *Ingegneria Ferroviaria*, April 2009, Rome, No 4, pp. 361–371.

(Table 13.2). This increase in speed does not affect the passenger's ride comfort; however, it has an impact on the track superstructure as it affects the geometrical positioning of the wheelsets on the track (bogies' curving ability) and, through this, an impact on the forces acting on the wheel-rail surface (Table 13.2).

Based on the above, tilting trains enable railway companies to reduce run times while at the same time they can maintain their old infrastructure assets, provided that those are in good condition.

13.2 TILTING TECHNIQUES AND SYSTEMS

Depending on how the tilting of the car body is achieved, two tilting principles are distinguished, namely passive tilting and active tilting (Profillidis, 1998).

13.2.1 Passive tilting

Tilting is activated by the vehicle's inertia forces and not by some mechanism. The trains' vehicles are manufactured to have a lowered centre of gravity, for the centrifugal forces exerted on horizontal curves to force them to tilt. This principle was developed during the first attempts to apply tilt technology on trains (TurboTrain, USA, 1968) and it was applied successfully in the Talgo Spanish trains (Talgo tilting system). Passive tilting allows for the tilting angle φ_t between the vehicle car body and the bogie to have a value between 3° and 5°.

13.2.2 Active tilting

The main issue regarding active tilting lies with the initiation of the tilting at the right moment, that is, just before the vehicle enters the transition curves, and not with obtaining/maintaining the desired tilting angle. More specifically, the car body must be located at the desired position when the vehicle passes through the circular segment of the curve as well as when it returns to the straight segment of the track. In the opposite case, passengers will encounter a violent lateral acceleration which, although short-term, will be annoying, if not unacceptable. The problem is even greater if the track alignment involves two successive reverse curves without an intermediate straight segment.

For this reason, the car body must have already obtained the desired tilting angle at the curve's transition curves. To achieve this, given that the time within which the tilting process must be completed is very limited (around 2 s), the car body's tilting must occur mechanically rather than by applying the principle of passive tilting.

Until today, two active tilting techniques have been developed, namely the European technique and the Japanese technique (Figure 13.4).

In the European technique, tilting is achieved by using equipment that is mounted solely on the vehicle and is operated hydraulically or electrically. A gyroscope placed on the head of the train 'recognises' the cant of the outer rail in curves and determines the tilting angle. At the same time, an accelerometer placed on the front bogie measures the lateral acceleration generated in curves and acts for the determination of the progressive tilting of the car body. These two equipment devices interact with a microcomputer unit that generates the necessary commands. This technique allows for a tilting angle between the car body and bogie of $\varphi_t = 8°$. It was adopted as a basic principle by various rolling stock manufacturers and was applied with variants (different technical versions). This technique is applied for the following tilting systems: Pendolino, Fiat SIG, Nuovo Pendolino, ASEA, Bombardier, CAF (SIBI), Adtranz, and Siemens (Chiara et al., 2008).

In the Japanese technique, tilting is achieved by using equipment that is placed both on the track and on the vehicle (Figures 13.4 and 13.5). The technique combines passive tilting, that is, the natural pendulum function using the centrifugal force, with a pneumatic pressure (air) mechanical system. The identification of the track is achieved via an electromagnetic system that detects the curve. The geometric characteristics of the track are recorded for the entire length of the route in an electronic format using a computer that is mounted on the vehicle. The position of the train on the track is calculated every moment in relation to its speed, with the aid of a system that enables continuous communication between the track and the train. This technique could be considered more efficient than the previous one

Japanese form Mechanism of passive and controlled pendulum-type tilting	European Form Mechanism of active and forced inclined link-type tilting
[1] Tilt rolling center [2] Bolster spring [3] Tilt beam	
[4] Bogie frame [5] Track surface [6] Cant	

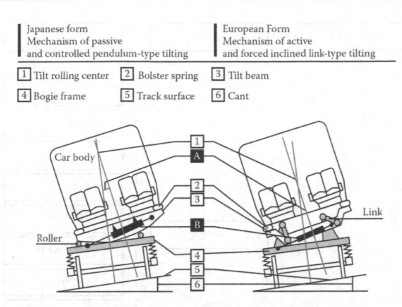

Figure 13.4 The two, active tilting techniques. [Adapted from Online image, available at: http://www.hita chirail.com/products/rollingstock/tilting/feature05.html (accessed December 2013). From Hitachi.]

Figure 13.5 Taiwan tilting train. [Adapted from Online image, available at: http://www.hitachi.co.jp/New/c news/month/2015/01/0109.html (accessed October 2015). From Hitachi.]

because of the improved lateral dynamic comfort experienced by the passengers when entering and exiting a curve. This comes as a result of the prediction of the exact point when the tilting commences. The above advantage is offset by the disadvantage of requiring the installation of equipment on the track. This system was manufactured by Hitachi and allows for a tilting angle of $\varphi_t = 5°$.

Figure 13.6 showcases the classification of tilting trains based on the tilting operation principle, the tilting technique, and the tilting system that is being used.

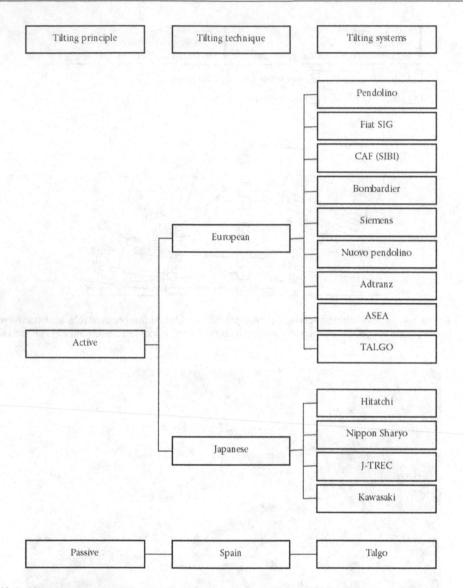

Figure 13.6 Classification of tilting trains based on the tilting operation principle, the tilting technique, and the tilting system.

13.3 MAIN CONSTRUCTIONAL AND OPERATIONAL CHARACTERISTICS OF TILTING TRAINS

13.3.1 Performances in terms of speed

Tilting trains are operated on new high-speed tracks where their maximum running speed can be V_{max} = 220–320 km/h; however, most commonly they are operated on upgraded conventional speed tracks where their maximum running speed is V_{max} = 150–200 km/h. On metric gauge tracks, the recorded running speed varies between V_{max} = 120 and 165 km/h, and the maximum speed is observed at the Australian railway network (V_{max} = 165 km/h).

Equation 13.1 is the mathematical expression of the tilting train's speed when running on curved segments of the horizontal alignment for standard gauge track (Chiara et al., 2008):

$$V_p = \sqrt{\left[(U + I + 2e_o \cdot \tan \varphi_t) \cdot R_c\right]/11.8} \qquad (13.1)$$

where:

V_p: Passage speed at curves (km/h)

R_c: Radius of curvature in the horizontal alignment (m)

U: Cant of the track (mm)

I: Cant deficiency (mm)

φ_t: Tilting angle (degrees)

$2e_o$: Theoretical distance between the running surfaces of the right and the left wheel when centred $\approx$ distance between the vertical axis of symmetry of the two rails

Equations 13.2 through 13.4 provide the mathematical expression of speed obtained by applying Equation 13.1 for conventional trains, trains with passive tilting, and trains with active tilting, respectively, considering the following track and rolling stock data:

- $2e_o = 1{,}500$ mm
- $U = 160$ mm
- $I = 105$ mm ($\gamma_{nc} = 0.7$ m/s^2)
- $\varphi_t = 0°$ for conventional trains
- $\varphi_t = 3.5°$ for tilting trains with passive tilting
- $\varphi_t = 8°$ for tilting trains with active tilting

$$V_p = 4.74\sqrt{R_c} \qquad (13.2)$$

$$V_p = 5.49\sqrt{R_c} \qquad (13.3)$$

$$V_p = 6.35\sqrt{R_c} \qquad (13.4)$$

Figure 13.7 illustrates a diagram that provides the variation of permitted passage speed V_p at curves in relation to the radius of curvature R_c for the three aforementioned train categories.

According to Equations 13.2 through 13.4 and Figure 13.7, tilting trains allow for an increase of running speeds by around 15% in the case of passive tilting and of around 35% in the case of active tilting. For the same geometrical design features, active tilting achieves higher running speeds by about 15% than passive tilting.

In practice, tilting trains have recorded increased running speeds in horizontal curves by 10–40%.

13.3.2 Tilting angle

Trains with active tilting achieve tilting angles of up to 8° (8° using the European technique and 5–6° using the Japanese technique). This results in an elevation of the car body in curves at the level of the external rail up to 200 mm for a standard gauge track.

Trains with passive tilting achieve tilting angles between 3° and 5° (3.5° under service conditions).

13.3.3 Track gauge

Tilting car-body trains can run on various track gauges (1,067, 1,435, 1,524, and 1,668 mm). In Spain, some Talgo trains are equipped with wheelsets that have variable axle lengths.

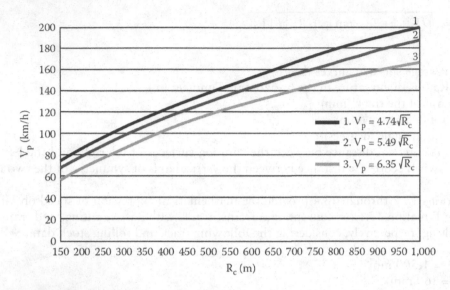

Figure 13.7 Variation of the permitted passage speed at curves in relation to the radius of curvature, for conventional trains and for tilting trains. (Adapted from Chiara, B., Hauser, G., and Elia, A., 2008, I Treni ad Asseto Variabile: Evoluzione, Prestazioni e Prospettive, *Ingegneria Ferroviaria*, July–August (7–8) pp. 609–648.)

13.3.4 Axle load

It usually ranges from 14 t to 17 t. For conventional high-speed trains, axle load reaches 17 t, while that of conventional-speed trains is higher.

13.3.5 Track superstructure

As mentioned in Section 13.1, the increase in speed affects the track's superstructure as it increases the value of lateral forces acting on the contact surface between the wheel and the rail (Kubata, 1992; UIC, 2003; Pyrgidis, 2009). Hence, the track should be able to offer increased lateral resistance. To meet these requirements, the track must have rails that are continuous welded with a minimum weight of 60 kg/m (UIC 60), concrete sleepers, and elastic fastenings. Slab track is the optimal solution. In the case of ballasted track, a ballast of great hardness and minimum layer thickness of 35 cm is needed (with an increased degree of compaction and increased ballast shoulder width). Last but not least, the track should be well stabilised.

13.3.6 Bogies' technology

All available bogies' technologies can be used for tilting trains [conventional bogies (Pendolino system), bogies with self-steering axles (ASEA system), and bogies with independently rotating wheels (Talgo system)].

13.3.7 Train formation

Tilting trains serve passenger trips exclusively. As regards their formation, tilting trains can be loco-hauled trains, railcars (multiple units), or push-pull trains.

13.3.8 Signalling

Tilting car-body technology is compatible with all signalling systems.

13.3.9 Traction

Tilting trains can be powered by either diesel or electricity. In the case of electric power, the electric power supply system requires adjustments as per the technical specifications of the tilting train.

13.3.10 Cost of rolling stock supply

The cost of tilting trains is around 15% higher than the cost of conventional trains (Profillidis, 2014).

13.4 REQUIREMENTS FOR IMPLEMENTING THE SYSTEM

Every railway company wishes to reduce run times and increase dynamic passengers' comfort for the network it manages. Operating tilting trains is an alternative to having a higher-quality infrastructure that ensures the desired performances described above.

Within this framework, the selection dilemma between conventional trains and tilting trains is directly related to the infrastructure on which they will be circulated.

More specifically, the following subsection will highlight cases in which tilting trains can be a choice.

13.4.1 Existing conventional-speed infrastructure

When the track alignment is characterised by a large percentage of horizontal curved segments and the run times need to be reduced without intervention on the track's layout geometric characteristics (radii, transition curves, and cant of track). As a prerequisite, the track's superstructure must be in good condition and it must have the necessary mechanical strength to be able to receive increased lateral loads exerted on the interface of the wheel flange and the rail. Meanwhile, the vehicles' new kinematic gauge, as a result of the tilting, must still lie within the limits of the current allowable minimum structure gauge.

Table 13.3 summarises the advantages and disadvantages of operating tilting trains on existing conventional-speed infrastructure.

The reduction in run time is directly related to the length of horizontal curved segments.

Table 13.3 Advantages/disadvantages of tilting trains in comparison with conventional trains – operation on existing conventional-speed infrastructure

Advantages	Disadvantages
• Reduction of run time on curved segments of the track • Reduction of the total run time • Zero cost of interventions on the track's alignment • Maintenance of mixed train traffic operation	• Additional stress and wear on the track's superstructure • Increased cost of rolling stock acquisition • Potential cost for the improvement of the track's lateral resistance and for widening the structure's gauge • More frequent maintenance of the track's superstructure • More complex maintenance of rolling stock

Table 13.4 shows:

- For routes where the curved segments of the horizontal alignment amount to 50%, 60%, 70%, and 80% of the total route length.
- Considering an increase of speed at curved sections of the track equal to 25%, by using tilting technology.

The percentage (%) reduction of run times obtained by using tilting technology trains compared to conventional ones.

In practice, the reduction in run times is around 12–20%.

Operating tilting trains allows for maintaining the mixed traffic of trains. More specifically, it allows for routing of passenger and freight trains on the same track, since no modifications on the track's alignment geometry are required (see Chapter 15).

When tilting trains equipped with conventional bogies run on tracks with small horizontal alignment radii (R_c = 300–500 m) and where flange contact is occurred, according to the literature (Pyrgidis, 2009):

- A 10% increase in passage speed V_p results in a 0.5 t increase of the guidance force F_{11} (front wheelset, left wheel) (Figure 13.8).
- A 20% increase in passage speed V_p results in a 1.0 t increase of the guidance force F_{11}.
- A 30% increase in passage speed V_p results in a 1.5 t increase of the guidance force F_{11}.

Table 13.4 Percentage reduction of run time by using tilting trains in comparison with conventional ones, for various percentages of horizontal curved segments' length in the total route length

Percentage of curved segments of the horizontal alignment in the total path length	50%	60%	70%	80%
Percentage reduction of run time by using tilting trains	10%	12%	14%	16%

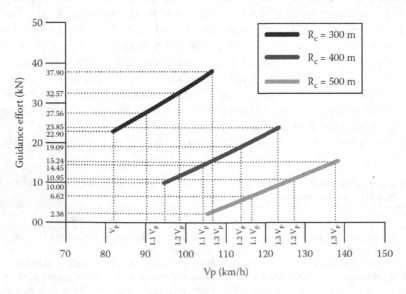

Figure 13.8 Motion of tilting trains with conventional bogies on horizontal curves with small radius – variation of the guidance force in relation to passage speed. (Adapted from Pyrgidis, C., 2009, Tilting trains/conventional trains – Comparison of the lateral forces acting on the track, *Ingegneria Ferroviaria*, April 2009, Rome, No 4, pp. 361–371.)

When tilting trains equipped with conventional bogies run on tracks with medium horizontal alignment radii (R_c = 1,200–2,000 m) where flange contact is avoided, an increase in energy consumption at the level of four wheels of each bogie is observed.

13.4.2 New conventional-speed infrastructure

When the terrain requires the adoption of many horizontally curved segments with small radii and the aim is to achieve better performance in terms of run times or dynamic passenger comfort, new conventional – speed infrastructure is needed.

The use of bogies with self-steering axles for tilting conventional-speed trains is the preferred choice for relatively low running speeds (V_{max} < 160 km/h) and for large numbers of curved segments of the horizontal alignment with small and medium radii (Pyrgidis and Demiridis, 2006; Pyrgidis, 2009).

13.4.3 New high-speed infrastructure

When the route is characterised by a high percentage of curved segments of the horizontal alignment and the aim is to further increase the lateral dynamical passenger comfort, new high-speed infrastructure is needed.

For high-speed tilting trains, it is essential to use conventional bogies or bogies with independently rotating wheels (Pyrgidis and Demiridis, 2006; Pyrgidis, 2009).

During the tests, and also during the up-to-date operation of tilting trains, the following malfunctions were recorded and resolved:

- 'Locking' of the car body in a tilted position on a straight path, that is, after leaving the curved segments.
- Problems during coupling of railcars (multiple units).
- Disabling of the tilting mechanism due to bad weather conditions.
- Compatibility problems with the signalling system.
- Problems with software.
- Very intense tilting of the car body.

13.5 HISTORIC OVERVIEW AND PRESENT SITUATION

Table 13.5 summarises the available data for tilting trains operating worldwide. All the data recorded relate to the end of the year 2019. The raw data were obtained from various available sources and cross-checked. Afterwards, they were further manipulated for the needs of this study.

Table 13.6 includes the number of new tilting train types per five years that was put in circulation until 2019. In total, since this particular technology was first implemented, 53 different types of such trains have been used. The highest number of tilting train types to be put in circulation was observed in the 1988–2002 period (30 types).

Of the total of 53 train types:

- 60% were of normal gauge and 30% of metric gauge.
- 34% circulated in Japan (18 types). This is followed by Germany, Italy, and Spain with 5 types each (9%).
- 15 train types developed high speeds (200–250 km/h) and 7 train types very high-speed (> 250 km/h).

Table 13.5 Tilting trains operating worldwide

Type	Operation launch year	Country	Tilting system	Max speed (km/h)	Track gauge (mm)	Train formation	Tilting angle (degrees)
TurboTrain	1968–82	Canada	UAC (Passive)	193	1,435	LC+TC	5
381 Series	1973	Japan	Hitachi	120	1,067	EMU	
Talgo Pendular	1974–93	Spain–Germany	TALGO	180–220	1,668 -1,435	LC+TC	3–3.5
ETR 401	1975	Italy	Pendolino	250	1,435	LC+TC	13
Class 370 APT	1975–86	United Kingdom	Pendolino	200	1,435	LC+TC	
LRC	1982	Canada	Bombardier	160	1,435	LC+TC	4–5
TRD 594	1982	Spain	CAF (SIBI)	160	1,668	DMU	6
ETR 450	1988–2015	Italy	Pendolino	250	1,435	EMU	8
2000 Series	1989	Japan	Hitachi	130	1,067	DMU	5
ICT	1989	Germany	Pendolino	230	1,435	EMU	8
X 2000	1990	Sweden	ASEA	210	1,435	LC+TC+DVT	8
Series 8000	1992	Japan	Hitachi	160	1,067	EMU	5
VT 610	1992–2014	Germany	Pendolino	160	1,435	DMU	8
8000 Series	1992	Japan	Hitachi	160	1,067	EMU	5
E351 Series	1993–2007	Japan	Hitachi	130	1,067	EMU	5
ETR 460	1994–1997	Italy	Pendolino	250	1,435	EMU	8
HOT 7000	1994	Japan	Hitachi	130	1,067	DMU	5
83 Series	1994	Japan	Hitachi	130	1,067	EMU	5
281 Series	1994	Japan	Hitachi	130	1,067	DMU	5
961 Series	1995	Japan	Hitachi	130	1,067	EMU	5
283 Series	1996	Japan	Hitachi	145	1,067	EMU + DMU	5–6
383 Series	1996	Japan	Hitachi	130	1,067	EMU	5
SM 220-SM3	1996–2003	Finland	Pendolino	220	1,524	EMU	8
ETR 470	1996–2015	Switzerland	Pendolino	200	1,435	EMU	8
VT 611	1997	Germany	Adtranz	160	1,435	DMU	8
ETR 480	1997	Italy	Pendolino	250	1,435	EMU	8
BM 73	1997	Netherlands	ASEA	210	1,435	EMU	8
QR Tilt Train	1998	Australia	Hitachi	165	1,067	EMU + DMU	5

(Continued)

Table 13.5 (Continued) Tilting trains operating worldwide

Type	Operation launch year	Country	Tilting system	Max speed (km/h)	Track gauge (mm)	Train formation	Tilting angle (degrees)
ALARIS 490	1999–2014	Spain	Pendolino	220	1,668	EMU	8
VT 612	1999	Germany	Adtranz	160	1,435	DMU	8
CP 400	1999	Portugal	Pendolino	220	1,668	EMU	8
ICN	2000	Switzerland	Adtranz	200	1,435	EMU	8
ACELA	2000	United States	Bombardier	240	1,435	DMU	4.2 (4–6)
SZ 310	2000	Slovenia	Pendolino	200	1,435	EMU	8
TALGO 350	2000	Spain	TALGO	350 (330)	1,435	LC + TC	3.5
BM 93 (Talent)	2001	Netherlands	Bombardier	140	1,435	DMU	7
Class 221 SuperVoyager	2002	United Kingdom	Bombardier	200	1,435	DMU	8
British Rail CLASS 390	2003	United Kingdom	FIAT SIG	225	1,435	EMU	8
CDT 680	2005	Czech Republic	Pendolino	230	1,435	EMU	8
MEITECHU Series	2005	Japan	Nippon Sharyo	120	1,067	EMU	5
N 700	2007	Japan	Hitachi + Nippon Sharyo	300	1,435	LC + TC	5
ETR 600	2008	Italy	Pendolino (New)	250	1,435	EMU	8
ETR 610	2008	Switzerland	Pendolino (New)	250	1,435	EMU	8
E5 Series Shinkasen	2011	Japan	Hitachi	300	1,435	EMU	5
E6 Series Shinkasen	2013	Japan	Hitachi	320	1,435	EMU	5
TALGO AVRIL	2013	Spain	TALGO	380	1,435	LC + TC	5
TRA	2013	Taiwan	Nippon Sharyo	150	1,435	EMU	5
TTX (Hanvit 200)	2013	South Korea	KRRI	200 (180)	1,435	EMU	8
ED250	2014	Poland	Pendolino	250	1,435	EMU	8
8600 Series	2014	Japan	Kawasaki	(200)	1,067	EMU	5
H5 Series Shinkansen	2016	Japan	Hitachi +	130	1,435	EMU	1.5
E353 Series	2017	Japan	Kawasaki	320	1,067	EMU	
2600 Series	2017	Japan	J-TREC	130	1,067	EMU	
		Japan	Kawasaki	120	1,067	DMU	

DMU, Diesel Multiple Units; EMU, Electrical Multipe Units; LC, LoComotive; TC, trailer vehicle; DVT, Driving Van Trailer.

Table 13.6 Number of new types of tilting trains put in circulation per five years until 2019

Time period	Number of new types of tilting trains put in circulation
1968–1972	1
1973–1977	4
1978–1982	2
1983–1987	0
1988–1992	7
1993–1997	13
1998–2002	10
2003–2007	4
2008–2012	3
2013–2017	9
Total	53

As it can be concluded from the above, tilting car-body train technology has been a research topic of interest for railway engineers for decades.

In more detail:

The 1930s

Experiments commence in the United States. The vehicles are named 'pendulum' and achieve the tilting of the car body through inertia forces.

The 1950s

In 1956, in France, an engineer named Mauzin builds and experiments with a single vehicle that uses passive (unassisted) tilting.

The 1960s
In 1967, the Italian company Fiat Ferroviaria manufactures the prototype tilting train YO160, named Pendolino.
In 1968, the United States launches the first high-speed tilting train with passive tilting (TurboTrain).
Active tilting trains are being studied in Germany for the first time.
The 1970s
The first experimental tilting trains using the active tilting feature in Japanese railways.
Tilting trains conquer Italy with Pendolino train at first and ETR trains later on.
Development of tilting trains' technology in Spain (Talgo).
The British railways build the experimental tilting train named APT (Advanced Passenger Train). The technology used for the construction of the train is similar to that of the Italian ETR and the Spanish Talgo. The British railways, due to technical problems, were never able to make this train reliable enough to be put into service.
ABB develops an alternative active tilting system.
The 1980s
In Italy, the commercialisation of tilting trains is intensified (ETR trains).
Operation of tilting trains in Japan and Canada (LRC).
The 1990s
In the early 1990s, the Swedish train X2000 (manufactured by ABB) is put into service.

In 1992, tilting trains enter the German market with the VT 610, followed by Finland and Switzerland (1996).

In 1996, the French National Railways (SNCF) and Alstom study the use of tilting trains on existing tracks and the conversion of a TGV into tilting TGV.

In 1997, the application of tilting technology on the connection Boston–Washington is studied.

The decade 1990–2000 is the peak period for this technology. During this decade, most of the tilting body train models were built. After 2000, the number decreases, yet it remains constant.

The 2000s

In 2000, Alstom purchases 51% and in 2002 the total of Fiat Ferroviaria's shares.

In 2001, ADtranz is purchased by Bombardier Transportation. During the time of purchase, ADtranz was the second manufacturer of rail rolling stock worldwide.

In 2004, Virgin Trains puts tilting trains named British Rail Class 390 into service on the mainline of the UK's west coast.

The 2010s

In December 2013, Talgo manufactures AVRIL (Alta Velocidad Rueda Independiente Ligero) train, with a design speed V_{rs} = 380 km/h.

In South Korea, the Korean Railroad Research Institute (KRRI) together with a team of Korean manufacturers built the Tilting Train Express (TTX), or Hanvit 200.

In Japan, Hitachi with Kawasaki, built in 2014–2015 the H5 series Shinkansen running on normal gauge tracks. Kawasaki (2014–2018), J-TREC (2015) and Kawasaki (2017) built the 8600 series, the E353 series, and the 2600 series, respectively running on metric gauge (1,067 mm).

In Italy, the circulation of tilting trains is limited (end 2019). There are still a small number of tilting trains in operation, particularly ETR460 (though operated without tilting), ETR610, and ETR 480/485.

REFERENCES

ABB. n.d., *The Fast-Train – Technology for Faster Passenger Services on Existing Railway Lines*, Leaflet, Manheim, Germany.

Chiara, B., Hauser, G. and Elia, A. 2008, I treni ad asseto variabile: evoluzione, prestazioni e prospettive, *Ingegneria Ferroviaria*, July–August (7–8), pp. 609–648.

Kubata, G. 1992, La technologie des véhicules à caisse inclinable – Perspectives et limites, *Rail International*, Brussels, 6–7, p. 163.

Online image, available online at: http://www.hitachi-rail.com/products/rolling stock/tilting/feature05.html (accessed December 2013).

Online image, available online at: http://www.hitachi.co.jp/New/cnews/month/2015/01/0109.html (accessed October 2015).

Profillidis, V.A. 2014, *Railway Management and Engineering*, Ashgate, Farnham.

Profillidis, V. 1998, *A Survey of Operational, Technical and Economic Characteristics of Tilting Trains*, Rail Engineering International, The Netherlands, 2, pp. 3–7.

Pyrgidis, C. and Demiridis, N. 2006, The effects of tilting trains on the track superstructure, *1st International Congress, 'Railway Conditioning and Monitoring' 2006*, IET, 29–30 November 2006, Birmingham, UK, pp. 38–43.

Pyrgidis, C. 2009, Tilting trains/conventional trains – Comparison of the lateral forces acting on the track, *Ingegneria Ferroviaria*, April 2009, Rome, No. 4, pp. 361–371.

UIC. 2003, FICHE 705, *Infrastructure Pour Les Trains à Caisse Inclinable*, 1st edition, August 2003, Paris, France.

Chapter 14

Metric track gauge intercity railway networks

14.1 DEFINITION AND DESCRIPTION OF THE SYSTEM

The term 'metric track gauge railway networks' describes networks with a track gauge of 900–1,100 mm.

Metric gauge appears in many countries, mostly in parts of their network; but in some cases, it is the exclusive track gauge.

The principal difference between normal and metric track gauges lies in the permanent way. Metric track gauge occupies less surface area, which means lower implementation cost and less expropriation. Moreover, a metric gauge line is generally quite flexible and adaptable to the landscape; this feature makes it ideal for mountainous areas.

The vast majority of metric tracks were designed and constructed many years ago. The following were mainly the reasons for choosing a narrower track gauge back then:

- *Economic*: It is estimated that a metric gauge track results in 30% less implementation cost than the track of normal gauge.
- *Strategic*: To ensure incompatibility with the networks of neighbouring countries and thus protection from a potential enemy invasion.
- *Operational*: In those days, traffic volumes and transport capacity requirements were quite low, and the performance of the rolling stock was limited.

In this framework, most metric lines are nowadays characterised by:

- Low running speeds, usually V_{max} < 120 km/h.
- Limited axle loads, usually Q_{max} < 16–18 t.
- Small transport capacity, passenger, or freight.

Nevertheless, there are lines that operate at a maximum running speed of V_{max} = 160 km/h (South Africa), axle loads of Q = 25–30 t (Australia, South Africa), and transport more than 6 billion passengers annually (Japan, JR East).

The tilting train of Queensland Rail in Australia holds the record of being the fastest train on metric track gauge in the world (V = 210 km/h).

Metric gauge systems are classified as follows:

On the basis of the exact track gauge:
- 914 mm (3 ft).
- 950 mm.
- 1,000 mm (metric).
- 1,050 mm and 1,055 mm.
- 1,065 mm and 1,067 mm (Cape gauge).

DOI: 10.1201/9781003046073-14

On the basis of the maximum running speed V_{max}:
- Low speed ($V_{max} \leq 80$ km/h).
- Medium speed (80 km/h $< V_{max} \leq 120$ km/h).
- High speed ($V_{max} > 120$ km/h).

On the basis of the maximum allowed axle load Q_{max}:
- Small axle load ($Q_{max} \leq 16$ t).
- Medium axle load (16 t $< Q_{max} \leq 20$ t).
- High axle load (heavy haul) ($Q_{max} > 20$ t).

Metric track gauge applies for nearly all railway systems (urban, suburban, regional, intercity, steep longitudinal gradients, etc.).

South Africa possesses the longest metric railway network (33,520 km), followed by Brazil (23,489 km), Japan (20,264 km), Australia (15,160 km), and Argentina (11,080 km). The majority of metric lines feature a gauge of 1,065 mm and 1,067 mm. The largest percentage of metric lines is in Africa. Many neighbouring countries use the same track gauge, which ensures interoperability.

The current trend dictates that metric lines are converted to normal gauge lines. On the contrary, normal gauge lines are not converted into metric anywhere in the world, while the ones already existing are only conditionally upgraded.

14.2 MAIN CONSTRUCTIONAL CHARACTERISTICS OF INTERCITY METRIC TRACK GAUGE LINES

14.2.1 Track alignment: differences between tracks of metric and normal gauge

The track geometry alignment design of a metric track (1,000 mm) is different from a normal track (1,435 mm) in the following aspects (Montagné, 1988; OSE, 2004):

- The distance between the vertical axis of symmetry of the two rails ($2e_o$) in the metric track is smaller (1,056 mm, for 31.6 kg/m rails) than in a normal track (1,500 mm) since it depends directly on the track gauge.
- For the same speed V and curve radius R_c of the horizontal alignment of a metric and a normal gauge track, the value of the theoretical cant U_{th} is about 30% smaller for the metric track than it is for the normal one.

For $g = 9.81$ m/s^2, $2e_o = 1,056$ mm and after appropriately converting the relevant units, the following equation is derived:

$$U_{th}(mm) = 8.306 \cdot \frac{V_{max}^2 (km/h)}{R_c(m)} \tag{14.1}$$

Equation 14.1 holds for rails that weigh 31.67 kg/m. In the case of rails that are of different quality, a different coefficient (than 8.306) is used. For instance, for UIC 50 and UIC 54 rail types, the coefficient is set to 8.416.

- For the same speed V and curve radius R_c of the horizontal alignment of a metric and a normal track, the value of the cant U is smaller in the case of the metric track.

- The maximum value of cant deficiency I for metric track gauge lines does not exceed 100 mm, while for the normal track gauge lines one can find values of up to 150 mm.
- The metric track, when considering the same daily traffic volume, speed and curve radius in the horizontal alignment, is designed with a smaller value of cant excess E_c than the normal gauge track. Specifically, for the metric track, the maximum allowed cant excess usually does not exceed 70 mm, while a normal gauge network can allow values of up to 110 mm (UIC Meter Gauge Group, 1998).
- The rate of change of cant deficiency for a metric track is theoretically smaller than for a normal gauge track, regardless of the applied speed.
- Metric track lines are usually designed with shorter transition curves than normal gauge lines; in some cases transition curves are completely omitted, leading to higher twist values. The same problem appears when a metric track applies the same value for change of cant per unit length as a normal track. Therefore, the maximum allowed twist in the metric track is lower than in the normal track (Rhätische Bahn, 1986; SNCF, 1998). In horizontal alignment, when designing successive compound or reverse curves, the intermediate straight segment can be shorter in the case of a metric track.
- In the vertical alignment, when designing successive gradient curves, an intermediate straight segment of minimum 20 m must be included in the metric track (30 m for the normal gauge track).

In general, there are no commonly accepted regulations or standards regarding track layout and track superstructure characteristics of a metric line.

Table 14.1 presents in summary, the main track alignment geometric characteristics, the boundary values, and standards used for the Greek metric railway network. These values originate from the 'Regulation for the Track Layout and Superstructure of a Metric Line', which was composed in 2006 to upgrade the network and increase the track design speed from V_{max} = 90 km/h to V_{max} = 120–140 km/h (OSE, 2004; Lambropoulos et al., 2005).

The minimum allowed horizontal curve radius for newly constructed main lines is set to R_{cmin} = 150 m. The minimum length for transition curves can be calculated from the following two equations (both must be fulfilled):

$$L_{k\,min} = 0.4 \cdot U \tag{14.2}$$

$$L_{k\,min} = \frac{U \cdot V_{max}}{125} \tag{14.3}$$

where:

$L_{k\,min}$: Minimum allowed length for a transition curve (m)
U: Cant of the track (mm)
V_{max}: Maximum running speed (km/h)

In parts of the line that are being renovated or upgraded, the constant 115 is used instead of 125, where possible. Moreover, the coefficients 0.04615 and 0.05728 of Table 14.1 adopted for the calculation of R_{cmin} are valid for the case of continuous welded rails that weigh 31.6 kg/m ($2e_o$ = 1,056 mm). For rails of type UIC 50 and UIC 54, the aforementioned coefficient values change to 0.04675 and 0.05804, respectively.

Table 14.2 provides the equivalent standards for the networks of Australia (QR), South Africa (Spoornet), Japan (JR East), Switzerland (RhB), and France (SNCF).

Table 14.1 Greek railway network of metric track gauge – track alignment geometric characteristics

Characteristics	Notation	Value
Track gauge	2e	1,000 mm
Distance between the vertical axis of symmetry of the two rails	$2e_o$	1,056 mm
Maximum cant	U_{max}	105 mm
Maximum cant deficiency	I_{max}	75 mm
Maximum cant excess	E_{cmax}	70 mm
Maximum rate of change of cant deficiency	$\Delta I_{max}/\Delta t$	40 mm/s
Maximum track twist	g_i	$\dfrac{125}{V_{max}}(mm/s)$
Highest permitted value for track twist	g_{imax}	2.5 mm/m
Maximum permitted residual lateral acceleration	γ_{ncmax}	0.07 g
Minimum horizontal alignment curve radius	R_{cmin}	$R_{c\,min}(m) = 0.04615 \cdot V_{max}^2 (km/h)$ $R_{c\,min}(m) = 0.05728 \cdot \left(V_{max}^2 - V_{min}^2\right)(km/h)$
Minimum vertical alignment curve radius	R_{vmin}	2,000 m – Convex curves 1,500 m – Concave curves 500 m – Auxiliary or siding lines
Maximum longitudinal gradient	i_{max}	1.6%
Cant (normal)	U	$U(mm) = 4.845 \cdot \dfrac{V_{max}^2 (km/h)}{R_c(m)}$
Theoretical cant	U_{th}	$U_{th}(mm) = 8.306 \cdot \dfrac{V_{max}^2 (km/h)}{R_c(m)}$

14.2.2 Track superstructure

Table 14.3 presents the characteristics of the track superstructure (track design speed, axle load, and track panel components) for several metric gauge networks around the world.

In most cases, the maximum permitted axle load lies in the range of Q = 14–18 t. However, some metric networks, such as in South Africa (Spoornet) and in Australia (Queensland Rail), can sustain axle loads that exceed 25 t, owing to the reinforced track structure.

The majority of small axle-load metric networks operate on rails of specific weight of 30–50 kg/m and length of 10–20 m.

For high speeds (V > 120 km/h) and axle loads Q > 14 t, it is generally necessary to use rails that weigh more than 31.6 kg/m, while as for the use of the lighter rail that weighs 31.6 kg/m, the good condition of the track's geometry is a prerequisite (IVT/ETH, 1994).

Regarding the sleepers, the use of wooden sleepers is widely applied in low-speed metric gauge networks, while the use of concrete monoblock sleepers applies in high-speed metric gauge networks. According to the UIC recommendations, reinforced concrete sleepers are to be used in cases of track design speeds of V_d = 120–160 km/h and axle load Q ≤ 13 t. Moreover, the sleepers should be placed at a distance of ≤ 650 mm from each other. The length of the sleepers in most metric gauge tracks of small axle load ranges from 1,800 mm to 2,000 mm, and the distance between sleepers varies between 600 mm and 700 mm.

Table 14.2 Metric gauge railway networks in Australia, South Africa, Japan, Switzerland, and France –
track alignment geometric characteristics

Network operator	QR, Spoornet, JR East	RhB	SNCF
Track gauge (2e)	QR: 1,067 mm Spoornet: 1,065 mm JR East: 1,067 mm	1,000 mm	1,000 mm
Minimum horizontal alignment curve radius (R_{cmin})	100 m	45 m	
Straight segment length between reverse curves	$\geq$ 10 m	$\geq$ 20 m	$\geq$ 30 m
Theoretical (U_{th}) and normal cant (U)		$U = \dfrac{4.55 \cdot V_{max}^2}{R_c}$	$U = \dfrac{8.5 \cdot V_{max}^2}{R_c}$
Maximum cant deficiency (I_{max})	50–60 mm	86 mm	70–90 mm
Maximum normal cant (U_{max})	100–110 mm	105 mm	100 mm
Highest permitted value for track twist (g_{imax})	QR: 3.33‰ Spoornet: 4.0‰ JR East: 2.5‰	2.5‰	2.5‰
Maximum longitudinal gradient (i_{max})	QR : 20‰ Spoornet: 25‰	70‰	25‰
Minimum vertical alignment curve radius (R_{vmin})	R_{vmin} = 2,000–4,000 m (V_{max} > 100 km/h) $R_{V_{min}}$ = 1,650 m (V_{max} > 60 km/h) Convex: $R_{V_{min}}$ = 525 m Concave: R_{vmin} = 300 m (V_{max} < 60 km/h)	Convex: $R_{V_{min}} \geq \dfrac{V_{max}^2}{2.5}$ $R_{V_{min}}$ = 1,500 m Concave $R_{V_{min}} \geq \dfrac{V_{max}^2}{4}$ $R_{V_{min}}$ = 1,000 m	$0.35 \cdot V_{max}^2$ R_{vmin} = 2,000 m

The use of sleeper anchors (the so-called 'safety caps') drastically increases the lateral resistance of the track panel; however, this also depends on how dense they are installed (i.e., every fourth, third, second, or even every sleeper). According to Lymberis et al. (2006), it is feasible to apply continuous welding of the track by using concrete sleepers; also feasible is the placement of anchors on each sleeper and track that weigh 31.6 kg/m at a radius of R_c = 100 m.

Most metric tracks employ ballasts of 240–300–mm thickness, but there are some cases where the ballast is constrained to 180–210 mm (SNCF, 1998).

14.3 ADVANTAGES AND DISADVANTAGES OF INTERCITY METRIC GAUGE LINES

14.3.1 Advantages

- The integration of a metric gauge line to the terrain requires smaller width for the permanent way when compared with the normal gauge line. This results in lower implementation and expropriation costs.
- It is relatively easy and also environmentally friendly to lay the track in areas with a difficult landscape.

Table 14.3 Track superstructure characteristics from different metric gauge networks around the world

Network operator	Country	Speed		Load	Rails			Type				Sleepers — Dimensions			
		Passenger trains (km/h)	Freight trains (km/h)	Per axle (t)	Weight (kg/m)	Length (m)	CWR	Wooden	Monoblock concrete	Twin block	Steel	Length (mm)	Width (mm)	Height (mm)	Distance in between (mm)
Spoornet	South Africa	160	80	30	57	36	Yes	Yes	Yes			2,200	300	258	650
QR	Australia	140	100	25	60		Yes	Yes	Yes			2,150	230	115	685
JR Hokkaido	Japan	140	100		50	25	Yes		Yes			2,000	240	240	640
JR West	Japan	130	110	18	50	25	Yes	Yes	Yes			2,000	240	170	568
JR East	Japan	130	110	16	50	25	Yes		Yes			2,000	200	140	625
JR Shikoku	Japan	130	95	16	50	25	Yes		Yes			2,000	240	201	661
JR Kyushu	Japan	130	95		50	25	Yes	Yes	Yes			2,100	200	140	550
SNCFT	Tunisia	120	65	16	46	18	Yes	Yes		Yes	Yes	1,711	274	200	600
SRT	Thailand	120	70	15	40	18	Yes	Yes				2,000	200	150	650
CJR	Japan	120			60	25	Yes		Yes			2,000	240	190	
SARCC	South Africa	110	80	22	48	36	Yes								700
JR Freight	Japan	–	110	17	50	25	Yes	Yes				2,100	200	140	694
Perumka	Indonesia	100	75	18	42	17	Yes	Yes	Yes			2,000	220	200	600
KTMB	Malaysia	100	72	16	40	12	Yes	Yes	Yes			2,000	280	235	600
IR	India	100	75	16	41	13	Yes	Yes			Yes				650
OSE	Greece	90	75	14	31	15	Yes	Yes				1,800	220	130	640
TAZARA	Tanzania	80	80	20	45	12	Yes	Yes	Yes			2,000	200	145	680
SITARAIL	Côte d'Ivoire	80	60	17	36		Yes			Yes	Yes	1,800	120	75	666
PNR	Philippines	75	45	15	37	20	Yes	Yes	Yes			2,134	203	127	600
KR	Kenya	72	65	17	40	12	Yes				Yes	2,057	279		500
URC	Uganda	65	55	14	40	12	Yes	Yes			Yes	1,990	280	90	700
RNCFM	Madagascar	60	45	16	30	12	Yes	Yes			Yes	1,900	220	170	600
CFL	Angola	50	45	16	30	10	Yes	Yes			Yes	1,850			700
CFRC	Cambodia	45	30	15	30	12	Yes	Yes			Yes	1,800	200	150	650

Source: Adapted from Paradissopoulos, I., and Paradissopoulou, F., 1998, Comparative analysis of track technical characteristics of metric gauge railway networks, Technika Chronika, TEE, I, I.

14.3.2 Disadvantages

- There is a higher risk of vehicles' overturning, which inevitably limits the allowed running speed (Chenuc, 1988). Reducing the speed results in the reduction of track capacity (Pyrgidis and Stergidou, 2011).
- The dynamic gauge of vehicles running on metric tracks is normally smaller than on normal tracks to ensure stability. This limits the capacity of freight wagons and imposes constraints on carrying large cargo and on the comfortable arrangement of the interior of passenger rolling stock.

14.4 REQUIREMENTS FOR IMPLEMENTING THE SYSTEM

As already mentioned, there is a trend toward converting metric gauge lines into normal ones (while no normal gauge line is being converted into a narrow gauge one).

The construction of a new metric line should be carried out when:

- Interoperability with the neighbouring networks is guaranteed. This condition explains why particular track gauges are widespread in discrete parts of the planet.
- It is considered to be an extension of an existing metric gauge line which is desired to be retained and upgraded.
- The regions served are environmentally sensitive areas and/or with a difficult landscape.
- It constitutes a section of an isolated network, thus making it economically more feasible.

In all cases, a transition from a metric track to normal gauge or the opposite is a costly and timely process. This can be achieved in several ways:

- By transferring passengers/cargo at the transition point. However, this has a negative impact on the cost and transport/travel time.
- By use of a mixed gauge track (Figure 14.1). This option provides the possibility of operating both metric and normal gauge bogies on the same track superstructure.

Figure 14.1 Mixed gauge track (normal and metric gauge tracks, Kiato, Greece). (Photo: A. Klonos.)

Figure 14.2 Varying-gauge wheelset. (Online image, available at: https://www.youtube.com/watch?v=U_L FlUkcPNM, Talgo 250 gage change – Animated (accessed 14 March 2015).)

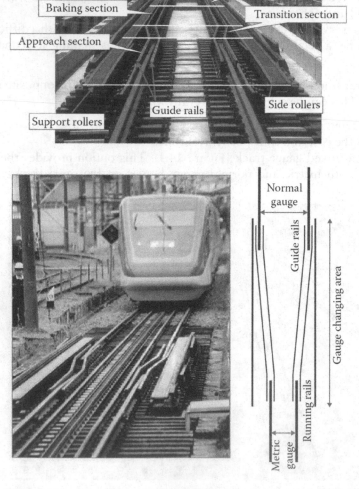

Figure 14.3 Track gauge change over equipment. (Adapted from Motoki, H., et al., 2001, *Developing a gauge-changing EMU*, available at: http://www.uic.org/cdrom/2001/wcrr2001/pdf/sp/1_1_1/413.pdf.)

Nevertheless, the implementation and maintenance costs, in this case, are significantly high (Alvarez, 2010).

- By changing bogies. In this method, the car body is lifted while the bogie underneath changes. The car body is then repositioned to the new bogie. Evidently, this approach is considerably time-consuming.
- By use of varying-gauge axles (automatic track gauge changeover). In this case, railway vehicles need to be equipped with wheelsets of varying gauges (Figure 14.2). The gauge of these axles' changes, following the transition as the train passes over a section of the track panel that is specifically equipped with a relevant track gauge changeover configuration (Figure 14.3). In this procedure, the wheels of the vehicles 'unlock' and move closer to each other – or further, depending on the case – and 'relock' accordingly, while the train runs over the specified section. The transition mechanism of the axle ensures the safe standstill of the wheel. Such systems are being used in several countries (Spain, France, and Japan) and mostly around cross-border railway passes of (e.g., Spain, France and Japan) with different track gauges (e.g., border pass Spain–France). Such a novel system based on independently rotating wheels has been developed in Switzerland (Prose AG, 2015). The system has been designed to allow operation on both Montreux Oberland Bernois Railway's (MOB) metre-gauge line and BLS standard-gauge infrastructure and has been tested successfully by Prose in Zweisimmen (Burroughs, 2019).

REFERENCES

Alvarez, A.G. 2010, *Automatic Track Gauge Changeover for Trains in Spain*, VíaLibre Technical Collection, Madrid.

Burroughs, D. 2019, GoldenPass variable-gauge system successfully tested, *IRJ*, available online at: https://www.railjournal.com/rolling-stock/goldenpass-variable-gauge-system-successfully-tested/ (accessed 6 December 2020).

Chenuc, C. 1988, Vitesse et sécurité sur ligne à voie métrique, *RGCF*, July-August, 7–8, pp. 13–17.

IVT/ETH. 1994, *Optimierung des Oberbaus bei Meterspurbahnen*, Zürich.

Lambropoulos, A., Lymberis, K. and Pyrgidis, C. 2005, The new regulation for the infrastructure of the hellenic, metric gauge, railway network – Application impacts, *2nd International Conference, 'Development of Railway Transportation Systems'*, Athens, 1–2 December 2005, pp. 133–144.

Lymberis, K., Pyrgidis, C. and Lambropoulos, A. 2006, L' utilizzo delle rotaie leggere per binario a scartamento metrico. Un approccio teorico all' impiego in curva di piccolo raggio, *Ingegneria Ferroviaria*, March 2006, Rome, No. 3, pp. 211–222.

Montagné, S. 1988, Les paramètres de tracé en voie étroite, *RGCF*, July–August, 7–8, pp. 7–12.

Motoki, H., Takahashi, A., Okamoto, I. and Oda, K. 2001, *Developing a Gauge-Changing EMU*, available online at: http://www.uic.org/cdrom/2001/wcrr2001/pdf/sp/1_1_1/413.pdf (accessed 07 August 2015).

Online image, available online at: https://www.youtube.com/watch?v=U_LFIUkcPNM, Talgo 250 gage change – Animated (accessed 14 March 2015).

OSE. 2004, Creation of a new regulation for the infrastructure of the metric gauge network of Peloponnesus, *Research Programme*, Aristotle University of Thessaloniki/National University of Athens.

Paradissopoulos, I. and Paradissopoulou, F. 1998, Comparative analysis of track technical characteristics of metric gauge railway networks, *Technika Chronika*, TEE, I, 1, pp. 37–48.

Prose AG. 2015, Factsheets, available online at: http://www.prose.ch/de/home/download/factsheets/? (accessed 14 March 2015).

Pyrgidis, C. and Stergidou, A. 2011, Investigation of the influence of railway system parameters on track capacity, *11th International Conference 'Railway Engineering-2011'*, 29–30 June 2011, London.

Rhätische Bahn. 1986, NormalienbuchUnterbau/Oberbau ChurR 30.1. 1975/1986.

SNCF. 1998, *Voies étroites: Particularités de pose et d'entretien*, Direction de l'Equipment, Paris, France.

UIC Meter Gauge Group. 1998, *Common Recommended Practices for Meter Gauge Rolling Stock*, Draft 5, November, Paris, France.

Chapter 15

Organisation and management of freight railway transport

15.1 PROVIDED SERVICES AND CLASSIFICATION OF FREIGHT RAILWAY TRANSPORTATION SYSTEMS

The term freight railway transport refers to any movement of goods by rail.

In freight railway transportation, there are differences in the provided services that can be attributed to the variety of customer demands, and most importantly, to the volume and type of the transported cargo (Alias, 1985).

The provided services fall into the following categories:

- Less-than-wagonload services.
- Wagonload services.
- Services of combined transportation.
- Services of heavy haul transport.
- Special services.

The less-than-wagonload services, also termed as part-load traffic, Less-than-Car-Load (LCL), and Less-than-Truck-Load (LTL) concern the transport of small packages under very strict time conditions and can be characterised as high-speed freight services (e.g., TGV postal high-speed trains, Mercitalia Fast) (Troche, 2005; Zanuy, 2013). In the provided rail freight services, also included are the transportation of small containers, pallets, and other forms of cargo that do not make up a full wagon, under given conditions (Zanuy, 2013).

Wagonload services are distinguished in Single Wagon Load (SWL) services and in Train Load services (TL).

SWL services involve cargo movements using single-wagon trains. In these traditional freight services, the wagons (each wagon containing loads of single or multiple shippers/consignees) start their journey from different origin points or end their journey at different destinations which necessitates that they pass through intermediary marshalling yards.

Regarding TL services cargo movement is performed by:

- *Unit trains* also termed *full trains*: These trains are formed of full wagons which start from the same origin and travel to the same destination. These trains make full use of the tractive power of the locomotives. They do not stop at marshalling yards and they run between two specific terminal stations. They have one shipper, one consignee, and one bill of lading, and serve more than one final receiver (Zanuy, 2013).
- *Block trains*[1]: Like unit trains, block trains run between two specific terminal stations without stopping at any marshalling yard. Their difference lies in the fact that they do not use all the power of their traction units while hauling.
- *Exclusive trains*[*]: They are unit trains or block trains that serve a single final receiver.

---

[1] The nomenclature varies. Block trains and exclusive trains are also usually termed 'unit trains'.

DOI: 10.1201/9781003046073-15

The wagons' assembling process in marshalling yards is time consuming and costly; hence during the scheduling of the transported freight distribution, particular concern is given so that most wagons are not subjected to the process of sorting more than once, or twice at the most.

Combined train transportation covers a wide range of services (see Section 15.5.1). Heavy haul transportation services use wagons of high transport capacity forming trains with an axle load of Q = 25–40 t (see Chapter 16).

Special services involve dedicated connections between mines and ports or mines and factories. The railway operates on a continuous basis ensuring a constant load flow, while special wagons are often used for transportation. A 'door-to-door' delivery can also be considered as a special service, since the railway company undertakes the receiving and delivering products from/to the user's 'door'.

Freight railway transport systems may be distinguished based on several characteristics. Specifically:

- *Based on load*:
  They may be classified among systems (trains) transporting conventional loads (under 25 t per axle), systems transporting heavy loads (25–40 t per axle) (See Chapter 16), and systems of 'mass transport' (See Section 15.5.2).
- *Based on traffic composition*:
  Depending on the percentage per type of trains running over a corridor, a classification may be made between systems running on a dedicated freight corridor, meaning the circulation of exclusively freight trains on the corridor, and systems running on a mixed traffic corridor, meaning the simultaneous circulation of freight and passenger trains along the same line. Moreover, there may also be the case of mixed traffic circulation, during which, however, a clear emphasis and prioritisation is given to one of the two types of traffic (corridor mainly used by freight or by passenger trains) (See Section 1.4.5).
- *Based on the type of goods that are being transported*:
  Transported goods may be classified between conventional and dangerous, which require, among others, specialised wagons and procedures (See Section 15.6.3).
- *Based on the geographical characteristics of the connection*:
  From a geographical perspective, freight railway transport may be classified as local, long-distance national, international and intercontinental transport.
- *Based on their complementarity with other modes of transport*:
  The use or not of other modes of transport besides railways at any point of the freight transport chain, meaning from the point of origin to the final destination, distinguishes railway transportation into combined transport or transport exclusively by rail.

15.2 SERVICE LEVEL OF FREIGHT RAILWAY TRANSPORT: QUALITY PARAMETERS

The basic actors that are involved and that influence a freight railway transport system are:

- *The dispatchers*: The cargo dispatchers are the ones that create the demand for freight transport and are responsible for the choice of transportation mode. Therefore, it is by them that the market share of railways is largely influenced.
- *The recipients*: The satisfaction of the recipients also largely affects the choice of transportation mode and therefore the market demand.

- *The railway operators – owners and operators of rolling stock*: Railway operators set the available in the market supply of freight transportation and their choices define the nature of the provided services (block trains, unit trains, etc.), the timetable frequency, the transport or not of dangerous goods, the availability of mass transport, and the involvement in combined transport.
- *The railway infrastructure managers*: They also define the supply and the quality of offered services. They are responsible for the construction, availability, and maintenance of tracks, signaling and electrification equipment, terminal stations, special installations, etc.
- *The European Union and National Authorities*: The institutional and legal framework in place largely affect the freight railway transport sector. European Authorities with the issuing of Directives and the granting of funds in a particular field of research or application affect to a great degree the nature of the offered services. National Authorities are responsible for the adaptation of Directives to National conditions and for their implementation.

The basic requirements of the various actors are presented in Table 15.1.

The level of service provided by freight railway transportation is evaluated according to the following quality parameters:

- Tariffs.
- Service frequency.
- Punctuality in the delivery of goods and regularity of itineraries.
- Transport possibilities provided by the railway system (e.g., the capability of transporting goods of large quantities, and bulk products).
- Facilities offered at freight stations (e.g., storage area, and appropriate equipment for loading and unloading).
- Flexibility and quick response to customer demands.
- Insurance during the transportation of goods (against theft, damage).
- Monitoring of cargo during its movement.
- Special services provision (transport of dangerous goods, combined transport, and specialised cargo transport).

Additional quality parameters for international freight railway transport are given below:

- Short delays at border stations.
- Ensuring interoperability.

Table 15.1 Basic requirements of the actors involved in freight railway transport (Boysen, 2012)

Actor	Requirements
Dispatchers	Mass transport, low transport cost, reliability, good wagon placement within train formation, safety during transport and waiting at marshaling yards, relatively short transport duration, regularity and frequency of dispatches, availability of special-use wagons
Recipients	Low transport cost, transport without damages to cargo (quality), service reliability, adherence to the timetable, last mile services
Railway Operators – Owners and Operators of rolling Stock	Low maintenance cost, equipment reliability, optimal use of the track capacity, safety, railway infrastructure of a good condition
Railway Infrastructure Managers	Use of infrastructure, use of network capacity, low maintenance costs, rolling stock of a good condition
National Authorities	Healthier competition conditions, environmental protection

Research on the requirements of dispatchers and recipients has concluded that a low transport cost is their primary requirement (Lundberg, 2006). However, in order for the transport service to be considered attractive and competitive other factors are also required. Dispatchers, for instance, consider punctuality as more important than transport duration, although for some special types of service (such as 24-h delivery), it may be considered as highly important for specific origin–destination pairs.

Some of the main features that characterise freight railway transport are:

- *Length of trains*: In Europe, the length of trains can reach up to 750 m (although exceptions exist, such as in Estonia, Denmark, Germany, and in specific corridors in France, where the length of trains may reach up to 800 or even 1000 m) (CER, 2016; Islam and Mortimer, 2017). In the United States, trains used for the transportation of conventional loads are larger (up to 2 km), whereas, in dedicated mine railways (such as those in Africa), their length can reach up to 4 km for heavy haul rail transport.
- *Weight of trains*: Trains routed in Europe usually weigh around 1,500–2,000 t. Only three countries, namely Russia, Sweden, and Norway, operate trains of 5,000 t. On the contrary, in the USA, where it is mainly freight trains that are being circulated and where the track gauge is wider, routed trains are of a significantly higher weight (normally 3,000–5,000 t). Heavy trains (5,000 t) are also routed in Australia, Canada, China, India, and South Africa.

  Both weight and length are important productivity factors for the railway freight industry and are influenced by various features of the system.

  The permissible length is dictated by station length and by the design of the brake system. On the other hand, the features of the brake system and the design of the coupling equipment affect the permissible weight.

  The adoption of the automatic coupler (Janney/Knuckle coupler in North America and China, Willison/SA3 in former the USSR) allowed for heavier and longer trains while the continued use of the screw coupler in Europe leads to shorter/lighter trains and thus less economical service provision.
- *Maximum train running speed*: It ranges between $V_{max} = 100$ and 120 km/h. The heavy haul rail transport requires lower running speeds ($V_{max} = 80–100$ km/h).
- *Axle load*: For the transportation of conventional loads, the maximum permissible axle load on normal gauge tracks is $Q = 22.5$ t, whereas for the transportation of heavy loads, it is $Q = 25–40$ t.
- *Longitudinal gradient*: Operation is highly influenced by the longitudinal gradient of the track. Gradients that exceed 1.6% result in a restriction of the train's maximum weight and speed while there is also an increase in the requirements for the rolling stock (available power, braking performance, etc.).
- *Reliability of services*: The acceptable delay time of a freight train depends on the length of the route. One could consider the acceptable delay time to be equal to 1 h per 500 km of length. Punctuality, in other words, minimisation of delays, constitutes one of the parameters that determine the level of quality that a freight railway system provides.
- *Terminal layout*: Freight transport requires stations to be equipped with special facilities for collecting and handling goods. The average distance between small marshalling yards in Germany is around 160–200 km, between medium marshalling yards 800–1,200 km and between large marshalling yards 3,000–3,500 km (Jorgensen and Sorenson, 1997).

15.3 ROLLING STOCK FOR FREIGHT

Freight railway transport is achieved with the use of diesel or electric loco-hauled trains. As mentioned in Section 1.2.2., the trailer vehicles that are used may be classified into two basic categories (Metzler, 1985 ; Pyrgidis and Dolianitis, 2021):

- General-use freight vehicles (or freight cars or goods wagons or trucks) intended to transport goods.
- Specific-use freight wagons, which are intended for the transportation of certain types of freight only.

General-use freight vehicles may be classified based on the type of covering and height of sides, into:

- Covered wagons (or boxcars).
- Open high sided wagons (or gondolas).
- Open low sided wagons (or gondolas).
- Open vehicles with no sides (or flat cars).

Specific-use freight wagons, may be classified based on the nature of products they transport into (indicatively):

- Tank cars (for petroleum products, chemical or food products, for liquified gas or cryogenic transport).
- Refrigerated wagons.
- Hopper cars (with central, bilateral, or pneumatic unloading).
- Road vehicles carriers (double deck or single deck).
- Cars for road/rail combined traffic (piggyback cars).
- Cement wagons.
- Special flat cars for containers and other specific goods.
- Center partition cars.

A more thorough classification of freight railway vehicles has been proposed by UIC (Schiffer, 2005).

15.4 SCHEDULING OF FREIGHT TRAIN SERVICES

Contrary to passenger railway services, freight railway services do not usually provide their customers with regular timetables (Lambropoulos, 2004).

The common practice is to provide certain routes in the form of reserved 'paths' which are performed only when there is sufficient load.

Another common practice is that there are no predetermined routes and the trains depart as soon as the load reaches a certain target (e.g., departure as soon as 1,500 t or 100 wagons are gathered). This is practiced by the American railway network. The adjusting and scheduling of services take place ad hoc on an everyday basis.

However, there is a trend to apply regular services similar to those of passenger trains. The paths are used with or without sufficient cargo and the customers' cargo is shipped with the first available train, the main objective being to increase the transportation speed and the reliability of the railway system.

15.5 THE TRENDS IN THE DOMAIN OF FREIGHT RAIL TRANSPORTATION

The term 'trends' in the domain of freight rail transportation refers to all the strategic choices and policies that either have been adopted in recent years and are being followed by almost the totality of railway companies or that are planned to be adopted in the near future. Such strategic choices and policies have been proven to contribute positively to the performance and economic efficiency of both the railway system as well as to contribute positively to the protection of the environment and to society in general. On this basis, they are supported financially at a research level and are closely associated with innovation.

The main trends include:

- Liberalisation of transportation.
- Interoperability.
- Combined transport.
- Mass transportation.
- The very long-distance transportation (e.g., Europe–Asia).
- Increased safety and security.
- Environmental protection.
- Reduced maintenance costs.
- Increased running speed.
- Reduced delays at border stations.
- Increased network capacity.
- Digitalisation
- Automation of train movement

Some of the above trends are discussed in more depth below. The last one (Automation of train movement) and cutting-edge technologies, in general, are discussed in Chapter 20.

15.5.1 Combined transport

The container revolutionised freight transport since it allows for easy transfers among all modes of transportation. At the same time, other techniques have been developed, allowing the transportation of cargo from one transport mode to another without handling the cargo itself. The common feature of all these techniques is the use of more than one transport modes in an attempt to make use of the unique advantages of each one of them. This is what characterises this type of transportation as *combined transport*.

The potential combinations of various techniques and transport modes are usually grouped into three categories:

- *Multimodal transport*: Transportation is achieved with the use of more than one mode. More precisely, one mode uses the other on a common route (i.e., one mode is loaded on the other, Figure 15.1).
- *Intermodal transport*: The major part of transportation takes place on a ship, whereas distribution/collection of the cargo in ports continues with other transport modes. Intermodal transportation sees each mode of transportation as a different contract. On the other hand, multimodal transportation maintains higher efficiency as the entire process comes under the supervision of one single carrier.
- *Combined transport*: The major part of transportation takes place on either railway or barge and collection/distribution continues on the road.

Figure 15.1 Multimodal transport – train loading in ship, Sassnitz hafen, Germany. (Photo: A. Klonos.)

Usually the term *combined* covers all categories, whereas the American term *intermodal* covers train/truck transportation. However, in the European Union (EU), the term *combined transport* refers to the transit of a trailer or semi-trailer truck (with or without tractor), a swap body or a container, using road transportation for the initial or final part of the route and ships, barges or railways for the main part of the route.

The EU has made great attempts to achieve the unimpeded and economically viable transportation of passengers and cargo throughout Europe at the least possible external costs (accidents, pollution, and traffic congestion). The combined transport technique perfectly responds to this policy. The 'door-to-door' cargo transfer can combine different means of transportation: flexible road transport for collection/distribution and – friendly to the environment – sea, inland water, and rail transport for the main transportation.

Aside from the above-mentioned trend, the EU is also facing the issue of alpine crossings. Switzerland and Austria pose restrictions to the passage of lorries from their land, either with general restrictions (restriction of trucks weighing more than 28 t) or with technical or economic barriers [tolls imposed according to the technology of the vehicle (emissions, sound pollution, and suspensions friendly to the road surface) restrictions in permits].

Generally, these policies aim to promote the use of railways for long-distance transport.

Combined transport constitutes a growing market for American railways since road transport is the country's third best customer after coal mines and power stations. In Europe, though railways receive great financial aid, their financial return is not very promising. In reality, although a significant part of the European rail freight transport is being handled with combined transport, revenue remains low.

Railways can offer four different 'products' of combined transport:

- *Transport of unified load* is the transport of containers and swap bodies with flat railway vehicles—-platforms. Containers are mostly used for sea transport, while swap bodies tend to be used for train–truck transport (Figures 15.2 and 15.3).
- *Piggyback transport* is the transport of a road trailer placed on special wagons – platforms without its tractor unit and its driver (Figure 15.4).
- *Rolling road or rolling highway transport* (ROllende LAndstraβe-ROLA): This is the transport of a whole truck on special flat, low-floor railway vehicles. The driver accompanies its truck on a special sleeping car, which is included in the train formation (Figure 15.5).
- *Roadrailer technique* is the technique in which railway bogies are adjusted on the basis of special road trailers (with their rubber-tyred wheels) which can be hauled like railway wagons. It is the least common technique in Eurasia.

Table 15.2 presents the advantages and disadvantages as well as the technical and operational requirements of the above techniques.

Figure 15.2 Swap-body loading to a railway wagon, Malmo, Sweden. (Photo: A. Klonos.)

Figure 15.3 Container loading to a railway wagon, Thriasio, Greece. (Photo: A. Klonos.)

Figure 15.4 Piggyback rail transport, Gavle, Sweden. (Photo: A. Klonos.)

15.5.2 Mass transport

The term 'mass transportation' is used to describe transport under the following transport conditions (most of which must be fulfilled for the term to apply) (Savy, 2006; Pyrgidis,1989):

- The transport of large quantities.
- The long-distance transportation.
- The high-speed transportation.
- The transportation with modes of great transportation capacity.
- The transportation with transport means of high occupancy.

Figure 15.5 Technique of rolling highway transport (ROLA), Bockstein, Austria. (Photo: A. Klonos.)

Table 15.2 Combined transport techniques: advantages/disadvantages

Techniques	Advantages	Disadvantages	Requirements
Unified loads (containers)	• Standardised technique • Global dissemination • Possibility of stowage in height • Robust construction, durability	• Large stowage area needed • Expensive handling equipment	• Special rail vehicles • Cranes for container handling • Applied mainly for sea transport
Unified loads (swap bodies)	• Large capacity • Technology affordable for end users • Easier handling	• Lighter structure • Large stowage area needed	• Cranes for container handling with special equipment • Applied mainly for land transport
Piggyback (transport of truck trailer placed on special wagons)	• Fast upload • Flexibility	• Large area needed • Enlarged loading gauge • High cost of equipment • Special management/ organisation	• Cranes for container handling • Special wagons [Trailer Of Flat Cars (TOFC)] • Applied mainly for land transport
Rolling road transport (ROLA)	• Simple procedures to be followed at stations • No additional investment by road transport companies is required	• Expensive and maintenance heavy rail vehicles • Low speeds • High operational cost • Enlarged loading gauge/ increased loads	• Special low-floor railway vehicles • The driver has to accompany the train • Applied mainly for land transport

The policy of mass transport somehow justifies the role of transhippers since they can achieve 'economies of scale' for a specific transfer. On the other hand, the additional profit obtained by transshipments allows consignors and consignees to pay a smaller fare for their cargo to be transported. Finally, the 'small' customers who fail to gather large loads can also join in mass transport, thus reducing the cost of transferring their cargo.

According to the above, consignees, consignors, and shippers have common interests in this policy of mass transportation. As a result, research has been conducted and money has been invested in means of transportation, techniques, and equipment suitable for massification of flows. This has resulted in the development of:

- Container ships.
- Cargo planes.
- Road trucks of heavy tonnage.
- Special trains for moving cargo (unit trains, block trains).
- Automated marshalling yards.

The railway is a suitable means for mass transportation. According to the volume of load it carries, the railway adopts different solutions in order to respond more successfully to the policy of mass transport (Figure 15.6) (Pyrgidis, 1989).

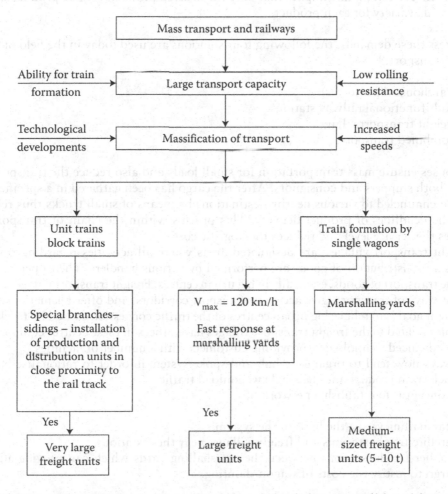

Figure 15.6 Mass transportation and railways. Feasible operational alternatives. (Adapted from Pyrgidis, C., 1989, Transport de mercaderies: Avantages i inconvenients de la massificacio de fluxos, Espais – revista del department de politica territorial i ombres publiques de Catalunya, No. 17, May-June 1989, pp. 26–31.)

More precisely:

a. When freight units are too big, cargo movement is performed with unit trains or block trains that are not passing through marshalling yards. For this purpose, special sidings are used to connect the railway line with the production centres or the production centres are installed/set up very close to the line.
b. When freight units are relatively big, expediting is done with single-wagon trains that move at high running speeds (V_{max} = 120 km/h) while their delay in marshalling yards is the least possible (automatic marshalling yards).
c. Finally, in the case of medium- and small-size freight units (5–10 t), expediting is done with single-wagon trains but the service level is low.

Nowadays, consignors and consignees have three main demands:

• Immediate order delivery, which requires frequent and flexible services.
• 'Stock' reduction; this implies that only the ordered products will be produced.
• A wide variety for each product.

To address these demands, the following four solutions are used today in the field of freight railway transport:

• Warehouses (platforms).
• Multifunctional railway stations.
• Freight transport villages.
• Combined transport.

Warehouses ensure mass transportation for small loads and also reduce the transportation cost for both shippers and consignors. After the cargo has been gathered in a specific place, it is then channeled to various nearby destinations by means of small trucks thus responding to the deadlines of product delivery. This process within the chain of transportation improves the service level and reduces the 'logistic cost'.

'Freight transport villages' are designated areas where all activities regarding transport logistics and distribution of goods are performed by various handlers. Their operation supports the transport of goods, especially in big urban centres. Freight transport villages connect different ways of transport, give access to transport corridors, and offer telematics services. Their main goal is to relieve big urban centres of the traffic congestion caused by trucks. The companies related to the freight transport villages have offices in these centres equipped with modern advanced technologies, providing customers with a number of facilities.

Railways now tend to organise a hub-and-spoke system in order to rationalise both the unit/block train product and the single wagonload traffic.

The concept is to establish a network of:

• Marshalling yards (the hubs of the system).
• Satellite freight stations and freight villages near these stations.
• Smaller freight stations between the marshalling yards which can provide adequate cargo to justify the costs of wagon shunting.

Between the hubs run mainly unit or block trains. Only a limited number of trains perform pick or leave wagons at intermediate stations.

The trains are marshalled at the yards and arranged into groups for the satellite stations, freight villages, or sidings.

The aim is to utilise the maximum available capacity of a train path (in terms of tractive power, length, and weight) between the marshalling yards and to reduce the delay times at intermediate stations.

15.5.3 Higher speeds

For several years, a coordinated effort has been taken towards an increase in the speeds of freight trains (Ferreira and Murray, 1997). This aims to increase track capacity and increase the share of railways in the emerging market of lightweight, time-sensitive, and high-value goods. Moreover, and based on relatively recent research (Spectrum Consortium, 2012, 2013), the railways are capable of attracting 10–15% of such loads that are as of now transferred via road means, if the offered services are improved.

For the time being, the maximum running speed of freight trains in Europe is 120 km/h, while the average running speed (commercial speed) ranges between 18 km/h and 39 km/h. This is mainly due to restrictions external to the railway system (such as shorting processes), delays, and the lack of a centralised management system. At the same time, the average running speed of road freight vehicles is 60 km/h (Liguori, 2018). A far better situation is encountered in the case of container freight trains for combined transport, for which the average speed is, according to operating entities such as DB Schenker, GBRF, and Fret SNCF, 50–60 km/h (Islam and Mortimer, 2017).

In order to alleviate this problem, nine international railway corridors that create an efficient network have already been built. Their infrastructure is subjected to centralised management, independent of the respective national networks. Thus, the average running speed of freight trains along these corridors reaches that of 50 km/h. A successful example of such a corridor is the connection of North Italy with the port at Hamburg (Liguori, 2018).

Another indication of the trend to increase speeds is the research into the creation of high-speed freight rail. The definition of such a network differs from the one for passenger trains and involves more characteristics, such as the nature of the cargo, the rolling stock, and the operating principles of the network. In terms of speed, however, a limit could be set at 160 km/h for characterising a freight railway corridor as high-speed. The operating principle that the creation of such a network is based on is: 'faster than road transport – cheaper than air transport' (Troche, 2005).

The creation of a fleet of high-speed freight trains gives rise to some problems, especially concerning the use of the conventional speed network. In any case, there have already been attempts at routing such trains. Characteristic examples may be found in the TGV Fret of the French SNCF and in the InterCargo Express (ICGE) of the German DB. The fastest freight train currently is the Mercitalia Fast of the Italian FS, which runs at an average speed of 180 km/h. The routing of high-speed trains is also being studied in Austria, Sweden, and the Baltic countries (connection between Talin in Estonia with Kaunas in Lithuania) (RailFreight.com, 2018).

Finally, and in order to achieve an increase in the speed of freight trains, it is first necessary to resolve the issues that this fact will give rise to. Specifically, higher speeds require increased energy consumption and create the need for improving the existing braking and wagon coupling systems as well as enhancing infrastructure to address the increase in dynamic forces. To that end, Shift2Rail has proposed the redesign of freight wagons to be compatible with maximum running speeds of 120–160 km/h (Shift2Rail, 2015).

15.6 THE MAIN DILEMMAS FOR RAILWAY COMPANIES IN THE DOMAIN OF FREIGHT RAIL TRANSPORTATION

15.6.1 The 'open' issues

The sector of freight railway transport is evolving by setting targets that typically adhere to the interests of the various involved actors. The choice of targets is to a large extent influenced by the up-to-date trends in the field of railway transport and by market needs, while their fulfillment is supported by specific actions.

The transportation of goods constitutes one of the most lucrative functions of a railway system. However, profitability depends not only on the volume of the transported goods, the distances over which they are transported, and the tariffs that are applied but also on other parameters that permeate freight railway transportation. The choice of such parameters is not an easy one for railway companies to make, since, firstly, they do not accurately know the positive or negative impact they will have on their financial efficiency and, secondly, many such choices are not solely up to them.

Some of the crucial questions posed by railway companies regarding what is more profitable are:

- Simultaneously routing passenger and freight trains (mixed exploitation) on the same railway corridor, or differentiating passenger from freight traffic (dedicated exploitation)?
- Routing trains of conventional or heavy loads?
- Providing single wagonload services or trainload services (block trains)?
- Long trains or short trains?
- Diesel trains or electrification?
- Should dangerous goods be transported by rail?

Some of these questions have been adequately explored and solutions that satisfy the requirements of the various actors have been proposed. On the contrary and for most of these questions, no univocal answer has been given (Figure 15.7). In this section, some of these dilemmas are documented. The benefits and weaknesses of each side are presented, while the conclusions of relevant research on which side is to be adopted are also presented.

15.6.2 Mixed traffic operation

15.6.2.1 Description and justification of the problem

Throughout the world, the vast majority of railway networks/corridors concern mixed train operation (Batisse, 1994, 1995a, b; UIC, 2007). Express and local passenger trains and freight trains are routed on the same track. For many years, this was the basic rule in the rail transport sector. On the one hand, this practice seems to achieve economies of scale, as most trains use the same railway infrastructure; on the other hand, it creates problems in the operation and maintenance of the network, as trains of different functionality circulate on the same track.

More specifically, many features of the freight wagons/trains differ substantially from those of the passenger cars/trains. As a result, sharing the same track affects the design, construction, operation, and maintenance of a railway system either directly or indirectly. Indicatively, the case of constructing either single or double track for mixed traffic is presented below:

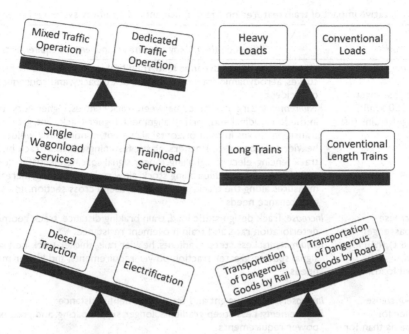

Figure 15.7 The main dilemma for railway companies in the domain of freight rail transportation and their adoption trend.

- Track alignment design is based on the track design speed, which is defined and imposed based on passenger trains.
- The construction of the superstructure takes into account both the speed of the passenger trains and the axle load of the freight trains.
- The maintenance policy for the track takes into account the daily traffic load that is derived from both passenger and freight trains.
- The timetable graph must incorporate trains that run at different speeds with whatever consequences this may have on track capacity.

Table 15.3 shows the qualitative impact of the features which differ substantially between freight and passenger vehicles/trains, on the components of a railway system.

A question which is a matter of concern to many railway companies is:

> What is more economically profitable for a railway company? The simultaneous routing of passenger and freight trains (mixed operation) on a railway network/corridor or the differentiation between passenger and freight traffic (dedicated operation)? Which are the criteria and basic principles that will lead railway operators to select the operational framework that they will adopt:
>
> - For a new railway connection which they are planning to construct?
> - For an existing railway connection of mixed traffic operation where demand is modified?

Mixed traffic operation networks satisfy primarily passenger transportation. This priority usually leads to resource inadequacy for freight trains, which are further delayed in favour of passenger trains. The needs of the freight transportation market differ from those of

Table 15.3 Qualitative impact of train features on the components of a railway system – requirements

Train/vehicle feature – variation	*Effects on the railway system's components – requirements*
Maximum running speed – increase Is greater for passenger trains (120–350 km/h) than for freight trains (60–120 km/h)	*Increase:* Track design dynamic load, train braking distance, centrifugal force in curves, aerodynamic train resistance, track capacity, and consequences in case of accident *Requirements:* Larger distance between track centres, higher curvature radius in the longitudinal and vertical alignment, higher track cant, greater length of transition curves in the horizontal alignment, continuously welded and heavier rails, concrete sleepers, elastic fastenings, thicker track bed layers, track fencing, electric signalling, longer signal spacing, electrification (for V > 160–180 km/h), specific rolling stock, slow train overtaking, increased safety measures along the track, bigger tunnel useful cross section, and higher maintenance needs.
Axle load – increase Is lower for passenger trains (12–18 t) than for freight trains (conventional loads: 16–25 t)	*Increase:* Track design static load, train braking distance, track geometry defects deterioration rate, and train movement resistance *Requirements:* Less steep gradients, heavier rails, thicker track bed layers, longer signal spacing, greater traction power requirements, and higher maintenance needs
Train weight – increase Is much smaller for passenger trains than for freight trains	*Increase:* Braking weight and train movement resistance *Requirements:* Less steep gradients, longer signal spacing, and greater traction power requirements
Train length – increase Is much smaller for passenger trains	*Decrease:* Track capacity *Requirements:* Longer tracks and platforms in stations
Daily traffic load – increase	*Increase:* Track maintenance needs and track geometry defects deterioration rate *Requirements:* Heavier rails, thicker track bed layers, and higher maintenance needs
Dynamic passenger comfort Concerns mainly the passenger cars	*Requirements:* In the case of passenger trains, lateral residual centrifugal acceleration < 1.0 m/s^2, greater curvature radius in the longitudinal and vertical alignment, higher track cant, and greater length of transition curves in the horizontal alignment In the case of mixed traffic operation, adoption of a track cant that satisfies both train categories
Punctuality	*Requirements:* Imperative for passenger trains – desirable for freight trains
Vehicle clearance gauge	*Requirements:* Variable depot and station dimensioning, larger distance between track centres and civil engineering structures height clearance
Terminal stations	*Requirements:* Totally different design and equipment for passenger and for freight stations. Passenger trains stop much more frequently (every 50–100 km) than freight trains (every 300–800 km)
Transported goods	*Requirements:* Different safety measures required for the infrastructure and the rolling stock when transporting passengers compared to cargo. Special safety measures in tunnels when carrying passengers and special safety measures on the 'open' line when carrying dangerous goods

Source: Adapted from Christogiannis, E., and Pyrgidis, C., 2013, An investigation into the relationship between the traffic composition of a railway network and its economic profitability, *Rail Engineering International*, 1, pp. 13–16; Pyrgidis, C., and Christogiannis, E., 2011, The problem of the presence of passenger and freight trains in the same track and their impact on the profitability of the railways companies, *9th World Congress on Railway Research 'Meeting the challenges for future mobility'*, 22–26 May 2011, Lille, France, Congress Proceedings (CD); Pyrgidis, C., and Christogiannis, E., 2012, The problem of the presence of passenger and freight trains on the same track, *Elsevier Procedia Social and Behavioral Sciences*, 48, pp. 1143–1164.

passenger transportation. This situation seems to enforce the gradual segregation of railway networks/corridors for passenger and freight transportation. Indicatively, it should be noted that:

- Large railway networks, particularly those of countries with significant industrial output, such as in China, Russia, India, etc., have constructed or are planning to construct dedicated passenger and/or freight railway corridors [OECD, 2002; ADB (Asian Development Bank), 2005; Woodburn et al., 2008; Dedicated Freight Corridor Corporation of India, 2015].
- The construction of dedicated passenger railway corridors in Japan has proven to be an especially successful and financially profitable investment, and, as such, the Japanese railway network is growing continuously, and the railway has secured a large part of the market.
- The abandonment of conventional long-distance passenger services in the United States and the shift by railway companies toward freight transportation has resulted in significant profits for these companies and has significantly increased the share of the railway in freight transportation.
- The trend toward a global economy without restrictions in freight movement is prompting large networks worldwide to design and create international dedicated railway freight corridors.
- The World Bank recommends that railway organisations in distress should proceed to the managerial separation of passenger and freight transportation activities.
- The competent Directorates of Transport of the European Union seek to establish criteria and basic principles by dint of which choices will be made in regard to the operational framework which will be adopted for upgraded or newly designed lines.

An essential prerequisite for the creation of long dedicated freight railway corridors would be the existence of connection amongst the networks that will be served and the ensuring of interoperability.

15.6.2.2 Contribution towards solving the dilemma – investigation of the impact of traffic composition on the economic profitability of a railway system

15.6.2.2.1 Data published by the railway networks

In a survey conducted in 2007, based on data publicly available from the networks themselves, traffic composition was associated with the profitability of a railway network (Christogiannis and Pyrgidis, 2013).

In particular, the following 20 networks were thoroughly examined (United States, Canada, Lithuania, Australia, Estonia, Latvia, Russia, Ukraine, China, South Africa, Poland, Austria, Romania, Germany, France, Spain, India, Italy, United Kingdom, and Japan).

As 'economic profitability' of a railway network/corridor is defined its capacity to produce significant transport volume, compared to other competitive transportation means and to be financially robust, that is, to show profit rather than deficits and debts.

The economic profitability of a railway system can be characterised by:

- The company's profits and debts owed on a fixed time basis (e.g., over a five-year period).
- The company's position in the competitive market.

- The organisation and structure of the company.
- The company's growth activities.

Based on their economic profitability railways systems (networks) can be classified in the following five categories:

A+: Very positive.
A: Positive.
AB: Neutral – balanced.
B: Negative.
B–: Very negative.

The methodology for the classification of railway networks and railway corridors in the aforementioned categories according to their economic profitability is outlined hereunder (Christogiannis and Pyrgidis, 2013, Christogiannis, 2012).

Figure 15.8 illustrates the results of this procedure. The following conclusions may be deduced from this figure:

- Networks belonging to traffic composition categories I and V, that is, networks with dedicated freight and passenger operation, respectively, belong to the category of economic profitability A+, that is, they present very positive economic profitability.
- The majority of networks belonging to the traffic composition category III, that is, with mixed traffic operation, present a negative or very negative economic profitability.

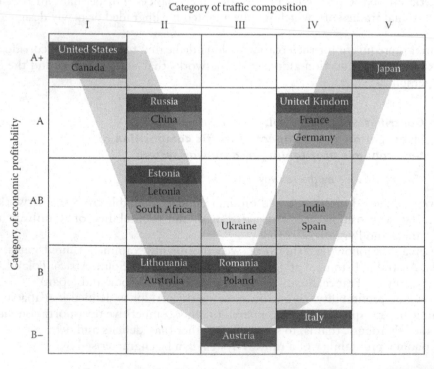

Figure 15.8 Correlation between traffic composition and economic profitability of a railway network. (Adapted from Christogiannis, E., and Pyrgidis, C., 2013, An investigation into the relationship between the traffic composition of a railway network and its economic profitability, *Rail Engineering International*, 1, pp. 13–16.)

- The majority of networks with emphasis on freight train traffic shows a positive balanced economic efficiency while no network shows a very negative profitability.
- Networks with emphasis on passenger train traffic, by a majority, do not show a negative profitability.

As can be seen in the translucent grey 'V' in Figure 15.8, the more dedicated (passenger or freight) the traffic composition is, the more the increase in the economic profitability of networks. On the contrary, the more mixed the traffic composition tends to be, the more the decline in their economic profitability.

15.6.2.2.2 Mathematical simulation: (a) Selection of the operational framework for a new railway corridor

The impact of traffic composition on the economic profitability of a new railway corridor has been investigated with the aid of a mathematical model as part of a PhD dissertation thesis (Christogiannis, 2012).

Within the framework of this research, the rail infrastructure manager is also the owner of the rolling stock and the operating company. Two financial indicators were considered to be the figures that characterise the economic profitability of a new railway corridor: (a) the Net Present Value (NPV) of the investment and (b) the Internal Rate of Return (IRR) of the investment.

The five operation scenarios of a new railroad connection presented in Table 15.4 were examined (Pyrgidis and Christogiannis, 2013a; Christogiannis and Pyrgidis, 2014, 2015).

The mathematical model that was developed allows, for various operation scenarios concerning the traffic composition, the following:

- Calculation of the expenses which are necessary for the implementation of the new railway link. The expenses calculated by the model for each of the five operation scenarios examined include the cost for the construction of the new track infrastructure

Table 15.4 New railway connection – operation scenarios under study – basic railway system design parameters

Schematic image of the scenario	Description	Basic railway system design parameters
Freight and passengers	Scenario S1: double track – mixed traffic operation	V_d = 200 km/h, Q_{max} = 22.5 t, i_{max} = 2%, V_{pas} = 200 km/h, V_f = 120 km/h, rails UIC 60, and γ_{ncmax} = 0.50 m/s^2
Passengers	Scenario S2: double track – passenger-dedicated operation	V_d = 200 km/h, Q_{max} = 22.5 t, i_{max} = 2.5%, V_{pas} = 200 km/h, rails UIC 60, and γ_{nmaxc} = 0.50 m/s^2
Freight	Scenario S3: double track – freight-dedicated operation	V_d = 120 km/h, Q_{max} = 25 t, i_{max} = 1.5%, V_{fr} = 120 km/h, rails UIC 70, and γ_{ncmax} = 1.0 m/s^2
Freight	Scenario S4: single track – freight-dedicated operation	V_d = 120 km/h, Q_{max} = 25 t, i_{max} = 1.5%, V_{fr} = 120 km/h, rails UIC 70, and γ_{ncmax} = 1.0 m/s^2
Freight and passengers	Scenario S5: single track – mixed traffic operation	V_d = 200 km/h, Q_{max} = 22.5 t, i_{max} = 2%, V_{pas} = 200 km/h, V_{fr} = 120 km/h, rails UIC 60, and γ_{ncmax} = 0.50 m/s^2

Source: Adapted from Christogiannis, E., 2012, Investigation of the impact of traffic composition on the economic profitability of a railway corridor – Fundamental principles and mathematical simulation for the selection of operational scenario for a railway corridor, PhD Thesis, Aristotle University of Thessaloniki, Thessaloniki, Greece.

V_d, track design speed; V_{pas}, V_{fr}, maximum speed of passenger and freight trains, respectively; Q_{max}, maximum axle load or vertical design axle load of a railway infrastructure; i_{max}, maximum longitudinal gradient, and γ_{ncmax}, maximum permitted residual centrifugal acceleration.

(substructure, civil engineering works, superstructure, stations, electrification and signaling systems, level crossings, expropriations, rolling stock maintenance facilities,
and studies), the cost for rolling stock acquisition, the operating costs, the maintenance cost, and the financial cost.
- Calculation of the revenue generated for the undertaking. Revenue is considered to
originate from passengers and freight, while the residual value of the project and revenue from loans are also included in the calculation.
- Calculation of the economic profitability of each exploitation scenario on the basis of
the method of the NPV and IRR of the investment, for the duration of the financial
lifespan of the investment.
- Comparison of the operation scenarios (reference scenarios) regarding their economic
profitability in general and, more specifically, regarding their individual cost elements
and other railway system parameters (e.g., infrastructure cost and maintenance cost).
- Study of the influence of various design, construction, operational, and financial
parameters of the railway system on the system's economic profitability.
- Selection, on the basis of demand for passenger and/or freight volume to be transported via rail on a connection, of the operation scenario that allows the highest economic profitability.

The conclusions drawn are grouped into the following three 'areas':

1. *Comparison of economic profitability scenarios*
   - Among scenarios S1, S2, and S3, which concern the construction and operation of
a new railway corridor consisting of a double track, the most economically advantageous solution appears to be scenario S3, that is, the case of dedicated freight
train operation, followed by scenario S2, that is, dedicated passenger train operation, followed by mixed train traffic operation.
   - Between scenarios S4 and S5, which concern the construction and operation of a
new railway corridor consisting of a single track, the most economically advantageous solution appears to be scenario S4, that is, the case of dedicated freight train
traffic. In fact, scenario S4 is almost equivalent to scenario S1.
2. *Influence of various railway system parameters on economic profitability*
   - As the length of the railway connection increases, the NPV of the investment and,
thus, the economic profitability of the railway system also increases. Scenarios S3,
S2, and S1 have an NPV equal to €11,000 M for connection lengths of S = 580,
760, and 1,120 km, respectively.
   - The economic profitability of all operation scenarios decreases drastically as the
topography of the landscape becomes harsher. In fact, for mountainous topography, it is negative in all cases. The change in the topography of the landscape from
average evenness to flatlands results in a significant increase in the NPV. More specifically, this increase is equal to 89% for scenario S1, 50% for scenario S2, 46%
for scenario S3, 69% for scenario S4, and 190% for scenario S5.
   - The change in the composition of traffic significantly affects the economic profitability of scenarios S1 and S5 (Figure 15.9). In the case of scenario S1, the curve
of the diagram of the ratio of passenger trains to the total number of routed trains
and of the NPV has a parabolic shape which points downwards. This curve reaches
its minimum in the case of mixed traffic (50% passenger – 50% freight trains). The
maximum values of the NPV in scenario S1 appear in the case of dedicated train
operation and are 32% higher in the case of dedicated freight train operation in
relation to the case of dedicated passenger train operation.

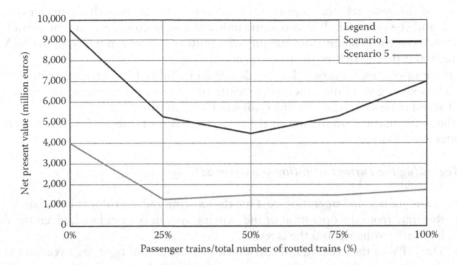

Figure 15.9 Variation of the NPV in relation to the traffic composition – scenarios S1 (double track) and S5 (single track). (Adapted from Christogiannis, E., and Pyrgidis, C., 2014, Investigation of the impact of traffic composition on the economic profitability of a new railway corridor, *Journal of Rail and Rapid Transit*, Proceedings of the IMechE, Part F, 228(4), pp. 389–401.)

3. *Selection criteria of rail corridor operational framework*

The basic criterion for the selection of an operation scenario for a new railway corridor concerns the characteristics of transportation demand and, particularly, what is being transported, passengers or freight, in combination with the volume being transported and the topography of the landscape.

- For average demand values (e.g., 9,000 passengers per day per direction/81,000 t per day per direction, 25,000 pas/25,000 t), scenario S1 (double track – mixed operation) presents the highest NPV.
- For high demand values (e.g., 18,000 passengers per day per direction/162,000 t, 50,000 pas/50,000 t), dedicated traffic scenarios (i.e., 0 pas/162,000 t, 50,000 pas/0 t) presents the highest NPV, both in the case of a single track and in the case of a double track. Furthermore, S3, the dedicated freight operation scenario presents a higher NPV in relation to S2, the dedicated passenger operation scenario.
- When the demand ratio of the number of passengers to the freight tons approaches 1/3 (traffic composition percentage 50–50%), this 'favours' mixed operation scenarios (scenarios S1 and S5).

By comparing the results deriving from the two approaches (published railway networks data and mathematical simulation), compatible conclusions are drawn. In particular, the curves of Figures 15.8 and 15.9 have exactly the same form and lead, at least qualitatively, to the same conclusions regarding the impact of the traffic composition on the economic profitability of the railway system.

15.6.2.2.3 Mathematical simulation: (b) Selection of the operational framework for an existing railway corridor

Pyrgidis and Christogiannis (2013b) and Christogiannis and Pyrgidis (2015) show the conditions under which it is economically viable to change the operating framework of an existing railway corridor from mixed traffic to dedicated traffic.

As part of this research, the railway infrastructure manager is both the owner of the rolling stock and the operator. Two economic indicators were considered as the main features that characterise the economic profitability of a railway corridor, namely (a) the NPV of the investment and (b) the IRR of the investment.

The three operating scenarios that are described in Table 15.5 were examined and compared. The existing rail link which is currently in operation is implemented with a single track of standard gauge, where traffic is allowed in both directions.

For the mathematical simulation of the three scenarios of Table 15.5, the steps outlined hereunder are followed:

Regarding the current situation (scenario S1)

- Maintenance and operation cost for the existing rail corridor is calculated.
- Revenue from the operation of the existing system is calculated, given the demand for traffic volumes and the fares.
- The NPV of the existing rail corridor for a time period $t_{fin}-t'$ (t_{fin}: year of the end of the corridor's economic life and t': year of change of the corridor's operating frame).

In the case of scenario S1 (mixed traffic operation), the ratio of the number of routed passenger trains to routed freight trains is 3:1 (75% of passenger trains and 25% freight trains). The saturation rate of the track's capacity is already equal to 70% and may not be further increased. On this basis, the transport volume, for both passenger and freight, is specified.

The line is considered to be saturated and it cannot respond to newly increased demand.

Regarding the new situation (S2 and S3 scenarios):

- For both considered scenarios of dedicated operation, the required changes to the existing infrastructure (replacement of track superstructure components, changes in the signaling plan, upgrading of railway stations, etc.) are recorded.
- For the new scenarios of dedicated operation, it is considered that the demand for passenger or freight transport initially remains the same in comparison with the existing situation and then it gradually increases by a specific percentage.

Table 15.5 Existing railway infrastructure – operation scenarios under consideration – basic railway system design parameters

Schematic image of the scenario	Description	Basic railway system design parameters
Passengers and freight	Scenario S1: Current situation; single track – mixed traffic operation	$V_d = 160$ km/h, $Q_{max} = 22.5$ t, $i_{max} = 20\%o$, $V_{pas} = 160$ km/h, $V_{fr} = 100$ km/h, electrification, and electric side signalling
Passengers	Scenario S2: New situation: single track – passenger-dedicated operation	$V_d = 160$ km/h, $Q_{max} = 22.5$ t, $i_{max} = 20\%o$, $V_{pas} = 160$ km/h, electrification, and electric side signalling
Freight	Scenario S3: New situation: single track – freight-dedicated operation	$V_d = 160$ km/h, $Q_{max} = 22.5$ t, $i_{max} = 20\%o$, $V_{fr} = 120$ km/h, electrification, and electric side signalling

Source: Adapted from Pyrgidis, C., and Christogiannis, E., 2013b, Mixed or pure train routing? The case of an existing railway corridor, *6th International Conference for Transport Research ICTR*, 17–18 October 2013, Thessaloniki, Conference Proceedings; Christogiannis, E. and Pyrgidis, C. 2015, Selection of the optimum exploitation scenario for an interurban railway corridor by the help of mathematical models, *Ingegneria Ferroviaria*, 5, pp. 427–448.

- The required changes in rolling stock in order to serve the new demand (additional rolling stock) are recorded.
- The required changes in the operating system (more scheduled trains daily, additional stations of a different category, changes in the number of required staff, etc.).
- The cost of implementing the above-required interventions, the maintenance cost and the operating cost of the new dedicated operation railway corridor are calculated.
- The revenue (for a given fare price) from the operation of the new system is calculated.
- The NPV of the railway corridor for a period $t_{fin}-t'$ is calculated.
- The NPVs of scenarios S2 and S3 are compared with the respective NPV of scenario S1 and with each other and the limit of the demand for which the dedicated operation scenarios are more cost-effective than the mixed operation on one is sought.

For both dedicated traffic scenarios and for the case of a new situation, it is considered that the track capacity saturation percentage may not exceed 70% while the train formation remains stable, which means that additional demand is treated by adding and routing extra trains. On this basis, for the new operating scenarios, the increased demand for passenger or freight transport is catered for by increasing the scheduled trains.

For scenario S2 (dedicated passenger operation), it is considered that the freight transport volume that will no longer be carried by rail will be carried by other modes of transport. The same applies to the passenger transport volume in the case of scenario S3 (dedicated freight operation).

The conclusions drawn from the above analysis are summarised as follows:

Scenario S1 (current situation – mixed operation) presents a low profit for a connection length S = 500 km. Profitability is approximately tripled when the connection length is equal to S = 1,000 km. The 70% saturation rate responds to the routing of 19 trains per day per direction at a ratio of passenger to freight trains that is equal to 3:1.

A comparison between the economic profitability of scenario S2 (dedicated passenger operation) and the economic profitability of scenario S1 (existing mixed operation) shows that:

- For a rail connection that has a length of S = 500 km, scenario S2 is more efficient than scenario S1 if the passenger volume is nearly doubled (i.e., if it is increased from 8,000 to 15,000 passengers per direction per day).
- For a rail connection that has a length of S = 1,000 km, scenario S2 is more efficient than scenario S1 if the passenger volume is increased by 20% or more (i.e., if it is increased from 8,000 to 9,600 passengers per direction per day).

A comparison between the economic profitability of scenario S3 (dedicated freight operation) and the economic profitability of scenario S1 (existing mixed operation) shows that:

- For a rail connection with a length of S = 500 km, scenario S3 is more efficient than scenario S1 if the passenger volume is nearly tripled (i.e., if it is increased from 11,000 to 30,000 tons per direction per day).
- For a rail connection with a length of S = 1,000 km, scenario S3 is more efficient than scenario S1 if the passenger volume is doubled (i.e., if it is increased from 11,000 to 30,000 tons per direction per day).

A comparison between the economic profitability of scenario S2 (dedicated passenger operation) and the economic profitability of scenario S3 (dedicated freight operation) shows that:

- For both rail connection lengths that are being considered and for a saturation rate of the line capacity that is greater than 40%, scenario S3 is more efficient than scenario S2.
- For a saturation rate of the line capacity that is equal to 68%, scenario S3 is more efficient than scenario S2 by approximately 19% for a rail connection which has a length of S = 500 km and by 7% for a rail connection which has a length of S = 1,000 km.

Regardless of the exploitation scenario, as the length of the rail connection increases, the NPV of investment and, therefore, the economic profitability of the railway system also increases. The increase rate for NPV in relation to the rail connection length is greater for scenario S3, followed by the respective rates for scenarios S2 and S1.

15.6.3 Transportation of dangerous goods

15.6.3.1 Description and justification of the problem

The term 'dangerous goods' covers materials and objects, the transportation of which is allowed only under certain conditions. These loads are categorised according to their physical and chemical properties. Fluid and solid fuel, gas, explosives, nuclear material, as well as polluting and corrosive materials are considered dangerous loads.

Based on the need to transport dangerous goods and the intensive competition between the alternative potential modes of transportation for such goods, the question arises which transport mode is better suited to provide this service based on criteria such as flexibility, economic efficiency, client requirement fulfillment, and most importantly, safety.

The railway, as mentioned in Chapter 1, has the following advantages:

- Providing mass and safe transport.
- Allowing the routing and movement of trains regardless of weather conditions.
- Being a friendly towards the environment mode.
- Having a single degree of freedom that facilitates the automation of several of a train's functions and its movement in general.

The main disadvantage of freight railway transport is the lack of door-to-door service. Moreover, important disadvantages, when compared with road freight transport, include the low number of scheduled routes and the low quality of the rolling stock.

On the contrary, the main disadvantages of road freight transport are the following (UIC and CER, 2012):

- Inability to support mass transport (especially in the case of liquid goods).
- Non-efficient vehicles in regard to the processes of specific services (including the washing of empty tanks).
- Increased chance of accidents due to the high traffic volumes in motorways and their non-ideal geometrical characteristics as well as, in some cases, due to adverse weather conditions.

Accidents during the transport of dangerous goods are usually catastrophic for all those involved, for the infrastructure, the wider area where they occur, and for the environment. As such, they are subject to intense regulatory specifications. In general, safety is a paramount concern, even more so than for conventional freight transport.

The various European railway organisations transfer over 100 million tons of dangerous goods every year (Energy efficiency in railways, 2009). Ninety-five percent of this activity concerns tanks of private companies (UIC and CER, 2012). Railway dangerous goods transport amounts to 25% of the relevant world market among all transport modes, excluding pipelines, while the potential of railways in this field is far larger (UIC and CER, 2012).

Transportation of the aforementioned loads is done with various specialised wagons (e.g., tank cars, covered wagons, and open high-sided wagons).

All types of freight trains are routed:

- Unit and block trains for the transport of chemical goods and gas.
- Exclusive trains for transport of liquid fuel.
- Combined transport for chemical goods that have to pass through environmentally sensitive areas that are subject to strict environmental protection laws.

15.6.3.2 Contribution towards solving the dilemma

In order for the railway to be efficient in regards to the transport of dangerous goods, the relevant policies should revolve around the following (Energy efficiency in railways, 2009; Journal de la Marine Marchande, 1994; Duche and Baspeyras, 1996; Pyrgidis and Basbas, 2000; Basbas and Pyrgidis, 2001).

1. Differentiation from the rest of freight transport services

    Organisation and execution of different activities associated with the transportation of dangerous loads should be examined and treated separately from the transport of conventional (non-dangerous) loads.

    Dangerous cargo transport differs from other freight activities in the following (OSE, 1998):
    - Procedures for product identification and completion of consignment.
    - Wagon loading and unloading.
    - Regulations for dangerous load trains parked in high-risk areas.
    - Signage of wagons.
    - Design and method of storage/packaging.
    - Emergency response.
    - Procedures for train shunting, train formation, and braking.
    - Cleaning of empty wagons.

        Activities related to transporting these products are legally established by international conventions and are conducted under strictly defined conditions of safety. The regulations applied to rail transport are COTIF/CIM/RID (www.cit-rail.org/en/rail-transport.../cotif/).

2. Creation of safe transport conditions

    In case of an accident, the extent of damage can be relatively higher than that caused by any other modes of transportation, because of the massification of railway transport. Therefore, the prevention of any incident is of the utmost importance.

    In this context, it is required that:
    - The condition of track superstructure used is really good (to avoid derailment).
    - There is a special maintenance programme and testing of the rolling stock (to avoid derailment, material leakage, etc.).
    - The maintenance and repairing area for vehicles transporting dangerous goods is different from that of the conventional wagons; special instructions for the

repairing of vehicles are made and special areas are provided for cleaning empty tanks (to avoid explosions, fainting, and fumes).
- Reception and distribution tracks specifically assigned to trains transporting dangerous goods are equipped with proper fencing, lighting, fire extinguishing, and sewage systems (protection of high-risk areas).
- Modern technologies are applied for the monitoring of loads during the whole route (for immediate intervention and effective treatment).

3. Special measures to protect the environment

An accidental leakage of the vehicle tanks or the transported goods' packages, or even the maltreatment of the material throughout the loading and unloading-transfer phase, involves a high risk of polluting the environment due to the nature of these products (see Section 19.4). Measures taken for accident prevention also ensure the safety and protection of the environment. However, there are additional measures, aiming exclusively at the protection of the environment. These are:
- The drainage system of waste at cleaning stations of empty tanks.
- The special design of the superstructure (e.g., slab track instead of ballasted track) at points of shunting tracks used by trains carrying dangerous loads, aiming at the protection of the groundwater in case of leakage.

The railway is a safe transport mode and as such it could be considered as ideal for the transport of dangerous goods (1 accident every billion ton-kilometers, 40 times safer than road transport, Eurostat, 2016). For railway transport and for every 1 billion ton-kilometers, only 0.34 accidents correspond to the transport of dangerous goods. For road transport, the corresponding ratio is about 15 accidents every 1 billion ton-kilometers (OSE, 1998; Pyrgidis and Basbas, 2001).

The railway is the safest mode for the transport of dangerous goods in the European Union based on relevant statistics with only 0.1 deaths ever 1 billion ton-kilometers. This number is 10 times lower than the mortality risk during the road transport of dangerous goods (OSE, 1998).

15.6.4 Long or short trains

15.6.4.1 Description and justification of the problem

A question that currently troubles several freight railway operators is:

'What is more cost-efficient? Routing over a railway network/corridor long or short freight trains?'

The optimisation of the length of a freight train, in order to best serve all transport requirements (mass transport, low transport cost, timetable reliability, low energy consumption, capability of trains to approach station platforms, flexibility in terms of timetable integration, etc.) is still an 'open issue'.

Adopting longer trains results in both advantages and disadvantages.

It is obvious that the adoption of longer trains contributes towards the creation of economies of scale, which is one of the main advantages of railways over competitive modes and which allows for mass transport. After all, research shown that the size of the dispatch is a crucial factor in choosing a mode of transport, either because larger dispatches are more likely to be transported via rail (Samimi et al., 2010) or because the choices between dispatch size and mode of transport mutually affect one another (Combes, 2012; Steer Davies Gleave, 2015).

Moreover, the current trend for transporting more, lighter, and more valuable goods in dispatches of greater volume calls for longer train formations.

A vertical increase in the dimensions of European trains, according to the model followed by trains in the U.S.A. with stacked containers, would require exceptionally costly and lengthy infrastructure investments (which would be likely to exacerbate the issues of interoperability). On the contrary, an increase in terms of train length is a far easier solution towards that direction and could result in better use of the track capacity of the current network and in a reduction of the cost of consumed energy (Zanui, 2013).

Some additional benefits to an increase in the length of trains are the following:

- Routing less dispatches for the transport of a given volume of goods (Atanassov and Dick, 2015). This results in a better use of the existing track capacity, especially for corridors of increased transport volume (Vogel, 2000; Vierth and Karlsson, 2013).
- Use of less personnel and dispatching of fewer locomotives for the transport of a given volume of goods (reduced operating cost) (Vogel, 2000).
- Substantial reduction in the cost of supply chain management (logistics) (Vierth and Karlsson, 2013).

On the contrary, an excessive increase in the length of trains could result in the following problems:

- Reduced track capacity (see Section 15.6.4.2).
- Uneven supply and demand (empty wagons).
- Compatibility issues with existing infrastructure.
- Time consuming coupling and shorting.
- Requirements for more effective coupling devices.
- Increase in the weight of trains.

15.6.4.2 Contribution towards solving the dilemma

The issue of utilising longer freight trains over European networks has been the focus of research both for specific routes as well as for European Union as a whole.

UIC funded in 2013 a research project to study the effects that the use of longer and heavier trains would have on railway infrastructure, track capacity, and cost-efficiency. This research concerned trains longer than 750 m and with an axle load heavier than 22.5 t, since those are the current upper limits of European railways. The analysis was conducted for various railway corridors and was based on information collected by both railway operators and infrastructure managers. The results indicated that the use of longer and/or heavier trains reduced the transport cost per ton. This reduction appeared to be usually greater for an increase in length rather than for an increase in weight (UIC, 2013).

Another relevant research (Vierth and Karlsson, 2013) was focused on an international railway corridor between Sweden and Germany. The researcher created and 'run' simulation models for three different lengths of trains and specifically for 750 m, 1,000 m, and 1,500 m. The results indicated that when there is an available amount of goods to be transported, an increase in train length results in an increase in ton-kilometers.

In 2011–2014, and under the supervision and funding of the European Union, a consortium of private European railway companies undertook the MARATHON (MAke RAil The HOpe for protecting Nature) project. In this endeavor, researchers coupled two 750 m long trains and created a 1,500 m long trainset with a secondary locomotive in the middle of the set, which was equipped with a number of innovative technologies. The results were overwhelmingly positive since a reduction in transport cost of over 30% was achieved (UIC, 2013).

Finally, the effect of an increase in train length on track capacity in terms of the maximum number of routed trains per day was also investigated (Christogiannis, 2012). It should be noted that track capacity is affected by the time headway of trains, which is affected by the length of blocks and the trains themselves. An increase in the length of trains from 500 to 1500 m, results in a decrease in track capacity from 88.3 to 81.9 trains/day.

When comparing the positive and negative effects and investigating the results of the aforementioned studies, it may be concluded that an increase in the length of trains would be potentially beneficial for European railways. However, more research is required in order to validate this conclusion and in order to define the optimum train length for current conditions. Finally, an investment in increasing train length maximises its potential only when combined with efforts to create better-surrounding conditions, such as the separation of freight and passenger traffic, an equal investment in infrastructure, and the automation of coupling and shorting processes.

REFERENCES

ADB (Asian Development Bank). 2005, *Railway Passenger and Freight Policy Reform Study, Technical Assistance – People's Republic of China*, November 2005, Mandaluyong..

Alias, J. 1985, *La voie ferrée – Exploitation technique et commerciale*, Lecture Notes, Vol. 3, ENPC, Paris.

Atanassov, I. and Dick, T. 2015, Delay and required infrastructure investment to operate long freight trains on single-track railways with short sidings, *IHHA 2015 Conference*, Perth.

Basbas, S. and Pyrgidis, C. 2001, Model development for rail transportation of hazardous materials in Greece, *Journal of Environmental Protection and Ecology*, Vol. 2(2), pp. 418–427.

Batisse, F. 1994, Les grandes tendances du trafic ferroviaire dans le monde, *Rail International*, January, pp. 15–22.

Batisse, F. 1995a, Le trafic de masse, atout du chemin de fer (1ère partie), *Rail International*, April, pp. 26–33.

Batisse, F. 1995b, Le trafic de masse, atout du chemin de fer (2ème partie), Rail International, June–July, pp. 23–29.

Boysen, H.E. 2012, *Genetal Model of Railway Transportation Capacity*, Computers in Railways XIII, WIT Press, pp. 335–348.

CER 2016, *Longer Trains: Facts & Experiences in Europe*, Bruxelles.

Christogiannis, E. 2012, *Investigation of the Impact of Traffic Composition on the Economic Profitability of a Railway Corridor – Fundamental Principles and Mathematical Simulation for the Selection of Operational Scenario for a Railway Corridor*, PhD Thesis, Aristotle University of Thessaloniki, Thessaloniki, Greece.

Christogiannis, E. and Pyrgidis, C. 2013, An investigation into the relationship between the traffic composition of a railway network and its economic profitability, *Rail Engineering International*, No. 1, pp. 13–16.

Christogiannis, E. and Pyrgidis, C. 2014, Investigation of the impact of traffic composition on the economic profitability of a new railway corridor, *Journal of Rail and Rapid Transit, Proc. IMechE, Part F*, Vol. 228(4), pp. 389–401.

Christogiannis, E. and Pyrgidis, C. 2015, Selection of the optimum exploitation scenario for an interurban railway corridor by the help of mathematical models, *Ingegneria Ferroviaria*, No. 5, pp. 427–448.

Combes, F. 2012, An empirical evaluation of the EOQ model of choice of shipment size in freight transport, *Transportation Research Record - Journal of the TRB*, pp. 92–98.

Dedicated Freight Corridors and High Speed Rail India's -Ultra Low Carbon Mega Rail Projects, Ministry of Railways, India 2015, available online at: http://en.wikipedia.org/wiki/Dedicated_Freight_Corridor_Corporation_of_India, 2015 (assessed 14 March 2015).

Duche, T. and Baspeyras, G. 1996, Le transport des marchandises dangereuses par chemin de fer, *RGCF*, April, pp. 45–53.

Energy efficiency in railways: Energy storage and electric generation in diesel electric locomotives, 2009, *20th International Conference and Exhibition on Electricity Distribution*, Prague, Czech Republic.

Eurostat. 2016, Available online at: https:// projects,shift2rail.org/S2r_projects.aspx

Ferreira, L. and Murray, M.H. 1997, Modelling rail track deterioration and maintenance: Current practices and future needs, *Transport Reviews*, Vol. 17, No. 3, pp. 207–221.

Jorgensen, M. and Sorenson, S. 1997, *Estimating Emissions for Air Railway Traffic, Report for the Project MEET: Methodologies for Estimating Air Pollutant Emissions from Transport*, Department of Energy Engineering, Technical University of Denmark, Lyngby, Denmark, available online at: www.inrets.fr/infis/cost319/MEETDeliverable17.pdf.

Journal de la Marine Marchande. 1994, *Les axes de progrès du transport ferroviaire des marchandises dangereuses*, No. 2308, 9 September.

Islam, D. and Mortimer, P. 2017, Longer, faster and heavier freight trains: Is this the solution for European railways? Findings from a case study, *Benchmarking: An International Journal*, Vol. 24, No. 4, pp. 994–1012.

Lambropoulos, A. 2004, *Design of Rail Freight Transport*, Lecture Notes, MSc Program "Design, organisation and management of transportation systems", AUTh, Faculty of Engineering, Thessaloniki, Greece.

Liguori, P. 2018, European railroads: How Europe's rail freight can connect ports to the hinterlands, joc.com, 26/02/2018, available online at: https://www.joc.com/rail-intermodal/international-rail/how-europes-rail-freight-can-connect-ports-to-hinterlands_20180226.html 6 April 2021.

Lundberg, S. 2006, *Freight Customers' Valuation of Factors of Importance in the Transportation Market*, Royal Institute of Technology, Stockholm.

Metzler. 1985, *Le matériel à voyageurs*, Lecture Notes, ENPC, Paris.

OECD. 2002, *Railway Reform in China – Promoting Competition*, Summary and recommendations of an OECD/DRC Seminar on Rail Reform in Beijing, 28–29 January.

OSE, 1998, *Transportation of Dangerous Goods in the Greek Railway Network*, Aristotle University of Thessaloniki/OSE, Greece, 1997–1998.

Pyrgidis, C. 1989, *Transport de mercaderies: Avantages i inconvenients de la massificacio de fluxos*, Espais – revista del department de politica territorial i ombres publiques de Catalunya, No. 17, May–June 1989, pp. 26–31.

Pyrgidis, C. and Basbas, S. 2000, Rail transportation of dangerous goods – Action plan development for Greek State Railways, *Rail Engineering International*, Netherlands, No. 3, pp. 8–11.

Pyrgidis, C. and Christogiannis, E. 2011, The problem of the presence of passenger and freight trains in the same track and their impact on the profitability of the railways companies, *9th World Congress on Railway Research 'Meeting the Challenges for Future Mobility'*, Lille, France 22–26 May 2011.

Pyrgidis, C., and Christogiannis, E. 2012, The problem of the presence of passenger and freight trains on the same track, *Elsevier Procedia Social and Behavioral Sciences*, Vol. 48, pp. 1143–1164.

Pyrgidis, C. and Christogiannis, E. 2013a, Freight dedicated railway corridors for conventional and for heavy loads – A comparative assessment of the economic profitability of the two systems, *3rd International Conference on Recent Advances in Railway Engineering – ICRARE*, Iran University of Science and Technology, Teheran, 30 Aprilto 1 May.

Pyrgidis, C. and Christogiannis, E. 2013b, Mixed or pure train routing? The case of an existing railway corridor, *6th International Conference for Transport Resarch ICTR*, Thessaloniki, 17–18 October.

Pyrgidis, C. and Dolianitis, A. 2021, *Rail Vehicle Classification, Encyclopedia of Transportation*, Elsevier.

RailFreight.com. 2018, High-speed freight train Italy hits the track on 7 November, 02/11/2018, available online at: https://www.railfreight.com/railfreight/2018/11/02/high-speed-freight-train-italy-hits-the-track-on-7-november/

Samimi, A., Mohammadian, A. and Kawamura, K. 2010, A behavioral freight movement microsimulation model: Method and data, *The International Journal of Transportation Research*, Vol. 2, pp. 53–62.

Savy, M. 2006, *Le Transport de Marchandises*, Eyrolles, Paris.

Schiffer, V. 2005, Generic types of freight cars (In German), available online at: http://home.wtal.de/gueterwagen/gatt-zde.htm

Shift2Rail Joint Undertaking. 2015, *Multi-Annual Action Plan*, Brussels.

SPECTRUM Consortium. 2012, *Market Based Operational Requirements for New Rail Freight Services*, Final Deliverable D1.2, Newcastle upon Tyne.

SPECTRUM Consortium. 2013, *Deliverable D2.4 Concepts*.

Steer Davies Gleave. 2015, *Freight on Road: Why EU Shippers Prefer Truck to Train*, European Union.

Troche, G. 2005, *High-Speed Rail Freight Sub-Report in Efficient Train Systems for Freight Transport*, KTH Railway Group Report 0512, Stockholm.

UIC. 2007, *Statistic Data*, available online at: www.UIC.org

UIC and CER. 2012, *Greening Transport: Reduce External Costs*, Paris, available online at: http://www.uic.org/spip.php?article1799

UIC. 2013, *Heavy and/or Long Trains*, UIC, Paris.

Vierth, I. and Karlsson R., 2013, Effects of longer lorries and freight trains in an international corridor between Sweden and Germany, *41st European Transport Conference*, Frankfurt.

Vogel, H. 2000, Longer freight trains: Possibilities and limitations, *Rail International*, Vol. 31, No. 4.

Woodburn, A., Allen, J., Browne, M. and Leonardi, J. 2008, *The Impacts of Globalisation on International Road and Rail Freight Transport Activity-Past Trends and Future Perspectives*, Transport Studies Department, University of Westminster, London, UK.

www.cit-rail.org/en/rail-transport…/cotif/

Zanuy, A.C. 2013, *Future Prospects on Railway Freight Transportation a Particular View of the Weight Issue on Intermodal Trains*, PhD, Technische Universität Berlin, available online at: https://www.railways.tu-berlin.de/fileadmin/fg98/papers/2013/PhD_Armando_Carrillo_Zanuy.pdf

Chapter 16

Heavy haul rail transport

16.1 DEFINITION AND GENERAL DESCRIPTION OF THE SYSTEM

The term 'heavy haul rail transport' describes any railway operation using trains of a minimum axle load equal to 25 t (Q = 25–40 t).

Trains that are intended for carrying heavy loads operate either in dedicated freight railway corridors or in corridors with mixed traffic operation. They serve the transport of conventional and hazardous goods. Railway networks intended for heavy haul transport satisfy both single wagonload services and trainload services (see Section 15.1) and usually have a broad or normal track gauge (Figure 16.1). The heavy haul rail transport requires reduced running speeds (V_{max} = 80–100 km/h).

Table 16.1 presents, for various heavy haul railway corridors, the number of tracks and the track gauge used, the length of the connection they serve, the allowed axle load, the total weight of the train, the traffic volume that they transport annually, the features that are selected for the track superstructure and the maximum running speed of freight trains.

In Section 16.3, constructional and functional characteristics of heavy haul rail transport systems that differ from those of conventional freight rail systems are identified. The heavy haul transport differs from other rail freight activities and should be considered and treated separately from the transportation of conventional loads.

16.2 THE INTERNATIONAL MARKET IN HEAVY HAUL RAIL TRANSPORT

Heavy trains are commonly used in America, South Africa, and Australia, where heavy haul rail transports have the largest share in the rail freight market. The train that carries the heaviest loads worldwide is also the longest one and is found in Australia. This train's payload is 82,000 t, its gross weight is 99,734 t, and is composed of 682 wagons and 8 diesel locomotives of 6,000 HP. The train's length is 7.2 km and it transports bulk minerals from one part of Australia to the other across thousands of kilometres, passing through uninhabited areas and deserts (Christogiannis, 2012).

In the EU, the average hauled weight of freight trains reaches 1,000 t, while in America trains that are 10–20 times heavier are used, the axle load of which is double than the respective axle load of trains at EU countries; trains operated in America also are 2–3 times longer than EU ones.

Moreover, in the EU, there are cases in which trains of high axle loads are dispatched over specific lines, such as in Hungary with trains of 4,000 t and Sweden with trains of 8,600 t. However, such cases amount to a minuscule percentage of the total European Network (Christogiannis, 2012; CER, 2016).

DOI: 10.1201/9781003046073-16

Figure 16.1 Heavy haul rail transport, convoy formation. (Adapted from Vossloh, 2014, Vossloh_Segment_Heavy Haul_09-2014_ENG.PDF.)

The tracks that are used for the heavy haul rail transport either belong to mixed traffic operation networks (India, Russia, etc.), in which the tracks are designed for a high design axle load, or to networks where freight trains have priority and which are specially constructed for heavy loads (the United States, Canada, etc.).

Countries in which such transportation is widely used are usually countries where the largest volume of transported goods are bulk products, which are related either to the needs of power production, for example, carbon, or to their manufacture, for example, iron ore. In the EU on the contrary, the transported products are mainly construction materials and chemicals. For such products, there is high competition between the railway and maritime and road transport.

16.3 DIFFERENCES BETWEEN CONVENTIONAL AND HEAVY HAUL FREIGHT RAILWAY NETWORKS

The operation of heavy haul freight trains on the one hand seems to achieve 'economies of scale' as these trains have a much higher transport capacity than the conventional freight trains. On the other hand, however, such an activity increases the implementation and maintenance cost of the railway track. The reason for this is that many features of the heavy haul freight trains/wagons differ substantially from the respective features of the conventional haul freight trains/wagons.

In columns (1)–(3) of Table 16.2, the features that differentiate between the two systems are presented.

Table 16.1 Main characteristics of railway lines intended for heavy haul rail transport (indicatively)

Country/line	Line length (km)	Number of tracks/ gauge (mm)	Axle load (t)	Train load (t)/ annual payload (million t)	Type of carried cargo	Rail's weight/sleepers type	Maximum running speed (km/h)
Australia/Pilbara–Rio Tinto	1,100	Single/1,435	32.5	29,500/220	Steel	68 kg/m/concrete	80
South Africa/ Sishen–Saldanha	861	Single/1,067	30	34,000/47	Steel	60 kg/m/concrete	80
North America/Burlington	43	Double/1,435	30	19,000/50	Coal	68 kg/m/wooden and concrete	75
Norway–Sweden/ Ofotbanen	43	Single/1,435	31	8,200/30	Steel	54–60 kg/m/wooden and concrete	50–90
Russia–Mongolia–Chin a–Kazakhstan/Trans-S iberian	9,244	Single/1,520	> 30	> 30,000/60	Containers	68–75 kg/m/wooden and concrete	80
USA/Wyoming Joint Line–Powder River Basin	165	Double/1,435	> 30	19,000/300	Coal	68 kg/m/concrete	80
China/ Datong–Qinhuangdao	653	Double/1,435	25	20,000/380	Coal	75 kg/m/concrete	80–100
Western Australia/ Fortescue	280	Single/1,435	40	32,000/55	Steel	68 kg/m/concrete	80
North-western Australia/ Hamersley	388	Single/1,435	30	26,000/55	Coal, minerals, iron, and steel	68 kg/m/concrete	75

Table 16.2 Features of the heavy haul freight wagons/trains that differ significantly from those of the conventional haul freight wagon/trains – effects (positive or negative) and requirements for the design, construction, operation, and maintenance of the railway system

Wagon/train characteristics that are different in the two railway systems (1)	Conventional freight trains values of the features (2)	Heavy haul freight trains values of the features (3)	Effects (positive or negative)/requirements for heavy haul railway networks (4)
Running speed (V_{max})	60–120 km/h (100 km/h)	50–100 km/h (80 km/h)	*Effects:* Lower track capacity, longer run time *Requirements:* Smaller curvature radius in the longitudinal and vertical alignment
Axle load (Q)	16–25 t	25–40 t	*Effects:* Higher track geometry defects deterioration rate, longer train braking distance, and higher transported volume of goods *Requirements:* Steeper gradients in vertical alignment, heavier rails, sleepers of higher mechanical resistance, thicker ballast layer, longer signal spacing, greater traction power requirements, higher maintenance needs, and wagons of higher transport capacity
Train weight (B_{tr})	1,500–3,000 t	5,000–35,000 t	*Effects:* Greater braking weight, higher transported volume of goods *Requirements:* Steeper gradients in vertical alignment, longer signal spacing, and greater traction power requirements
Daily traffic load (T_f)	10,000–100,000 t	100,000– 300,000 t	*Effects:* Higher track geometry defects deterioration rate, and higher transported volume of goods *Requirements:* Heavier rails, sleepers of higher mechanical resistance, thicker ballast layer, and higher maintenance needs
Train length (L_{tr})	400–800 m	1,000–4,000 m	*Effects:* Lower track capacity, higher transported volume of goods *Requirements:* Longer tracks and platforms in stations
Vehicle clearance gauge	Normal	Enlarged	*Effects:* Larger rolling stock gauge *Requirements:* Differentiates depot and station dimensioning, distance between track centres, and height clearance under civil engineering structures
Type of transported goods	All types	Mainly steel, ore, and coal	

Source: Adapted from Pyrgidis, C., and Christogiannis, E., 2013, Freight dedicated railway corridors for conventional and for heavy loads – A comparative assessment of the economic profitability of the two systems, 3rd International Conference on Recent Advances in Railway Engineering – ICRARE 2013, Iran University of Science and Technology, 30 April–1 May 2013, Teheran, Congress Proceedings.

16.4 IMPACTS OF HEAVY HAUL RAIL OPERATIONS AND MAIN DESIGN PRINCIPLES

Column (4) of Table 16.2 lists the impacts, both positive and negative, which cause differentiations between the two systems, as well as the requirements imposed on their design, construction, and operation (Pyrgidis and Christogiannis, 2013).

The axle load is directly or indirectly involved in the analytical expressions of all forces acting on the wheel–rail contact surface and influences the behaviour of both the rolling stock and the track (Pyrgidis, 2009).

The presence of high axle loads creates many technical problems. The transporting of heavy loads results in increased stresses applied on the rails and transferred by the features of the track superstructure to the subgrade. This has serious implications on the rails (such as cracks, damage to the weld points and breakage due to fatigue) and on the sleepers (e.g., wear). Moreover, damage is observed at the areas of switches and crossings. Stresses on the track bed layers and the formation layer are likely to exceed the permissible values. Therefore, the maintenance cost and the frequency of track inspections are increased.

In order for the heavy haul rail transport to become financially efficient, it is required: (i) for heavy loads' lines that are already in operation to be frequently checked concerning defects of the track superstructure and (ii) for lines that are under construction, all features of the infrastructure to be specially designed. More specifically, as discussed below, the main requirements are the use of heavier rails, the use of sleepers that are of greater mechanical strength, the increased density of the sleeper layout, the special design of the ballast regarding its thickness and the type of materials used, the provision of specifically designed formation layer, and the improvement of the soil if it is of poor quality.

For heavy haul rail transport (Q > 30 t), the specifications and technical solutions that are described hereunder should be adopted.

16.4.1 Selection of track infrastructure components

16.4.1.1 Selection of the track's alignment geometric characteristics

In the case of broad gauge track, the same track infrastructure can serve much heavier loads than in the cases of normal and metric gauge track.

In practice, an increase in the axle load (up to a certain degree) is not treated by widening the gauge but by increasing the mechanical strength of the features of the track's superstructure (e.g., heavier rails).

The increase of the axle load leads to the adoption of lower longitudinal gradients. The adoption of larger longitudinal gradients in many cases reduces the tunnelling works and the construction of bridges considerably; however, it requires locomotives of greater traction power (Geo-Technical Engineering Directorate, 2007).

16.4.1.2 Selection of rails

When the axle load is increased, an increase of the cross section of the rail is required, that is, rails that are heavier and of greater mechanical strength are required (UIC 60, UIC 72 grade steel 90 kp/mm^2).

Based on Equation 16.1 (Esveld, 1989), it is deduced that axle loads of 20, 25, and 30 t require rails of minimum weight equal to 50, 60, and 70 kg/m, respectively. An increase of the axle load by 25% requires a 20% increase of the rail weight:

$$B_r = 2.25Q + 3 \qquad (16.1)$$

where:
   B_r: Rail weight per metre (kg).
   Q: Axle load (t).

For axle load up to 25 t rails of 60 kg/m 90 UTS (Ultimate Tensile Strength) is sufficient. For higher axle loads, heavier rails are required.

For a load Q = 30 t, rails of 68.5 or 71 kg/m must be used.

Regarding the stress that is developed on the rail (and therefore regarding the rail's wear), it increases as the quantity $(Q_o)^v$ increases (Alias, 1977), where v is exponent, the values of which are between 3 and 4, and Q_o is the wheel load.

Longer wavelength irregularities, usually known as 'waves', which appear on the rail's rolling surface, are due to fatigue of the wheel–rail contact surface. During the heavy haul rail transport, the pressure on the contact surface is very high, and this results in the development of corrugations troughs (with gross plastic flow), which have a wavelength of 200–300 mm and a frequency of 30 Hz, for average traffic speed (Grassie and Kalousek, 1993; Grassie and Elkins, 1998). The longer wavelength corrugations have a particularly big impact on the maintenance, as they increase costs by up to 30%.

In any case, it is evident that the increase of axle loads intensifies the phenomena of fatigue and their consequences.

Increased axle load results in wear and fatigue during the early stages of the line's operation (Lichtberger, 2005).

Regarding the frequency of the required track maintenance work, the influence of vertical loads is catalytic. The deterioration of the track's quality is proportional to the third power of the value of the axle loads. An increase in the axle load by 10% reduces the intervals between two consecutive maintenance works by 30% (Liu, 2005).

Finally, with regard to the quality of steel, it is required that the steel used be harder and heat-treated, in order to be able to bear the increased loads, to have increased resistance against wear and fatigue, and to ensure increased lifetime for the rails.

16.4.1.3 Selection of the type of sleepers and the distances between them

The tracks on which high axle loads are expected to be imposed are usually constructed using sleepers made of prestressed concrete. The sleepers' density is 1,540 sleepers per km or 1,660 sleepers per km. The use of 60 kg/m rails and prestressed concrete sleepers placed at distances of 43 cm between them is suitable for the operation of axle loads that are equal to 30 t.

16.4.1.4 Selection and dimensioning of track bed layer features

By applying Equation 16.2 (Esveld et al., 1989; UIC, 2006; Profillidis, 2014), the impact of the axle load on the thickness of the track ballast and the sub-ballast layers can be assessed as:

$$e_{bt} = e_b + e_{sb} = E_b + a_b + b_b + c_b + d_b + f_b \qquad (16.2)$$

where:
   e_b: Ballast layer thickness (m).
   e_{sb}: Sub-ballast layer thickness (m).
   e_{bt}: Total thickness of ballast and sub-ballast layers (m).
   E_b: Parameter that depends on the quality category of the soil and the bearing capacity of the substructure.
   a_b: Parameter that depends on the classification of the track in the UIC classes.
   b_b: Parameter that depends on the sleepers' length and material.
   c_b: Parameter that depends on the volume of the required track maintenance work.

d_b: Parameter that depends on the maximum axle load.

f_b: Parameter that depends on the track design speed and the bearing capacity of the substructure.

By using Equation 16.2, the contribution rate of the parameter d_b on the total thickness e_{bt}, and, thus, the influence rate of the axle load Q on it can be derived graphically (Figure 16.2). As shown in Figure 16.2, the influence rate of the axle load on the total thickness of ballast and sub-ballast layers is negligible for passenger trains (Q ≤ 20 t), while for freight trains it can increase the thickness by 10–21%.

The thickness of the sub-ballast can range between 15 cm and 75 cm, depending on the quality of the soil material of the substructure.

For an axle load equal to Q = 30 t, the required thickness of the ballast layer is 25 cm and the required thickness of the sub-ballast layer is 15 cm for speed up to 100 km/h. However, a clean ballast layer with a thickness of 300 mm may prove to be the best solution.

The gravel should be of high hardness. Frequent monitoring of the ballast with the use of mechanical means and appropriate techniques is required.

The existence of high axle loads in combination with a weak subgrade requires that a sub-ballast layer which is no less than 1 m thick be placed between the ballast and the subgrade.

16.4.1.5 Construction principles of the formation layer

The introduction of an axle load of 30 t requires the provision of a formation layer of sufficient thickness in order to improve the bearing capacity just below the sub-ballast. It is obvious that a weak subgrade will lead to a rapid deterioration of the track's geometry, which will render the operation of heavy axle-load trains unsafe, thereby imposing an additional requirement for increased and more frequent maintenance (Dingqing and Chrismer, 1999).

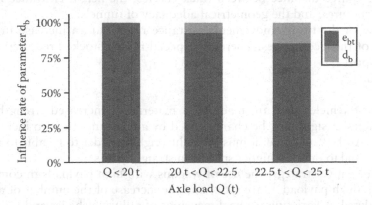

Figure 16.2 Influence of the parameter d_b on the total thickness of ballast and sub-ballast layers e_{bt}. (Adapted from Christogiannis, E., 2012, Investigation of the impact of traffic composition on the economic profitability of a railway corridor – Fundamental principles and mathematical simulation for the selection of operational scenario for a railway corridor, PhD thesis (in Greek), Aristotle University of Thessaloniki, Thessaloniki, Greece; Pyrgidis, C., and Christogiannis, E., 2012, The problem of the presence of passenger and freight trains on the same track, *International Congress TRA (Transport Research Arena) 2012, 'Sustainable mobility through innovation'*, 23–26 April 2012, Athens; Pyrgidis, C., and Christogiannis, E., 2011, The problem of the presence of passenger and freight trains on the same track and their impact on the profitability of the railway companies, *9th World Congress on Railway Research 'Meeting the challenges for future mobility'*, 22–26 May 2011, Lille, France.)

For heavy haul rail transport, the following are required:

- Stabilisation of the substructure's soil with suitable mechanical means during construction.
- Improvement of the foundation soil in case it is considered to be of poor quality.

16.4.1.6 Dimensioning of bridges

Bridges require special design in the case of heavy haul rail transport. Since sometimes cracks are developed on the concrete along the bridges, it is necessary that the substructure of the bridge be strengthened (see Section 2.2.3.4).

16.4.1.7 Dimensioning of the signalling system

The heavy haul rail transport results in an increase of the total weight of the train, thus increasing significantly the braking distance (which depends on the total weight of the train, the speed of the train, the total resistance of the train, and the longitudinal gradient of the line). For this reason, there is an urgent need for a special study on the signalling system.

16.4.2 Effects on the rolling stock

The increase of the axle load results in:

- An increase of the static gauge of the rolling stock.
- An increase of the train's movement resistance.

The effect of the rolling stock's static gauge on the components of the railway system mainly concerns the required distance between track centres, the height clearance from the civil engineering structures, and the geometrical adequacy of tunnels.

The increase of the train's movement resistance results in an increase in the required engine power of the locomotives. Therefore, special rolling stock is required for the heavy haul rail transport.

More specifically:

- The use of vehicles that are made by a material of increased strength is required. The vehicles' design must be characterised by a constant effort to increase transport capacity. At the same time, it must be lightweight in order to be able to move at steep gradients and to develop higher speeds on straight paths.
- As an alternative strategy, the use of wagons with high payloads in comparison with their tare (high payload – tare ratio) and the increase of the number of axles can also be considered. A variation of the dimensions of railway vehicles and more specifically the reduction of the wheels' diameter is essential.
- The use of 3-axle bogies increases the transport capacity, keeping the forces applied on the track within the permissible limits.

Special probes for the impact of high load on the wheels (Wheel Impact Load Detector – WILD) can provide control and monitoring capabilities regarding the effects of heavy rail vehicles on the track, thereby constituting a valuable tool for the study of a heavy haul rail system.

16.4.3 Effects on the operation

The heavy haul rail transport regards trains of large length, thereby reducing the track capacity.

In addition, the heavy haul rail transport usually concerns the transport of goods over long distances, that is, it is implemented at lines that are of long length. A typical example is Trans-Siberian line, with a length of 9,244 km, which crosses two continents, namely Europe and Asia.

The distributed power, or in other words the use of more than one locomotive in the train formation (at the front and the back, at the front and the middle, and at the front, the middle, and the back), makes the transport operation much more effective. This option reduces stress on couplers and buffers, enables longer and heavier trains, improves force distribution in curves, and enables quicker brake response (Hoffrichter, 2014).

The increase of the axle loads from 20 t to 30 t is expected to result in an increase in the track's maintenance cost by three times, depending on the quality of the infrastructure. It is still unclear as studies show that after the rails are subject to grinding processes and lubrication, the observed increase in maintenance cost is only 3%.

According to the literature (Christogiannis, 2012), in the case of two railway tracks with axle-load values of Q = 16 t and Q′ = 22.5 t, respectively, and assuming that the speed is equal in both cases, the following mathematical equation applies:

$$\frac{C_o'}{C_o} = 1.406^{2/\pm} \tag{16.3}$$

where:

    C_o, C_o': The respective track maintenance cost.

    α, β: Coefficients that are empirically determined (Esveld, 1989; Lichtberger, 2005), depending on the type of the superstructure wear.

The ratio C_o'/C_o is calculated between 1.41 and 2.78 (the ratio β/α is calculated between 1 and 3.5), that is, the maintenance cost in the case of a track for which Q′ = 22.5 t is between 41% and 178% greater than the respective cost for a track for which Q = 16 t.

16.5 ECONOMIC EFFICIENCY OF HEAVY HAUL RAIL TRANSPORT

A question faced by many railway operators nowadays is: 'What is more economically efficient for a railway company? Operating of conventional load freight trains or heavy haul freight trains on a new railway corridor which is dedicated for freight operation?' (see also Chapter 15).

The issue of economic viability of rail networks that are used for the heavy haul transport over long distances contains several uncertain factors that affect the cost. The increased axle load initially seems like a profitable factor; however, a more realistic assessment is deemed necessary. The increased costs caused by the increased energy consumption, the increased investment in rolling stock, the requirements for the components of the infrastructure, and the more frequent and increased wear of the track and the rolling stock constitute factors that may change the facts.

A technological solution is to increase the maximum allowable payload for a given axle load by improving the net percentage of tare.

Moreover, the use of multiple axle vehicles, which are more popular for road transport, could be a solution for railways as it increases the transport capacity. On the contrary, the strength and safety of bridges should be examined from scratch.

Mathematical models for decision-making regarding the transfer from a conventional load freight system to a heavy load system have been developed. In a relevant survey that was carried out (Christogiannis, 2012; Pyrgidis and Christogiannis, 2013), the economic efficiency of heavy haul rail transport rail (axle load of 30 t) was compared with the economic efficiency of rail freight transport of conventional loads (axle load 22.5 t) with the aid of mathematical models. The comparison concerns the implementation and operation of a new single track of standard gauge dedicated for freight traffic and is implemented by considering various values of freight workload demand (10,000–130,000 t per day per direction) and of connection lengths (S = 500 and 1,000 km).

Within the framework of this research, the rail infrastructure manager is also the owner of the rolling stock and the operating company. The financial indicator that has been considered to characterise the economic profitability of the new railway corridor is the Net Present Value (NPV) of the investment.

The following steps were methodologically followed:

- Initially, a minimum daily freight load value was considered to be equal to 10,000 t per direction for both corridors. It was considered that this demand:
  - As concerns the conventional network, is served by ten trains per direction which are formed of a number of locomotives and a number of wagons that can meet the above requirement. All wagons have an occupancy ratio of 80%.
  - As concerns the heavy haul corridor, is also served by ten trains per direction which are formed of a number of locomotives and a number of wagons and can meet the above demands given the same occupancy ratio (80%).

In both operation scenarios, a necessary prerequisite is that, in accordance with the UIC method, the track capacity saturation ratio should not exceed 70%. Assuming a connection length equal to S = 500 and 1,000 km, the NPV is thus estimated for both systems.

- The value of the daily freight load increased by 100% (20,000 t) for both corridors. Thus in order to meet this demand, it was considered that:
  - As concerns the conventional rail corridor, the number of wagons is initially increased (maximum value of 28 wagons) and thereafter, if demand cannot thus be met, the number of scheduled trains is increased. The occupancy ratio of the wagons remains constant and the track capacity saturation ratio does not exceed 70% of the practical capacity of the line, in accordance with the UIC method. In each case, the number of locomotives required is calculated.
  - As concerns the heavy haul rail corridor, the number of wagons is initially increased (maximum value of 85 wagons) and thereafter, if demand cannot thus be met, the number of scheduled trains is increased. The occupancy ratio of the wagons remains constant and the track capacity saturation ratio does not exceed 70% of the practical capacity of the line. In each case, the number of locomotives required is calculated.

Assuming the connection length to be equal to S = 500 and 1,000 km, the NPV is thus calculated for both corridors:

- The value of the daily freight load is gradually increased by steps of 10,000 t, and the same procedure is repeated.
- After being suitably recorded, the results are compared and evaluated.

Table 16.3 indicatively presents, for a connection length of S = 1,000 km and for the different freight volume values under examination, for both exploitation scenarios:

- The formation of the train (locomotive and wagons).
- The number of daily services per direction.
- The saturation ratio of track capacity.
- The NPV for each of the two scenarios.

It is noted that the initials EXCA (EXceeded CApacity) indicate that 70% of the practical capacity of the track is exceeded and, for this reason, the financial indicator is not recorded.

The diagram in Figure 16.3 shows the change in NPV in relation to freight volume demand for both operation scenarios examined and for both connection lengths considered.

By examining all the combinations of demand and connection length, the following conclusions have been reached:

- The conventional load freight-dedicated corridor can serve up to around 40,000 t daily for each direction, while the heavy haul rail corridor can cater for roughly three times that volume.
- Both systems have a negative NPV for a daily freight of up to approximately 20,000 t for each direction.
- For daily freights per direction of up to 40,000 t, which can be served by both systems, conventional load corridors are economically more profitable.

Table 16.3 Application of the mathematical model – results for connection length S = 1,000 km

	Conventional axle-load line				Heavy axle-load line			
Demand (t/day/ direction)	Trains scheduled per day per direction	Train formation (locomotive + wagons)	Track capacity saturation ratio (max permitted = 70%)	NPV (€M)	Trains scheduled per day per direction	Train formation (locomotive + wagons)	Track capacity saturation ratio (max permitted = 70%)	NPV (€M)
10,000	10	1 + 18	29%	−7,205	10	1 + 13	34	−8,986
20,000	13	2 + 28	37%	−2,374	10	1 + 26	34	−4,398
30,000	19	2 + 28	54%	1,976	10	1 + 38	34	−58
40,000	25	2 + 28	71%	6,311	10	1 + 51	34	4,534
50,000	−	−	EXCA	−	10	1 + 63	34	8,785
60,000	−	−	EXCA	−	10	2 + 76	34	13,233
70,000	−	−	EXCA	−	11	2 + 85	38	19,412
80,000	−	−	EXCA	−	12	2 + 85	41	22,461
90,000	−	−	EXCA	−	14	2 + 85	48	28,433
100,000	−	−	EXCA	−	15	2 + 85	52	31,462
110,000	−	−	EXCA	−	17	2 + 85	59	37,447
120,000	−	−	EXCA	−	19	2 + 85	66	43,525
130,000	−	−	EXCA	−	20	2 + 85	69	46,461

Source: Adapted from Pyrgidis, C. and Christogiannis, E. 2013, Freight dedicated railway corridors for conventional and for heavy loads – A comparative assessment of the economic profitability of the two systems, 3rd International Conference on Recent Advances in Railway Engineering – ICRARE 2013, Iran University of Science and Technology, 30 April–1 May 2013, Teheran, Congress Proceedings.

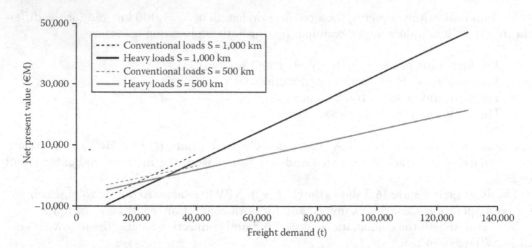

Figure 16.3 Variation of NPV in relation to the freight demand for a conventional axle-load line and for a line for heavy axle loads – length of connection S = 500 and 1,000 km. (Adapted from Pyrgidis, C., and Christogiannis, E., 2013, Freight dedicated railway corridors for conventional and for heavy loads – A comparative assessment of the economic profitability of the two systems, 3rd International Conference on Recent Advances in Railway Engineering – ICRARE 2013, Iran University of Science and Technology, 30 April–1 May 2013, Teheran, Congress Proceedings.)

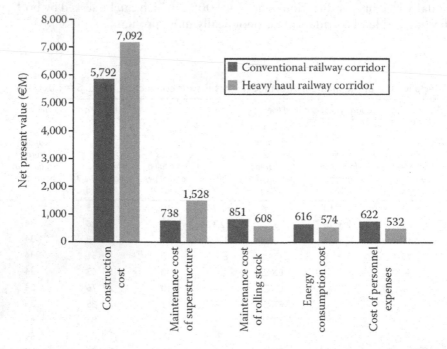

Figure 16.4 Construction, maintenance, and operational costs for conventional and heavy axle-load lines – length of connection S = 1,000 km, demand for freight = 30,000 t per day per direction.

- For heavy haul rail corridors with a daily freight greater than around 30,000 t, the increase in the connection length results in a significant increase in profitability as it translates to an approximate doubling of the NPV. Similar conclusions also apply for conventional freights; however, the point where it becomes profitable is at around 25,000 t.

The histogram in Figure 16.4 presents the different costs incurred for the two exploitation scenarios examined, for daily freight volumes per direction equal to 30,000 t and for connection length S = 1,000 km.

The intermediate calculations showed that in the case of the heavy haul rail corridor in comparison with the conventional freight corridor:

- The total construction cost of the infrastructure (studies, expropriations, civil engineering works, superstructure, substructure, track systems, and facilities) is approximately 18.5% more.
- The construction cost of the superstructure is 15% more.
- The maintenance cost of the superstructure is about 52% more.

On the contrary, the cost for the maintenance of the rolling stock, the energy consumption cost, and the personnel cost are lower.

REFERENCES

Alias, J. 1977, *La Voie Ferrée*, Editions Eyrolles, Paris.
CER. 2016, *Longer Trains: Facts & Experiences in Europe*, Bruxelles.
Christogiannis, E. 2012, *Investigation of the Impact of Traffic Composition on the Economic Profitability of a Railway Corridor – Fundamental Principles and Mathematical Simulation for the Selection of Operational Scenario for a Railway Corridor*, PhD thesis (in Greek), Aristotle University of Thessaloniki, Thessaloniki, Greece.
Dingqing, L. and Chrismer, S. 1999, *Soft Subgrade Remedies under Heavy Axle Loads*, TTCI, Pueblo, CO.
Esveld, C. 1989, *Modern Railway Track*, MRT-Productions, West Germany, Duisburg.
Esveld, C., Jourdain, A., Kaess, G. and Shenton, M. 1989, The consequences on track maintenance of increasing axle loads from 20 to 22.5 t, *Rail Engineering International*, Vol. 2, pp. 20–24.
Geo-Technical Engineering Directorate. 2007, *Research Designs and Standards Organisation*, Lucknow, India, July.
Grassie, S.L. and Elkins, J.A. 1998, Corrugation on North American transit lines, *Vehicle System Dynamics*, Vol. 28(Suppl.), pp. 5–17.
Grassie, S.L. and Kalousek, J. 1993, Rail corrugation: Characteristics, causes and treatments, *Proceedings of the IMechE, Part F: Journal of Rail and Rapid Transit*, Vol. 207F, pp. 57–68.
Hoffrichter, A. 2014, *Heavy Haul Freight*, Lecture Notes, Master in Railway Systems Engineering and Integration, University of Birmingham, UK.
Lichtberger, B. 2005, *Track Compendium – Formation, Permanent Way, Maintenance, Economics*, Eurail Press, Hamburg.
Liu, Z. 2005, Track deterioration and countermeasure after running 10000 ton heavy haul fleet on Daqin coal line of China, *International Congress, Railway Engineering 2005*, 29–30 June 2005, University of Westminster, London.
Profillidis, V. A. 2014, *Railway Management and Engineering*, Ashgate, Farnham.
Pyrgidis, C. 2009, Transversal and longitudinal forces exerted on the track – Problems and solutions, *10th International Congress Railway Engineering 2009*, 24–25 June 2009, London.
Pyrgidis, C. and Christogiannis, E. 2011, The problem of the presence of passenger and freight trains on the same track and their impact on the profitability of the railway companies, *9th World Congress on Railway Research 'Meeting the Challenges for Future Mobility'*, 22–26 May 2011, Lille, France.
Pyrgidis, C. and Christogiannis, E. 2012, The problem of the presence of passenger and freight trains on the same track, *International Congress TRA (Transport Research Arena) 2012, 'Sustainable Mobility through Innovation'*, 23–26 April 2012, Athens.

Pyrgidis, C. and Christogiannis, E. 2013, Freight dedicated railway corridors for conventional and for heavy loads – A comparative assessment of the economic profitability of the two systems, *3rd International Conference on Recent Advances in Railway Engineering – ICRARE 2013*, Iran University of Science and Technology, 30 April–1 May, 2013, Teheran.

UIC. 2006, *Earthworks and Track Bed for Railway Lines*, CODE 719R, 3rd edition, Paris, France.

Vossloh. 2014, Vossloh Segment_Heavy Haul 09–2014_ENG.PDF. Vossloh, Werdohl.

Chapter 17

Operation of railway systems under specific weather conditions and natural phenomena

17.1 SPECIFIC WEATHER CONDITIONS/NATURAL PHENOMENA AND THE RAILWAY SYSTEMS

Today, extreme weather conditions (strong crosswinds, frost/heavy snowfall, and high temperatures) and natural phenomena (heavy leaf fall, sandstorms, and earthquakes) occur that have an impact on both the rolling stock and the infrastructure and, thus, the operation and safety of railway transportation systems.

Their occurrence is likely to increase in the future due to climate change and, thus, existing railway systems will increasingly have to operate under conditions beyond their initial design specifications. Moreover, the railways are expanding into parts of the globe in which intense weather conditions are present either permanently or for extended periods of time (deserts, coastlines, high altitudes, etc.).

As it is hard to predict when and where extreme weather conditions and natural phenomena will occur, effective measures may not always be in place on existing railway lines to counteract or manage their impact (Rossetti, 2017). However, when introducing new railway lines in areas that are known to be prone to extreme weather conditions and natural phenomena, it is desirable that these meet new and stricter specifications that would need to be decided upon. In this framework, it would be necessary to reconsider the way in which all railway lines are designed, constructed, operated, and maintained in the future, in order to counteract the impact of extreme weather conditions and natural phenomena; a fact that is already being addressed by the International Union of Railways (UIC, www.ariccs.org). In addition, a uniform management system to counteract the severe weather conditions and natural phenomena is called for, which would need to be included in the track supervision regulations of the various railway organisations.

As a result of the above, several actors and responsible parties around the world have over the past decade started to study the effects of weather conditions on railway operations. An indicative example may be found in a 2007 study, the questionnaire of which was completed by representatives of over 40 railway networks (UIC, 2016).

Long delays, closed line segments for extended periods, and a higher maintenance cost are among the common consequences of the special conditions that are investigated. At the same time, and in cases where the phenomena are particularly intense, safety issues arise (vehicles overturning, derailments).

Mitigation measures exist and are applied in all cases. However, taking into account that the actors involved in preventing and dealing with the consequences are varied, the interfaces with the railway system are numerous, and the impacts on the system's operation are high, it would be beneficial to develop for each special weather condition a complete and integrated management system.

DOI: 10.1201/9781003046073-17

This chapter deals with the impact that these external parameters may have on train operation, as well as possible preventive/counteractive measures that could be implemented to alleviate their impact. The categories of systems that are investigated include high- and very high-speed networks and conventional speed intercity/regional networks (of either exclusively passenger or mixed traffic operation) (Pyrgidis et al., 2017; Dolianitis et al., 2017).

17.2 SPECIFIC WEATHER CONDITIONS

Weather conditions that may have an impact on train operation include:

- Strong crosswinds.
- Frost and heavy snowfall.
- High temperatures.

17.2.1 Strong crosswinds

17.2.1.1 Interfaces with the railway system: impacts

Strong crosswinds may have an impact on the dynamic behaviour of a train as a result of:

- *Lateral forces*: These forces affect the lateral dynamic behaviour of the entire train, in that they cause lateral vehicle instability, which may lead to a derailment.
- *Lifting forces*: These forces, which result from air intrusion in the empty space between vehicle-body and track superstructure, may in certain circumstances lift the vehicle from the track, which may lead to a derailment due to wheel climb, as well as to ballast flying that may cause vehicle damage.
- *Drag forces*: These forces, the extent of which depends on the running speed and the aerodynamic shape of the vehicle, may lead to an increase in energy consumption and aerodynamic noise.

When trains are operating under strong crosswind conditions, in combination with heavy rainfall, especially in exposed areas, such as bridges or high embankments, the impact may accumulate and the risk of vehicle overturning may increase (Dios Sanz Bobi et al., 2009; Imai et al., 2002).

When comparing lateral wind conditions in combination with rain with those without rain, the critical overturning speed of the train is reduced by 10–20%. Thus, the train may overturn more easily.

The critical overturning speed takes various values depending on the angle under which wind forces are applied and is increased as the trains move at lower speeds (Zheng, 2009; Baker et al., 2010).

In Europe, specifications as regards vehicle overturning risks are laid down in Technical Specifications for Interoperability (TSIs).

17.2.1.2 Possible mitigation measures

To combat the three types of forces that result from strong crosswinds, the following measures could be applied:

- *Lateral forces*: These could be counteracted by:
  - Placing anemometers at strategic locations along the track, in order to monitor and record wind speed.

- Implementing speed restrictions which, however, may increase run time and, thus, cause train delays.
- Installing wind barriers, which are usually composed of dense wire meshes or perforated metallic foil and have a height of 2–3 m.
- Increasing the height of existing noise barriers.
- Lifting forces: These could be counteracted by:
  - Installing an aerodynamically shaped shield under the nose of the train.
  - Installing protective covers underneath the vehicle.
  - Placing the sleepers higher or by lowering the ballast layer, thus creating a larger void between the bottom of the rail and the top of the ballast, in order to allow any compressed air that is trapped between the vehicle and the track to escape, thus reducing aerodynamic pressure. Ballast is more prone to being picked up when it is lying on the surface of the sleepers. Lowering the ballast profile by 2–3 cm is a mitigation strategy that is sometimes adopted to combat ballast flying. However, in some cases, this has led to an increase in tamping demand, possibly due to the lower resistance of the track to lateral displacement that is associated with a lower ballast layer (US Department of Transportation, 2015).
- *Drag forces*: These could be counteracted by implementing measures that reduce the aerodynamic vehicle resistance, such as by:
  - Placing covers over the equipment of the train.
  - Covering the entire vehicle-body except the bogies.
  - Adopting bogie skirts.
  - Adopting under-bogie covers.

17.2.2 Frost/heavy snowfall

17.2.2.1 Interfaces with the railway system: impacts

Frost and heavy snowfall, which usually occur during the winter months, may cause (Eddowes et al., 2003):

- Problems as regards diesel engine operation.
- Rail breakages.
- Malfunctioning of points due to the presence of ice.
- Snow coverage of track and/or vehicles.
- A lower wheel/rail adhesion – increased braking distance.
- An increase in track loading.
- Malfunctioning of track and vehicle equipment due to snow intrusion.

The most serious consequence could be a derailment due to point failure or rail fracture. Also, significant timetabling problems may arise due to train delays. Furthermore, increased wear occurs throughout the railway system, either due to braking issues or due to the operation of the system's infrastructure elements outside their specifications. All the above can lead to a demand for maintenance.

17.2.2.2 Possible mitigation measures

The problems that can be caused by frost and heavy snowfall may be counteracted by (Eddowes et al., 2003):

- Using point heating systems, in order to prevent (Heat Trace Ltd, n.d):
  - The moving rail of the switch from freezing to the stock rail.
  - The moving rail of the switch from freezing to the supporting slide plates.
  - The build-up of snow, sleet, or hail between the switch rail and the stock rail that could jeopardise the proper functioning of the point system.
- Removing snow from the track by means of appropriate vehicles/equipment.
- Implementing a winter timetable, also embracing the running of empty trains during the night to prevent snow from accumulating.
- Implementing protection measures against avalanches, such as snow fences, avalanche protection forests, avalanche wedges, etc. (Noguchi and Fujii, 2000).
- Implementing temperature condition monitoring systems.
- Fitting trains with specialised equipment (snow ploughs, steam jets, etc.).
- Designing and running trains that are suitable for operation in low-temperature conditions. This mainly refers to how the surfaces of the rolling stock are designed – they should not be flat but curved so that snow and ice can fall off, and the body, shell, and various openings of the vehicles should be adequately insulated.
- Ensuring that adequate track drainage is in place. When temperatures rise above 0°C, snow melts, and the melting water seeps through the ballast bed layers. Also, if melted snow is not effectively removed from the track, it may cause damage to various track components when it refreezes.

17.2.3 High temperatures

17.2.3.1 Interfaces with the railway system: impacts

High temperatures can have a negative effect on a railway system, in that the resulting increase in thermal loading of the various structural elements may lead to (Eddowes et al., 2003; Baker et al., 2010):

- An increased risk of track buckling.
- A drying out of the earthworks.
- An increased demand for vehicle air-conditioning.
- An increased demand for ventilation, especially in tunnels.
- An increased growth of vegetation along the track which, in turn, may lead to an increase in leaf fall.

The most serious consequence that could occur is a derailment, due to either track buckling or earthworks failure. Also, when vehicles are not properly air-conditioned, high-temperature conditions may lead to serious passenger discomfort while energy consumption is also increased.

17.2.3.2 Possible mitigation measures

Mitigation measures that may be implemented to prevent/counteract the impact of high-temperature conditions include (Eddowes et al., 2003):

- Implementing speed restrictions either (Bruzek et al., 2015):
  - In advance (also to allow timetable adjustments) when it is expected that, on a given day, the air temperature may exceed a certain threshold. Although simple in its implementation (it does not require the use of special equipment), this measure

is not particularly accurate, as speed restrictions may also affect track sections where there is no risk of track buckling; or

- In real time, this requires the use of remote temperature monitoring equipment. Although this measure is relatively accurate, its adoption is costly and does not always allow timetables to be adjusted in time.

- Constantly monitoring and maintaining the resistance of the track to lateral displacement.
- Installing sprinkler systems at switches and crossings, in order to lower the temperature of the rails and, thus, ensure their proper functioning.
- Painting the rails of specifically vulnerable track sections white, so that they absorb less heat. A study has shown that, depending on the coating material used, a reduction in temperature of as high as 10°C can be achieved using this method (Ritter and Al-Nazeer, 2014).
- Installing slab track instead of ballasted track or using heavier rails.
- Using effective ventilation and air-conditioning systems.
- Using platform screen doors.
- Reducing through various means the energy that is consumed by the railway system (reduced running speed, utilisation of lighter vehicles, regenerative braking, and more efficient lighting systems).

17.3 NATURAL PHENOMENA

Natural phenomena that may have an impact on train operation include:

- Sandstorms.
- Heavy leaf fall.
- Earthquakes.

17.3.1 Sandstorms

17.3.1.1 Interfaces with the railway system: impacts

A sandstorm, which entails the flow of air and solid particles, may occur when strong winds are present in intensely sandy desert areas, in conjunction with the presence of a dry climate (Zheng, 2009). Sandstorms, which occur in many countries, such as in China, India, Iran, Saudi Arabia, the United Arab Emirates, the United States affect the lateral dynamic behaviour of a train, especially when running at high and very high speeds (Xiong et al., 2011).

The sand particles that are transferred by the wind may:

- Accumulate in switches and crossings and obstruct or even prohibit their functioning.
- Intrude into the ballast layer (ballast contamination) (Zakeri, 2012; Zakeri et al., 2012; Zakeri et al., 2011) and affect track stiffness.
- Get into the electrification and signalling equipment, as well as into the mechanical/electrical components of the rolling stock, causing damage and an impediment to their functioning.
- Be deposited on the running surface of the rail and contribute to an increase in rail defects.
- Accumulate in the track drainage systems, leading to blockages (Zakeri, 2012).
- Lead to a decrease in critical train speed, in the case of specific crosswind speeds, resulting in an increased risk of vehicle overturning (Xiong et al., 2011).

Sandstorms are distinguished based on typical values of sand concentration and wind speed into six levels (Table 17.1).

A sandstorm reduces the safe speed limit of trains for a given lateral wind speed. Specifically, sand concentration increases the lifting forces by 50%, the lateral forces by 33.55%, and the overturning moment (Xiong et al., 2011).

Furthermore, sandstorms may lead to line closures for extensive periods of time due to the presence of sand on the track, delays due to speed restrictions, track component failure (sleepers, rails, fastenings, elastic pads) due to the erosive properties of sand, as well as to an increase in maintenance demand (removal of sand from the track, ballast cleaning) and, thus, costs. Also, when sandstorms are very strong, there could be safety risks as regards train operation (vehicle overturning, derailment), as well as passenger discomfort.

17.3.1.2 Possible mitigation measures

Mitigation measures that may be implemented to prevent/counteract the impact of sandstorms include:

- Adopting an appropriate horizontal alignment when constructing a railway track in sandy desert areas, which takes into account sand-flow corridors (Zakeri et al., 2011).
- Appropriate design of the leeward sides of the track, so as to not be subject to wind corrosion (Kollmann, 2013).
- Deploying devices that measure wind speed and sandstorm conditions (Voestalpine, 2017).
- Implementing speed restrictions based on the intensity (wind speed) of prevailing sandstorms (Xiong et al., 2011).
- Conducting inspections to detect the presence of any sand on the track, for instance, by the train crew or by cyclic visual inspections and condition monitoring.
- Creating and maintaining a data base to record the occurrence of sandstorms.
- Implementing (Cheng and Xue, 2014):
  - Sand-retaining walls.
  - Sand-stabilising measures.
  - Sand-flow guiding measures.
- Removing sand from the railway track by means of sand removal machines (Kollmann, 2013).
- Adopting either a conventional slab track or a humped slab track (Figure 17.1). In the latter case, the rails are in a higher position in that they rest on reinforced-concrete

Table 17.1 Wind speeds and sand concentrations for different sandstorm levels

Sandstorm class	Total suspended particulate (mg/m³)	Volume fraction of sand	Wind speed (m/s)
Floating dust	0.4	1.6×10^{-10}	5
Blowing dust	1.2	4.8×10^{-10}	10
Weak sandstorm	6	2.4×10^{-10}	15
Medium sandstorm	30	1.2×10^{-10}	20
Srong sandstorm	90	3.6×10^{-10}	25
Particularly strong sandstorm	270	1.08×10^{-10}	30

Source: Adapted from Xiong, H., Yu, W., Chen, D., and Shao X., 2011, "Numerical Study on the Aerodynamic Performance and Safe Running of High-Speed Trains in Sandstorms", *Journal of Zhejiang University – Science A: Applied Physics & Engineering*, Volume 12, Issue 12, 2011.

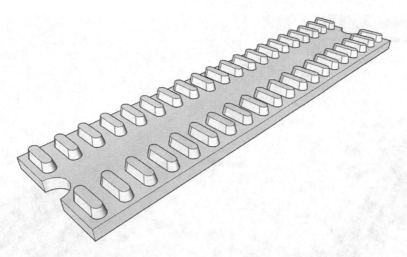

Figure 17.1 Humped slab track. (Adapted from Zakeri, J. A., and Forghani, M., 2012, "Railway Route Design in Desert Areas", *American Journal of Environmental Engineering*, Number 2, 2012.)

Figure 17.2 Tubular modular track. (Adapted from Gräbe, P. J., Vorster, D. J., Shaw, F. J., and Van Haute, S. C. A. 2011, The Use of Ballastless Track Structures for Rail Roading in Desert Conditions", *Proceedings of the International Heavy Haul Association (IHHA) Conference*, Calgary, Canada, 19–22 June 2011.)

'humps' that protrude from the slab, which allows sand to flow underneath the rails (Zakeri and Forghani, 2012).

• Adopting a tubular modular track (Figure 17.2), in which case the rails would rest on longitudinal reinforced-concrete beams that are placed in modules with a length of approximately 6 m each; galvanised steel bars ensure that the respective track gauge is maintained (Gräbe et al., 2011).

Figure 17.3 Humped sleepers. (Adapted from Zakeri, J. A., Esmaeili, M., Fathali, M., 2011, "Evaluation of Humped Slab Track Performance in Desert Railways", *Proceedings of the Institution of Mechanical Engineers, Part F: Journal of Rail and Rapid Transit*, Volume 225, Issue 6, 2011.)

- Adopting humped sleepers (Figure 17.3) which, with the rails being positioned 15 cm higher than when using conventional sleepers, allow sand to pass underneath the track, thus preventing sand from accumulating (Zakeri et al., 2011).
- Using hydraulic point systems and point components that are covered to prevent sand intrusion (Kollmann, 2013).
- Using head-hardened rails.
- Placing specialised brushes along the track and in parallel to it, particularly at switches and crossings (Osborn, n.d.).
- Utilising vibration reduction measures at the source (railway track) as well as along the propagation path and the final recipients (Zakeri et al., 2012).

17.3.2 Heavy leaf fall

17.3.2.1 Interfaces with the railway system: impacts

On many railway networks, the presence of leaves on the track is a recurring problem. The wheels compress the leaves, transforming them into a black, shiny, concentrated substance that may coat both the rails and the wheels. This coating is particularly difficult to remove as opposed to initial leaf concentrations which may be easily removed (Arias-Cuevas, 2010).

There are two main consequences of heavy leaf fall on railway systems. Specifically (Arias-Cuevas, 2010; Lewis and Dwyer-Joyce, 2009):

- Rolling issues.
- Improper functioning of track circuits.

The presence of leaves on the running surface of the rail leads to a reduction in wheel/rail adhesion, especially when there is humidity, which may increase the occurrence of wheel slip and slide, thereby affecting the braking and acceleration performance of the train. Further, the presence of leaves on the track may impede the proper functioning of the track circuits that are needed to detect track occupancy. When these are not working properly, a track section may seem to be free of traffic, whilst in fact it may be occupied; this may have serious consequences. Thus, the presence of leaves on the track may lead to delays or even cancellations of trains and, in the case of an impediment of the proper functioning of the track circuits, to dangerous situations.

17.3.2.2 Possible mitigation measures

Mitigation measures that may be implemented to prevent/counteract the impact of leaf fall include (RSSB, 2007):

- Track related interventions, such as:
  - Regularly cutting back any vegetation that is present along railway lines.
  - Increasing the level of wheel/rail adhesion of the rail running surface, for instance, by sanding.
- Rolling stock related interventions, such as:
  - Removing leaf deposits by means of either pressurised water or a mixture of sand and gel (sandite) by means of train-mounted equipment.
  - Removing leaf deposits by means of lasers that are mounted directly in front of the wheels, so as to burn the compressed leaf coating whilst the train is passing. Using this method, the rails are dried, deterring further leaf deposits. In case the laser beam loses its aim, it is temporarily deactivated for safety reasons. Using this method, substances with a thickness of up to 20 mm can be removed at speeds of up to 80 km/h (Gray, 2014).
  - Removing leaf deposits by means of sanding, either activated by the train driver or automatically, in places where there is a need. There are also sanding devices whereby the operation is automatically adjusted based on the running speed of the train (Thommesen et al., 2014).
  - Adopting devices that prevent wheel slip, in that they automatically detect wheel slide and then ensure that braking is applied in a controlled manner (Thommesen et al., 2014).
  - Adopting electromagnetic braking, which operates independently from wheel–rail contact and is therefore less affected by any wheel–rail adhesion or friction problems (Thommesen et al., 2014).
  - Adopting eddy-current braking, which uses a magnetic field that induces electrical eddy current in the railhead. In turn, these produce a resisting force on the moving magnets of the braking system, and thus braking is applied. Since there is no contact with the railhead, this system is not affected by any wheel–rail adhesion problems (Thommesen et al., 2014).
  - Implementing measures aimed at the rail infrastructure manager and the train drivers (special training for drivers, instruction manuals, early driver warning, and autumn timetabling).

In summary, the presence of leaves on the rolling surface of the rails (as a consequence of heavy leaf fall) reduces adhesion and thus impedes the traction, braking, and acceleration performances of trains. At the same time, it may impede the function of the electric signalling equipment (track circuits).

17.3.3 Earthquakes

17.3.3.1 Interfaces with the railway system: impacts

Earthquakes may have an impact on the track, the rolling stock, as well as the civil engineering structures and, depending on its severity, may lead to a derailment and vehicle overturning. Earthquakes may lead to a derailment due to either seismic vibrations or the damages sustained by the track (buckling, structural failure, etc.).

Damages sustained by the railway infrastructure may include:

- Changes in the condition of ballast layer and track formation.
- Structural failures of bridges, tunnels, and stations.
- Damage of track components, switches and crossings, slab track, as well as electrification and signalling equipment.

Thus, earthquakes may cause damage to various components of the railway system and, thus, impede its proper functioning. Depending on the magnitude of the earthquake, these consequences may range from short train delays to full cancellation of train services for extensive periods of time. Cancellations would be needed to allow repair work to be conducted (removal of derailed vehicles, replacement of damaged rails, sleepers, fastenings, cables and wires of overhead line systems, etc.), which involves inherent costs. Earthquakes may also cause damage to the rolling stock, which may need repair or replacement. Besides material damage, derailments due to earthquakes may cause injuries or loss of life, the extent of which depends on the magnitude of the earthquake and the rescue action plan that is in place.

17.3.3.2 Possible mitigation measures

Mitigation measures that may be implemented to prevent/counteract the impact of earthquakes include the following actions:

- Adopting early warning systems and immediately halting rail traffic. A seismometer could detect a primary seismic wave which, depending on its location, may be followed by more destructive secondary waves within a few seconds. Since electrical signals travel at a significantly faster speed than the secondary waves, this may allow power to be shut down and emergency braking to be applied in time, i.e., before the secondary waves reach the railway lines.
- Reinforcing bridges, buildings, and overhead line masts.
- Using, in the case of slab track, modular pre-fabricated concrete slabs, which allow a faster and less costly maintenance or replacement in case of an earthquake (www.max -boegl.de, n.d.).
- Installation of guard rails, which prevent lateral wheel movement. However, they do not prevent wheel hunting (Horioka, 2013).
- Stabilising the ballast using anchors, which may help prevent lateral track displacement during an earthquake (www.rtri.or.jp, n.d.).
- Implementing evacuation maps, improved evacuation routes, and the use of specially trained personnel.

REFERENCES

Arias-Cuevas, O. 2010, *Low Adhesion in the Wheel-Rail Contact – Investigations towards a Better Understanding of the Problem and Its Possible Countermeasures*, Doctoral Thesis, TU Delft, The Netherlands, 8 September 2010.

Baker, C.J., Chapman, I. and Dobney, K. 2010, Climate change and the railway industry: A review, *Proceedings of the Institution of Mechanical Engineers, Part C: Journal of Mechanical Engineering Science*, Vo. 224, No. 3, pp. 519–528.

Bruzek, R., Trosino, M., Kreisel, L. and Al-Nazer, L. 2015, Rail temperature approximation and heat slow order – Best practices, *Proceedings of the ASME Joint Rail Conference*, San Jose, CA, 23–26 March 2015.

Cheng, J. and Xue, C. 2014, The sand-damage prevention engineering system for the railway in the desert region of the Qinghai-Tibet plateau, *Journal of Wind Engineering and Industrial Aerodynamics*, Vol. 125, pp. 30–37.

Dios Sanz Bobi, J., Suarez, B., Núñez, J.G. and Vázquez, J.A.B. 2009, Protection high speed trains against lateral wind effects, *Proceedings of the ASME International Mechanical Engineering Congress and Exposition*, Lake Buena Vista, FL, 13–19 November 2009.

Dolianitis, A., Pyrgidis, C. and Spartalis, P. 2017, Operation of railway systems under special weather conditions, *8th International Congress ICTR 2017 'The Future of Transportation: A Vision for 2030'*, 28–29 September 2017, Thessaloniki, Greece.

Eddowes, M.J., Waller, D., Taylor, P., Briggs, B., Meade, T. and Ferguson, I. 2003, *Railway Safety Implications of Weather, Climate and Climate Change: Final Report*, Rail Safety and Standards Board (RSSB), England, 2003 (www.rssb.co.uk).

Imai T., Fujii, T., Tanemoto, K., Shimamura, T., Maeda, T., Ishida, H. and Hibino, Y. 2002, New train regulation method based on wind direction and velocity of natural wind against strong winds, *Journal of Wind Engineering and Industrial Aerodynamics*, Vol. 90, pp. 1601–1610.

Noguchi, T. and Fujii, T. 2000, Minimizing the effect of natural disasters, *Japan Railway & Transport Review*, March 2000.

Gräbe, P.J., Vorster, D.J., Shaw, F.J. and Van Haute, S.C.A. 2011, The use of ballastless track structures for rail roading in desert conditions, *Proceedings of the International Heavy Haul Association (IHHA) Conference*, Calgary, Canada, 19–22 June 2011.

Gray, R. 2014, End of the line for autumn train delays? Dutch begin trials to zap leaf litter from the tracks with lasers, *MailOnline*, 4 December 2014 (www.dailymail.co.uk).

Horioka, K. 2013, Clarification of mechanism of Shinkansen derailment in the 2011 great East Japan earthquake and countermeasures against earthquakes, *JR East Technical Review*, No. 27, 2013.

Kollmann, J. 2013. *Railway Operations under Harsh Environmental Conditions Sand, Dust & Humidity Problems and Technical Solutions / Mitigation Measures*. Presented at Berlin, Voestalpine.

Lewis S.R. and Dwyer-Joyce, R.S. 2009, Effect of contaminants on wear, fatigue and traction, *Wheel-Rail Interface Handbook*, Woodhead publishing, Sawston, 2009.

Osborn. n.d., *Solutions for Turnout Protection against Sand and Snow*, available online at: http://www.osborn.com/media/PDF/literature-de/Osborn-TurbFly-English.pdf. 1 April 2021.

Pyrgidis, C., Dolianitis, A. and Spartalis, P. 2017, The impact of specific weather conditions and natural phenomena on train operation, *Rail Engineering International*, Number 4, 2017, pp. 5–8, The Netherlands.

Ritter, G.W. and Al-Nazeer, L. 2014, Coatings to control solar heat gain on rails, *Proceedings of the AREMA 2014 Conference*, Chicago, IL, 28 September–1 October 2014.

Rossetti, M.A. 2017, Analysis of weather events on U.S. railroads, *Proceedings of the 23rd Conference on Interactive Processing Systems (IIPS) for Meteorology*, Oceanography and Hydrology, San Antonio, TX, 15–18 January 2007.

Thommesen, J., Duijm, N.J. and Andersen, H.B. 2014, *Management of Low Adhesion on Railway Tracks in European Countries, Kgs.*, DTU Management Engineering, Lyngby, Denmark, 2014.

UIC. 2016, *Adaptation of Railway Infrastructure to Climate Change*, available online at: http://unfccc.int/files/adaptation/application/pdf/iuc_further_information_101110.pdf. 06 April 2021.

US Department of Transportation. 2015, Federal Railroad Administration, Identification of high-speed rail ballast flight risk factors and risk mitigation strategies – final report, Washington, DC, April 2015.

Voestalpine. 2017, *Professional System for Wind and Sandstorm Monitoring*, available online at: http://www.sst.ag/produkte/diagnosesysteme/mistral-desert/.

www.max-boegl.de, Documents on slab track.

RSSB. 2007, *Characteristics of Railhead Leaf Contamination: Summary Report (T354 Report)*, Rail Safety and Standards Board (RSSB), England, 2007 (www.rssb.co.uk).

www.rtri.or.jp, 2015, *2014–15 Annual Report 2014–2015*, Railway Technical Research Institute, Japan.

www.ariccs.org.

Heat Trace, 2019, www.heat-trace.com, *A Guide to Rail Network HEAT Tracing Applications*, Heat Trace Ltd., England.

Xiong, H., Yu, W., Chen, D. and Shao X. 2011, Numerical study on the aerodynamic performance and safe running of high-speed trains in sandstorms, *Journal of Zhejiang University – Science A: Applied Physics & Engineering*, Vol. 12, No. 12, pp. 971–978.

Zakeri, J.A., Esmaeili, M. and Fathali, M. 2011, Evaluation of humped slab track performance in desert railways, *Proceedings of the Institution of Mechanical Engineers, Part F: Journal of Rail and Rapid Transit*, Vol. 225, No. 6, pp. 566–573.

Zakeri, J.A. 2012, Investigation on railway track maintenance in sandy-dry areas, *Structure and Infrastructure Engineering Maintenance, Management, Life – Cycle Design and Performance*, Vol. 8, No. 2, pp. 135–140.

Zakeri, J.A., Esmaeili, M., Mosayebi, S., and Abbasi, R. 2012, Effects of vibration in desert area caused by moving trains, *Journal of Modern Transportation*, Vol. 20, No. 1, pp. 16–23.

Zakeri, J.A. and Forghani, M. 2012, Railway route design in desert areas, *American Journal of Environmental Engineering*, No. 2, pp. 13–18.

Zheng, X. 2009, *Mechanics of Wind-Blown Sand Movements*, Springer Science & Business Media, Berlin.

Chapter 18

Railway safety

18.1 TYPES OF RAILWAY INCIDENTS AND DEFINITION OF RAILWAY SAFETY

18.1.1 Types of railway incidents

Railway incidents are classified into three groups: accidents, events, and failures. Relevant definitions are given below (Pyrgidis and Kotoulas, 2006; ERA, 2009):

- *Accident*: All undesired or unexpected sudden incidents or a specific chain of similar incidents that have (or had) undesirable consequences for the railway system (railway infrastructure, rolling stock, and railway operation) and the environment.
- *Event*: Every incident not characterised as an accident that concerns the operation of the trains and also that affects their safety. An event may be the cause of an accident.
- *Failure*: Failure represents a specific category of an event. It can be defined as any technical malfunction of the railway infrastructure and of the rolling stock that affects the safety of the operation of the whole system. Accordingly, a failure may be the cause of an accident.
- *Incident*: It is a unified definition of an accident, an event, or a failure.

18.1.2 Definition of railway safety

The safety that a railway system provides to its users can be defined with the aid of the two following approaches.

18.1.2.1 Based on risk level

This approach suggests a qualitative assessment of safety. In the case of a railway system, the term 'safety' describes the guarantee, through the constituents and the components of the railway system, that during operation the risk level is not described as 'non-permissible' (European Standard EN50126-1, 2000; EC, 2009).

The classification of the risk is uniquely accrued by the combination of the frequency and the severity of an event. This correlation defines the following four risk levels (Table 18.1):

- *Non-permissible*: Accidents of this category must be eliminated. It represents the most significant category and necessitates urgent safety measures by the services responsible, regardless of the financial and operational cost.
- *Non-desirable*: Accidents of this category can be accepted only in case of inability to contain their consequences and always upon the relevant approval of the authority in charge.

Table 18.1 Risk levels based on the frequency and severity of accidents

		Accident severity			
	Risk levels	Catastrophic	Severe	Minor importance	Negligible
Accident frequency	Frequent	Non-permissible	Non-permissible	Non-permissible	Non-desirable
	Possible	Non-permissible	Non-permissible	Non-desirable	Permissible
	Occasional	Non-permissible	Non-desirable	Non-desirable	Permissible
	Unusual	Non-desirable	Non-desirable	Permissible	Unimportant
	Rare	Permissible	Permissible	Unimportant	Unimportant
	Unlikely	Unimportant	Unimportant	Unimportant	Unimportant

Source: Adapted from European Standard EN50126-1, 2000, *Railway Applications: Reliability, Availability, Maintainability and Safety (RAMS)*, Part 1, CENELEC European Standards (European Committee for Electromechanical Standardization).

- *Permissible*: It corresponds to a generally acceptable safety level, without excluding further improvements, if it is feasible.
- *Unimportant*: The incidents of this category are acceptable, provided that there is approval from the competent authority.

Regarding the classification of accidents according to the severity of their consequences, various attempts have been made in Europe. The EU has introduced a common approach for such efforts. More specifically, the following definitions are suggested:

- *Catastrophic*: Fatalities and/or multiple severe injuries and/or severe environmental impact and/or extensive property damage.
- *Severe*: One fatality and/or serious injury, and/or significant environmental impact, and/or limited severe property damage.
- *Low severity*: Minor injury, and/or significant threat (or low impact) on the environment, and/or limited damage.
- *Negligible*: Possible minor injury and/or minor property damage.

To the classification stated above, an additional definition could be introduced, considering, more specifically, 'events and failures that did not lead up to an accident'.

The EU by adopting standards and by forming the appropriate legislative framework has quantified some of the consequences mentioned above. For example, extensive damages refer to 'those for which the Accident Investigation Body can directly estimate that a minimum of €2,000,000 are required for their restoration' (Directive 2004/49).

However, there is no precise definition of the consequences, and therefore the above classification is applied at the initiative and discretion of railway safety stakeholders. Thus, according to the British Railtrack, the equivalence between fatalities and injuries is defined as follows:

- 1 fatality = 10 severe injuries.
- 1 severe injury = 20 minor injuries.

On the contrary, as far as incident frequency is concerned, there are as of yet no standards clearly defining the borderlines between the various classifications (possible, circumstantial, etc.), and this renders the application of Table 18.1 difficult.

In Section 18.9, a methodology for the classification of accident frequency is suggested.

18.1.2.2 Based on incident 'indicators'

This approach suggests a quantitative assessment of safety. The safety that a railway system provides is evaluated by the incidents that occurred during a specific time period (e.g., 1 year) and had consequences on the track, the rolling stock, the passengers, the cargo, and the environment. In this context, indicators based on incidents that have occurred over a given time period are being used (ERA, 2013).

Worldwide various countries use specific indicators to assess the safety of their railway networks. These indicators differ very little among the various countries (FRA, 2003; Australian Government, 2012; Indian Government, 2013; Japan Transport Safety Board, 2013).

To apply the EU Directive 2004/49 for safety and its revision 2009/149/EC (EC, 2009), ERA (European Railway Agency) proposed a series of indicators (ERA, 2013) concerning rail incidents, their impact in relation to human life, economic impact, technical impact, etc. More emphasis is placed on human life, as any incident has direct consequences upon it. Consequences may include fatality, serious injury, and light injury. These consequences combined with the number of accidents and the economic impact form the values of corresponding indicators. These indicators are necessary for further decisions on the prevention measures that need to be adopted.

These indicators are as follows:

Indicators related to accidents (per year):
- Total number of serious accidents (number).
- Relative number of serious accidents (number/train-kilometre).
- Distribution of accidents per accident category.
- Fatality risk indicator: Death toll as a result of train accidents per million train-kilometre.

Taking into consideration all fatalities from rail accidents (excluding suicides), the EU 'fatality risk indicator' for the period of 2009–2011 had a value of 0.31 fatalities per million train-kilometre (ERA, 2013).

- Total number of deaths and serious injuries per accident category (number).
- Relative number of deaths and serious injuries per accident category (number/train-kilometre).
- Breakdown of accidents according to different users/stakeholders of the railway system.

Indicators for the financial impact assessment of accidents:

- Total cost (in €).
- Unit costs (€/train-kilometre) for the number of fatalities and serious injuries, the cost of environmental impact, the cost of damage to rolling stock or infrastructure, and the cost of delays resulting from accidents.

The downgrading of a risk level (e.g., from 'non-desirable' to 'permissible') requires additional safety measures, which inevitably increase the investment cost, as well as the operational and maintenance cost of the railway system.

The measures taken by the railway authority, aiming at reducing the probability of incident occurrence, are called *preventive measures*.

The measures that must pre-exist in order to reduce the impact of an incident and to make rational actions following the incident (e.g., in case of a train which is immobilised on the track due to breakdown, evacuation of the train, and process of removal of passengers from the area of the incident) are classified as *management measures* (consequence containing measures and escape and rescue measures).

18.2 SIGNIFICANCE OF SAFETY IN RAILWAY SYSTEMS AND DIFFERENCES IN ROAD SAFETY

18.2.1 Significance of safety in railway systems

Each railway company tries to optimise the processes and output of its production, maintaining at the same time a high safety level for all the activities involved.

More specifically, a transportation system can benefit from the safe production of transport services in the following areas:

* In the level of service offered to the users.
* In the economic efficiency of the system.
* In the environment surrounding the system.

Furthermore, in what concerns the railway system, safety is one of the major inherent advantages over its main competitor (i.e., private cars), according to the official international accident records. In this context, a railway company needs to focus on safety in order to obtain a significant share in the transportation market.

Railway safety constitutes a first priority target for the EU within the framework of the efforts targeted at the revitalisation of the railway sector. During the last 20 years, the EU published a series of directives specified on safety issues (the most representative is the Directive 2004/49) (EC, 2004a). Besides, the EU established the ERA which is, among others, committed to initiating methods for the evaluation and the monitoring of the safety of railway systems in Europe in a unified and integrated manner (EC, 2004b).

18.2.2 Distinctions between railway and road safety

As a transportation system, the railway differs from the road vis-à-vis its three main constituents, namely railway infrastructure, rolling stock, and railway operation. Consequently, there are distinct differences in terms of safety between the road vehicle and the train, both in terms of the characteristics of the incidents (type, severity) and in terms of the safety measures taken for their prevention and management.

Indicatively:

* The railway is constrained to one degree of freedom. Owing to the impossibility of a train performing manoeuvres while moving, braking is the only option when faced with the risk of two trains colliding or a train colliding with an obstacle. However, because of the relatively low adhesion between the wheel and rail (steel/steel contact) and the greater braking load, the braking distance of a train is much greater than that of a road vehicle (see Tables 1.5 and 1.6). Therefore, braking rarely prevents a collision. Hence it is of great importance for the railway to 'prevent' such accidents by taking the necessary measures in order to avoid a collision.
* The railway's operational and constructional features increase the impact of the aerodynamic effects that are developed during train movement (high speed, long length, and large frontal cross section). These phenomena may have negative consequences on

the rolling stock, the passengers, and other users of the system (e.g., passengers waiting or moving on the platforms) as well as on the staff working by the track. At the same time, due to the train's large lateral surface, it bears greater transversal wind loading, thereby making it more susceptible to overturning, owing to crosswinds.

- The conventional road transport means cannot use the railway track because of its structure (rails/sleepers/ballast). Moreover, very often the landscape, in which the layout of the track is integrated, is inaccessible to road transport. Consequently, if a train is immobilised on the track, either due to a fault or due to an accident, the evacuation of passengers from the site of the incident and the provision of first aid is in many cases a particularly challenging operation.
- The rolling surface of roads, contrary to that of the railway track, is almost impermeable to water; therefore, driving in icy conditions or during heavy rainfall is hazardous (risk of sliding or aquaplaning).
- A collision between two trains, or a collision between a train and an obstacle, is inevitably violent due to the train's large inertia. To mitigate this, engineers choose high safety factors during the dimensioning of the rolling stock. The vehicle's frame is designed with higher resistance and front vehicles are equipped with reinforced bumpers. Furthermore, several auxiliary mechanisms and automated systems ensure the smooth rolling and operation of the train in case of failure of certain functions and inform the driver of potential problems.

18.3 CLASSIFICATION OF RAILWAY INCIDENTS

In general, the railway incidents can be classified into ten main categories, as illustrated in the second column of Table 18.2 (Pyrgidis and Kotoulas, 2006).

Within each main category, there are subcategories. The classification adopted here takes well into account the classification suggested by the European Directive for Safety.

According to the line segments on which they may occur, rail incidents are classified into the following categories:

- Incidents at civil engineering structures.
- Incidents at stations/stops.
- Incidents on the 'open track'.
- Incidents at Railway Level Crossings (RLCs).

18.4 CAUSES OF RAILWAY INCIDENTS

When it comes to railway incidents, special attention must be paid to the clear determination of the causes behind an event. The infrastructure manager or the railway company is expected to examine what triggered a particular incident in order to take corrective measures and subsequently mitigate the risk.

To that aim, the causes of railway incidents are grouped into three different levels, with regard to the 'source' event that triggered the chain.

Table 18.3 presents the various causes, according to the three aforementioned levels (Pyrgidis and Kotoulas, 2006).

A first-level cause might be one of the three main constituents of the railway system and/or a combination of them, as is the usual case.

Table 18.2 Main categories and subcategories of railway incidents

	Main incident category	Incident subcategory
1	Vehicle derailments	• Derailment on the main track • Derailment on secondary tracks • Derailment on turnouts • Derailment at civil engineering structures
2	Train collisions	• Collision between trains (head-on, rear-on, head to flank, and broadside) • Collision between train and obstacle on the track
3	Separation of a train formation (splitting of coupled vehicles)	• Splitting of a passenger/freight train formation
4	Pedestrian/animal getting hit by the rolling stock	• Pedestrian/animal drifts on track
5	Fires/explosions	• Fire on the train/in the tunnel/next to the track/in facilities • Explosion on the train/in the tunnel/next to the track/in facilities
6	Incidents at RLCs	• Collision between train and road vehicle • Collision between train and truck • Collision between train and motorcycle • Pedestrian drifts at RLC area
7	Work accidents	• Incidents during working hours • Incidents during the return
8	Incidents involving hazardous goods	• Derailment/collision/splitting of a train formation carrying hazardous goods • Explosion in a wagon carrying hazardous cargo on route/during the process of maintenance works/during the cleaning process • Leak of toxic liquids/toxic vapours/radioactive fumes from the wagons
9	Other miscellaneous incidents	• Vandalism • Sabotage • Terrorist actions • Occupation of the track (by strikes) • Suicides and suicide attempts • Occupation of the track by waters/flooding of the facilities • Passenger(s) fall(s) from the platforms
10	Events able to cause one of the aforementioned incidents 1–9	• Violation of incorrect application of the regulations by the staff • Technical failures in the railway infrastructure and/or in the rolling stock

Source: Adapted from Pyrgidis, C., and Kotoulas, L., 2006, An integrated system for the recording and monitoring of railway incidents, *6th World Congress in Railway Research (WCRR)*, 4–8 June, Montréal, Canada, Congress Proceedings.

First-level causes can also originate from external factors; these cases are classified in Table 18.2, under the group label *other miscellaneous incidents*. Typical examples include natural phenomena such as large-scale earthquakes, heavy and continuous rainfall or snowfall, and extreme weather conditions in general (see also Chapter 17). Furthermore, terrorist attacks, vandalism, sabotage, suicide attempts and suicides, level crossing accidents (if the accident is caused by the road vehicle), etc. are also included.

Thus, the first-level causes might be:

• The railway infrastructure (track, civil engineering structures, track systems, track facilities, and premises).
• The rolling stock (locomotives and trailer vehicles).
• The railway operation.

Table 18.3 Classification of causes of railway incidents by origin, in first, second, and third level

Cause of incident – first level	Cause of incident – second level
1 Railway infrastructure	• Failure in the track superstructure • Failure in the track bed • Failure in tunnels/bridges • Failure in the earthworks/retaining walls • Failure in the signalling/electrification/telecommunications systems • Failure in the level crossings • Failure in the track fencing
2 Rolling stock	• Failure in the car body/bogies • Failure in the axles/wheels • Failure in the doors/windows • Failure in the braking system • Failure in the electrical/electronic equipment • Failure in the coupling/buffering system • Failure in the motors
3 Operation	• Violation of the regulations • Misapplication of the regulations (e.g., red light violation, excessive speed, wrong switch setting, erroneous radio communication, wrong vehicle loading, and sudden braking/accelerating of the train) • Unprofessional condition of the employee (tired, drunk, etc.)
4 External causes	• Railway employee's sickness • Car driver negligence at a level crossing • Car driver inability due to alcohol or drug use • Car driver loss of control due to weather conditions • Terrorist attack • Vandalism • Sabotage • Suicide attempts • Suicides • Extreme weather conditions and natural phenomena (earthquake, storm flooding, strong winds, snowfall, ice, etc.)

Cause of incident – third level

Failures

In the track superstructure	• Rail cracks • Rail defects • Track geometry defects • Sleeper cracks • Error in the track alignment geometry (insufficient cant, insufficient length of transition curves, etc.) • Soft ballast • Insufficient ballast • Ice • Slackened or inadequate fastenings • Bad condition of track switches and crossings
In the track bed	• Ballast disintegration • Collapse of substructure
In tunnel	• Lateral displacement • Water flooding • Narrow (inadequate) structure gauge • Insufficient aerodynamic behaviour • Structural failure
On bridges	• Structural failure • Pillar displacements • Aggrading • Rusting of metallic elements • Undermining of the pillars

(Continued)

Table 18.3 (Continued) Classification of causes of railway incidents by origin, in first, second, and third level

Cause of incident – first level	Cause of incident – second level
In earthworks and retaining walls	• Embankment settlement • Embankment slope failure (collapse, large transverse movements, slope material loss) • Falling rocks
In the signalling system	• Burnt lamps • Cable failures • Relay failures/malfunctions • Power supply failure • Track circuit failures • Faults in tele-commanding • Defects related to rail cracks • False signal indication • Miscellaneous external causes
In the electrification system	• Insulator failures • Defective station equipment • Short circuits • Power failure • Other external causes
At level crossings	• See Table 18.13
Along the track fencing	• Destroyed track fencing • Inadequate fencing in terms of design and structural characteristics
In the car body	• Fracture/distortion of the vehicle frame • Fractured or damaged container • Fractures/cracks in wagons with hazardous content
In the bogies	• Fracture of bogie frame • Distortion of suspension springs • Inappropriate design of suspension system
In the wheelset	• Fracture of axles • Axle-box overheating • Wheel cracks (flange, tread) • Wheel wearing (flange, tread)
At the doors	• Door malfunction
At the windows	• Broken windows
In the coupling system	• Damaged/broken couplers • Poor coupling
In the motors	• Motor failure
In the vehicle's electrical/ electronic equipment	• Cab-signal failure/defects • GSM-R failure/errors in the radio communication
In the braking system	• Braking system malfunction/failure

- Combination of the above.
- Causes external to the railway system.

18.5 SAFETY IN CIVIL ENGINEERING STRUCTURES

18.5.1 Railway civil engineering structures and related incidents

The term 'civil engineering structures' in rail transportation systems describe all the structures built along the track, with the purpose of integrating the track layout in areas with difficult topography and/or sensitive environment. They must ensure the safe rolling of trains while being harmoniously integrated into the existing environment.

Accidents taking place on civil engineering structures are usually the gravest and have the worst consequences, including several fatalities, due to the profound difficulty for all escaping or rescuing operations (on bridges, in tunnels, etc.) as well as the high cost of the applied mitigation measures.

Virtually, all incidents taking place in the 'open track' can also occur on civil engineering structures. Table 18.4 shows the incidents involved which require special handling from the operator of the railway system (Pyrgidis et al., 2005; Pyrgidis and Kehagia, 2012).

18.5.2 Safety at railway bridges

The main cause of accidents on bridges is crosswind (Fujii et al., 1999). A characteristic example is the accident on the Amarube Bridge in Japan, in 1986, where seven railway vehicles were lifted by the wind and pushed over the bridge, causing six deaths.

It has been proven that when the speed of the wind exceeds 25 m/s, the transversal and vertical acceleration of the bridge deck increase alarmingly, rendering its crossing by a train extremely unsafe (Li et al., 2005; Xia et al., 2006, 2008).

The thicker the bridge deck, the greater the transversal force coefficient.

In the event of an untoward incident (collision of trains, derailment, and immobilisation of train on a bridge), the presence of a civil engineering structure above ground level, or

Table 18.4 Incidents on rail civil engineering structures requiring special management

Civil engineering structures	Railway track
Bridges	• Train derailment and falling from the bridge (for various reasons) • Pedestrians dragged along by rolling stock • Train derailment due to strong crosswinds and falling from the bridge • Workers dragged along by rolling stock (due to aerodynamic phenomena) • Train immobilised on a bridge
Tunnels	• Fires or emission of toxic gases and smoke inside the tunnels • Work accidents • Passengers' discomfort with regard to noise • Shattering of window pane on train carriage • Train immobilised inside a tunnel
Road overpasses	• Loss of road vehicle control and falling from the road deck on the track • Object falls from the road on the track
Embankments	• Train derailment due to strong crosswinds and overtopping from the top of the embankment • Train derailment due to failure of embankment slopes (collapse, loss of material) • Violation of permitted limit of track defects due to embankment settlement – train derailment • Train immobilised on a high embankment
Cuttings	• Collision between train and obstacle on the track • Accident due to landslide • Rock fall • Train immobilised in deep cutting
Fencing	• Animals getting hit due to drifting onto the track • Humans getting hit due to gaining access to the track (negligence, or to commit sabotage/suicide) • Accident on the side road network leading to an intrusion of a road vehicle onto the track

water, as well as the narrow and constrained space, further complicate the evacuation of the passengers and the access of rescue services and increase the severity of the incident.

In all cases, the placement of anti-derailment protective checkrails along the track should be implemented (Figures 18.1 and 18.2).

Table 18.5 illustrates the safety measures used on railway bridges.

18.5.3 Safety in railway tunnels

The most severe accidents that can take place inside tunnels involve fire and, consequently, emission of smoke and toxic gases. Additional problems are generated by the presence of aerodynamic pressures or reduced ventilation (in the case of diesel locomotives). When entering tunnels, high and very high-speed trains suffer from a sudden shift in the pressure conditions around them, and passengers experience a sudden reduction in their acoustic comfort. Moreover, cracking of windowpanes may take place due to this pressure. In such cases (train

Figure 18.1 Anti-derailment protective check rails. (Photo: A. Klonos.)

Figure 18.2 Anti-derailment protective check rails. (Photo: A. Klonos.)

Table 18.5 Mitigation measures implemented on railway bridges

Category of measures	Railway bridges
Preventive measures	• Wind barriers and drapes • Anemometers • Footways for workers
Consequence containing measures	• Anti-derailment protective checkrails
Escape and rescue measures	• Footways for evacuation of passengers in case of an emergency • Construction of emergency exits for evacuating to a safe place • Safety manholes

Table 18.6 Relative risk value for various tunnel configurations compared with the risk of a single-bore double-track tunnel without auxiliary tunnel

Tunnel configurations	Relative risk value
Single-bore double-track tunnel	100
Single-bore double-track tunnel + auxiliary tunnel	80
Twin-bore tunnel	50–60
Twin-bore tunnel + auxiliary tunnel	40
Three single-track tunnels	< 40

Source: Adapted from Diamantidis, D., Zuccarelli, F., and Westha, A., 2000, Safety of long railway tunnels, *Reliability Engineering and System Safety*, 67c, 135–145.

collisions, fire, immobilisation of train inside tunnel, etc.), the presence of a civil engineering structure below ground increases the severity of possible consequences. Especially in long tunnels – where rescue services are faced with the urgency to access the incident site as quickly as possible and initiate fast evacuation – the situation can get really difficult.

Two construction types – regarding railway tunnels – have prevailed internationally: the single-bore double-track tunnel (one tunnel with two tracks) and the twin-bore tunnel (two tunnels with single track).

The main advantage of a twin-bore track tunnel is that there is no risk of a head-on collision and there are no aerodynamic problems occurring from trains moving in opposite directions; moreover, there is a high degree of protection in case of a fire event. On the contrary, the single-bore double-track tunnel has the major advantage of lower construction cost and clearly reduced aerodynamic effects (Maeda, 1996).

Table 18.6 presents the relative risk values for various tunnel configurations compared with the risk value of a single-bore double-track tunnel. Evidently, the three single-track tunnels and the twin bore with an auxiliary tunnel present a lower risk but bear a significantly higher construction cost.

Table 18.7 presents the safety measures taken in railway tunnels.

18.5.4 Safety at road overpasses

Accidents at road overpasses and railway underpasses usually occur as a result of objects falling from the road bridge. The most severe incidents occur when the pillars supporting the road overpass are not adequately reinforced and protected against the impingement of a derailed train.

Table 18.8 presents the safety measures taken at road overpasses constructed along a railway line.

Table 18.7 Mitigation measures implemented in railway tunnels

Category of measures	Rail tunnels
Preventive measures	• Control for surveillance of tunnels • Hot-box detection devices placed at tunnel entrance (Figure 18.3) • Avoidance of switches and crossings inside tunnels
Consequence containing measures	• Fire-resistant materials and structures • Automatic fire, smoke, and toxic gas detection systems • Installation of fire extinguishing system • Ventilation system to control heat and smoke • Water supply • Emergency power supply • Measures to reduce aerodynamic effects
Escape and rescue measures	• Escape routes and emergency exits • Safety and evacuation lighting • Emergency exits leading to ground (Figure 18.4) • Emergency contacts • Auxiliary tunnel for single-bore double-track tunnel • Cross-connections between tunnel tubes • Staff refugees

Figure 18.3 Hot-box detection devices. (Adapted from online image available at: http://www.mermecgroup
.com/inspection-technology/hot-box-detector-/418/1/hot-box-detector-.php (accessed 12 June
2015).)

Figure 18.4 Escape exit, Kallidromos Tunnel, Greece.

Table 18.8 Mitigation measures implemented at road overpasses

Category of measures	Road overpasses (rail underpasses)
Preventive measures	• Proper design of alignment geometry and cross section of the road • Pedestrian facilities on the road bridge • Signing and road markings on the road
Consequence containing measures	• Safety vertical parapets and barriers • Protective wall for road bridge supporting columns • Horizontal protective grid along the bridge (outside the bridge deck) (Figure 18.5)

Figure 18.5 Vertical and horizontal protection along the road bridge (outside the bridge deck), Vienna, Austria. (Photo: A. Klonos.)

18.5.5 Safety on high embankments

A considerable portion of the railway incidents is solely due to the presence of high embankments.

By their very nature, embankments impose an altitudinal difference between the rolling surface of the track and the natural ground level. Due to this fact, the impacts of an embankment failure are exacerbated (e.g., a fall from a significant height after derailment). At the same time, high embankments make it difficult, in case of accidents or trains immobilised on the track, to remove passengers and to allow access to rescue crews.

The two main causes of railway embankment failure are slope failure and settlement. These types of failure may be the result of inadequate compacting during the construction of the embankment or due to the presence of water.

In the construction of embankments for civil engineering structures, particular attention must be paid to their height (lower height relative to road embankments) (Pyrgidis et al., 2005) and to their compactness and coherence so as to avoid settlement effects (Selig and Waters, 1994). The geometric track defects (cyclic top, twist, etc.) that may occur along the track are particularly dangerous for the movement of trains and, in any case, they are dangerous for the operation of the railway system.

Table 18.9 presents the safety measures taken on high rail embankments.

18.5.6 Safety in deep cuttings

The most common accidents on the railway cuttings are caused by rocks that fall from the slopes of cuts on the track. In this case, there is a high risk of collision of the passing train with the obstacle on the track, given that the presence of the obstacle may not be perceptible promptly.

Table 18.10 presents the safety measures used in deep rail cuttings.

18.5.7 Safety in fencing

Fencing prohibits the crosswise traverse of the railway line by humans or animals and in general their access to the occupied area. Their prime role is the reduction of interfaces between the railway system and the surrounding environment and any activities that are taking place in it. To this end, it aims to:

- Reduce the chance of accidents that involve railway vehicles hitting pedestrians or animals.
- Protect the fauna ecosystem and large mammals in particular.

Table 18.9 Mitigation measures implemented on high rail embankments

Category of measures	High rail embankments
Preventive measures	• Drainage, protection from groundwater and rain water • Reinforcement of earth embankment's foundation • Monitoring of movements/displacements by optic fibres • Wind barriers and anemometers • Anti-derailment protective checkrails (in case of very high embankments)
Escape and rescue measures	• Splitting the railway track into 'safety zones'. Connecting 'safety zones' with the road network. Contingency plans for evacuating to safe areas

Table 18.10 Mitigation measures implemented in deep rail cuttings

Measure type	Deep rail cuttings
Preventive measures	• Protection from rock fall (fences and catch nets, protective gullies, rock-trap ditches, and retention walls) • Protection against slope slip • Track guard presence (foot patrols)
Escape and rescue measures	• Splitting the railway track into 'safety zones'. Connecting 'safety zones' with the road network. Contingency plans for evacuating to safe areas

A large portion of railway incidents is attributed to the lack of fencing (see Table 18.4) or to the improper functioning of the fencing in place.

The following types of incidents are attributed to the improper functioning of fencing:

• Animals/humans gaining access to the track due to damaged fencing.
• Animals/humans gaining access to the track due to inadequate design and/or structural characteristics of the fencing.

The following actions and mitigations measures improve the proper function of the fencing along the track:

• Increase of the height of the fencing or double fencing.
• Addition of barbed wire.
• Specialised escape routes.
• Adequate lighting.
• Closed Circuit TeleVision (CCTV) surveillance systems.
• Movement tracking systems.
• Guard patrols.
• Regular maintenance.
• Specialised reflectors.
• Small openings of a single direction.
• Sound emitting devices.
• Warning signs.
• Animal tracking systems.

The obligation to construct adequate fencing along the railway track is, in most countries, derived from appropriate legislation.

In most cases, the construction of such a project is at the discretion of the infrastructure managers.

Fencing is obligatory along the entirety of the length of railway corridors serving high or very high-speed trains. Moreover, it is also deemed necessary for tracks that pass through urban and residential areas.

Finally, and with regard to the characteristic of the wider area that they exclude, fencings may be distinguished into five categories. Each of these categories has different characteristics, as follows:

1. Fencing in an urban/residential area to prevent the illegal access of people. Concerns all railway systems.
2. Fencing to distinguish railway ownership (mostly in rural areas). Concerns all railway systems.

3. Fencing to separate the road network from the railway tracks (when they have parallel alignments). Concerns all railway systems.
4. Fencing to prevent the access of domesticated animals to the tracks in areas that constitute pastures. Concerns intercity railways of high and very high speeds.
5. Fencing to prevent the access of wild animals in protected areas and forests. Concerns intercity railways of high and very high speeds.

18.6 SAFETY AT RAILWAY STATIONS

When a train runs through a railway station at a certain speed, the generated airflow can disturb people standing on the platforms (passengers, station personnel, etc.). Reducing the speed of the train when moving next to platforms or keeping a minimum distance from the track when standing on the platforms can mitigate the problem.

For instance, the British Railways apply a minimum passenger safety distance of 1.50 m from the edge of the platform when the trains' passage speed V_p exceeds 200 km/h. The respective distance for the station personnel is at least 2.00 m.

Table 18.11 presents the maximum permitted passage speed for a train, in case of human presence at a distance of 1 and 2 m, respectively.

A lot of platform-related accidents involve baby carriages, strollers, and other wheeled equipment, susceptible to moving after the wind flow caused by a passing train. There are specific measures that counter these phenomena; the most common is installing ventilation systems above the platforms in order to weaken the generated wind flow and reduce its force and speed.

18.7 SAFETY ON THE 'OPEN' TRACK

18.7.1 Potential risks

All the incidents recorded in Table 18.2 can also take place in the 'open track'. Table 18.12 presents the potential risks entailed within train operation in the 'open track' as well as the applied mitigation measures.

The term 'potential risks' includes all the incidents that can occur during the operation, as well as the contingency actions that follow such incidents.

Table 18.12 does not include the following:

- All incidents that can be avoided if the technical specifications of the project are completely fulfilled (e.g., derailments due to insufficient geometrical design with regard to the track design speed).
- All incidents that can be avoided if track maintenance is properly carried out (e.g., derailments due to geometric track defects).

Table 18.11 Maximum permitted passage speed for trains in case of human presence at distances of 1 and 2 m, respectively

Wind force criterion	Maximum permitted train passage speed for distance of 1 m (km/h)	Maximum permitted train passage speed for distance of 2 m (km/h)
Beaufort scale level 5 (for passengers)	80–118	98–146
Beaufort scale level 7 (for personnel)	127–188	156–232

Table 18.12 Potential risks in the 'open track' and respective safety measures

Incidents	Safety measures
• Pedestrian/animal drifts on track	Fencing along the track, overpasses
• Collision between train and obstacle on the track	Fencing along the track, overpasses
• Train is derailed due to axle-box overheating	Hot axle-box detection
• Train is derailed and subsequently enters the road network	Metal beam crash barriers, reinforced concrete walls, etc.
• Train is derailed (overturned) following strong crosswind	Wind barriers and speed reduction
Actions	
• Passengers must evacuate the train and the incident site	Adequate space along the track at each side providing safe and efficient flow of passengers and personnel
	In case the track lies on high embankments or next to a cliff edge, lateral protection of the track can be ensured by railings, barriers, etc. Emergency routes and exits should also be clearly communicated to passengers through proper signs
• Passengers need to relocate to areas where first aid can be provided	Emergency exits leading from the track occupation zone to locations that are connected with the road network and which allow for medical and first-aid agencies to operate
	Easy-to-access emergency exits (In case of embankments, appropriately designed stairways are required to facilitate the passengers' descend from the track level. In case of earth cuts, appropriately designed stairways are required to allow passengers to ascend from the track to the fencing level)
• Ambulances should be able to get as close as possible to the track	Emergency exits leading from the track occupation zone to locations connected with the road network
• Emergency personnel should be able to quickly and easily access the point on the track where the train was halted	Special track – road vehicles
• Planning for crossovers in the track layout in order to account for maintenance activities as well as optimisation of the operation in anticipated emergency cases	Crossovers

- All incidents that are caused by possible malfunctions in the electromechanical installations and the related equipment.
- Incidents at RLCs (they are examined in Section 18.8).

18.7.2 Safety measures

The entire track should be split into subsections called 'safety zones'. This approach offers several possibilities:

- Immediate detection of the incident's corresponding track section.
- Quickly determining the nearest available emergency exits in all cases.
- Operational and contingency planning and structuring for the railway system operators.

Some of the suggested safety measures of Table 18.12 are analysed hereunder:

- *Emergency exits:* The locations of emergency exits are chosen so as to optimally satisfy the following criteria:
  - They must lead to a safe place, away from the original source of risk, as well as from other ensuing consequent risks. Therefore, emergency exits and routes are usually placed next to the road network.
  - They must ensure that these locations can be accessed by multiple directions. Thus – if present – locations near road junctions, upper and lower crossings of intersecting roads, etc. are usually preferred.
  - They need to facilitate the deployment of rescue teams for passengers and personnel, as well as medical and technical services to remove the halted rolling stock, provide first aid, and support in general.

The distance between consecutive emergency exits is determined by the system operator.
- *Markings:* Emergency exits should be properly marked along the line by signs communicating the direction to the nearest exit as well as the relevant distance.
- *Road and rail corridors separation measures:* Sometimes the distance between road and rail is particularly small, and this creates the need for additional protective measures in order to safeguard the system. These measures include:
  - Protective security barrier (0.75 m high)
  - Reinforced concrete walls (1.15 m high)

The selection and configuration of the required separating elements depend on the altitudinal difference as well as on the horizontal distance between the transport corridors. Horizontal distance is defined as the distance between the outer rail and the edge of the security barrier or the auxiliary lane (or the traffic lane of the secondary road). The altitudinal difference represents the height difference between the rolling surface of the rail and the road running surface level.

18.8 SAFETY AT RLCS

The causes of accidents at RLCs may be distinguished into five basic categories, as follows (Woods et al., 2008):

- Causes attributed to the human factor and derived from the road network (these refer to accidents that are due to the intentional or unintentional actions of road vehicle drivers).
- Causes attributed to the human factor and derived from the railway network. (These refer to accidents that are due to oversight or mistakes of railway employees, such as crossing guards and drivers.)
- Causes attributed to issues of a technical nature derived from the road network. (These refer to accidents that are due to technical issues of the road vehicles, such as the improper function of brakes.)
- Causes attributed to issues of a technical nature derived from the railway network. (These refer to accidents that are due to malfunctions in the RLC equipment, such as signage and barriers.)
- Other causes. (These refer to accidents that may not be attributed to any of the previous four categories, e.g, accidents due to animal crossings.)

The vast majority of RLC accidents may be attributed to a human factor derived from the road network (Woods et al., 2008). This fact constitutes a focal point in exploring solutions in terms of RLC risk management.

Table 18.13 presents the causes of RLC incidents, grouped into three levels according to the 'source' event that triggered the chain.

In a railway level crossing collision incident, regardless of who provokes it – namely the road vehicle or the train – it is the vehicle driver's life that is being threatened; what is more, only she/he can actually do something to prevent the accident from happening. Therefore,

Table 18.13 Classification of causes of railway incidents at RLCs, into first, second, and third level of source

Causes of incidents at the first level	Causes of incidents at the second level
Railway infrastructure	• Failure of RLC equipment • Poor geometry/construction of the RLC
Railway operation	• Liability of the train driver • Liability of the level crossing guard
Causes external to the railway system	• Liability of the driver of the road vehicle • Liability of the pedestrian • Breakdown of the road • Act of terrorism • Vandalism • Sabotage • Extreme weather conditions

Causes of incidents at the third level	
Failure of RLC equipment	• Broken barriers • Failure of the train announcing system • Burnt signal lamp • Power cut-off • Mechanical breakdown of barriers, sound, and signal light activation
Faulty geometry/construction	• Reduced visibility due to obstacles • Insufficient lighting – improper lighting • Problematic geometry of intersecting road • Lack of road or railway signposting • Poor state of signposting • Poor state of road surface • Poor state of the 'surface' of the level crossing
Liability of the train driver	• Wrong manoeuvre • Violation of signalling rules
Liability of the level crossing guard	• Wrong manoeuvre • Negligence, recklessness
Liability of the driver of the road vehicle	• Violation of the traffic code • Poor judgement • Incapacity due to the influence of alcohol or drugs • Tiredness, distraction • Halted vehicle on the track
Liability of the pedestrian	• Distraction, not paying attention • Indisposition • Bad judgement • Incapacity to react due to the influence of alcohol or drugs • Fatigue

Source: Adapted from Pyrgidis, C., and Sarafidou, M., 2007, Railway level crossing management system, in *9th International Congress Railway Engineering 2007*, 20–21 June 2007, London, University of Westminster, Congress Proceedings.

she/he is expected not only to adhere to the traffic regulations but also to double check carefully the crossing, even if the signals show 'go'.

RLCs constitute a location where rail and road traffic flows intersect. Given that their density in conventional speed railway networks around the world remains high, the RLCs are a neuralgic point of the railway system, both in terms of traffic safety and operation.

For these reasons, level crossings should be examined separately, since they constitute a major cause of railway accidents, and all railway companies should employ a system to manage the safety level provided to their customers.

Managing accident risk at RLCs has been an important issue for both railway and highway infrastructure managers and operators. The term 'RLC management' refers to the procedure, methodology, and mitigation measures that are adopted, the equipment, staff, and funds that are available, the scientific 'tools' that are used, the regulations that are in place, etc. in order to achieve the following goal:

- Limiting the chance of an accident at RLCs to a minimum.
- Securing the passage of trains through the RLC location at the desired speed.
- Maintaining the cost of operation of the RLC and of the maintenance of its equipment at acceptable levels.

Figure 18.6 illustrates an integrated system for the management of the RLCs. This system includes the following managing tools (Pyrgidis and Sarafidou, 2007):

- Recording system of the existing situation of the network's RLCs (i.e., of their functional and structural characteristics)
- System for evaluation of the existing situation of RLCs (equipment, structural, and functional characteristics).
- Recording system of the incidents at RLCs.
- Monitoring system (through indicators) of the incidents.

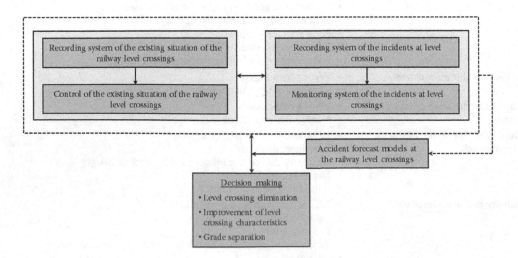

Figure 18.6 Integrated RLCs management system. (Adapted from Pyrgidis, C., and Sarafidou, M., 2007, Railway level crossing management system, in *9th International Congress Railway Engineering 2007*, 20–21 June 2007, London, University of Westminster, Congress Proceedings.)

The interaction of these four 'management tools' can help the infrastructure manager to take decisions regarding the interventions to be made on the network's RLCs, in order to minimise the occurrence of accidents and to help the operation of the network.

Those interventions are essentially three (Mallet, 1987; George, 1999):

- Elimination of RLCs.
- Improvement of RLC characteristics.
- Grade separation of the level crossings that are characterised as dangerous.

The removal of an RLC and the shifting of road traffic flow to the next possible level crossing through the adjacent road means essentially shifting the risk to the next crossing and results in increased run time for the road users.

Improving the construction and operational characteristics of an RLC includes:

- Conversion of the RLC from passive to active.[1]
- Upgrading the existing equipment of an active RLC.
- Systematic maintenance of the equipment of the RLC.
- Removal of the elements (e.g., advertising boards) that constitute potential distraction for the road vehicles' drivers and trains' drivers.
- Installation of CCTV.
- Installation of automatic obstacle detection system.
- Improvement of the visibility conditions.
- Improvement of the lighting.
- Improvement of the road (pavement) and level crossing surface.

Finally, the conversion of an RLC to a non-level crossing is performed by constructing underpasses or flyovers (overpasses).

The cost of the equipment of an RLC is characterised by significant variations among countries, mainly due to the diversification of the technical specifications adopted in each country and the differentiation of the individual characteristics of each network (e.g., single or double track and train speed). For example, the cost of automatic gates may be as low as €100,000 but it may also exceed €900,000.

Table 18.14 provides the cost values for various interventions at a passive RLC (Ioannidou and Pyrgidis, 2014).

When selecting the optimal solution, the following should/must be taken into account:

- Number of RLCs per line kilometre.
- Daily traffic moment (see Section 18.9).
- Statistical data related to accidents.
- Possibility to rearrange the road network in the broader area of the RLC.
- Land uses in the proximity of the crossing.
- Cost of intervention.

The final selection will result from a feasibility study (financial and socioeconomic analysis). In many cases, eventual social pressure might be a determining factor.

---

[1] *Passive level crossing*: level crossing without any form of warning system or protection activated when it is unsafe for the user to traverse the crossing. *Active level crossing*: level crossing where the crossing users are protected from or warned of the approaching train by devices activated when it is unsafe for the user to traverse the crossing.

Table 18.14 Cost values for various alternative interventions at a passive RLC (2014 data)

Type of intervention	Installation cost (in €)	Annual maintenance cost (in €)
Installation of semi-automatic barriers at the RLC	370,000	2,300–5,700
Installation of automatic barriers at the RLC	570,000	2,300–5,700
Removal of the RLC	50,000–70,000 (plus the construction of road connection in parallel and near the track)	–
Conversion to overpass	3,200 per m²	3,500

Source: Adapted from Ioannidou, A. M., and Pyrgidis, C., 2014, The safety level of railway infrastructure and its correlation with the cost of preventive and mitigation measures, *International Journal of Railway Research*, 1(1), 19–30, 2014.

At this point it should/must be emphasised that the application of such a system must always be accompanied by an organised effort from all bodies involved, with respect to information and to raising public awareness.

18.9 THE TRAFFIC MOMENT OF AN RLC

A parameter that significantly influences the risk level of RLCs is the daily traffic moment. (The number of trains moving on the track in both directions per 24 h multiplied by the number of passing road vehicles of all types in both directions of the crossing during the same 24-h period.)

It is included among the operational characteristics of an RLC and is the main (if not the only) criterion in many railway networks for the conversion of a passive crossing to an active one.

In Figure 18.7, the number of rail accidents depending on the traffic moment (M_{RLC}) of RLCs is given for:

- The railway network of North Greece.
- The time period 2004–2013.
- A sample of 69 accidents on RLCs of which data were available.
- Various categories of protection.

Figure 18.7 shows that the number of accidents increases in accordance with the increase of traffic moment. It is worth mentioning that in RLCs with a low traffic moment ($M_{RLC} < 2,000$), a higher number of accidents is observed than in crossings with $2,000 < M_{RLC} < 10,000$, because there are many unguarded crossings in this category ($M_{RLC} < 2,000$).

Figure 18.8 provides, for two active RLCs of the examined railway network with the same constructional features but different traffic moment, the:

- Total number of accidents that occurred in the decade 2004–2013.
- Number of deaths.
- Number of injuries.

Based on Figure 18.8, it is concluded that the increase in traffic moment is accompanied by an increase in accidents, especially fatal ones.

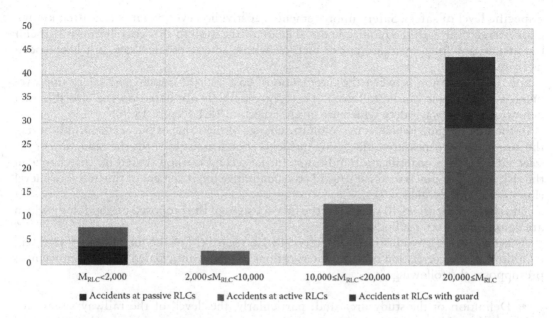

Figure 18.7 Railway network of North Greece – 2004–2013 period – number of rail accidents depending on the traffic moment and for various categories of protection of RLCs (Pyrgidis et al., 2016).

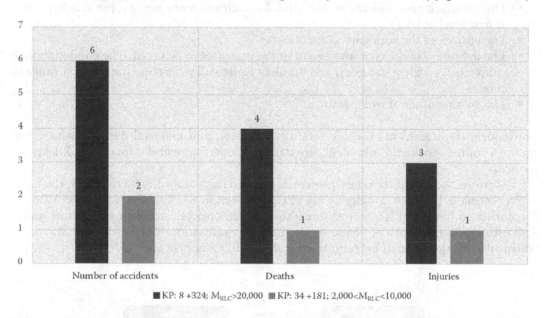

Figure 18.8 Railway network of North Greece – 2004–2013 time period – RLC – KP: 8 + 324 and KP : 34 + 181 – Accident data.

18.10 CORRELATION BETWEEN THE COST OF INTERVENTIONS AND THE SAFETY LEVEL IMPROVEMENT

18.10.1 General approach

One of the main issues that railway companies have traditionally dealt with is the amount of money they need to invest initially or during the system's operation in order to ensure a specific level of safety. Safety improvement is costly; however, what is not often known is its correlation with the required cost. The quantification of this correlation is difficult because it is defined by a number of factors, whose characteristics have not been specified yet.

For the correlation between the interventions' cost and the anticipated safety improvement, two methods can be followed. These two methods are based on the two different definitions of railway safety which are given in Section 18.1 (Figure 18.9).

In the first method (which is based on the change of the value of the accident indicators), the aim is that the measures addressing incidents should assist towards the reduction of the selected accident's quantification indicator. In the second method (based on the change in the risk level), the aim is to assist towards the qualitative improvement of the initial risk level (Ioannidou and Pyrgidis, 2014).

The chart of Figure 18.10 illustrates the first six steps of the proposed methodology which are common to both methods.

As seen in the chart illustrated in Figure 18.10, regardless of the methodology that will be followed, the correlation between interventions' cost and anticipated safety improvement presupposes the following:

- Definition of the study area and, particularly, the 'level' of the railway system for which accidents are assessed (e.g., whole network, railway corridor, track section, and railway system constituent).
- The approach per accident category and, for each accident category, per accident cause at first level at least.
- The costing of the accidents' consequences.
- The definition of the type and extent of the measures to be taken. The combination of study area, accident category, and accident cause will determine the relevant range of choices.
- The costing of the above measures.

Consequences of accidents include fatalities, injuries, and material damage which covers both rolling stock and infrastructure damages, environmental damage, and delays of service.

Concerning the cost of the consequences, it is stated that, according to the ERA, the Value of Preventing a Fatality/Casualty (VPF/VPC) amounts to €1,500,000, while the Value of Preventing an Injury (VPI) amounts to €200,000 (2008 prices). According to a more up-to-date estimation (2013 prices) made by the British organisation Rail Safety and Standards Board (RSSB), the Value of Preventing a Fatality (VPF) amounts to €2,230,000.

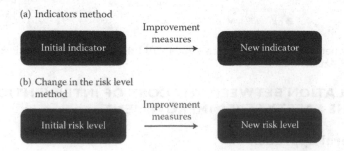

Figure 18.9 (a) and (b) Methods for the correlation between interventions' cost and anticipated safety level improvement.

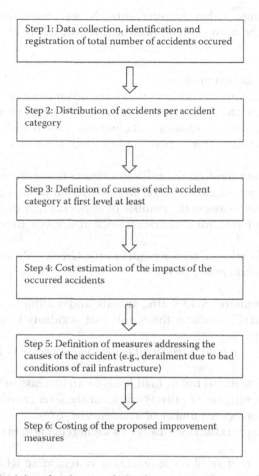

Step 1: Data collection, identification and registration of total number of accidents occured

Step 2: Distribution of accidents per accident category

Step 3: Definition of causes of each accident category at first level at least

Step 4: Cost estimation of the impacts of the occurred accidents

Step 5: Definition of measures addressing the causes of the accident (e.g., derailment due to bad conditions of rail infrastructure)

Step 6: Costing of the proposed improvement measures

Figure 18.10 Proposed methodology for the correlation between interventions' cost and anticipated safety improvement – first common steps for the two methods.

18.10.2 The change in the value of accident indicators

The proposed methodology uses an indicator which, depending on the incident, can be one of the indicators proposed by ERA, such as 'fatality risk indicator' or an indicator that involves the number of accidents for a specific accident category per vehicle-kilometre (e.g., the number of derailments per train-kilometre and the number of collisions per train-kilometre).

The correlation between the cost of interventions and safety improvement lies with the calculation of the amount of money that should be invested in order to reduce the current value of the indicator by a specific percentage or to set a new target value (i.e., the average rate applicable for EU countries for this incident category).

The first six steps that are common to both approaches (see Figure 18.10) are followed by the steps outlined hereunder:

Step 7: Assessment of the impact that the intervention's implementation has on the parameters that form the numerical expression and, as a result, the value of the indicator.

Step 8: New situation – calculation of new indicator's value.

Step 9: Correlation between the change in the indicator's value and the cost of interventions.

The assessment of the impact that the intervention's implementation has on the change of the indicator's value is the most difficult task. It can potentially be addressed in one of the three ways, namely:

- By appropriate prediction models.
- By recording the number of incidents that have taken place or will take place at a particular constituent of the railway system for at least 5 years after the implementation of preventive and mitigation measures and comparing them with the previous situation.
- Based on statistics from other networks with similar functionality.

Concerning the prediction models, the railway infrastructure managers have at their disposal, according to the international literature, prediction models for railway incidents at RLCs, which they can use to assess the number of accidents that may occur at an RLC with the given functional and structural characteristics at a given time period (Coleman and Stewart, 1976).

For instance, the application of such a model to the Greek railway network led to the following findings (Morfoulaki et al., 1994):

- Installation of automatic road traffic signals and audible warnings in combination with automatic barriers reduces the number of accidents by up to 50% (upgrade of passive RLC to active).
- The increase of 50% in road traffic load leads to an increase in the number of accidents by 15%.
- The increase of 50% in rail traffic load leads to an increase of 12% in accidents.
- The increase in the number of railway tracks at the level crossing (from single track to double track) increases the number of accidents by 10%.
- The reduction of road traffic lanes leads to a slight reduction in the number of accidents.

Moreover, the installation of an obstacle detection system at an RLC with semi-automated gates results in the reduction of accidents by 30% (Woods et al., 2008), while the installation of lighting results in the reduction of accidents by 45% (Varma, 2009).

On the basis of experience from the installation of CCTV on the road network, it can be assumed that the installation of such a system at an RLC would result in a reduction of intentional infringements by 50% (Woods et al., 2008).

Table 18.15 provides the improvement that is brought about on the security level of the RLC specific measures (Saccomanno et al., 2006; Washington and Oh, 2006).

18.10.3 The change in the risk level

In this method, the correlation between the cost of interventions and safety improvement lies with the assessment of the money that must be invested in order to change the current level of risk of a railway system to a lower one or to a desired level. This change can only be made by changing the frequency of accident occurrence, by altering the severity of accidents, or, finally, by a simultaneous change of both.

The first six steps that are common to both approaches are followed by the steps outlined hereunder:

Step 7: Classification of the accidents' frequency per accident category and the source of cause. For this process, a specific methodology is proposed in Section 18.9.3.1.

Table 18.15 Effectiveness of safety measures at RLC as a percentage of the reduction of accidents

Safety measures	Effectiveness (%)
'Speed humps' (measure for the reduction of speed on the road network)	36–40
Warning signage	0–50
Reduction of the road gradient at the location of the crossing	39–47
Increase of the distance from which the crossing is visible by the road network	0–50
Increase of the distance from which the crossing is visible by the railway network	10–41
Construction of a pedestrian crossing	0–50
Installation of lighting at the road network	15–45
Stop signs	35–46
Conversion of the RLC to non-level railway crossing	100
Road and railway network lighting	44
Removal of prohibition of the use of train audible warning system	53
Improvement of visibility	34
Improvement of the condition of the road pavement	48
Reduction of the speed limit for road vehicles	20

Source: Adapted from Washington, S., and Oh, J., 2006, Bayesian methodology incorporating expert judgment for ranking countermeasure effectiveness under uncertainty: Example applied at grade railroad crossings in Korea, *Accident Analysis and Prevention*, 38, 234–247.

Step 8: Classification of the accident's severity per accident category. For this process, a specific methodology is proposed in Section 18.9.3.2.

Step 9: Definition of risk level for each accident in combination with the frequency and severity, as defined in steps 7 and 8.

Step 10: Assessment of the intervention's impact on the accident's frequency and severity. Classification of the new accident's frequency and severity.

Step 11: New situation – calculation of the new risk level.

Step 12: Correlation between the results of the cost-benefit analysis and the new risk level.

The approach that is based on the change in the risk level is related with more problems in comparison with the approach that is based on the change in the value of incident indicators. These problems relate to the following:

1. The quantification of six categories as proposed in Table 18.1 regarding the frequency of incident occurrence. The key questions raised are the following:
   - What is the value of each frequency category, what are its measurement units, and which time period does it refer to?
   - Is the value of the frequency that characterises each frequency category the same for all accident categories?
   - Is there a distinction depending on the cause of the accident?
   - Is there a distinction depending on the category of railway system (metro, train, high-speed trains, suburban trains, etc.)?
2. The quantification of four categories proposed in Table 18.1 regarding the severity of incidents. The key questions raised are the following:
   - How is each category of severity defined?
   - Do the various accident categories belong uniquely to a particular category of severity?

3. The assessment of the impact that the application of specific measures has on the value of their frequency of occurrence. This assessment can be done for the three measures already outlined and for the case of the indicators methodological approach.
4. The assessment of the impact that the application of specific measures has on the value of severity of the incidents' impacts.

18.10.3.1 Characterisation of the frequency of a particular incident

The quantification of frequency is an issue that remains under investigation. A literature review (Ioannidou and Pyrgidis, 2014) attempts to approach this topic. More specifically, a methodology is proposed in order to set quantitative limits of frequency categories as listed in Table 18.1 for each accident category. The main indicator used in order to set the values for frequency categories is the average number of accidents per accident category which have occurred at a large number of representative networks (e.g., the EU countries). It is expressed in different measurement units depending on the accident category and has a specific year as a reference. The average is considered to be the value for the 'occasional' frequency category while the values for the other categories are based on the average, by increasing or decreasing it. An indication of the recommended percentages is provided in Table 18.16.

On the basis of the above, the steps of the proposed methodology for a specific accident and for a given year are as follows:

1. Assignment of the accident to the appropriate category.
2. Collection of the necessary data so as to allow for the calculation of the average number of accidents.
3. Calculation of the average for the accident's category, by using the measurement unit that corresponds to the particular accident category.
4. Setting the average as value for the 'occasional' frequency category.
5. Calculation and setting of the values of other frequency categories using Table 18.16 for the specific accident category.
6. Determination of the accident's frequency category on the basis of the position of its average value in Table 18.16.

In many cases, the proposed methodology should be further specialised in order to address a particular cause of occurrence for each incident (e.g., derailment (incident) and infrastructure (cause)).

Table 18.16 Indicative percentages that form the values for the various frequency categories

Frequency category	Indicative percentage so as to set the values
Frequent	40% increase of the average
Possible	20% increase of the average
Occasional	Average number of accidents per accident category
Unusual	20% decrease of the average
Rare	30% decrease of the average
Unlikely	40% decrease of the average

Source: Ioannidou, A. M., and Pyrgidis, C., 2014, The safety level of railway infrastructure and its correlation with the cost of preventive and mitigation measures, *International Journal of Railway Research*, 1(1), 19–30, 2014.

Table 18.17 Definition of severity based on incident category

Severity categories	Accident categories
Catastrophic	Derailment of trains
	Collision of trains
Severe	Accidents at level crossings
	Accidents to persons involving rolling stock in motion
Low severity	Fires in rolling stock
Negligible	Other accidents

Source: Adapted from Ioannidou, A. M., and Pyrgidis, C., 2014, The safety level of railway infrastructure and its correlation with the cost of preventive and mitigation measures, *International Journal of Railway Research*, 1(1), 19–30, 2014.

18.10.3.2 Characterisation of the severity of a particular incident

Ioannidou and Pyrgidis (2014) attempt to approach the topic of incident severity in two ways.

The first approach involves the incident category (as defined in Section 18.3, Table 18.2) with its usual consequences. More specifically, it is considered that some incident categories have, in most cases, catastrophic consequences, such as loss of human lives. The classification of severity proposed in this research based on the incident categories is presented in Table 18.17.

Derailments and collisions are two incident categories which, regardless of the cause of their occurrence, in many cases, cause fatal accidents and/or multiple severe injuries and/or severe environmental impact, and/or extensive material damage; hence, they are classified into the 'catastrophic' severity category. Incidents at RLCs and incidents caused by rolling stock in motion usually cause those impacts recorded in the 'severe' incident category and are classified under the 'severe' category. For the same reasons, fires are classified under the 'low severity' category although in some cases their consequences can be catastrophic.

The second approach regarding severity involves its quantification based on the actual consequences of incidents that have occurred. A key indicator in order to form the values for each severity category of each incident category is the average of their consequences (at the national or international level). Depending on the average of their consequences, incidents are classified under one of the four categories, namely catastrophic, severe, low severity, and negligible as already defined in the above.

18.10.4 Case studies

18.10.4.1 Individual passive RLC – conversion to active RLC

Incident type: Accident.
Accident category: Accident at passive RLC.
Special accident category: Collision of a train with a road vehicle.
Cause of accident: First level: Railway infrastructure; Second level: Poor visibility.
Used indicator: Number of fatal accidents (each with at least one fatality); hence the number of fatalities in the long term of 25 years = 10 fatalities = 0.40 fatalities per year.
Intervention: Improvement of RLCs' constructional characteristics.
Measure: Installation of automatic barriers.

Intervention cost: €570,000 (installation of automatic barriers) + €5,000 (annual maintenance cost).

Impact of measure implementation: Reduction of fatal accidents and, therefore, of the number of fatalities by 50% (Morfoulaki et al., 1994).

Indicator's new value: 5 fatalities over 25 years – 0.20 fatalities per year.

Cost of fatalities: €836,000 × number of fatalities + €760,000 per year (fixed premiums) (ERA, 2009).

Economic life period of barriers = 25 years.

Results of cost-benefit analysis: cost-benefit ratio = 4.761069 >> 1 (25-year assessment period, 5.5% discount rate).

18.10.4.2 Individual passive RLC – conversion to overpass

Incident type: Accident.

Accident category: Accident at passive RLC.

Special accident category: Collision of a train with a road vehicle.

Cause of accident: First level: Railway infrastructure; Second level: Poor visibility.

Used indicator: Number of fatal accidents (each with at least one fatality); hence the number of fatalities in the long term of 25 years = 10 fatalities = 0.40 fatalities per year.

Intervention: Grade separation of the level crossing.

Measure: Construction of an overpass.

Intervention cost: €3,200,000 + €3,500 (annual maintenance cost).

Impact of measure implementation: Reduction of fatal accidents and, therefore, of the number of fatalities by 100%.

Indicator's new value: 0 fatalities over 25 years – 0.00 fatalities per year.

Cost of fatalities: €836,000 × number of fatalities + €760,000 per year (fixed premiums) (ERA, 2009).

Economic life period of barriers = 25 years.

Results of cost-benefit analysis: cost-benefit ratio = 1.661069 > 1 (25-year assessment period, 5.5% discount rate).

Internal Return Ratio (IRR) = 12%.

18.10.4.3 Passive level crossings at railway network level

Incident type: Accident.

Accident category: Accident at passive RLC.

Special accident category: Collision of a train with a road vehicle.

Cause of accident: First level: Railway infrastructure; Second level: Poor visibility.

Total number of fatalities per year: 10.

Total length of track: 2,500 km.

Number of passive RLCs = 500.

Used indicator: Number of fatalities as a result of accidents at passive RLCs per RLC/track-km = 0.000008 fatalities per year.

Assumption: The accidents occur in 50 RLCs only (out of 500). Half of these (25) present a total number of 6 accidents during the 25-year period each and the rest (25) present a total number of 4 accidents each.

Intervention: Improvement of RLCs' constructional characteristics.

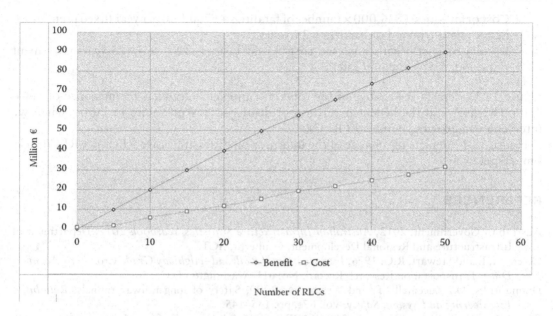

Figure 18.11 Change of the total cost and the benefit when we intervene in a growing number of RLCs (Pyrgidis et al., 2016).

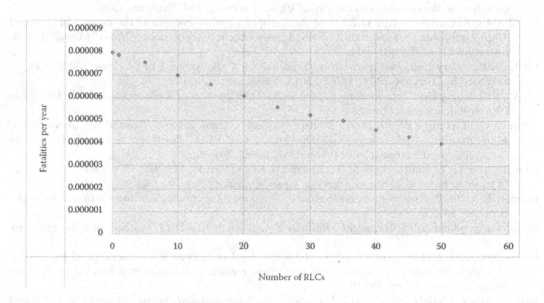

Figure 18.12 Change of the indicator 'fatalities at passive RLCs per RLC/track-km' when we intervene in a growing number of RLCs (Pyrgidis et al., 2016).

Measure: Installation of 50 automatic barriers.

Intervention cost: €570,000 (installation of automatic barriers) + €5,000 (annual maintenance cost).

Impact of measure implementation: Reduction of fatal accidents by 50% (i.e., 5 per year).

Indicator's new value: 0.000004.

Cost of fatalities: €836,000 × number of fatalities + €760,000 per year (fixed premiums).
Economic life period of barriers = 25 years.
Results of cost-benefit analysis: cost-benefit ratio = 2.81 > 1 (25-year assessment period, 5.5% discount rate).

Figure 18.11 depicts the change of the total cost (implementation + maintenance of automatic barriers) and the benefit (reduction of deaths – Present Values in Euros) when we intervene in a growing number of RLCs.

Figure 18.12 depicts the change of the indicator fatalities at passive RLCs per RLC/track-km, respectively.

REFERENCES

Australian Government. 2012, *Australian Infrastructure Statistics Yearbook 2012*, Department of Infrastructure and Regional Development, Canberra, ACT.

Coleman, I. and Stewart, R.G. 1976, *Investigation of Railroad–Highway Grade Crossing – Accident Data*, Transportation Research Record, No. 611, Washington D.C.

Diamantidis, D., Zuccarelli, F. and Westha, A. 2000, Safety of long railway tunnels, *Reliability Engineering and System Safety*, Vol. 67c, pp. 135–145.

EC. 2009, Commission Directive 2009/149/EC, *Official Journal of the European Union*, 28 November.

EC. 2004a, Directive 2004/49/EC of the European Parliament and of the council of 29 April 2004 on safety on the community's railways, *Official Journal L 164*, European Council, 0044–0113.

EC. 2004b, Regulation (EC) 881/2004 of the European Parliament and of the council of 29 April 2004 establishing a European Railway Agency (Agency Regulation), *Official Journal L 164*, European Council, 0001–0043.

ERA. 2009, *Safety Unit, Implementation Guidance for CSIs*, Annex 1 of Directive 2004/49/EC as amended by Directive 2009/149/EC, 2013, Valenciennes, France.

ERA. 2013, *Safety Unit: Intermediate Report on the Development of Railway Safety in the European Union*, Valenciennes, France.

European Standard EN50126-1. 2000, *Railway Applications: Reliability, Availability, Maintainability and Safety (RAMS)*, Part 1, CENELEC European Standards (European Committee for Electromechanical Standardization),.Bruxelles

Fujii, T., Maeda, T., Ishida, H., Imai, T., Tanemoto, K. and Suzuki, M. 1999, Wind-induced accidents of trains/vehicles and their measures in Japan, *Quarterly Report of RTRI*, Vol. 40, pp. 74–78.

George, B. 1999, Passages à niveaux: Comment améliorer la sécurité?, *Rail International*, Brussels, June, pp. 24–27.

Indian Government. 2013, *Statistical – Railway Year Book 2012 – 2013*, Ministry of Railways, New Delhi.

Ioannidou, A.M. and Pyrgidis, C. 2014, The safety level of railway infrastructure and its correlation with the cost of preventive and mitigation measures, *International Journal of Railway Research*, Vol. 1(1), pp. 19–30.

Japan Transport Safety Board. 2013, *Statistics – Railway Accident*, Japan Transport Safety Board Secretariat, Tokyo.

Li, Y., Qiang, S., Liao, H. and Xu, Y.L. 2005, Dynamics of wind-rail vehicle-bridge systems, *Journal of Wind Engineering and Industrial Aerodynamics*, Vol. 93, pp. 483–507.

Maeda, T. 1996, Micro-pressure wave radiating from tunnel portal and pressure variation due to train passage, *Quarterly Report of RTRI*, December 1996, Vol. 37(4), pp. 199–203.

Mallet, P. 1987, Passages à niveaux: 15 ans de suppressions, *Revue Générale des Chemins de Fer*, Vol. 106, pp. 19–23.

Morfoulaki, M., Papaioannou, M. and Pyrgidis, C. 1994, Accident prediction at railway grade crossings: application to the Greek railway network, *Rail Engineering International*, Netherlands, Vol. 23(4), pp. 9–12.

Online image, available online at: http://www.mermecgroup.com/inspection-technology/hot-box-det ector-/418/1/hot-box-detector-.php (accessed 12 June 2015).

Pyrgidis, C. and Kotoulas, L. 2006, An integrated system for the recording and monitoring of railway incidents, *6th World Congress in Railway Research (WCRR)*, 4–8 June, Montréal, Canada.

Pyrgidis, C. and Kehagia, F. 2012, Safety measures on rail and road engineering structures – Comparative assessment, *2nd International Conference on Road and Rail Infrastructure (CETRA)*, 7–9 May 2012, Dubrovnic, Croatia, pp. 1051–1058.

Pyrgidis, C., Mouratidis, A. and Tzavara, S. 2005, Railway structures – Peculiarities/differences in relation with road structures, *8th International Congress Railway Engineering – 2005, 29–30 June 2005*, London, University of Westminster.

Pyrgidis, C. and Sarafidou, M. 2007, Railway level crossing management system, *9th International Congress Railway Engineering 2007, 20–21 June 2007*, London, University of Westminster.

Pyrgidis, C., Papacharitou, E. and Elefteriadis, A. 2016, Risk management at railway level crossings: Proposal for a decision support system, *6th International Congress Transport Research Arena (TRA)*, 'Moving forward: Innovative solutions for Tomorrow's Mobility', 18–21 April 2016, Warsaw, Poland.

Saccomanno, F., Park, J. and Fu, L. 2006, Analysis of countermeasure effects for highway – Railway grade crossings, *9th International Level Crossing Safety and Trespass Prevention Symposium*, 6–8 June, Montreal, Canada.

Selig, E.T. and Waters, J.M. 1994, *Track Geotechnology and Substructure Management*, Thomas Telford, London.

U.S. FRA. 2003, *Guide for Preparing Accident/Incident Reports*, Federal Railroad Administration Office of Safety Analysis, May.

Varma. 2009, *Level Crossing Safety Management Information System*, Chief Engineer, Indian Railways, New Delhi.

Washington, S. and Oh, J. 2006, Bayesian methodology incorporating expert judgment for ranking countermeasure effectiveness under uncertainty: Example applied at grade railroad crossings in Korea, *Accident Analysis and Prevention*, Vol. 38, pp. 234–247.

Woods, M.D., MacLauchlan, I., Barrett, J., Slovak, R., Wegele, S., Quiroga, L., Berrado, A. et al. 2008, Safer European level crossing appraisal and technology (SELCAT). *D3 – Report on Risk Modelling Techniques for Level Crossing Risk and System Safety Evaluation*. Report SELCAT-WP2-D2-V2. Rail Safety and Standards Board (RSSB).

Xia, H., Guo, W.W., Zhang, N. and Sun, G.J. 2008, Dynamic analysis of a train–bridge system under wind action, *Computers and Structures*, Vol. 86, pp. 1845–1855.

Xia, H., Zhang, N. and Guo, W.W. 2006. Analysis of resonance mechanism and conditions of train–bridge system, *Journal of Sound and Vibration*, Vol. 297, pp. 810–822.

Chapter 19

Railway and the natural environment

19.1 NATURAL ENVIRONMENT OF THE RAILWAY

Rail transport systems can move at grade, underground, and above the ground (elevated). In this context, the constituents of the natural environment which these systems' infrastructure and operation affect are air, the ground surface, and the subsoil.

Specifically, the air is affected by:

- The pollution caused mostly by air pollutants, which are emitted during the operation of diesel trains (air pollution) and during the production of electric power (electric trains).
- The noise emitted during the trains' operation (acoustic annoyance).

The ground surface, as well as anything that exists or moves on it (humans, flora, fauna, constructions, etc.), in general, are affected by:

- The noise and vibrations which are caused during the running of trains on the rails and are transmitted through the ground to neighbouring constructions (ground-borne noise and vibrations).
- The impacts caused to other means of transport, due to the land take by the railway infrastructure.
- The changes caused to the landscape, and, especially for the case of urban means of transport, to the urban space (integration into the topographic relief/surrounding built environment).
- The resulting change in the aesthetics of the surrounding area, resulting from the presence of the new means of transport (visual annoyance).
- The disturbance of various activities of local residents and particularly for the urban means of transport, the disruption of continuity in the urban space.
- The disturbance of the ecosystems.

The subsoil and the underground water are affected by:

- The pollution, which is caused by waste and pollutants released by railway vehicles during their movement in the main traffic lines, or derived from various activities that take place in the facilities and premises, the waste of which penetrate the soil (soil and water pollution).
- The vibration that is transmitted to the soil through the track superstructure.

DOI: 10.1201/9781003046073-19

Finally, all three constituents of the natural railway environment, and, in general, the whole planet are affected by the climate change caused by the so-called greenhouse gases emitted by the means of transport.

In the broader context of environmental sustainability in addition to the above-mentioned effects, energy consumption is of particular importance.

All the above essentially compose the interface between the railway and the natural environment and are described in more detail in the following sections. The land take, the ground-borne noise and vibrations, the disruption of continuity of the urban space, the changes in land uses and land values, and the integration of the railway infrastructure into the urban environment can be described as the interface between the railway and the so-called built-up environment. The land take has been examined in Sections 1.5.1 and 1.5.3.

19.2 ENERGY CONSUMPTION

19.2.1 Definition: units expressing energy consumption

Energy is defined as the capacity of an object or a system to produce work. Railway systems move by using mainly two sources of energy: oil, in the form of diesel fuel (diesel trains), and electricity (electric trains). Diesel locomotives convert thermal energy to kinetic energy, while electric locomotives convert electrical energy to kinetic energy.

The quantities commonly used for the expression of energy consumption of the railway are displayed in Table 19.1, and the relationships that link the expression units with each other in Table 19.2.

19.2.2 Energy-consuming railway activities

The energy-consuming activities of the railway system are the circulation of trains and the railway facilities/premises (stations, depots, etc.).

With regard to the operation of trains, energy is required to meet the six basic operations, and specifically:

- Acceleration of the train.
- Traction of the train.

Table 19.1 Units expressing energy consumption for passenger and freight railway transport

Passenger transport	Freight transport
kW h/passenger-seat	–
kW h/passenger-kilometre (kW h/pkm)	kW h/tkm
MJ/passenger-seat	–
MJ/passenger-kilometre (MJ/pkm)	MJ/tkm
Petrol (L)/passenger-seat	Petrol (L)/tkm
BTU/passenger-kilometre (BTU/pkm)	BTU/tkm

Table 19.2 Relationship between units that are used to express energy consumption

1 MJ = 0.2778 kW h
1 BTU = 2.931×10^{-4} kW h
1 L petrol = 9.7 kW h

- Train movement on track segments with longitudinal gradient.
- Supply of the control systems.
- Lighting, heating, cooling, and ventilation of the vehicles.
- Transmission of power through the electric network to the driving wheels, in the case of electrification.

With regard to the operation of the stations, it should be highlighted that in railway networks with a large number of stations, the energy consumed for the operation of lifts, escalators, as well as lighting and heating of the stations, should comprise up to 20% of the total energy consumed by the railway system (UIC, 2008).

Finally, energy is consumed during both the production and the distribution of the final energy that is being used for the operation of the various activities of the railway system.

19.2.3 Special features of each railway system category

Figure 19.1 displays the energy consumption per category of passenger train, and for different traction systems.

It should be noted that regional trains consume more energy than intercity trains and that electric trains consume much less energy than diesel trains (approximately 1/3).

The chart of Figure 19.2 displays the relationship between energy consumption and speed for various train categories.

By evaluating the above data, it can be stated that conventional-speed trains, running at speeds V < 200 km/h, consume 0.050–0.130 kW h/pkm (passenger-kilometre).

According to Network Rail, 2009, energy consumption for European high and very high-speed trains ranges between 0.034 kW h/pkm and 0.041 kW h/pkm, while speed ranges between 180 km/h and 350 km/h. The high increase in the speed of trains does not result in a corresponding increase in energy consumption. One of the reasons behind this is the fact that the average distance of intermediate stops is lower than that of the conventional-speed railway.

Concerning urban railway systems, the metro is an energy-consuming system, since the train movement, as well as the operation of the auxiliary train and station facilities equipment, is performed through the supply of electrical energy.

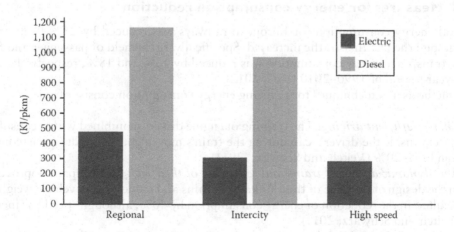

Figure 19.1 Energy consumption of passenger railway transport per train category and traction type, 2005. (Adapted from UIC, 2012, *Energy Consumption and CO$_2$ Emissions, Railway Handbook*, 2012.)

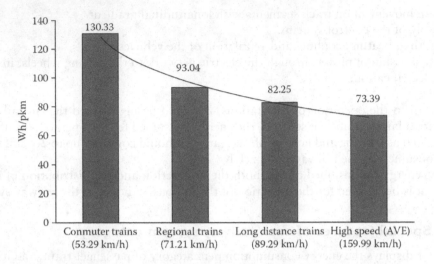

Figure 19.2 Energy consumption for various categories of railway systems per ascending order of average speed. (Adapted from Garcia, A., 2010a, *High Speed, Energy Consumption and Emissions*, UIC, 21 December 2010.)

Table 19.3 Range of fluctuation in the values of energy consumption for various categories of railway systems

Interface with the environment	High and very high-speed railway	Conventional-speed railway	Metro	Tramway
Energy consumption (kW h/pkm)	0.035–0.100	0.050–0.130	0.015–0.055	0.008

The tramway is characterised by comparatively lower energy consumption, with about 50% being lost during braking. The use of energy-storage systems saves energy. Lighter vehicles allow for lower energy consumption (UIC, 2008).

Table 19.3 displays the range of fluctuation in the values of energy consumption for various categories of railway systems.

19.2.4 Measures for energy consumption reduction

The total energy consumption for European railways was reduced by 25% from 1990 to 2010, despite the fact that traffic increased. Specifically, in the field of passenger and freight railway transport, energy consumption was reduced by 13% and 18%, respectively, during the 20-year period of 1990–2010 (UIC, 2012).

The methods and techniques for reducing energy consumption consist of:

- *Energy-efficient driving*: The training of engine drivers, combined with a consultation system inside the driver's cab during the train's movement, can reduce fuel consumption by 5–20% (Veitch and Schwarz, 2011).
- *Aerodynamic design of trains and reduction of their weight*: In Japan, improvements in the design of the nose of the Shinkansen trains and reductions in vehicle weight have resulted in the reduction of energy consumption by 40%, although speed has increased (Veitch and Schwarz, 2011).

  The reduction of weight is of particular importance in urban railway systems (metro, tramway, and suburban railway) that make a lot of stops. The large weight of the trains, combined with frequent start-ups, leads to high energy consumption. Carbon fibres are the new composite material that has been increasingly applied recently and contributes towards the construction of lighter vehicles.

- *Systems for the storage of the energy that is diffused during braking*: Energy storage can be performed by batteries, supercapacitors, flywheels, and superconducting magnetic energy-storage systems.
- *LED eco lighting*: This is a very low-energy lighting solution, which consumes up to three times less energy than conventional systems.

19.3 AIR POLLUTION

19.3.1 Definition: units expressing air pollution

Air pollution is defined as any condition where there are substances in the ambient air, in much higher concentration than the normal, which may cause measurable effects to humans and, in general, to organisms, vegetation, or constructions/materials.

The impacts of the railway, as well as of transport systems in general, on air pollution are divided into two main categories:

- Local pollution, which includes all forms of air pollution that affect the quality of ambient air and health on a local scale.
- General pollution (of the planet), which involves the destruction of the ozone layer, and the 'greenhouse effect'.

The main pollutants emitted by various means of transport and producing local pollution are carbon monoxide (CO), nitrogen oxides (NO_X), hydrocarbons (HC), solid particles (TSP), sulphur dioxide (SO_2), and various mineral traces.

The main gases of the atmosphere that cause general pollution are carbon dioxide (CO_2), methane (CH_4), nitrogen oxide (N_2O), chlorofluorocarbons (CFCs), tropospheric ozone, and vapours at high altitude.

The quantities commonly used in the expression of local air pollution caused by the railway are displayed in Table 19.4.

General air pollution is measured based on the generated carbon footprint. The carbon footprint is defined as *the total sets of greenhouse gas emissions caused by an organisation, event, product, or person* and is measured in emitted tons of carbon dioxide equivalent (t CO_2e) (Carbon Trust, 2012).

19.3.2 Railway activities causing air pollution

Air pollution is caused during the operation, as well as during the implementation, of a railway system.

Table 19.4 Indices for the expression of local air pollution caused by railway systems

Passenger transport	Freight transport
kgr or gr CO_2/pkm or seat	kgr or gr CO_2/tkm
gr NO_x/pkm or seat	gr NO_x/tkm
gr NMHC/pkm or seat	gr NMHC/tkm
gr SO_2/pkm or seat	gr SO_2/tkm

During the operation of a railway system, there are air pollutants only in the case of diesel traction, whereas in electrification, ambient air in the area close to the railway network is affected only by the use of polychlorinated biphenyl in the power converters of the electrical equipment of the rolling stock and the substations. In contrast, air pollutants are produced in the area of electrical power production.

During the construction of railway projects, the sources of air pollution are the following:

- Emissions of air pollutants from a variety of machinery used in various construction operations.
- Dust from excavations. The quantities of dust emissions from roads and non-asphalt surfaces range widely (1–10 kg/vehicle-kilometre) (Maropoulou, 2009).
- Additional emissions from the road traffic that serve the specific project.

19.3.3 Special features of each railway system category

According to:

- A research conducted in 2006 by the Centre for Neighborhood Technologies, high-speed trains release into the atmosphere 30–70 CO_2 gr /pkm (CNT, 2006).
- Data from DB AG TREMOD (Kettner, 2008), the emission of CO_2 in long-distance passenger railway transport is 49 gr/pkm.
- A study conducted in 2010 by JR Central in the Tokyo–Osaka connection, with a length of 515 km, the emissions of CO_2 for the N700 'Nozomi' train were estimated to be 8.54 gr/pkm (JR Central, 2010).

Figure 19.3 displays the emissions of CO_2 per passenger train category in Europe and for different traction systems.

By evaluating the above data, we can state that high and very high-speed trains emit carbon dioxide (CO_2) in the range of 10–50 gr/pkm.

The conventional-speed railway emits 29% more pollutants per passenger than the high and very high-speed railway (Garcia, 2010b).

With regard to the tramway, the only ambient air pollutants that it may cause are the particles derived from the friction between wheels and rails; however, due to their small quantity,

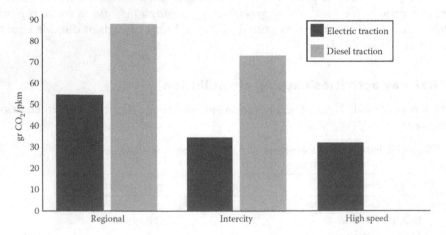

Figure 19.3 CO_2 emissions per train and traction type in Europe, 2005. (Adapted from UIC, 2012, *Energy Consumption and CO_2 Emissions, Railway Handbook*, 2012.)

these are considered negligible, compared with other pollutants within a city. According to studies conducted in Melbourne, the emissions of carbon dioxide that correspond to each passenger-kilometre are estimated for a tramway at an average of 60 gr CO_2/pkm.

Table 19.5 displays the range of fluctuation of the values of air pollution for various railway system categories.

19.3.4 Measures for air pollution reduction

The railway is the only means of transport whose share with regard to carbon dioxide (CO_2) emission has been reduced since 1990, while the respective shares of all other transport means have increased. The total CO_2 emissions from railway operations in the EU were reduced by 39% from 1990 to 2010. More specifically, CO_2 emissions were reduced by 27% per passenger-kilometre and 41% per train-kilometre (UIC, 2012).

Beyond the application of electrification, in order to deal with air pollution, a variety of new techniques, constructions, and environmentally friendly materials have been used, such as:

- *Hydrogen trains*: The term 'hydrogen train' (hydrail) implies all types of railway vehicles, large or small, which use hydrogen as a source of energy to power their locomotives. These trains convert the chemical energy of hydrogen to mechanical energy.
- *Natural gas, biofuels.*
- *Cement sleepers with the addition of slag*: Cement causes significant environmental impacts because of the CO_2 that is emitted during its production. The construction of this environmentally friendly sleeper has been performed by replacing part of the high-durability Portland cement with granulated slag (ground-granulated blast furnace slag), and part of the fine-grained aggregates with oxidising slag (electric-arc furnace oxidising slag).
- *Brakes with pads made of natural fibre*: Many railway companies want to abolish the use of porous metals in brakes, which results in the diffusion of heavy metal particles into the environment. According to a research by Sustainable Technologies Initiative (STI), natural fibres can replace the costly aramid fibres used for brake pads without any reduction in performance, and with clearly less impacts to the environment (reduction of dust contained in these materials, and which is released to the environment, as pads wear).
- *Renewable energy sources (solar, aeolic, and geothermal)*: An example of the use of solar energy is the railway tunnel (cut-and-cover tunnel) near Antwerp, Belgium, with a total length of 3.4 km, on top of which 17,820 solar panels have been installed, generating 3,300 MW h of electrical power every year (Figure 19.4). The Blackfriars railway station in London, which has been constructed on the bridge bearing the same name, hosts 4,400 photovoltaic panels on its roofs. In Japan, solar panels are installed on stations and facilities, while in Italy the installation of wind power generators on bridges is being investigated (aeolic bridges) (Pikal, 2011).

Table 19.5 Range of fluctuation of the values of air pollution for various railway system categories

Interface with the environment	High and very high-speed trains	Conventional-speed trains	Freight transport (UIC, 2012)
Air pollution (during system operation)	10–50 gr CO_2/pkm (30 gr CO_2/pkm)	Electrification 15–65 gr CO_2/pkm (40 gr CO_2/pkm) Diesel traction 75 gr CO_2/tkm	Electrification 18 gr CO_2/pkm Diesel traction 28 gr CO_2/pkm

Figure 19.4 Solar panels on top of railway tunnel, Belgium. (Adapted from Gifford, J., 2011, *High-Speed Rail Line Installation Powers First Solar Train*, available online at: http://www.pv-magazine.com/news /details/beitrag/high-speed-rail-line-installation-powers-first-solar-train-_100003253/#axzz3X y0A1P8R (accessed 30 April 2015).)

19.4 SOIL AND WATER POLLUTION

19.4.1 Definition: measurement methods of soil and water pollution

Soil pollution is defined as any unwanted change in the physical, chemical, and biological properties of soil, which under conditions is, or may become, harmful to humans and other plant and animal organisms.

Soil and water pollution may be defined in the following ways:

- In the event of fuel leak, by identifying the soil content of insoluble hydrocarbons, polycyclic aromatic hydrocarbons, and volatile aromatic compounds that are toxic substances contained in oil.
- By measuring the pH of soil and water.
- By measuring the organic substances ending in underground and surface waters, using the BOD (BODu) index (Biochemical Oxygen Demand).
- By measuring the concentration of foreign materials in the soil, in terms of mass per volume unit.
- By measuring the concentration of foreign materials in water, in terms of mass per volume unit.

19.4.2 Railway activities causing soil pollution

The soil pollution that is caused as a result of the presence of a railway system is due to the (Donta, 2010):

- Sealing of the soils (i.e., when the soil surface is covered by watertight material, e.g., concrete).

- Release of inorganic and organic substances to the environment by the passing trains. These substances may end into the underground and surface waters, after permeating the track bed layers and passing in the foundation soil.
- Combating vegetation on the track superstructure, by using pesticides that are hard to degrade.
- Disturbance of the flow of underground waters during the construction of the railway system.
- Pollution resulting from an accident, during the transportation and storage of fuel, hazardous goods, and other materials, causing their leak into the ground and underground waters.
- Pollution due to liquid waste, emerging from the activities that take place in the area of the depot.

19.4.3 Special features of each railway system category

The problem of soil and water pollution involves mainly the operators of urban railway systems and systems of railway transportation of hazardous goods.

Specifically, tramway vehicles possess a liquid waste cleaning unit which is installed on the train itself, while they normally use a superstructure which is embedded in the pavement. The areas with a high likelihood of soil degradation are mainly the railway facilities/premises.

With regard to the railway transportation of hazardous goods, due to the nature of the products in case of an accident, leaks from vehicle tanks or from packages, or just poor handling by the staff during loading/unloading or transfer of materials, the risk of soil pollution is high. Further to the accident prevention measures that, apart from the safety, also guarantee the protection of the environment, there are also measures that aim exclusively at the protection of the soil, such as the following:

- Waste drainage system in the washing stations of empty tanks.
- Special configuration of the track superstructure (e.g., slab track instead of ballasted track) in the siding lines of reception and forwarding of trains transporting hazardous goods.

The various activities associated with the movement of these products are regulated by international conventions and are performed under strictly defined safety conditions. For railway transportation, the provisions of COTIF/CIM/RID apply (Europa, 2012).

19.4.4 Countermeasures against the pollution of soil due to the presence of the railway

The most significant countermeasures against the pollution of soil and waters due to the presence of the railway are the following:

- Regular monitoring of the quality of the soil and water.
- Maintenance of facilities and fuel storage tanks.
- Placement of special surface layers (covers) and drainage systems at places where the train stops, such as, before the lighting signals at the stations, and in warehousing and maintenance areas.
- Placement of special oil collectors to prevent pollution from leaking oil.
- Equipping vehicles with composting toilets.

19.5 VISUAL ANNOYANCE

19.5.1 Definition: measurement methods of visual annoyance

Visual annoyance is divided into visual obstruction and visual nuisance.

Visual obstruction is defined as the rate of an observer's field of vision that is covered by the infrastructure and the facilities/premises of the railway system. It is estimated through the angle formed between the point of observation and the relevant railway structure element.

Visual nuisance is defined as the discomfort caused to the observer by the presence of the infrastructure of the railway system.

Visual nuisance constitutes a design and construction parameter of a transport system and it entails the system's aesthetics as a whole, since that determines the degradation or improvement of the landscape in which the system is to be integrated into.

By studying the current approaches for evaluating visual nuisance, as such evaluations are conducted in large transport projects today, two major shortcomings have been identified:

- The lack of an objective and quantitative method for evaluating visual nuisance. On the contrary, there exist a number of guidelines and best practices from various institutes, which cannot however constitute but a generalised and theoretical basis.
- The lack of an objective supervisory/regulatory authority that evaluates the visual nuisance of projects.

In recent literature (Lagarias, 2015; Pyrgidis and Barbagli, 2019; Pyrgidis et al., 2018, Pyrgidis et al., 2020), a relatively objective and numerical method for evaluating the visual nuisance caused by a tramway and a monorail system as a whole is proposed. The methodology is based on assigning points to each of the structural elements of these systems and then evaluating the entire system based on its overall score (Lagarias, 2015). More specifically the following steps were considered:

1. The structural elements of a tramway and a monorail system that contribute to visual nuisance were identified and recorded (Table 19.6).

Table 19.6 Structural elements of a tramway and of a monorail system that contribute to visual nuisance – weighting factor per element

Structural element	Tramway	Monorail
Exterior image of the rolling stock	$w_i = 3.0$	$w_i = 2.7$
Interior image of the rolling stock	$w_i = 1.3$	$w_i = 1.9$
Stops/stations	$w_i = 2.7$	$w_i = 2.8$
Electrification system	$w_i = 2.8$	n.a
Superstructure covering materials	$w_i = 2.7$	n.a
Corridor separation techniques	$w_i = 2.3$	n.a
Signalling equipment	$w_i = 1.4$	n.a
Depot(s)*	$w_i = 1.4$	n.a
Pillars	n.a	$w_i = 3.1$
Guidance beams	n.a	$w_i = 2.7$
Emergency escape ways	n.a	$w_i = 2.3$
Sum	**17.6**	**15.5**

*Monorails depots, even though they may provoke visual nuisance, were excluded from the structural elements to be assessed. This choice was based on the arguments that are situated in a single point along the route and are usually located outside the core urban area of a city or in areas of lower aesthetic significance (e.g., industrial zones).

2. The different available aesthetic solutions for each of the structural elements were identified and recorded (Tables 19.7 and 19.8; column 1).
3. These aesthetic solutions were then categorised qualitatively into five aesthetic categories: O, A, B, C, D (Tables 19.7 and 19.8; column 2). In order to rank the aesthetic solutions in the five aesthetic categories, the following criteria were taken into account:
   • The concealment of the structural element from the line of sight of observers.
   • The limitation in the number or size of the different parts of the structural element.
   • The providence during the design of the structural element for the reduction of its visual nuisance, meaning a design that considers the aesthetics of the element as a priority.
4. A score of Visual Nuisance Points VNPs was attributed to each of these aesthetic categories (0 to 4) (Table 19.9).
5. A weighting factor w_i concerning the contribution to visual nuisance of each of the structural elements to the entire tramway/monorail system was identified (Table 19.6). For this purpose, a series of interviews were conducted with different experts (16 in total) (Lagarias, 2015; Pyrgidis and Barbagli, 2019). Each of the structural elements was given a grade from 3 (least impact) to 10 (most impact) by the interviewees based on the impact that particular element has on the visual nuisance caused by the system. The mean grade for each structural element was calculated. The mean grade was then divided by a reference value (the minimum grade of 3) and rounded to the first decimal point.
6. A formula for estimating the overall level of Visual Nuisance VN that is caused by the operation of a tramway/monorail system was proposed (Equation 19.1) (Lagarias, 2015).
7. The overall VN score that is derived from the application of this formula may be used to evaluate a tramway/monorail system as a whole (Table 19.10).

The level of visual nuisance VN that is caused by the operation of a tramway/monorail may be evaluated using the following formula:

$$VN = \frac{\sum_i (w_i \times VNP_i)}{\sum_i w_i} \tag{19.1}$$

where:

VNP$_i$: is the Visual Nuisance Points of every structural element i of the tramway/monorail. They are dependent on the qualitative category of the aesthetic solution chosen to reduce the visual nuisance caused by structural element i.

w_i: is the weighting factor that defines the level of influence every structural element i has on the visual nuisance caused by the system as a whole. It is dependent on the size of the structural element, its construction site, its final location, and the influence it exerts on the perception of observers.

Taking into account, for the two examined types of urban transport systems:

• The weighting factors w_i per structural element are presented in Table 19.6.
• The different available aesthetic solutions for each of the structural elements in combination with the aesthetic category in which they belong (O, A, B, C, D) are presented in Tables 19.7 and 19.8.

Table 19.7 Available aesthetic solutions per structural element of a tramway and their association with a specific aesthetic category

Structural element – aesthetic solution	Aesthetic category
Exterior image of the rolling stock	
• Modern vehicles that are designed exclusively for the system they are made for, while taking into account the existing character of the urban area that the system will be located in.	A
• Modern vehicles with an innovative design that does not take into account the existing character of the urban area.	B
• Conventional vehicles without any distinctiveness in their design.	C
• Use of trams that are longer than 65 m.	Degrade by one category
Interior image of the rolling stock	
• Innovative design, with large open spaces and no advertisements.	A
• Large windows and adequate open spaces at eye level, limited use of advertisements.	B
• Small windows or obstruction of the passengers' sight with many elements at eye level or extensive use of advertisements.	C
Stops	
• Mainly small and discreet stops. Limited size of structural elements and main use of glass or thin metallic parts.	A
• Stops with a distinctive design that is integrated in the urban area in which they are constructed. The design takes into account the reduction of visual nuisance through the use of transparent or thin parts.	A
• Stops with a distinctive architectural design, with the use however of large structural elements that hide part of the sky or the urban area.	B
• Conventional stops with large structural elements. No effort to reduce visual nuisance.	C
• Placement of advertisements on the surfaces of the stop (over 50%).	D
Electrification system	
• No use of catenary wires and electrification poles. Ground level electrification (free catenary system).	O
• Effort to limit the amount of catenary wires per track.	B
• Effort to limit the amount of electrification poles, use of existing buildings to support catenary wires.	B
• No effort to reduce the amount of catenary wires or electrification poles.	C
Tramway superstructure covering materials	
• Use of cover elements (turf or coloured stones) that have a visual continuity with the surrounding landscape, meaning that the cover materials appear to be a continuation of the surrounding ground.	A
• Use of cover elements or coloured stones, without, however, taking into account the visual continuity with the surrounding landscape.	B
• Tramway corridors with no covering materials.	C
Tramway corridor separation techniques	
• Separation (from the other means of transport) with the use of small structural elements that are designed to improve the area's aesthetics (vegetation or well-designed elements).	A
• Separation with the use of small structural elements, poles, or fences.	B
• Separation with the use of large structural elements, poles, or fences or other solid non-transparent elements that are over 1 m in height.	C
• Placement of advertisements on the structural elements used for separation.	D
Signalling equipment	
• Effort to limit the use of signalling equipment or use of a distinctive design of poles and signs.	B
• Use of conventional signalling poles.	C
Depot(s)	
• The depot is placed outside of the urban area.	O
• The depot is placed within the urban area but there is consideration for its aesthetics.	B
• The depot is placed within the urban area but there is no consideration for its aesthetics.	C

Table 19.8 Available aesthetic solutions per structural element of a monorail system and their association with a specific aesthetic category

Structural element – aesthetic solution	Aesthetic category
Exterior image of the rolling stock	
• Modern vehicles that are designed exclusively for the system they are made for, while taking into account the existing character of the urban area that the system will be located in.	A
• Modern vehicles with an innovative design that does not take into account the existing character of the urban area.	B
• Conventional vehicles without any distinctiveness in their design or any association with the characteristics of the urban landscape.	C
• Use of long train sets (> 70 m).	Degrade by one category
Interior image of the rolling stock	
• Innovative design, with large open spaces and no advertisements.	O
• Large windows and adequate open spaces at eye level, limited use of advertisements.	A
• Small windows or obstruction of the passengers' sight with many elements at eye level or extensive use of advertisements.	B
Stations	
• Integration of stations within existing buildings without significant alterations to their face.	O
• Stations with the use of transparent or thin parts in combination with a concurrent limitation to their size to one minimally satisfying their functional requirements as well as a placement near ground level so as to avoid hiding part of the sky.	A
• Stations with use of transparent or thin parts or with a limitation to their size to one minimally satisfying their functional requirements or with a placement near ground level so as to avoid hiding part of the sky.	B
• Conventional stops with large structural elements. No effort to reduce visual nuisance.	C
Pillars	
• Limited size of pillars with the use of thinner cross sections.	B
• Adoption of one or even a combination of the following ways (covering the pillars with some form of vegetation or with reflective panels).	B
• No consideration for reducing the visual nuisance caused by pillars.	C
• Construction of pillars along higher altitude areas that are visible throughout the city (in case that this solution is not chosen for panoramic view reasons).	Degrade by one level
Choice of route	
• If a route offering scenery of a particular value to the passengers is chosen. (Choice of a route offering scenery of a particular value to the passengers)	Reduction of 0.5 VNP points from the final value of VN factor
Escape ways	
• No escape way present.	O
• Escape way existing but concealed within the guidance beam overall gauge.	A
• Escape way existing and not concealed within the guidance beam overall gauge.	B
Guidance beams	
• Use of beams with reduced width and height (lower than 0.7 m and 1.4 m, respectively).	B
• Conventional monorail beam.	C

Table 19.9 Visual Nuisance Points per aesthetic category

Aesthetic category	VNP
Solution Category O Concealment of the structural element from any observers	0
Solution Category A No option for concealment – effort to limit the size of the element that is observable and simultaneous significant effort for an aesthetically pleasing design of the element	I
Solution Category B No option for concealment – either an effort to limit the size of the element that is observable or significant effort for an aesthetically pleasing design of the element	2
Solution Category C No option for concealment – no effort to limit the size of the element that is observable and no effort for an aesthetically pleasing design of the element	3
Solution Category D Increase of visual nuisance through the use of larger structural elements or use of additional elements that are not functionally necessary and obstruct the sight of an observer (advertisements, large non-transparent elements)	4

- The special circumstances, such as choice of route or length of vehicles, under which aesthetic categories are downgraded or upgraded, are presented in Tables 19.7 and 19.8.

From Equation 19.1, the following maximum and minimum achievable Visual Nuisance Points per system may be derived:

- Tramway: Max 3.36 Min 0.84
- Monorail: Max 3.10 Min 0.42

Based on these limit values, the overall Visual Nuisance score may be used to evaluate a tramway/monorail system as a whole using three qualitative categories (I, II, or III) as shown in Table 19.10.

The lower the value of the Total VN the better the aesthetics of the systems and, consequently, the lower the visual annoyance to the residents of the area where the system operates and to its users.

The findings of this research may be applied during the bidding process for or at the design stage of a new tramway/monorail system, at the evaluation of an existing system, or finally for the evaluation of corrective interventions aimed at upgrading an existing system. They are mainly of interest to the assessors of urban railway systems, as the proposed methodology can replace their qualitative decision process with a quantitative approach for the evaluation of visual nuisance. They are also of interest to designers as it provides them with a list of best practices for the reduction of visual nuisance. The proposed methodology is also useful in order to provide decision-makers with simple and solid quantitative data. In any case, it constitutes a scientific tool to support the applicability verification of environmental impacts of any specific urban guided mass transport system (see Chapter 21).

In intercity railway networks, with regard to visual annoyance, more considerations are involved, which describe the level of attraction of the landscape and, to a lesser extent, the observer's visual nuisance.

The aesthetics of the civil engineering structures and the trains, as well as the level of attraction of the landscape, before and after the project, constitute elements that can be graded subjectively by experts and the public.

In any case, the correct estimate of the visual outcome of a railway project can be performed only after the completion of its construction. Environmental design can provide for

Table 19.10 Evaluation of the total visual nuisance caused by a tramway and a monorail system

Total VN value Tramway	Total VN value Monorail	System evaluation
0.84–1.68	0.42–1.31	Visual Nuisance Qualitative Category I The tramway/monorail system has reduced to a large extent the visual nuisance it causes. It has taken this parameter into consideration at the design level and has chosen effective solutions in partially or totally concealing the structural elements from observers. At the same time, a priority has been given to its tasteful design. It has a low negative impact on the image of the urban area while at the same time it includes visually pleasant elements.
1.68–2.52	1.31–2.20	Visual Nuisance Qualitative Category II The tramway/monorail system has partially reduced the visual nuisance it causes. It has limited the size and intrusiveness of some elements and has improved their aesthetics. It has a medium negative impact on the image of the urban area. There might be a need for individual corrective actions in some of the areas in which it operates.
2.52–3.36	2.20–3.10	Visual Nuisance Qualitative Category III The tramway/monorail system has taken few or no measures in reducing the visual nuisance it causes. Its structural elements limit the line of sight of observers to a large extent, while their design is neutral or unpleasant. The railway system has a high negative impact on the image of the urban area and is in need of corrective actions to limit the visual nuisance it causes.

(or prevent) the potential visual annoyance, based on the existing experience in relevant projects.

19.5.2 Railway activities causing visual annoyance

Visual nuisance is a consideration throughout all stages of a railway system's lifecycle and specifically at the construction stage, at the operational stage, and at the stage after the end of operations.

In the case of construction of the railway infrastructure of urban railway systems, strong visual annoyance is caused by the facilities of the construction sites and the fencing.

During the operation of railway systems, visual annoyance is caused by (see also Table 19.6):

- The railway infrastructure, which includes all elements of the track superstructure, the civil engineering structures, and the lineside systems. Such structural elements are the electrification system installations (overhead catenary system, substations, and electrification masts), the signalling system installations (masts), the presence of high embankments, the presence of noise barriers, the depots that are constructed in urban areas, and mainly the elevated infrastructure if exists (pillars, guidance beams, elevated stations).
- The rolling stock and specifically its exterior image (design taking into account the existing character of the urban area/city in which the system will be located).

19.5.3 Special features of each railway system category

For intercity systems, visual nuisance is more easily quantified as the effects it causes on the landscape are relevant to the space the system occupies, while for urban systems visual nuisance is a more complex issue, since it includes a series of interactions.

The main parameters of visual annoyance caused by the tramway are the exterior image of the rolling stock and the overhead catenary system (see also Table 19.6). The aesthetic annoyance caused by the wires is significant when the tramway lines are integrated into highly populated and visited areas, and particularly in sensitive regions (historic centres, traditional residential settlements, monuments, etc.). In general, however, the tramway's profile is particularly attractive to the citizens, while in many cases the tram itself comprises an attraction to the city.

The metro usually does not affect the aesthetics of cities, except only during the construction stage, due to the fact that it operates underground. With regard to underground railway stations, and in many cities, they are exceptionally decorated, and their aesthetics create a pleasant atmosphere for the user.

In the case of an urban elevated railway system (monorail and elevated metro), the main parameter of visual annoyance is the elevated infrastructure. The structural elements of their infrastructure have generally large dimensions, while at the same time several structural elements are present throughout or for a large section of the route. Most such elements are elevated, thus creating a second level of construction on top of the existing urban infrastructure and hindering further visibility towards the sky and in general the upward visual range of any observer and at any point in time. Concurrently, the mass of the structural elements of these elevated systems hides large portions of the urban landscape thus hindering visibility towards the city itself.

In the case of a monorail, visual nuisance is less impacting than that caused by an elevated metro operating on heavier infrastructure. Additionally, in the case where the monorail serves movement within amusement parks, zoos, etc., or transportation of passengers through areas that are of particular interest in terms of view, the innovative/futuristic factor of the monorail can, contrary to the norm, be considered as a suitable and pleasant presence. In these specific cases, it does not constitute a disturbance but rather has a desired positive impact. Of all the structural elements of a monorail system, the pillars play the most vital role in how intrusive a monorail system is to the aesthetics of an observer (see also Table 19.6) (Pyrgidis and Barbagli, 2019).

19.5.4 Countermeasures against visual annoyance caused by the presence of the railway

During the construction of any railway system, the architectural design should constitute an integral part of the total design. Its goal should be the construction of facilities that are of high aesthetics while, at the same time, functional and the delivery of user-friendly railway systems.

For the tram, the most effective solution to the aesthetic annoyance caused by the aerial cables is to perform power supply through the ground or via energy-storage devices (see Chapter 20), at least for those segments of the network that are environmentally sensitive. Other measures used in tramway networks for the reduction of visual annoyance are:

- Covering the tram traffic corridor with grass (Figure 19.5). This solution, on top of the aesthetic enhancement of the landscape, prevents the effect of temperature increase in city centres and provides a permeable layer that filters the precipitation waters, thereby limiting the pollution of the subsoil.
- The aesthetics of trains and stops.

Figure 19.5 Covering a tram corridor with grass, Montpellier, France. (Photo: A. Klonos.)

Figure 19.6 Covering a pillar of a monorail system with some form of vegetation, Sentosa 'Express' mono-rail, Singapore. (Adapted from Yoshitaka Hirabaya, Hitachi, Japan. On line image. Available at: https://www.monorails.org/tMspages/CnstSentosa01a.html) (accessed 25 February 2021).)

For the monorail, different architectural solutions for pillars may, to an extent, reduce the visual nuisance related to them. Some of the ways in which this may be achieved are the following:

- Covering the structural elements with some form of vegetation (Figure 19.6).
- Depicting scenes and adding pleasing graphical designs and colour schemes over the surfaces of the structural elements.
- Covering the structural elements with reflective panels (Figure 19.7).

Figure 19.7 Covering a pillar of a monorail system with reflective panels (Adapted from Online image. Available at: https://www.monorails.org/tMspages/enviro.html (accessed 25 February 2021).)

The emergency escape ways have a negative impact on aesthetics also, as their presence heavily affects the visual lightness of the guidance beam.

Finally, a factor called *slenderness ratio* defined as the width of the vehicle divided by the overall height of the vehicle above the beam could, perhaps, be an additional aspect that affects aesthetics, since this gives a measure of the perceived stability or awkwardness of the train.

In intercity railway systems, the usual measures used towards the reduction of visual annoyance are:

- Planting of embankment slopes.
- Fencing with plants.
- Transparent noise barriers and noise barriers with planting.
- The aesthetics of the civil engineering structures.

19.6 INTEGRATION OF THE TRACK INTO THE LANDSCAPE

19.6.1 Definition: measurement indices of integration

A railway line is considered optimised in terms of the track alignment, when it consists exclusively of straight sections and lies throughout its length on a horizontal plane. The integration of such an optimised track layout into the landscape requires earthworks and civil engineering structures, which alter the topographic relief and increase the implementation cost. The adoption of curved sections in the horizontal alignment and the adoption of longitudinal slopes in the vertical alignment reduce these interventions and smooth out the integration into the landscape.

The quality of integration of the railway network infrastructure is characterised by:

- An excess of the alterations performed in the topographic relief.
- Alteration in the drainage of surface waters.
- The level of reinstatement of the landscape.
- The land take.

Earthworks, and specifically, embankments and cuttings, constitute interventions that alter the topographic relief.

The indices usually used for the estimation and evaluation of the impacts of the earthworks of a railway system are the height of the embankments and the depth of the cuttings, as well as the volume of the soil that is removed or deposited for the requirements of the railway civil engineering structures.

The relatively small right-of-way (land take) of the railway track causes a much lesser problem to the natural environment than the motorway, resulting in the requirement for lower-width cuttings and embankments.

The reinstatement of the landscape to its original form or finding an accepted aesthetic solution for the newly formed situation is also of particular importance. The issue of the restoration of landscape has begun to gain significance during the last 35 years. Currently, the budgets of most projects allow for 1–2% to be spent on the reinstatement of the landscape.

19.6.2 Railway activities causing a change of landscape

The change of ground relief is mostly due to the execution of earthworks, and specifically due to the construction of embankments and cuttings.

Embankments and cuttings are constructed in order to ensure a smooth track alignment (straight sections where possible and small longitudinal slopes) in areas of a difficult landscape.

Embankments are used as an alternative to bridges, for access to areas of high-altitude difference, provided that their height is not prohibitive. Low-height embankments are also used, for the protection of the track superstructure from flooding.

Cuttings are used as an alternative to tunnels, for track layout within mountains, provided that their height is not prohibitive.

Deep cuttings may cause landslides on the slopes, while high embankments may result in the blocking of access to certain areas.

19.6.3 Special features of each railway system category

Regardless of the category of the railway system, the height of embankments should be, for reasons of traffic safety, relatively low, while the soil should have been compacted well (see Sections 18.5.5 and 18.5.6). Indicatively, it is stated that embankment heights of more than 20 m should be avoided in railway systems.

The application of very high speeds also involves the adoption of large curvature radii in the horizontal alignment, and a greater distance between track centres. This leads to a more difficult integration into the topographic relief.

The integration of the metro systems depends on the selection of the construction depth of the stations (see Section 5.4.2). The typical depth is 18–20 m from ground level; however, this fluctuates depending on the conditions.

19.6.4 Measures for smooth integration of the railway into the landscape

Some of the measures used for smooth integration of the railway systems into the topographic relief of an area are:

- The selection of tunnel boring machines after evaluating the geologic and geotechnical conditions, in order to minimise ground surface settlement.
- The construction of tunnels, rather than deep and long cuttings.
- The construction of bridges, rather than high and long embankments (Figure 19.8).
- The reinstatement of the landscape.

19.7 ECOSYSTEM DISTURBANCE

19.7.1 Definition: indices of expression of ecosystem disturbance

The integration of a railway line has a direct impact on the biological resources of the area through which the line passes:

- Alteration or destruction of the natural environment and the reproduction areas of rare or endangered species.
- Loss of a significant number of species. Endangered species under higher risk are large mammals (deer, boars, and bears), migratory and prey bird species, and bats.

Figure 19.8 Bridge construction for smoother integration into the topographic relief, Karya, Greece. (Photo: A. Klonos.)

- Impacts or measurable degradation of the protected natural environment, the sensitive natural vegetation, the wetlands, and other environmentally protected areas.
- Conflict with the provisions of the existing environmental regulations on a local, regional, or national level.

The restriction of access to the area can be assessed based on the impact observed on the fauna of the area (decrease or increase in the number of specific species, and, in general, the impact on the lifecycle of endemic species in the area, before the construction and during the operation of the railway network).

19.7.2 Railway activities causing ecosystem disturbance

Ecosystem disturbance is caused by:

- The land take by the railway.
- The track fencing.
- The noise emitted by the railway.

The compulsory fencing of main lines in high and very high-speed networks constitutes an important protection measure towards the reduction of animal mortality rate, and in many cases it may act negatively by reducing their likelihood of survival, such as in the case that railway track crossing from one side to the other has become impossible.

19.7.3 Special features of each railway system category

The problem relates mostly to high and very high-speed networks, to which fencing is imposed, and in general to all tracks passing through habitats.

19.7.4 Reduction measures of ecosystem disturbance

The most effective measure towards the reduction of ecosystem disturbance is the correct selection of the track alignment in the first place, and specifically the avoidance of passing of the railway track through areas characterised as sensitive.

The mitigation measures used for the reduction of ecosystem disturbance are:

- Ensuring the continuity of natural habitats by using wildlife crossings. Wildlife crossings may include underpass tunnels, viaducts, and overpasses (green bridges); amphibian tunnels, fish ladders, tunnels and culverts; and green roofs (http://en.wikipedia.org/wiki/Wildlife_crossing, 2015) (Figures 19.9 and 19.10).

The factors that define the effectiveness of overpasses for animals are:

- Proper vegetation in their entrance points.
- Their dimensions, the presence of natural lighting, and the level of noise.
- Their proper location.
- Systematic monitoring of animal movements and behaviour.
- The exchange of knowledge and cooperation among researcher engineers, biologists, and ecologists.

Figure 19.9 Wildlife overpass on a railway track, TGV Atlantique, Lavare, France. (Adapted from Vignal, B., 1990, SNCF Médiathèque.)

Figure 19.10 Wildlife overpass on a railway track, TGV Atlantique, France. (Adapted from D'Angelo, J. J., 1995, SNCF Médiathèque.)

The monitoring of overpasses reveals that animals become familiar with them and use them. Nowadays, the number of this specific type of overpass in Europe is fairly high, while in the last years it has been attempted to increase their number in America.

- Special protection of endangered flora, implantations of trees and shrubs, and special reforestation campaigns.

- Emission of sounds before the passage of trains in order to move animals away from the track. Moving dead animals that attract birds, away from the tracks.
- The prevention of forest fires.
- Use of a camera on the train assisted by a computer program in order to detect weeds along the track.

19.8 DISTURBANCE OF LOCAL RESIDENT ACTIVITIES: ACCESS RESTRICTION AND DISRUPTION OF URBAN SPACE

19.8.1 Definition: measurement indices of disturbance on local resident activities

The permanent way of a railway system cuts off and isolates the area from where it passes. The isolation caused has impacts on the communication and movement of humans on either side of the railway track, resulting in the disturbance of several of their activities.

When a line passes through urban areas, the continuity of the urban space is disrupted. This results in a reduction in the accessibility of various destinations which may be of interest to both pedestrians and road vehicles.

In the case of the passage of the railway infrastructure through non-urban areas, the agricultural and livestock farming activities are obstructed. With regard to agricultural farming, the operation is literally split into two pieces. This constitutes an important issue, and there are many examples of cases where it has not been solved effectively. It usually comprises the main cause of discord between railway organisations and farmers.

The right-of-way, the length of the line, the segregation level (e.g., fencing and insufficient side road network) and the characteristics of the area themselves determine the volume of disturbance caused to the activities of local residents.

19.8.2 Railway activities causing disturbance to local resident activities

The disturbance of activities resulting from the presence of the railway is due to the fact that the railway uses an exclusive traffic corridor. The track fencing imposed on high and very high-speed networks, as well as on conventional-speed networks which, for part of their length, are integrated into urban and suburban areas, intensifies the problem.

Accessibility problems are also caused during the construction of urban railway systems.

19.8.3 Special features of each railway system category

With regard to metros, there is no disruption arising in the urban space because a metro moves under the ground. Its stations in fact increase, to a great extent, access to nearby areas, permitting urban regeneration. At the same time, concentrated development around the stations reverses the spatial dispersion of activities caused by the use of private cars.

On the other hand, a tramway, being a means of transport that moves at grade, is characterised by the permanent occupancy of a specific part of the road's width, and the necessity to receive priority against other means of transport at intersections. Therefore, the traffic capacity of the road is reduced, while the delay of road vehicles at intersections is increased. Its impacts on the various activities and access are directly linked to the category of the corridor used, and the way it integrates along the width of the road arteries.

19.8.4 Measures for the reduction of disturbance caused to local residential activities due to the presence of railway infrastructure

The maintenance of the residential structure of the surrounding area and the protection of the quality of life, mostly for the population located within a small distance from the track (0–500 m), constitute a primary concern towards the preservation of the cultural environment. Furthermore, the maintenance of agricultural activity in areas of high sensitivity, in terms of agricultural use, and of course the maintenance of continuity of archaeological sites, whether officially declared as such or not, are the main features that are taken into account in the process of the track alignment.

The measures used in order to reduce the disturbance of activities caused by the passage of a railway track are:

- *The construction of overpasses/underpasses at sufficient distance:* On the basis of the type or the functionality of the transport means which intersect at different levels, unlevelled crossings are divided into the following categories:
  - Railway–railway.
  - Railway–road.
  - Railway–pedestrian bridge (Figure 19.11).
  - Railway–animal passage (Figures 19.9, 19.10, and 19.12).

    Railway overpasses/underpasses have to be constructed at a specific location along the line, based on traffic flow and environmental study. In each case, they are required for the rehabilitation of the local road network, in order for movement and communication on either side of the track to run smoothly, where possible.
- *The construction of level crossings at sufficient distance:* Level crossings are part of lineside systems and constitute a basic structural and functional component of all conventional-speed railway networks (V < 200 km/h).
- Finally, during the construction of urban railway networks, and specifically of underground railway stations, traffic arrangements are required.

Figure 19.11 Pedestrian bridge, Neoi. Poroi, Greece. (Photo: A. Klonos.)

Figure 19.12 Animal underpass, TGV Sud-Est, Saint-Laurent-d'Andenay, France. (Adapted from Henri, M.

19.9 ACOUSTIC ANNOYANCE

19.9.1 Definition: units expressing acoustic annoyance

Acoustic annoyance is proportional to the acoustic energy received by the human ear. Acoustic energy is a function of the noise level and the duration of exposure to that noise.

Noise could be defined in two ways, as given below:

- Any irregular, non-periodic noise, the instant value of which fluctuates in a random way.
- Any unwanted noise.

Noise is directly linked to human health and well-being.

The energy indices commonly used for the assessment of railway noise, and its impacts on humans, are the equivalent energy noise level $L_{eq,T}$, the index L_{dn} (day–night equivalent noise level), the index L_{den} (day–evening–night equivalent noise level), the Sound Exposure Level (SEL), and the maximum noise level L_{max}. All these indices are measured in dB (A) (http://www.eea.europa.eu/data-and-maps/indicators/traffic-noise-exposure-and-annoyance/noise-term, 2001; Tsohos et al., 2001; UIC, 2010).

Table 19.11 displays the permissible noise levels for humans.

19.9.2 Railway activities causing acoustic annoyance

Depending on their source, five types of noise are distinguished during the operation of railway transport. These include:

The *mechanical noise* caused by the electromechanical equipment of the train, primarily the power vehicles. It originates from various sources, which are indicatively power vehicle motors, the braking system, air compressors, the air conditioning, etc. This

Table 19.11 Permissible noise levels for humans

Noise level (dB (A))	Annoyance description
≥ 81	Unacceptable situation
80–78	Very noisy situation
77–75	Noisy situation
74–72	Nearly bearable situation
71–69	Good situation
≤ 68	Comfortable situation

type of noise is perceived and gains more importance, mostly during movement at low speeds (30–40 km/h).

The *rolling noise* (Kitagawa, 2009) caused by the 'wheel–rail' system. This noise is generated during the contact of wheels with rails, and, secondarily, by car body oscillations. The causes for the generation of this noise are indicatively the discontinuities of the track (e.g., the presence of rail joints, switches, and crossings), the wheel flange–rail contact, the hunting of the railway wheelset due to geometric track defects, etc. (Schweizer Norm. SN 671 250a, 2002). In general, the rolling noise plays a significant role in medium speeds, where it comprises the basic source of noise. The rolling noise generated during the contact between wheel flange–inner rail edge (negotiation of bogies in curved horizontal alignment sections of very small radius as in metro, tram, and depot tracks) is particularly annoying and is called squeal noise.

The *aerodynamic noise*, which is prevalent at very high speeds (V ≥ 250 km/h), is perceptible at high speeds (200 km/h ≤ V < 250 km/h), while it is considered negligible at conventional speeds (V < 200 km/h). It is due to the increase of aerodynamic resistance during the movement of the train; it depends significantly on the aerodynamic shape of the head and the rear of the train, its cross-sectional surface, the condition of its lateral surface, its length, the number of vehicle bogies, the aerodynamic protection of pantographs, etc.

The *arcing noise*, which is caused by electrical traction, is generated in the event of discontinuity in the contact of pantographs with the aerial power supply cables. This noise is very similar to that generated by trolleybuses.

The *ground-borne noise*, which is perceived by the occupants of the buildings near the railway track as a very low-frequency noise. It originates from the vibrations caused during the movement of the railway vehicles, either in a surface track or in an underground track (inside a tunnel), and which are transferred through the ground to adjacent buildings. Ground-borne noise is examined in Section 19.10.

The *elevated structure noise*, which is attributed to railway systems operating on an elevated guideway.

Finally, the *construction site noise*, generated by the construction site machinery during the construction stage of a railway project.

19.9.3 Special features of each railway system category

Acoustic annoyance concerns all railway systems. Table 19.12 displays the correlation between the type of noise, as defined in Section 19.9.2, and the various categories of railway systems.

The noise emitted during the passage of a very high-speed train increases by up to 300 km/h as a function of the third power of speed, while for speeds higher than 300 km/h, it increases as a function of the sixth to eighth power of speed (Relié, 1989; Maeda et al., 2010).

Table 19.12 Correlation between the type of noise and the various categories of railway systems

	High and very high-speed railway	Conventional-speed railway	Metro	Tramway
Mechanical noise	+	+	+	+
Rolling noise	+	+++	+++	+++
Squeal noise		+	++	+++
Aerodynamic noise	+++	+		
Arcing noise	+++	+ (electrification)	++	++
Ground-borne noise	+	+	+++	++
Construction site noise	+	+	+++	+++

⁺: Correlation.
⁺⁺: Strong correlation.
⁺⁺⁺: Very strong correlation.

Table 19.13 Measurement results of rolling noise of high and very high-speed trains in Europe (UIC)

Passage speed (km/h)	Rolling noise values measured at a distance of 25 m (dB (A))
250	86–90
300	89–92
320	91–93
350	94.5–97

Source: Adapted from Eadie, D.T., Kalousekb, J., and Chiddicka, K. C., 2002, Wear, 253(1–2), 185–192.

Noise limits in the high and very high-speed railway are clearly defined in the technical specifications for interoperability of the trans-European high-speed railway system (EC, 2008). According to these specifications, the external noise emitted by the rolling stock of the high-speed railway system is divided into noise at stop, start-up noise, and rolling noise.

Table 19.13 displays the measurement results of high and very high-speed trains' rolling noise in Europe (UIC).

An increase in the roughness of wheels may induce an increase of noise by 3–4 dB (A) (Eadie et al., 2002).

The chart of Figure 19.13 displays the relationship between the running speed of the tram and the noise caused.

Finally, with regard to freight trains, according to Kurer, a typical train formation emits 90 dB (A) (ECMT, 1993).

19.9.4 Countermeasures against acoustic annoyance

For the reduction of acoustic annoyance caused by railways, countermeasures are taken along the path of noise transmission and at the source of the noise. Almost all measures used at the source of noise are also used towards the reduction of ground-borne vibrations and noise (see Section 19.10).

19.9.4.1 The path of noise transmission

The measures used towards the reduction of railway noise along the path of its transmission are noise barriers (Figure 19.14) and fencing with planting (Figure 19.15). These measures

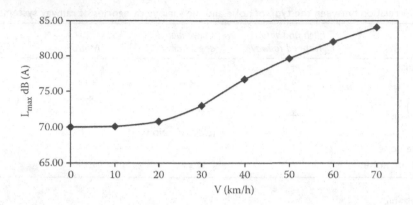

Figure 19.13 Relationship between external noise of the tram and running speed. (Adapted from Oikonomidis, D., Triantafyllopoulos, P., and Paidousi, M., 2003, Noise protection programme during the operation of the tram of Athens, *International Conference Contemporary Tram and LRT Systems*, 19–20 May 2003, Patras.)

Figure 19.14 Noise barriers made of reinforced concrete, TGV Atlantique, France. (Adapted from Olivain, P., 1989, SNCF Médiathèque.)

(overall noise-reduction potential 5–15dB; UIC, 2010) are effective only at a local level, since for the protection of longer sections of the railway networks, high investment is required.

The level of acoustic annoyance for the railway is higher compared with that of a road network; therefore, the construction material of the noise barriers should be characterised by higher sound-insulation performance, compared with that of a road network. Specifically, pursuant to regulation SN 671 250a, the noise barriers in the railway should ensure the reduction of the railway noise, at least by 10–15 dB.

Table 19.14 displays the values of the noise-reduction index R_s (an index which expresses the sound-insulating capacity of the construction material of noise barriers, having dB as the unit of measure), for specific frequencies.

Figure 19.15 Noise barriers made of concrete with planting bays. (Adapted from Online image, available at: http://www.tucrail.be/FR/media/Site_of_the_month/PublishingImages/2013/chantier_du_mois _06_2013_3.jpg (accessed 30 April 2015).)

Table 19.14 Values of noise-reduction index R_s for various frequencies

Frequency (Hz)	100	125	160	200	250	315	400	500	630
R_s (dB)	10	12	14	16	18	20	22	24	26
Frequency (Hz)	800	1,000	1,250	1,600	2,000	2,500	3,150	4,000	5,000
R_s (dB)	28	30	31	32	33	34	35	36	36

Source: Adapted from Schweizer Norm. SN 671 250a., 2002, *Schweizerischer Verband der Strassen – und Verkehrsfachleute* (VSS), May 2002.

Table 19.15 Values of sound-absorption coefficient α_s for specific frequencies

Frequency (Hz)	100	125	160	200	250	315	400	500	630
α_s	0.10	0.15	0.20	0.25	0.30	0.45	0.60	0.75	0.80
Frequency (Hz)	800	1,000	1,250	1,600	2,000	2,500	3,150	4,000	5,000
α_s	0.85	0.90	0.90	0.90	0.90	0.90	0.90	0.90	0.90

Source: Adapted from Schweizer Norm. SN 671 250a., 2002, *Schweizerischer Verband der Strassen – und Verkehrsfachleute* (VSS), May 2002.

To improve the efficiency of noise barriers on both sides, and for a height ≥ 2 m, adequate sound absorption should be ensured. Sound absorption, measured by coefficient α_s (sound-absorption coefficient), should, at least and at specific frequencies, take the values that are displayed in Table 19.15.

The maximum height distance between the noise barriers and the rolling surface of the rails is commonly 2 m, in order for the landscape aesthetics as well as the attractiveness of the adjacent inhabited areas and the visibility of the train passengers to be preserved (Bontinck, 1997).

When the height of 2 m is not sufficient for adequate sound protection from railway noise, noise insulated windows are placed in the affected buildings (overall noise-reduction potential 10–30 dB; UIC, 2010). The placement of even higher noise barriers is selected only when there are serious reasons.

Low noise barriers at a close distance from the tracks are avoided due to bad acoustic and low safety level. However, they are the last resort, in case there is no other alternative.

19.9.4.2 The source of noise

Measures taken at the source can reduce the noise throughout the railway system, provided that they are applied widely. Such measures are:

- *The use of brakes made of composite materials*: The replacement of wheel pads made of cast iron with wheel pads made of composite material (retrofitting with K-blocks or LL-brake blocks) ensures that the rolling surfaces of the wheels remain smooth; this results in a reduction of the rolling noise of up to 10 dB (UIC, 2010).
- *The use of resilient wheels* (Figure 19.16): Thanks to their special configuration, they reduce the ground-induced vibration through the reduction of unsprung masses and the noise through increased wheel damping. An additional beneficial effect is that resilient wheels absorb more roughness excitation than a monobloc wheel (because of the lower radial stiffness) and thus reduce the noise radiated by the rail. It has been observed that by using resilient wheels, the noise level is reduced by 5–6 dB. This technique is used mostly in tram vehicles (obviously resilient wheel could reduce even 10–15 dB of squealing noise).
- *Rail vibration dampers.*
- *The use of exclusively continuous welded rails and elastic fastenings.*
- *The aerodynamic design of pantographs.*
- *The absence of aerial wires.*
- *The proper maintenance of the rolling stock (reprofiling of wheels with wheel flats) and of the track (rail roughness can be kept low through acoustic grinding).*
- *Wheel dampers* (Betgen et al., 2012).

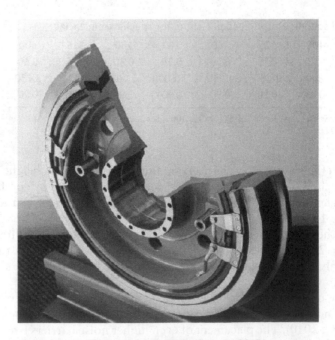

Figure 19.16 Resilient wheels. (From BONATRANS GROUP, A. S., 2015.)

19.10 GROUND-BORNE NOISE AND VIBRATIONS

19.10.1 Definition: measurement units of ground-borne noise and vibrations

One of the most important environmental impacts of railway systems, and specifically of urban railway systems moving on underground track sections, are the vibrations caused by the train movement which are borne and transmitted through the ground to its surface and the overlooking adjacent buildings (Figure 19.17).

The ground-borne vibrations are perceptible by the residents through the vibrations of the floors and the walls of the buildings. They can be annoying, since they can cause not only the movement of various objects inside the buildings, but also a secondary noise that originates mostly from the screeching of glass panels and cooking utensils. Such noise (ground-borne noise) is caused by vibrations in frequencies usually in the range of 40–80 Hz, because of the resonance of the building constructions in the tunnel vicinity. It becomes perceptible as a low-level muffled noise (rumble).

Given that the minimum typical limit of the human acoustic capacity is about 20 Hz, vibrations under this frequency become perceptible by the residents of the buildings as vibrations (ground-borne vibrations), while over this frequency they become perceptible both as vibrations and as sound.

Inside the railway vehicles, vibrations reduce the vertical dynamic comfort of passengers.

The size of the vibrations can be measured based on the:

- Displacement d_o (mm).
- Velocity v_o of the vibrations (mm/s).
- Acceleration α_o of the vibrations (mm/s²).

Vibrations are primarily expressed in the form of velocity or acceleration. They can, however, be expressed also in the form of noise levels, in dB.

The following conversion relationships apply:

$$L_{\alpha o(level)} = 20 \log_{10}(\alpha_o/\alpha_o')(dB) \tag{19.2}$$

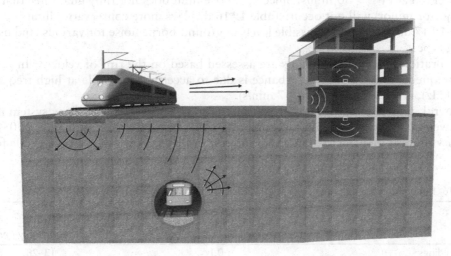

Figure 19.17 Transmission of vibrations in the area around railway lines. (Adapted from Loy, H., 2014, available online at: http://innorail.hu/en/vibration-mitigation-with-under-sleeper-pads/.)

where:

α$_o$: Acceleration of vibration in m/s^2.

α′$_o$: Reference level of the acceleration of vibration, 10^{-6} m/s^2 or 10^{-5} m/s^2.

The product $20\log_{10}(\alpha_o/\alpha'_o)$ (dB) is defined as VAL (Vibration Acceleration Level).

$$L_{v(level)} = 20\,\log_{10}\left(v_o\,/\,v'_o\right)(dB) \tag{19.3}$$

where:

v$_o$: Velocity of vibration.

v′$_o$: Reference level of the velocity of vibration, 10^{-9} m/s.

The product $20\log_{10}(v_o/v'_o)$ (dB) is defined as VVL (Vibration Velocity Level).

$$L_{d(level)} = 20\,\log_{10}\left(d_o\,/\,d'_o\right)(dB) \tag{19.4}$$

where:

d$_o$: Displacement.

d′$_o$: Reference displacement, 10^{-11} m.

An index used for the description of the amplitude of vibrations (as well as sound waves) is the root mean square (rms) of vibration velocity or acceleration (rms). It results from the summary of the square values of velocity or acceleration at each instant, and the calculation of the mean duration of a period. The rms of the velocity or acceleration is the square root of this mean value of time.

The impacts of vibrations on damage caused to the constructions are estimated through comparisons based on Peak Particle Velocity (PPV). In general, the vibration levels required to cause damage to the buildings are much higher than those considered acceptable by humans themselves (dynamic comfort).

The basic safety limit for avoiding damage is that the PPV value should not exceed 50 mm/s. The total of major damage to the buildings, as well as 94% of smaller damage, has been observed at PPV > 50 mm/s. Since the above limit does not fully guarantee that damage (major or minor) will not occur, Table 19.16 displays more conservative limits.

Table 19.17 displays the permissible levels of ground-borne noise for various land use and building types.

The vibration impacts on humans are assessed based on the rms of velocity. In general, at low frequencies (1–10 Hz) disturbance is due to acceleration, while at high frequencies (10–100 Hz), it is due to velocity (< 7 mm/s).

In general, the range of vibration frequencies for an underground railway system fluctuates between 0 Hz and 200 Hz (Eisenmann, 1994; Esveld, 2001; Lichtberger, 2005). The range of vibration frequencies that are of interest inside tunnels is 0–500 Hz, while for the

Table 19.16 Recommended maximum vibration velocity limits during construction

	Recommended vibration limits	
	Weighted acceleration (mm/s^2)	Equivalent velocity (mm/s)
Other buildings	0.1	13–28
Special-use buildings and monuments	0.05	1.3

Table 19.17 Permissible levels of ground-borne noise

| | Ground-borne noise (dB (A)) | | |
Space type	Detached house	Block of flats	Hotel
Low population density	30	35	40
Average residential density	35	40	45
High residential density	35	40	45
Commerce	40	45	50
Industry-intercity network	40	45	55

Source: Adapted from Vogiatzis, K., 2011, *Information Bulleting of the Hellenic Association of Transportation Engineers*, Issue No 176, April–May–June 2011.

ground surface, and therefore for the buildings, is 1–80 Hz. Vibrations in buildings usually peak at 50–63 Hz.

The critical condition emerges when the primary vibration frequency of the ground coincides with the natural frequency of the building. While the primary transversal vibration frequency of the building is 1–10 Hz, its individual structural elements can have higher natural frequencies.

The critical distance between tunnels and buildings, for ground-borne vibrations and for a ballasted track, fluctuates between 15 m and 25 m (Eisenmann, 1994).

Ground-borne noise in buildings should be measured in the middle of a room. The ground-borne and the secondary noise (rattling) should not be included in the measurements. It is expressed by the Sound Pressure Level (SPL) and is measured in dB (A) (Equation 19.5) (Eitzenberger, 2008). The frequency range that is of interest is between 16 Hz and 250 Hz:

$$L_p = 20 \log_{10}\left(p_o / p_o'\right)(dB) \tag{19.5}$$

where:

p_o: The mean noise pressure in N/m^2.

p_o': The relative mean reference pressure, usually equal to 2×10^{-5} N/m^2 for transmission through the air and 0.1 N/m^2 for any other means of transmission.

The ground-borne noise can be estimated if the ground-borne vibrations are known.

19.10.2 Railway activities causing and affecting ground-borne noise and vibrations

Ground-borne noise and vibrations are caused essentially during train movement at constant speed and acceleration.

The parameters that affect the transmission of vibrations and the noise caused through them are:

In terms of the source of a noise (wheel–rail interface, track superstructure, and rolling stock) (Nelson, 1996; Cox, 2002; Busch et al., 2005; Fujii et al., 2005):
1. *The train running speed*: For speed between 25 km/h and 115 km/h, an increase in the speed by 100% results in an increase of index VAL by 4–6 dB.
2. *The train axle load*: Doubling the axle load results, on tunnel level, in an increase of index VAL by 2–4 dB, in a frequency region between 40 Hz and 250 Hz.
3. *The sleepers.*

4. *The train type and length*: Trains with vehicles of uniform load cause higher vibration levels.

5. *The ballast thickness (in case of ballasted track)*: A fluctuation of ballast thickness from 30 cm to 70 cm under the sleeper does not have any effect.

   A slab track, compared with a ballasted track, has a higher noise level. Specifically, it reflects the rolling noise instead of absorbing it. It is estimated that the level of the emitted noise in the case of a slab track (without resilient fastenings and elastic pads) is 10 dB (A) higher than that of a ballasted track.

6. *The wheel–rail contact surface quality (conditions and quality of rolling)*: The roughness of wheel–rail comprises the major factor that stimulates the rolling noise. Increasing the roughness alone results in an increase of index VAL by 3–10 dB.

   According to measurements performed in Germany, France, and Japan, eliminating the wave corrugations of the rails by grinding may result in the reduction of the rolling noise by 6–14 dB, within a frequency range of 500–2,000 Hz. Measurements performed in the metro of Naples have shown that eliminating the wave corrugations at frequencies between 0.8 Hz and 80 Hz results in a reduction of index VAL, on the level of tunnel walls, by 9 dB, while at frequencies between 20 Hz and 250 Hz, by 17 dB.

   The elimination of wheel flats may lead, as recorded by measurements, to noise reduction by 3 dB (A).

   Bad rolling conditions in total (loose fastenings, wave corrugations of rails, etc.) may cause an increase of index VAL by 10–20 dB.

   The presence of Continuous Welded Rails (CWR) reduces the vibration levels, as it eliminates discontinuities on the rolling surface. An increase of index VAL by 5 dB has been recorded, compared with the jointed track.

7. *The braking system of the vehicles*: Vehicles with disc brakes cause less noise than those with cast-iron pads.

8. *The stiffness of the primary suspension of vehicles*, and their dynamic behaviour in general, as well as the parameters affecting it.

   The low values of vertical stiffness of the primary vehicle suspension system reduce vibrations. Indicatively, it is stated that in the event of vehicles with a stiff suspension, vibration increases of up to 10–15 dB were measured.

In terms of the path (tunnel walls, soil, and earthworks):

1. *Construction of the tunnel's bearing structure*: Doubling the width of the tunnel ring results in the reduction of index VAL (inside the tunnel) by 5–18 dB.

   The vibration levels inside the tunnel, in case the tunnel is constructed within rock, are higher by 12 dB (in high, audible frequencies), and lower by 5 dB (in low frequencies), than in the case where the tunnel is not constructed within a rock.

   The cut-and-cover sections (with a large cross section, close to the ground surface) generate higher noise levels than the sections constructed in tunnels with circular cross section.

In terms of the final receptor (buildings and humans):

1. *The foundation system of buildings*: In the case of foundation with slabs on grade, the reduction for frequencies higher than the natural frequency of the concrete slab is zero. The same applies to light constructions and buildings founded directly on a rock.

   In the case of foundation of a building on pillars, the vibration levels are higher down to the ground, close to the pillars, than on the ground surface.

For other cases of foundation, there is a reduction by 2–15 dB, depending on the foundation system and frequency of vibration.

- *The distance of the building from the tunnel*: Space waves attenuate by a rate of 6 dB (50% in amplitude) as the distance of the source of vibration doubles, without the presence of a damping material in the ground.
- *The depth of the tunnel*: Underground constructions close to the ground surface generate higher vibration levels than deep underground constructions.
- *The properties of the soil and the rock mass*: The velocity of wave propagation is higher in a dense and solid rock and lower in a less dense and solid one. In general, in wetlands with muddy soils, the vibrations attenuate fast, in sandy soils less fast, while in contrast, high vibrations are observed in hard soils near the track.
- *The construction material of the building's frame*: The generated vibrations affect more the constructions that are made of light wood and of metal.
- The number of floors of the buildings.

The acceleration level of vibrations is reduced as we ascend to the upper floors of a building. This attenuation fluctuates around 3 dB and is higher at high frequencies.

19.10.3 Special features of each railway system category

Ground vibrations caused by the passage of a high-speed train are similar to those caused by the passage of a conventional-speed train. The vibration level, however, is relatively lower, due to the high standards observed during the construction, operation, and maintenance of a high-speed railway track (Nelson, 1996).

The values of vibration of high-speed trains (V = 240 km/h) and in a distance of 30.5 m were estimated in the range of 75–80 V dB (VVL).

In trams, the accumulation of debris on the track superstructure surface causes vibrations of a fairly high-frequency range (10–200 Hz).

19.10.4 Countermeasures against vibrations and ground-borne noise

The impacts of ground-borne noise and vibrations are limited when:

- They are restricted to the initial vibrations.
- The range of frequencies within which they are generated is far from the natural frequency of the system which they affect (the building frame, the track bed system, etc.). Solutions that guarantee frequencies at least $\sqrt{2}$ times the natural frequency of the system are considered effective. Indicatively, it is stated that the floating slab systems (mass spring systems) act as a barrier against vibrations, with a frequency that is $\sqrt{2}$ times higher than the natural frequency.

In any case, the measures taken at the source, and specifically at the railway track, are the most economical and effective.

One of the basic principles applied is the increase of the mass of the track which is supported elastically. In this way, the natural frequency is reduced, and thus the efficiency of the system is increased, in terms of insulation against vibrations.

The technical solutions that can be applied for the reduction of vibrations and ground-borne noise are (Wettschureck, 1995, 2002; Eadie et al., 2002; Liu, 2005; Vogiatzis, 2006; Eitzenberger, 2008; Kuo et al., 2008) given below.

In all track bed cases:

- Elimination of the geometrical track defects.
- Elimination, mainly, of the wave corrugations of rails (by grinding the rails).
- Use of high-resilience pads and fastenings, for the absorption of vibrations.
- Elimination of wheel flattening.
- Use of resilient wheels (Figure 19.16).
- The increase of the width of the tunnel ring.
- The increase of the mass of the track superstructure supporting system (mass of concrete slab or sleepers).
- Wheel slide protection systems.
- Lubrication of the inner side of the rails in curved sections of the horizontal alignment that have a small radius (reduction of squeal noise).
- The increase of the strength of the rail and the use of heavier rails.
- Vehicle bogies with low vertical stiffness springs on the level of the primary suspension vehicles with light unsprung and semi-sprung masses.
- Placement of vibration-damping materials in the rail web (by welding) (web dampers). With this technique, a reduction of vibrations (of VAL) up to 2–5 dB can be attained.
- Continuous welded rails and movable point frogs on track crossings.
- Placement of switches and crossings, where possible, away from land use that is sensitive to vibrations.
- Reduction of the train passage speed.

In cases of ballasted track:

- Maintaining the ballast thickness, in all cases, to more than 30 cm.
- Interference of layers of resilient or elastomer materials under the ballast layers (sub-ballast mats).
- Placement of elastic pads between the sleepers and the ballast. A reduction of VAL by 15 dB at 125 Hz was recorded.
- Foamed ballast. This refers to stabilising the ballast, by using special polyurethane foam, towards the increase of its durability.
- High vibration-damping sleepers.
- High-width sleepers.

In cases of slab track:
  Two techniques may be distinguished, namely:
- *The solution of floating slab*: This technique was described in Section 5.3.2.
- Other solutions that attach a more resilient vertical behaviour to the superstructure and its support on the concrete slab. Such solutions (Nelson, 1996; Eadie et al., 2002; Cox, 2003; Vanhonacker and Leuven, 2005) are the following:
  - Resiliently supported twin-block sleepers.
  - Resilient direct fixing fastenings of rails.
  - Resilient pre-load APT-ST direct fixing system.
  - Super-resilient pre-load APT-BF direct fixing system of rails (Vanhonacker and Leuven, 2005).

19.11 IMPACTS ON LAND USE

Railway systems, like all other transport systems, ensure access conditions among various land uses and serve the mobility needs arising from the interaction of the activities taking place within them. This way the railway systems influence:

- The location of land uses in relation to proximity access to the transport network.
- The composition of land uses in the sense of relationships and interactions that are created through the links provided by the transport network.
- Land value, which is formed by the combined effect of accessibility and composition of land use.

The type and characteristics of the effect on the location and the composition of land uses and land values depend on the operational characteristics of the transport system. Thus, at the level of urban development, the areas around the main stations of the urban railway transport modes are developed featuring high density and mixed land use, combining housing, services, trade, and recreation, while the regions that are mainly served by the private car showcase trends towards suburbanisation, urban sprawl and zoning of land uses, such as industrial zones and major shopping centres (Newman and Kenworthy, 1999). On the contrary, the proximity of households or businesses to train stations may be considered a competitive advantage leading to increased land values (Debrezion et al., 2006).

19.12 COMPARATIVE ASSESSMENT OF THE IMPACTS OF VARIOUS MEANS OF TRANSPORT ON THE NATURAL ENVIRONMENT

19.12.1 Methodology approach

In order for a rational comparison to be made, between railway systems and other means of transport, it should relate to transport systems of similar functionality and performance, that is, systems that can be considered as competitive, and provide roughly the same level of service to their users.

In this context, the following comparisons take place in this section:

- For very long-distance trips (500–1,500 km), the aeroplane and the high-speed railway are compared with each other.
- For urban transport, the metro, the tram, the private cars, and the urban bus are compared with each other.
- For very high-speed transport, the aeroplane, the very high-speed train, and the magnetic levitation train are compared with each other.
- For the modes involved in freight transport, the freight train and the road truck are compared with each other.

19.12.2 Long distances: comparison between the aeroplane and the high-speed train

On the basis of a study conducted in 2010 by JR Central on the Tokyo–Osaka line, with a length of 515 km, the high-speed train (Shinkansen Series N700 'Nozomi') consumes about 1/8 of the amount of energy consumed per passenger/seat by an aeroplane (B777-200) and emits about 1/12 of the CO_2 emissions of the aeroplane per passenger/seat (Garcia, 2010a).

In a study conducted by the Institute for Energy and Environmental Research (IFEU), a comparison was performed of the Frankfurt–Hannover connection, between aeroplanes and high-speed trains, in terms of emission of pollutants. Although in Germany more than 50% of the required energy for train operation is generated by fossil fuels, a high-speed train, because of its high efficiency, emits much less pollutants, as is demonstrated in Table 19.18 (Kettner, 2008).

From Table 19.18 it is concluded that, for the specific connection, the high-speed train emits 4 times less CO_2 and 16 times less NO_x per passenger.

Table 19.18 Emission of pollutants in the Frankfurt–Hannover line

Data per passenger	Aeroplane	High-speed train
Energy consumption (equivalent in petrol litres)	32.8	11.1
Carbon dioxide (kg)	77.1	19.2
Greenhouse gases		
Global warming		
Particulate particles (gr)	2.1	1.0
Sulphur dioxide (gr)	43.4	19.5
Nitrogen oxides (gr)	268.3	17.2
Hydrocarbons, excluding methane (gr)	20.8	1.1

Source: Adapted from Kettner, J., 2008, *Future Challenges of Transport and Environment: The Role of Railways to Reduce Climate Gas Emission*, Deutsche Bahn AG.

Table 19.19 Summary of very long-distance trips

Comparison parameter	Comparison quantity	Relationship between high-speed train/aeroplane
Carbon dioxide	g/pkm	1/7
Nitrogen oxides	g/pkm	1/9
Hydrocarbons	g/pkm	1/18
Energy consumption	Equivalent fuel in L/pkm	1/4

Parameters of comparison and ratio.

Table 19.19 displays the relationship of the most significant ambient air pollutants for a high-speed train and an aeroplane. These results refer to the implementation of eight connections, which were investigated using computer software available at www.ecopassenger .org. For this calculation, the following assumptions were made:

- Regarding the train, it was assumed to be full of passengers.
- Regarding the aeroplane, the gas emissions of the modes that are used for the transfer from and to the airport were also taken into account.

As it is also demonstrated in Figure 19.18, the noise caused by an aeroplane is the most annoying, followed by that of a road, and, finally, the noise from a railway system. More specifically, from the people exposed to a noise level of 55 dB, which is critical, according to the World Health Organization (WHO), 30% considered aeroplane noise as annoying, 20% considered the noise produced by road means of transport as annoying, while only 10% were annoyed by the noise of the railway (Den Boer and Schroten, 2007).

The data held by the EU are consistent with the aforementioned, since it is concluded that the highest percentage of disturbance is derived by the noise of aeroplanes, at 25%, whereas the lowest percentage is assigned to railways, at 10%.

The noise caused by high-speed trains is a short-term annoyance, which affects the environment locally, when the train crosses a specific area. In contrast, the noise caused by aeroplanes lasts longer, since high-noise levels are generated in the areas around airports, both during landing and takeoff, and when an aeroplane flies at a low altitude.

19.12.3 Urban transport: comparison of the metro, the tram, the urban bus, and the private car

The bar chart of Figure 19.19 graphically displays the distance (in km) travelled by various means of urban transport, based on the consumption of 1 kgr of fuel equivalent.

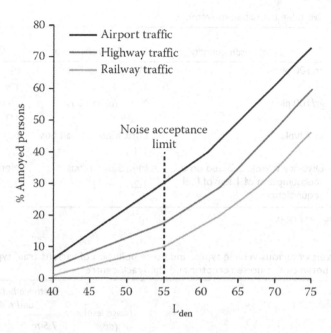

Figure 19.18 Percentage of annoyed persons from the traffic noise caused by various means of transport. (Adapted from Den Boer, L. C., and Schroten, A., 2007, Traffic noise reduction in Europe, *CE Delft Report*, 2007.)

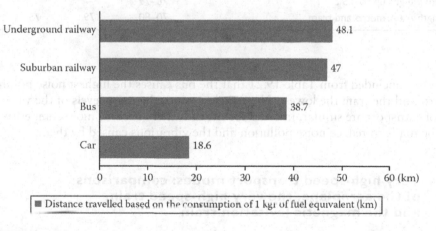

Figure 19.19 Energy efficiency of urban means of transport.

As demonstrated in Figure 19.19, the suburban railway and metro cover almost the same distance, over twice (2.5 times) that of the private car and 1.2 times that of the urban bus.

In the United States, a light metro vehicle with 55 passengers is estimated to require 0.19 kW h/pkm, an urban bus with 45 passengers consumes 0.2 kW h/pkm, and a private car with a driver and three passengers requires 0.33 kW h/pkm. In contrast, a private car carrying only his driver consumes about 1.34 kW h/pkm.

Table 19.20 displays the parameters compared for the four means of urban transport, and the relationship among them, for each of these parameters.

Table 19.21 shows the noise level of various vehicle types and the noise pollution of various train types, for a specific distance between the track centre and the receptor.

Table 19.20 Comparative table for urban movements

Comparison parameter	Comparison quantity	Relationship among	Means of transport
Nitrogen dioxide (NO$_2$)	gr/100 pkm	Metro/urban bus 1/3	Metro/private car 1/4
Carbon monoxide (CO)	gr/100 pkm	Metro/urban bus 1/189	Metro/private car 1/311
Energy consumption	kW h/pkm	Light metro/urban bus 1/1.1	Light metro/private car 1/7
Energy consumption	Distance travelled, based on the consumption of 1 kgr of fuel equivalent	Metro/urban bus 1.2/1	Metro/private car 2.5/1

Parameters of comparison and ratio.

Table 19.21 Noise levels of various vehicle types, and noise pollution of various train types, for a specific distance between the noise receptor and the track centre

Vehicle type	Noise level (dB)	Distance between noise receptor and railway track axis		
		7.5 m	15 m	25 m
Passenger cars with a maximum capacity of five seats	77	–		
Passenger transport vehicles (buses) with more than nine seats and weighing more than 3.5 t	80–83	–		
Buses with weight up to 3.5 t	78–79	–		
Short-length train, metro, and tram	70–80	79	75	72

It can be concluded from Table 19.21 that the bus causes the highest noise pollution, and the metro and the tram the lowest. Nevertheless, since the noise levels of the various urban means of transport are similar, in order for the railway to become more competitive, efforts should be made to reduce noise pollution and the vibrations caused by them.

19.12.4 Very high-speed transport modes: comparisons of the aeroplane, the very high-speed train, and the magnetic levitation train

With regard to the comparison of very high-speed means of transport, the comparison results are displayed in Table 19.22.

19.12.5 Freight transport: comparison of freight trains and road trucks

Table 19.23 displays the comparison parameters, as well as the comparison results.

Table 19.23 shows that a freight train has 4.7 times less carbon dioxide emissions than the equivalent number of road trucks. Furthermore, it consumes 3.3 times less energy.

In freight transport, the value of CO_2 emissions for the railway is 24 gr/tkm, while for road trucks, the respective value is 88 gr/tkm. Compared with the road truck, the CO_2 emissions of the railway is almost four times less.

Table 19.22 Comparative table for very high-speed movements

Comparison parameter	Comparison quantity	Relationship among means of transport		
Carbon dioxide		Magnetic levitation train/aeroplane 1/5.5	Magnetic levitation train/ very high-speed train 1/1.5	Very high-speed train/aeroplane 1/3.7
Energy consumption	BTU/passenger-mile	Magnetic levitation train/aeroplane 1/2.25	Magnetic levitation train/ very high-speed train 1/1.39	Very high-speed train/aeroplane 1/1.62
Noise pollution	dB (A)	Magnetic levitation train/aeroplane 1/1.2	Magnetic levitation train/ very high-speed train 1/1.125	Very high-speed train/aeroplane 1/1.06

Parameters of comparison and ratio.

Table 19.23 Comparative table for freight movements

Comparison parameters	Comparison quantity	Relationship of freight train/road truck
Carbon dioxide	100 tkm	1/4.7
Nitrogen oxides	100 tkm	1/19.3
Energy consumption	kW h/100 tkm	1/3.3

Parameters of comparison and ratio.

REFERENCES

Betgen, B., Bouvet, P., Thompson, D.J., Demilly, F. and Gerlach T. 2012, Assessment of the efficiency of railway wheel dampers using laboratory methods within the STARDAMP project, *Proceedings of the Acoustics 2012 Nantes Conference*, 23–27 April 2012, Nantes, France.

Bontinck, W. 1997, Les protections antibruit le long de la ligne a grand vitesse Belge, *Rail International*, Brussels, April, pp. 32–35.

Busch, T., Gendreau, M. and Amick, H. 2005, Vibration and noise criteria used to evaluate environmental impacts of transportation projects on sensitive facilities, *Proceedings of SPIE Conference 5933: Buildings for Nano Scale Research and Beyond, San Diego, CA*, 31 July–1 August 2005, San Diego, CA.

Carbon Trust. 2012, *Carbon Footprinting*, UK, March 2012, Carbon Trust, London.

CNT. 2006, *High Speed Rail and Green House Gas Emissions in the U.S.*, Center for Neighborhood Technology, Center for Clean Air Policy, January 2006.

Cox, S. 2002, Pandrol VANGUARD track fastening system. Reducing track-induced vibration levels in tunnels, *Rail Engineering International*, Vol. 2002(2), pp. 7–10.

Cox, S. 2003, Installation and testing of Pandrol VANGUARD base plates on MTRC, Hong Kong, *Proceedings of the International Conference and Exhibition Railway Engineering 2003*, 30 April–1 May 2003, London, UK.

D'Angelo, J.J. 1995, SNCF Médiathèque, Paris.

Debrezion, G., Pels, E. and Rietveld, P. 2006, The impact of rail transport on real estate prices: An empirical analysis of the Dutch housing markets, *Tinbergen Institute Discussion Paper*, Vol. 31(3), pp. 1–24.

Den Boer, L.C. and Schroten, A. 2007, Traffic noise reduction in Europe, *CE Delft Report*, 2007.

Donta, E. 2010, *European Legal Framework for Ground Protection*, dissertation thesis, Department of Biology, Aristotle University of Thessaloniki, Thessaloniki, Greece.

Eadie, D.T., Kalousekb, J. and Chiddicka, K.C. 2002, The role of high positive friction (HPF) modifier in the control of short pitch corrugations and related phenomena, *Wear*, Vol. 253(1–2), pp. 185–192.

EC. 2008, Commission decision of 21 February 2008 concerning a technical specification for interoperability relating to the "rolling stock" subsystem of the trans-European high speed rail system (notified under document number E).

ECMT. 1993, Transport growth in question, *12th International Symposium on Theory and Practice in Transport Economics*, Lisbon, Portugal, 1992.

Eisenmann, J. 1994, Vibrations and structure-borne noise by underground railways, *Rail Engineering International*, Vol. 1994(4), pp. 6–8.

Eitzenberger, A. 2008, *Train-Induced Vibrations in Tunnels – A Review*, Lulea University of Technology, Department of Civil, Mining and Environmental Engineering, Division of Mining and Geotechnical Engineering, Sweden, Lulea.

Esveld, C. 2001, *Modern Railway Track, Book*, MRT-Productions, West Germany, Duisburg.

Europa. 2012, Rail transport, *Summaries of EU Legislation*, available online at: http://europa.eu/legislation_summaries/transport/rail_transport/tr0006 (accessed 20 May 2012).

Fujii, K., Takei, Y. and Tsuno, K. 2005, Propagation properties of train induced vibration from tunnels, *Quarterly Review of RTRI*, Vol. 46(3), pp. 194–199.

Garcia, A. 2010a, *High Speed, Energy Consumption and Emissions*, UIC, 21 December 2010, Paris.

Garcia, A. 2010b, Energy consumption and emissions of high speed trains, *Transportation Research Record*, Vol. 2159, pp. 27–35.

Gifford, J. 2011, *High-Speed Rail Line Installation Powers First Solar Train*, available online at: http://www.pv-magazine.com/news/details/beitrag/high-speed-rail-line-installation-powers-first-solar-train-_100003253/#axzz3Xy0A1P8R (accessed 30 April 2015).

Henri, M. 1980, SNCF Médiathèque, available online at: http://en.wikipedia.org/wiki/Wildlife_crossing, 2015 (accessed 12 June 2016).

Hirabaya, Y., Hitachi, Japan, online image, available online at: https://www.monorails.org/tMspages/CnstSentosa01a.html) (accessed 25 February 2021).

http://www.eea.europa.eu/data-and-maps/indicators/traffic-noise-exposure-and-annoyance/noise-term-2001 (accessed 12 June 2016).

JR Central. 2010, *Central Japan Railway Company Environmental Report 2010*. JR Central, Tokyo.

Kettner, J. 2008, *Future Challenges of Transport and Environment: The Role of Railways to Reduce Climate Gas Emission*, Deutsche Bahn AG, Germany.

Kitagawa, T. 2009, The influence of wheel and track parameters on rolling noise, *Quarterly Review of RTRI*, Vol. 50(1), pp. 32–38.

Kuo, C., Huang, C. and Chen, Y. 2008, Vibration characteristics of floating slab track, *Journal of Sound and Vibration*, Vol. 317, pp. 1017–1034.

Lagarias, A. 2015, *The Aesthetic Integration of Railway Systems in Urban Space: A Methodology for Evaluating Visual Nuisance*, Thesis of final studies, Civil Engineering Department, Aristotle University of Thessaloniki, Greece.

Lichtberger, B. 2005, *Track Compendium, Formation, Permanent Way, Maintenance, Economics, Book*, Eurail Press, Hamburg, Germany.

Liu, L. 2005, Pandrol VANGUARD track fastening system installed on Guangzhou metro, China, *Rail Engineering International*, Vol. 34(4), pp. 7–10.

Loy, H. 2014, Available online at: http://innorail.hu/en/vibration-mitigation-with-under-sleeper-pads/ (accessed 30 April 2015).

Maeda, T., Gautier, P.-E., Hanson, C., Hemsworth, B., Nelson, J.T., Schulte-Werning, B., Thompson, D. and Vos, P. 2010, Noise and vibration mitigation for rail transportation systems: Recent studies on aerodynamic noise reduction at RTRI, *Proceedings of the 10th International Workshop on Railway Noise*, 18–22 October 2010, Nagahama, Japan.

Maropoulou, E. 2009, *Investigation of the Environmental Impacts for the Thessaloniki-Eidomeni Railway Line*, Alexandreio Technological Educational Institution, Department of Civil Infrastructure Engineers, Greece.

Nelson, J. 1996, Recent developments in ground-borne noise and vibration control, *Journal of Sound and Vibration*, Vol. 193(1), pp. 367–376.

Network Rail. 2009, *Comparing Environmental Impact of Conventional and High Speed Rail*, New Lines Program, London.

Newman, P. and Kenworthy, J. 1999, *Sustainability and Cities: Overcoming Automobile Dependence*, Island Press, Washington, DC.

Oikonomidis, D., Triantafyllopoulos, P. and Paidousi, M. 2003, Noise protection programme during the operation of the tram of Athens, *International Conference Contemporary Tram and LRT Systems*, 19–20 May 2003, Patras.

Olivain, P. 1989, SNCF Médiathèque,.Paris.

Online image, available online at: http://www.tucrail.be/FR/media/Site_of_the_month/PublishingImages/2013/chantier_du_mois_06_2013_3.jpg (accessed 30 April 2015).

Online image, available online at: http://www.hmmh.com/cmsdocuments/FTA_Ch_07.pdf (accessed 17 April 2015).

Online image, available online at: https://www.monorails.org/tMspages/enviro.html (accessed 25 February 2021).

Pikal, S. 2011, *Cool Bridge Made of 27 Wind Turbines*, available online at: http://www.mobilemag.com/2011/02/09/cool-bridge-made-of-27-wind-turbines/ (accessed 30 April 2015).

Pyrgidis, C., Lagarias, A. and Dolianitis, A. 2018, The aesthetic integration of a tramway system in the urban landscape – Evaluation of the visual nuisance, *4th Conference on Sustainable Urban Mobility – CSUM2018*, Skiathos Island, Greece, 24–25 May.

Pyrgidis, C. and Barbagli, M. 2019, Evaluation of the aesthetic impact of urban monorail systems, *11th International Monorail Association Annual Conference (Monorailex 2019)*, Chiba, Japan 24–26 November.

Pyrgidis, C., Lagarias, A., Garefallakis, I. and Spithakis, I. 2020, Evaluation of the aesthetic impact of urban mass transportation systems, *5th Conference on Sustainable Urban Mobility – CSUM2020*, Skiathos Island, Greece, 17–19 June.

Relié. 1989, *TGV L'Atlantique à 300 km/h*, La Vie du Rail, October 1989.

Schweizer Norm. SN 671 250a. 2002, *Schweizerischer Verband der Strassen – und Verkehrsfachleute* (VSS), May 2002.

Tsohos, G., Pyrgidis, C. and Demiridis, N. 2001, Contribution to the railway noise prediction, *Technika Chronika*, Scientific Edition of the Technical Chamber of Greece Vol. 21(3), pp. 143–156.

UIC. 2008, *Process, Power, People – Energy Efficiency for Railway Managers*, Association of Train Operating Companies (ATOC), Paris.

UIC. 2010, *Railway Noise in Europe, A 2010 Report on State of the Art*, Paris, France.

UIC. 2012, *Energy Consumption and CO_2 Emissions, Railway Handbook*, 2012, International Energy Agency, Paris, France.

Vanhonacker, P. and Leuven, A. 2005, Super resilient rail fixation systems to reduce squeal noise, vibration and rail corrugation, *Proceedings of the International Congress Rail Engineering 2005*, 29–30 June, London.

Veitch, A. and Schwarz, H. 2011, *Rail and Sustainable Development*, UIC, April 2011, Paris.

Vignal, B. 1990, SNCF Médiathèque, Paris.

Vogiatzis, K. 2006, Investigation of noise and vibration transmitted through the ground with the application of elastic rail consolidation and floating slab in parts of tram network in Athens, *Technika Chronika*, Scientific Edition of the Technical Chamber of Greece, Vol. I(3), Athens.

Vogiatzis, K. 2011, Noise and vibrations of fixed route transport systems transmitted through the ground criteria – Propagation standards, *Information Bulletin of the Hellenic Association of Transportation Engineers*, No. 176, April–May–June 2011.

Wettschureck, R. 1995, Vibration and structure-borne noise insulation by means of cellular polyurethane (PUR) elastomers in railway track applications, *Rail Engineering International*, Vol. 2, pp. 7–14.

Wettschureck, R. 2002, Long term properties of Sylomer ballast mats installed in the rapid transit railway tunnel near the Philharmonic Hall of Munich, *Rail Engineering International*, Vol. 4, pp. 4–11.

Chapter 20

The research in the railway domain
Cutting-edge technologies in railways

20.1 THE RESEARCH IN THE RAILWAY DOMAIN IN EUROPE

The European railway network constitutes a particularly complex and important transport system since, besides its size, it connects many different countries with a huge number of regular passenger and freight services. The high diversity in terms of rolling stock, infrastructure, operation systems, and national legal frameworks means that a potential system upgrade and a proper implementation of evolving cutting-edge technologies are particularly difficult and require the cooperation and concurrent actions of multiple actors.

Taking these particularities into account, the European Union founded in 2014 the Shift2Rail Joint Undertaking with the ultimate goal of optimising research in the railway domain across Europe and strengthening the European railway industry. Shift2Rail is a public–private partnership in the railway sector that in addition to the European Union, includes 8 founding members (Alstom, Ansaldo STS, Bombardier, Construcciones y Auxiliar de Ferrocarriles (CAF), Siemens, Thales and infrastructure managers, Trafikverket, and Network Rail) and 19 associated members.

Shift2Rail follows a holistic approach and aims to optimise the system by considering it as a whole rather than focusing on specific constituents and structural elements as was the case in the railway research domain prior to its founding (Shift2Rail, 2015 ; Shift2Rail, 2019a; https://shift2rail.org).

As it is mentioned in its Multi-Annual Action Plan, the added value of the Shift2Rail project will come from (Shift2Rail, 2015):

- Innovative ideas/materials/technologies that will be checked at a high Technology Readiness Level (TRL), meaning that they will be closer to the real needs of the market.
- Examining the railway system as a whole rather than looking just at its subsystems (constituents/elements).

The producer is the railway industry, the final product is the railway transport systems, the clients are the system's users, the operators, the managers of the infrastructure, and society as a whole.

As already mentioned, the unified and combinatorial approach to technological development and demonstration allows for a holistic approach to innovation in railway systems and not just for individual structural elements. Thus, it allows for the fastest possible evaluation of the compatibility among the various solutions under examination. This means that at an early stage, the practical value of developing technologies is validated and as such research efforts are focusing mostly on 'cutting-edge technologies' (see Section 20.2) (Shift2Rail, 2015).

DOI: 10.1201/9781003046073-20

The Shift2Rail is structured around five Innovation Programmes (IPs) and encompasses all the construction, technical, and operational procedures of the various railway subsystems. The above IPs are implemented gradually through research projects that Shift2Rail specifies, invites for, and funds (calls for members as well as open calls).

These IPs and the basic fields of research that they cover are shown below (Shift2Rail, 2015, 2019b):

- IP1: Cost-efficient and reliable trains:
  - New braking systems that allow for higher brake rates and lower noise emissions.
  - Innovative doors of lighter materials that improve safety and reliability levels.
  - Car-body shells that use composite or other lightweight materials (hybrid structures).
  - Modular train interior that will improve the transport capacity of the vehicle and boarding/alighting times.
  - New bogie technologies – improved dynamic behaviour of vehicles – optimal design of bogies depending on the railway system.
  - Lighter and smaller traction equipment.
  - Coupling systems.
  - Wireless train control and communication systems.
- IP2: Advanced traffic management and control systems:
  - Adaptable communications for all railway systems (quality of service, interface to signalling).
  - Railways network capacity increase (ATO up to GoA4 – UTO).
  - Track capacity increase through fluid moving block.
  - Advanced fail-safe train positioning (focus on satellite technology).
  - Onboard train integrity (Safety Integrity Level (SIL4)).
  - Zero on-site testing (control command in lab demonstrators).
  - Formal methods and standardisation for smart signalling systems.
  - Virtually coupled trainsets and smart switching and crossing.
  - Traffic management system.
  - Smart radio connected all-in-all wayside objects.
  - Cyber-security (including key management systems).
- IP3: Cost-efficient and reliable infrastructure:
  - Enhanced permanent way system (track configurations – e.g., switches and crossings, rails, fastenings, elastic pads, sleepers, trackbed systems – calculation/design methods, materials).
  - Increased lifecycle of bridges and tunnels (inspection, maintenance, repair methods, noise, and vibration reduction).
  - Future stations (increased capacity and security in larger stations and standardised design of smaller stations).
  - Smart systems for distributed energy resource management.
- IP4: IT solutions for attractive railway services:
  - Interoperability framework.
  - Travel shopping.
  - Booking and ticketing.
  - Trip tracker.
  - Travel companion.
  - Business analytics platform.
  - Integrated Technology Demonstrators (TDs).

- IP5: Technologies for sustainable and attractive European rail freight:
  - Improved wagon designs that will allow for higher speeds, lower track wear, and easier and shorter maintenance.
  - New freight propulsion concepts.
  - Fleet digitalisation and automation.
  - Marshalling yards, terminal stations, and branch lines with innovative equipment (e.g., intelligent video gates).
  - Real-time management of marshalling yards and track/network capacity.
  - Coupling systems – longer trains.
  - Strategic design – provided services – traffic composition.

The five above IPs are supported by Cross-Cutting Activities (CCAs) in areas and subjects that have increased importance for the goals of the master plan (such as the long term needs, socioeconomic research, safety and security, interoperability, and energy sustainability) and take into account the interrelations between the five IPs and the various subsystems (Shift2Rail, 2015).

Each IP consists of subcategories of innovative actions and solutions, the aptly named Technology Demonstrators (TDs). TDs are tested in a laboratory setting or on specific prototype trains so as to evaluate the performance of the new technologies involved. Subsequently, they are translated into Integrated Technology Demonstrators (ITDs) that also refer to tests and evaluations at a laboratory setting and, finally, they are implemented in an integrated environment that simulates real operating conditions (System Platform Demonstrators). The evaluation of the various TDs is quantified with the use of Technology Readiness Levels (TRLs) (Horizon, 2020).

In addition, to taking into account the progress in terms of digitalisation of the rail systems, a new series of projects dedicated to a novel concept of operations and functional system architecture has been set up and launched in 2018: this will provide the basis for the evolution of the technology transfer of solutions to operations, integrating by design interoperability and market uptake.

A large number of projects have already been completed or are underway (Shift2Rail, 2020). Table 20.1 includes, per IP, the projects that fall under the Shift2Rail framework.

By 2021, the European Commission is planning to establish a new European Partnership on Rail Research and Innovation in accordance with the Treaty on the Functioning of

Table 20.1 Shift2Rail projects that have been completed or are underway per IP

Innovation Programme (IP)	Abbreviation
IP1: Cost-efficient and reliable trains	PIVOT, CONNECTA, CONNECTA-2, PINTA, PINTA-2, Mat4Rail, RUN2RAIL, SAFE4RAIL, SAFE4RAIL-2, PIVOT2, CARBODIN, NEXTGEAR
IP2: Advanced traffic management and control system	X2Rail-1, X2Rail-2, X2Rail-3, X2Rail-4, VITE, ASTRAIL, ETALON, MOVINGRAIL, GATE4RAIL, EMULRADIO4RAIL, 4SECURAIL, OPTIMA, CYRAIL, MISTRAL
IP3: Cost-efficient, sustainable, and reliable high-capacity infrastructure	In2Track, IN2SMART, In2Stempo, Fair Stations, IN2DREAMS, S-CODE, MOMIT, ASSETS4RAIL, FUNDRES, In2Track2, IN2SMART2
IP4: IT solution for attractive railways services	CO-ACTIVE, ATTRACKTIVE, CONNECTIVE, COHESIVE, My-TRAC, SPRINT, SHIFT2MAAS, MaaSive, RIDE2RAIL, GoF4R, ST4RT
IP5: Technologies for sustainable and attractive European freight	ARCC, FFL4E, FR8HUB, FR8RAIL, FR8RAILII, INNOWAG, OptiYard, SMART, M2O, LOCATE, SMART2, FR8RAILIII, Dynafreight
IPX	FLEX-RAIL, TER4RAIL, B4CM, MVDC-ERS, RAILS, TRANSLATE4RAIL
CCAs: Cross-Cutting Activities	FINE1, IMPACT-1, IMPACT-2, LINX4RAIL, PLASA-2, GoSAFERAIL, OPEUS, SMaRTE, TRANSIT, FINE2, PLASA, NEAR2050, DESINATE

the European Union (TFEU), building upon the current Shift2Rail Joint Undertaking. To this end, the European Technology Platform, European Rail Research Advisory Council (ERRAC) sets out a new programme of technical and operational innovation that can transform railways' contribution to mobility in Europe, addressing the needs of railway users, the economy, and society and protect natural resources and the environment (Rail Strategic Research and Innovation Agenda, 2020).

The programme must have strong links with relevant stakeholders (including end users and staff) at European, national, regional, and sectoral level and with European Partnerships developing relevant technologies, such as related to hydrogen, batteries, and digital technology and to cyber-security. It focuses on autonomous vehicles, the Internet of Things and Artificial Intelligence. This will be made possible by focusing on a new telecommunications infrastructure that makes the best use of 5G technology or modern satellite communication, which is being developed across the entire rail industry.

20.2 DEFINITION AND CLASSIFICATION OF CUTTING-EDGE TECHNOLOGIES

The term 'cutting-edge' technology implies a technology which adheres to the following:

- Is fairly recent (last 5–10 years).
- Is innovative (sustaining/disruptive).
- Greatly improves system performance.
- Has the potential to or has started to be widely implemented.
- Is the subject of extensive research.

They may be distinguished into categories based on:

- The constituent(s) of the system that they concern (infrastructure, rolling stock, operation).
- The type of railway system to which they are applicable (high-speed, metro, freight, etc.).
- The stage in which their development is (under testing, under research, in use).

Based on the analysis of Chapter 15 concerning rail freight transport, one may state that there is, in all railway transportation systems, a dynamic correlation between dilemmas (open 'issues'), trends, and innovations that may not, however, be considered ideal. The dilemmas remain almost constant over time, since for most of them no univocal answer has been given. International current conditions, such as the imperative need to protect the environment, the energy crisis, or the market and the rail industry themselves indicate the strategic choices and policies that contribute positively to the performance and economic efficiency of the railway system, the environment, and the society in general. These trends are supported financially at a research level and are closely associated with innovation. On the other hand, innovation is based on technological progress and the opportunities that it offers to influence or even overturn the basis on which the various actors formulate the strategy to be followed for specific dilemmas.

The inverse, however, relation, meaning that the strategic development of innovations is guided by the dilemmas (after an answer to them has been given), is seen in certain cases to be absent. Specifically, research shows that in the domain of freight, trainload services are more lucrative than single wagonload services. However, it may be observed that an increased amount of funding has been given to research on the automation of marshalling yards and the revitalisation of single wagonload services. Moreover, even though networks of dedicated traffic composition (freight or passenger) show higher financial efficiency, technologies that lessen the negative impacts of mixed traffic composition are being developed, without however addressing specifically the topic of separating traffic or focusing on technologies that maximise the gains for a dedicated freight railway corridor.

A perhaps more rational approach would be for railway actors to set – through research – the most lucrative choices and decide collectively and rationally on future strategies so as to result in actions that are more beneficial for the sector and direct their funds in relating research fields. In that way, the best possible conditions for competition in the railway transport market are ensured.

Table 20.2 records, indicatively, for the three railway constituents (infrastructure, rolling stock, and operation) and maintenance activities, technologies that may (based on the aforementioned criteria) be considered as cutting-edge (Dutta, 2020; Duffin, 2020; Young, 2020; Shift2Rail, 2019b). The following sections describe in more detail some of them.

20.3 SMART WINDOWS

The operation of 'smart' windows is based on the use of innovative synthetic glasses that permit automatic adjustment of their optical and thermal properties, depending on external conditions. These windows absorb solar radiation and re-emit it inside the vehicle, ensuring natural lighting and heating, and allowing, at the same time, the user to control the amount of light and heat that permeates the glass (Figure 20.1).

Smart windows make use of various technologies:

- Thermotropic.
- Photochromic or photochromatic.
- Suspended Particle Devices (SPDs).
- Electrochromic.
- Reflective hydrides.

Thermotropic and photochromic technologies cannot be controlled manually. Thermotropic materials become darker by reacting directly with the sunlight and, as a result, their use in winter is not recommended, when it is required to heat the interior of the vehicles. Photochromic windows react with the heat (become darker) and, as a result, during the summer months, passengers have a limited view.

The last three technologies are the most interesting ones and are currently evolving.

SPD technology allows the window glass to convert its surface from transparent to non-transparent and also to any intermediate condition, automatically or manually, with a switch (Bonsor, 2010; Katanbafnasab and Abu-Hijleh, 2013).

Windows with the above properties were used for the first time in a test train route in Japan, on 20 April 2014 (Research Frontiers, n.d.; Railway Gazette, 2014a).

Electrochromic windows use special materials, which, along with an electrode system, may change colour when electric current permeates them. In essence, electrical power causes a chemical reaction that changes the properties of these materials. In specific electrochromic materials, this change relates to many colours. In electrochromic windows, however, the changes are between coloured, which reflects the light of a colour, and transparent, which reflects no light at all.

In an electrochromic window, electrical power is only required to implement the first change in transparency. The maintenance of a specific colour shade does not require constant

Table 20.2 Cutting-edge technologies per railway constituent (Indicatively)

Infrastructure	Rolling stock	Operation	Combination of constituents	Maintenance
• Mechatronic switches • Enhanced permanent way system • Enhanced switches and crossings • Low cost high-speed bridges • Long performing structures	• Smart windows • Hydrogen powered trains • Active suspension, actively steering • Active pantograph • Interactive train windows • New light carbodies • New door functions • Freight automatic coupler	• Digitalisation • Train automation • Big data • Digital twin models • The internet of trains • High-speed biometric and microchip ticketing systems • Energy-efficient train control/energy-efficient driving strategies	• Power supply systems with energy storage devices • Ground-level power supply systems • Obstacle Detection System	• Condition based and predictive maintenance • Laser railhead cleaner system • Hydrogen fuel cell rail milling machine • Integrated measuring systems • High-pressure waterjet cutting

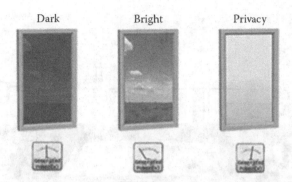

Figure 20.1 Operation of smart windows. (Adapted from Online image, available online: http://www.gree npacks.org/2010/03/15/smart-energy-glasses-generate-energy-stay-smart/ (accessed 7 April 2015; From Peer+B.V., 2015).)

voltage. Minimum voltage is required to implement and also reverse the initial change. This is also the major element that makes it very efficient in terms of energy. Depending on the construction parameters, full colouration and discolouration may need from 2 s to 10 min.

Reflective hydrides may be described as electrochromic materials; however, they behave in a different way. Instead of absorbing light, they reflect it. Thin films made of a nickel-magnesium alloy are able to change a transparent condition to a non-transparent one, and vice versa. This change can be supplied with high-voltage electric current (electrochromic technology) or the diffusion of hydrogen and oxygen gases (aerochrome technology). These materials are considered to be more efficient in terms of energy than other electrochromic materials (Bonsor, 2010).

Below are presented the advantages (with +) and disadvantages (with –) of smart windows, compared with windows using conventional technology (Bonsor, 2010):

- (+) Control, either automatic or manual, of the intensity of natural lighting. Adjustment by the staff and also by the passengers.
- (+) Lower weight (by 30%) and size.
- (+) Better soundproofing of the interior of the vehicle.
- (+) Lower installation and maintenance cost. Reduction of the operating cost (cost of heating, lighting, air conditioning, and abolish the use of sunscreens).
- (+) Higher thermal insulation capacity.
- (+) Reduction of reflections.
- (+) Quick change from transparent to non-transparent status.
- (+) Protection against ultraviolet radiation, and therefore, protection of the materials of the vehicle's interior space against wear and discolouration.
- (+) Resistance against a high range of temperatures.
- (+) High lifespan.
- (–) Significant initial investment.
- (–) Use of electric charge.

20.4 LASER RAILHEAD CLEANER SYSTEM

An issue encountered in many railway networks is the presence of 'leaves on the track'. Fallen leaves on the track superstructure cause wheel sliding that may lead to long delays,

Figure 20.2 The LRC system. (From Structon Rail, 2015.)

and even cancelled services. According to the British company Network Rail, leaves on tracks caused 4.5 million hours of passenger train delays in 2013 (Gray, 2014).

Wheels compress leaves, causing a layer of black slippery and hard substance like Teflon on the rolling surface of the railhead. The reduction of adhesion affects the rolling, braking, and signalling systems.

The solutions used nowadays towards this issue are either high-pressure water jetting or jetting a mixture of sand and gel called sandite, by ejectors placed on the trains. Both methods have disadvantages, since trains have to be supplied on a constant basis with the necessary materials, while severe corrosion can be caused on rails.

The first attempts to use a laser in order to clean railway lines are dated in 1999 in Great Britain, without, however, delivering the desired results.

In November 2014, the Dutch started retesting a system that burns leaves with a laser. Lasers are placed right in front of the wheels and target low, so that, as the train passes, they burn the accumulated mass of the leaves. Moreover, they dry the rails, preventing the accumulation of new leaves.

The LRC (Laser Railhead Cleaner) system (Figure 20.2) has been installed on a test basis on train DM-90 of the Dutch Railways. The wavelength used is absorbed only by the leaves and other organic materials that may be found on tracks.

The laser is deactivated temporarily in case it loses its target. So far, researchers have improved the system, so that it can clean up to 20 mm of substances on the rail, on both sides of the track, and at speeds that may reach 80 km/h (Daily Mail, 2010; Brown, 2014; Structon Rail, n.d.).

20.5 CATENARY-FREE POWER SUPPLY SYSTEMS OF TRAMWAYS

The power supply systems of a tramway are divided into four major categories: overhead power supply systems, ground-level power supply systems, induction systems, and onboard energy storage systems (Table 20.3).

These systems can be used on their own for the operation of a tramway system (as autonomous systems) or can be combined with each other (mixed power supply or hybrid systems).

In Table 20.4, all the potential combinations are presented. The potential combinations are expressed with the symbol $\sqrt{}$.

Table 20.3 Classification of power supply systems used in tramway networks

Major systems	Subcategories
Overhead supply systems	• Trolleybus type • Catenary type
Ground-level supply systems	• Conventional systems of third and fourth rail • APS • TramWave
Induction systems	• PRIMOVE
Onboard energy storage systems	• Supercapacitors • Batteries • Flywheels

Table 20.4 Potential combinations of tramway systems power supply

	Catenaries	APS	TramWave	PRIMOVE	Super-capacitors	Batteries
Catenaries	√	√	√	√	√	√
APS	√	√			√	Inability to recover energy from braking
TramWave	√		√		√	√
PRIMOVE	√			√	√	√
Super-capacitors	√	√	√	√	√	√
Batteries	√	Inability to recover energy from braking	√	√	√	√ (in a small percentage of line length)

The above data, as well as the data recorded and analysed in the following sections, relate to the year 2019. The raw data were obtained from various available sources and cross-checked. Afterwards they were further manipulated for the needs of this chapter.

In the following section, the systems of ground-level power supply and energy storage which are considered as cutting-edge technologies in the railway sector, as they meet the requirements of Section 20.2, are described in detail. All these technologies are in service. Flywheels are in a very primary stage. However, flywheels show a series of disadvantages that prevent their extensive use in railway applications, while they have a relatively high weight (Siemens, 2012; CCM, 2014).

20.5.1 Ground-level power supply systems

20.5.1.1 The APS system

The APS system (Alimentation Par le Sol – electrical power supply of trains on ground-level) was developed by Alstom, for the supply of electricity through a 'third rail' to trams. The third rail is literally a carrying conductor integrated into the ground and placed between the two main rails of the track.

The system operates with electrified segments at a voltage of 750 V, and a length of 8 m, separated by neutral segments of 3 m length (Figure 20.3).

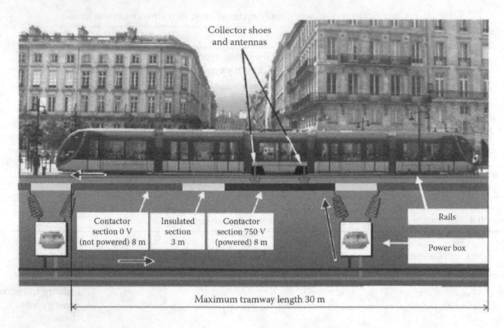

Figure 20.3 Operating principle of the APS system. (Adapted from Novales, M., 2010, Overhead wires free light rail systems, *90th TRB Annual Meeting 2011*, Washington, DC, 23–27 January, available online at: http://assets.conferencespot.org/fileserver/file/32564/filename/16668k.pdf) (accessed 11 April 2015; From Alstom Transportation, 2015.)

The required energy for the movement of the tram is ensured in the following way: each train (articulated tram) has two electricity collectors (collector 'shoes'), in the middle car, close to which is placed an antenna, which sends radio signals to 'power boxes' that supply the conductor with electric current only when the tram passes just over it, resulting in the elimination of the risk of electrocution to pedestrians, cyclists, animals, etc. (Figure 20.4). Power boxes are placed every 22 m. At the same time, the batteries placed on the roof of the vehicles are charged in the track sections where there is an overhead power supply and use the stored energy when the power supply from the ground is out of order due to 'power box' failure, and also in the areas of intersection with roads. The energy that is stored in the batteries permits autonomous movement of up to 50 m (Novales, 2010; Cooper Bussmann, 2012).

The minimum required length of line, in order to apply the APS power supply system, is 30 m. Conduction segments of 8 m length are activated and deactivated thanks to an isolation switch, which is located inside the embedded power box.

The specific technical characteristics of the APS system are summarised below:

- The tram collects power from the pairs of contact shoes placed under the central car of the tramway. The shoes are 3.2 m away from each other, which means that a tram cannot be decoupled in the neutral segment of the line of 3 m.
- The process of lowering the pantograph for the movement of the train by the APS technique is performed at specific stops during boarding/alighting of passengers.
- At road intersections, insulations are placed on rails. The total length of the insulated track sections at crossings is 5 m or more; therefore, when a tram is decoupled from the catenaries at a crossing, it should be supplied by the batteries in order to proceed to the next electrified track segment.

Figure 20.4 The APS system is totally safe for pedestrians, cyclists, and road vehicles. (Adapted from Finn, B., 2007, *Tram system in Bordeaux report on the tram system and underground power supply (aps) for Dublin city business association,* December 2007, available online at: http://www.dcba.ie/wp-con tent/uploads/2012/04/Bordeaux-Report-07.12.07.pdf (accessed 11 April 2015).)

- Electric current ceases to flow through the conductor after the passage of the tram. Therefore, if more than two consecutive electrified track segments are identified, the automatic power switch cuts off electrification in both segments, and its electrification cannot be recovered unless manually (activation normally takes about 20 min after isolation) (Mott MacDonald Group, 2008).

The APS system has been in practice for more than 15 years and the issues that occurred initially have been overcome (Finn, 2007). Specifically, it was applied for the first time in 2003 to the tram network of the city of Bordeaux, France, in 13.6 km out of the 44 km of the network, and in particular within the historic centre of the city (Figure 20.5).

Nowadays, the APS system is used in ten tramway networks worldwide (Table 20.5). The tram of Dubai, the construction of which was completed in November 2014, is the only tram in the world to have, throughout its length, the APS (autonomous system) ground-level power supply system.

The tram of Rio de Janeiro in Brazil uses the APS system in combination with supercapacitors. In the other eight systems, the APS system operates only in parts of the network, in conjunction with the conventional overhead catenary system (see Table 20.4).

The advantages of the APS system against other conventional solutions with overhead wires are:

- Its superiority with respect to aesthetics, as it relieves from visual annoyance.
- The ability to perform all activities that were prevented before due to the presence of overhead wires.
- The performance of trains with regard to speed and dynamic comfort of passengers, which are relevant to those with conventional power supply systems.

Figure 20.5 The tram of Bordeaux, where the APS system is used within a part of the network. (Adapted from Pline, 2008, Online image available at: https://commons.wikimedia.org/wiki/File:XDSC _7576-tramway-Bordeaux-ligne-B-place-des-Quinconces.jpg (accessed 8 August 2015).)

Table 20.5 Tramway systems that use APS technology (2019 data)

City (country)	Total line length (km)	Cost/km million (€) (2014 data)	Operation starting year	Percentage of line length equipped with overhead catenary system (%)	Line length with APS (km)/percentage of line length equipped with APS (%)
Tours (France)	15	27.8	2013	86.7	2/13.3
Orleans (France)	12	30.8	2012	82.5	2.1/17.5
Angers (France)	12	32.9	2011	87.5	1.5/12.5
Reims (France)	12	34.1	2011	83.3	2/16.7
Bordeaux (France)	44	35.7	2003	69.1	13.6/30.9
Dubai (UAE)	10.6	66.8	2014	0	10.6/100.0
Cuenca (Ecuador)	10.5		2016	88.6	1.2/11.4
			2018	21.5	22.7/68.5
Lusail (Qatar)	33.1		2019	87.5	1.5/12.5
			2016	0	APS + supercapacitors
Sydney (Australia)	12				
Rio de Janeiro (Brazil)	28				

Source: Adapted from Railway Gazette, n.d., available online at: http://www.railwaygazette.com/ (accessed 12 July 2014); Wikipedia, 2015a, Fenbahn, available online: http://de.wikipedia.org/wiki/Stra%C3%9Fenbahn (accessed 15 July 2014); Wikipedia, 2015b, Tramway, available online: http://fr.wikipedia.org/wiki/Tramway (accessed 22 July 2014); Light Rail Now, n.d., available online: http://www.lightrailnow.org/ (accessed 20 August 2014); Guerrieri, 2019, Catenary-Free Tramway Systems: Functional and Cost-Benefit Analysis for a Metropolitan Area, Urban Rail Transit, (4):289–309, available on line : https://doi.org/10.1007/s40864-019-00118-y (assessed on 2 January 2021).

Its disadvantages against conventional solutions with overhead wires are:

- The potential reduction of the performance in case of accumulation of rainwater.

    The requirement for immediate removal of water in case of flooding results in a significant increase in the maintenance and operating costs of the network, whereas the requirement for placement of protective covers and insulating materials on the current-carrying conductor, for protection against moisture, increases the infrastructure implementation cost.

    The requirement for a high level of tightness in the isolation equipment of a segment, which does not permit their placement in a soil with a high aquifer level, for example, on the banks of a river.

    In terms of their protection, their latest version is based on the EN 60529 standard and they can be kept submerged in 1 m of water for 15 days.
- When the tram is driven by the APS system, for safety reasons, braking with energy recovery is not possible (non-compatible with the regenerative-braking technology). Alstom investigates the improvement of energy efficiency through the on-vehicle installation of supercapacitors (SYSTRA, 2012).

    This technology has been implemented so far by Alstom in the Citadis 302 and 402 type vehicles.
- Problems in the presence of snow, frost, and sand.
- A design that should take into account the mechanical stress caused by road traffic. In the case of application to the city of Bordeaux, the railway line shares infrastructure with road traffic, and as a result, it should endure charging for years. In other French cities, the APS system is mostly installed in a protected corridor, with the exception of course of intersections, where the corridor is necessarily common.
- Finally, the higher total implementation cost of a tramway system that uses the APS system, compared with the cost of a tramway system, runs exclusively on a conventional power supply.

    Specifically, the use of APS technology, instead of the overhead catenary system, increases the following three components of the total project cost as follows:
    - The installation cost of the traction system

        The cost of the power supply equipment placed on the line in the case of using APS technology is seven times higher per metre of length than the cost of installation of an overhead catenary system.
    - The cost of rolling stock

        By convention, an increase should be taken into account in the region of 10–15% of the cost of conventional rolling stock, when an alternative power supply is used from the ground. The installation in retrospect of APS equipment on conventional power supply vehicles requires modifications (installation of a collection device, installation of batteries on the roof, installation of antennas under the frame of the bogies, and modification of the traction circuit). The layout of the substations is identical for both APS and conventional power supply.
    - The cost of works

        This is related to the superstructure, the substructure, and the restoration of the road infrastructure.

According to Alstom Transportation, the increase of the overall implementation cost (rolling stock + power supply + track + civil engineering) in comparison to the conventional solution (overhead catenary system) is estimated to (2019 data):

- 3–4% in case of a double-track line of 12 km length with 2 km APS (18.5%).
- 10% in case of a double-track line of 11 km length with almost all with APS (100%) (tram of Dubai).

20.5.1.2 The TramWave system

The TramWave system has many similarities to the APS system. It has been developed by Ansaldo. It supplies a voltage of 750 V through a continuous tubular conductor encased in the ground, which is placed between the two main rails of the track (Figure 20.6).

The power supply line is implemented by a series of current-carrying conductor segments that are insulated from each other, each of which has a length of 3 or 5 m. The segments are supplied with current only when the train passes over them. The presence of trams is detected by gravity sensors and other electrostatic means.

The energy collector is placed under the vehicles, whereas the power collector is placed in the ground within the power supply conductor. A ferromagnetic zone in the power supply conductor on the ground allows electrical energy to be transmitted to the train. The track segment with electric charge is very small – just 1.5 m – in order to maximise safety. As soon as the train passes from the specific segment, the electricity supply is cut off.

The TramWave can be adjusted in many types of vehicles, including those with rubber-tyred wheels, and operates simultaneously with the conventional overhead catenary system.

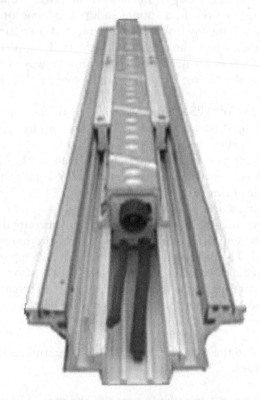

Figure 20.6 The conductor of the TramWave system. (Adapted from ANSALDO STS., n.d., TramWave ground-level power supply system (nooverheadlines), Italy, available online: http://www.ansa ldo-sts.com/sites/ansaldosts.message-asp.com/files/imce/tramwave_eng.pdf (accessed 11 April 2015).)

When the overhead system is in operation, a safety system sets the power collector placed in the ground to 'off' mode.

The TramWave can be combined with Onboard Energy Storage Systems (OESS), and the supply of electricity can be used in order to recharge it. In this way, the vehicle can be disconnected from the power supply cable and continue autonomously with the energy stored on the vehicle.

Ansaldo asserts that this system has lower maintenance requirements than conventional overhead catenary systems because any type of malfunction can affect only one tram, which is detected automatically by the diagnostic system and can be replaced within just 30 min. Furthermore, the immobilisation of a tram does not prevent other trams from circulating on the lines.

Two tramway systems operate actually with the TramWave system (see Table 20.6).
The differences between the TramWave and the APS system are the following:

- TramWave is a technology that has been applied only to two lines in practice.
- It can be adjusted to many vehicle types, including those with rubber-tyred wheels.
- The return current is transferred through the power supply conductor. In this way, the TramWave eliminates the effects of leakage currents, avoiding the requirement for insulation of the track infrastructure.
- There is the ability of energy recovery during braking (compatible with regenerative-braking technology).

The equipment of the TramWave system has been installed only on Sirio-type vehicles of Ansaldo. Upgrading the conventionally supplied trams for the operation of the TramWave system requires modifications on the vehicle.

20.5.1.3 The PRIMOVE system

This system has been developed by Bombardier. Like the other two systems, it has power supply equipment placed in parallel to the two main rails of the track. However, in this system, the ground power supply is performed inductively and not by contact (induction system). The power supply is used for either the operation of electric motors or charging the storage devices that may be placed on the vehicle.

In the PRIMOVE system, the tramway infrastructure has, in every 8 m, a separate segment of loops of inductive coil of 8 m length, which transfers power energy inductively to the train (Figure 20.7). These inductive coils (primary inductive loop) are fully embedded in the pavement. Their covering is performed with various materials, such as asphalt or concrete. Below the underground induction coil, there is a magnetic shield that prevents electromagnetic interference. Inverters are placed along the line and are supplied with a

Table 20.6 Tramway systems which use TramWave technology

City (country)	Line length (km)	Operation starting year	Percentage of length operating with overhead catenary system (%)	Line length with Tramwave system (km)/percentage of line length operating with the TramWave system (%)
Zhuhai (China)	8.7	2015	0	8.7/100
Beijing (China)	9.4	2015	57.5	4/42.5

Source: Adapted from Guerrieri, M., 2019, Catenary-Free Tramway Systems: Functional and Cost-Benefit Analysis for a Metropolitan Area, Urban Rail Transit, (4):289–309, available on line : https://doi.org/10.1007/s40864-019-00118-y (assessed on 2 January 2021).

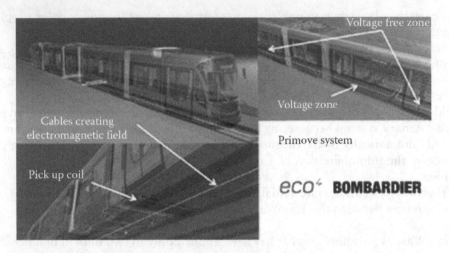

Figure 20.7 The PRIMOVE system. (From Bombardier Transportation, 2015.)

direct current of 750 V DC, and convert it to alternating current which can be used in the PRIMOVE system. When a high-frequency alternating current flows through a segment of induction coil loops, it causes an electromagnetic field. The segment is activated by the detection and control cable of the segment of induction coil loops (Vehicle Detection and PRIMOVE Segment Control (VDSC) cable), which detects the presence of a vehicle over it.

The vehicles have induction receivers under the floor (secondary inductive loop), which capture the inductive power that is transferred by the primary inductive loop and a compensation condenser (the PRIMOVE power receiver system), which converts the electromagnetic field to alternating current. On-vehicle inverters convert alternating current to direct current used for its power supply. They also have, on the roof, energy storage devices, and specifically, the MITRAC energy saver, which uses a double layer of capacitors. The capacitors store the released energy during braking and also from PRIMOVE's infrastructure, both during train movement and at stops for future use. It can also be combined with Li-ion batteries, leading to the optimisation of the system's energy efficiency and also the reduction of the required infrastructure. The detection of cable segments is performed by the detection antenna of the vehicle and control of the induction coil loop segment (VDSC antenna) on the vehicle, which coordinates their activation/deactivation.

Just like in the TramWave or the APS, the cable segments are supplied with power only when the train is over them, which makes the PRIMOVE system safe for pedestrians and road vehicles.

The system is able to supply 100–500 kW of power. It can be used for the operation of trams of 30–42 m length, on a ground slope of up to 6%, and for speeds of up to 90 km/h, with 270 kW of power (Brecher and Arthur, 2014).

Since the power supply is performed without contact, the system operation is not affected by snow, ice, or sand.

The PRIMOVE system was installed on a test basis in Augsburg, Germany, in 2011, in a segment of 0.8 km length. The pilot application was completed successfully in June 2012.

The differences of the PRIMOVE system, compared with TramWave and APS systems, are the following:

- Reliability even under adverse weather conditions (snow and ice), since the power supply is performed inductively.
- Compatibility with all types of pavements.

- There is also, like in the TramWave system, the ability of energy recovery during braking (compatible with regenerative-braking technology).

It is a technology that has not been applied in practice yet, and hence there are not any credible data about its cost and reliability.

20.5.2 Onboard energy storage systems

The main disadvantage of overhead and ground-level power supply systems is the fact that they require the installation of fixed infrastructure, with subsequent impact on the implementation and maintenance cost (Novales, 2010). An alternative solution is the use of Energy Storage Systems (ESS) on the vehicle (Onboard Energy Storage Systems (OESS)). The initial goal of this technology was not to avoid the use of overhead wires but to improve the energy efficiency of light railway systems.

Indeed, a way to reduce the energy consumption of a tram is through regenerative braking, which has been expanded widely to railway systems. Regenerative braking, in order to be effective, required the simultaneous presence of other vehicles in close distance from the braking vehicle for the absorption of surplus electrical energy, which would otherwise return to the overhead catenary system, especially when the power supply was performed with direct current. Overheating of the overhead wires caused by the total energy generated by the motors during the braking operation used to be a frequent event.

To prevent such an event, trams started to get equipped with integrated ESS, for recovery and using braking energy when necessary, such as in case of acceleration. Very shortly, technicians realised that this technology could be used in order to avoid the use of overhead wires in network segments.

The process of charging can be accomplished with three different methods:

- By the overhead catenary during conventional operation.
- By the regenerative-braking process.
- By the installation of charging infrastructure in stations.

ESS can be divided into three types of applications:

- *Mobile storage applications:* Mobile storage applications consist of ESS usually located on the vehicle roof. Every system works independently and the recovered energy is directly sent to the storage system placed on the vehicle. When the vehicle accelerates, energy is used in priority from the ESS to propel the vehicle.
- *Stationary storage applications:* Stationary storage applications consist of one or several ESS placed along the tracks. These devices recover the exceeding energy when no other vehicle is receptive meanwhile.
- *Stationary back to the network applications:* Stationary back to the grid applications whose main difference with the previous applications is that they do not store the recovered energy but send it to the main electrical network for other consumers such as lighting, escalators, administrative, and technical buildings or potentially sold back to the energy provider.

Charging of the storing devices can be achieved by the following methods:

- *Regenerative-braking process.*
- *Charging from power supply cables*

In this case, the batteries placed on the vehicle are charged by the overhead catenary system during the stage of conventional power supply, in order to store the necessary amount of energy for autonomous movement in the areas where the overhead catenary system has been abolished to achieve an improved aesthetic result. There is also the possibility of charging from a ground power supply, as it applies in the case of PRIMOVE.

- *Charging from power supply cables installed at stops*

The charging process is implemented at stops through an overhead catenary system which has been installed only in the area of stops. In this case, there is the requirement for rapid charging of storage devices (which is an advantage of supercapacitors over batteries). Charging can also be attained from the infrastructure which is installed inside the rails.

20.5.2.1 Supercapacitor charging/ESS (supercapacitors or ultracapacitors)

Supercapacitors comprise an improved and more competitive version of capacitors used in applications of means of transport. Supercapacitors are identical energy storage means to electrochemical cells, that is, batteries, with energy however being stored as electric charge, instead of chemical reactors, as it happens in batteries. High-capacity batteries are able to store large quantities of energy; however, their charging takes long. Not only are supercapacitors charged faster than batteries, but they also last longer, because they do not suffer from natural wear of charging and discharging, which exhausts the batteries. The charging of supercapacitors is implemented within a few seconds (20 s), which allows them to be charged at stops, during the boarding and alighting process.

Supercapacitors allow braking energy recovery, resulting in energy saving percentages of up to 30%.

In Table 20.7, the major differences between batteries and supercapacitors are listed.

The biggest disadvantage of supercapacitors, compared with lithium-ion batteries, is their much smaller energy density. Lithium-ion batteries may store up to 20 times the energy of supercapacitors for a given weight and size.

Supercapacitors are placed on tram roofs. This solution implies the increase in vehicle weight and cost, but it is offset by the saving of energy (Gonzalez-Gil et al., 2013).

New materials, such as graphene, could ensure high energy density in capacitors and fully replace lithium-ion batteries.

The purchase cost of supercapacitors is very high; however, it is expected to be reduced by the dissemination of technology.

Table 20.7 Major differences between batteries and supercapacitors for their use in tramway systems without overhead wires

Features	Batteries	Supercapacitors
Lifespan	20,000 charging cycles	100,000 charging cycles
Braking energy recovery	Very low	Yes
Indicative cost	20% of vehicle cost + replacement cost in the end of lifespan	20% of vehicle cost + replacement cost in the end of lifespan
Charging time	High	Low
Required time between two consecutive charges	High	Low

Source: Adapted from Global Mass Rapid Transport, 2014, *Catenary-free Trams: Technology and Recent Developments*, April 1, 2014.

The benefit enjoyed from the use of supercapacitors, as well as of batteries, urges manufacturers to adopt hybrid systems that use both technologies (see Section 20.5.2.2).

In Table 20.8, the tramway systems are presented which use supercapacitors.

The company of CSR ZELC from China presented in mid-2014, in its statement, one of the first trams to move exclusively through supercapacitor technology, and not use any aerial conductors in contact, for its power supply (Figure 20.8). The trains carry supercapacitors on their roofs, which are sufficiently charged when trains make their scheduled stops. CSR adds that the vehicles draw sufficient power during a stop of 10–30 s, which allows them to operate independently for up to 4 km. The supercapacitors provide power to the traction system for the acceleration of the tram, and they also draw energy from the 'braking energy recovery' system, which can recover up to 85% of the braking energy to be reused. The

Table 20.8 Tramway systems which use supercapacitors exclusively or in combination with other systems

System/vehicles	Company	City	Power supply
ACR Freedrive/Urbos	CAF	Zaragoza, Spain (2013)	OCS + supercapacitors + batteries
ACR Freedrive/Urbos	CAF	Granada, Spain (2017)	OCS + supercapacitors + batteries
ACR Freedrive/Urbos	CAF	Luxembourg (2020)	OCS + supercapacitors + batteries
ACR Evodrive/Urbos	CAF	Seville, Spain (2011)	OCS + supercapacitors
ACR Evodrive/Urbos	CAF	Kaohsiung, China (2015)	OCS + supercapacitors
Siemens Sitras – ES (Maxwell)			
	CSR ZELC	Guangzhou, China (2014)	Supercapacitors (they are charged at stops from ground power supply)
Sitras-HES/AVENIO	Siemens	Doha, Qatar (2016)	Supercapacitors + batteries (Ni-MH)
Voith/Dolphin	CNR	Shenyang, China (2013)	(they are charged at stops from
SRS with Ecopack	Changchun	Nice, France (2018)	overhead power supply)
CITADIS	Alstom		OCS + supercapacitors
			Supercapacitors + batteries
Sitras-HES	Siemens	Almada-Seixal (2008)	OCS + supercapacitors + batteries

Figure 20.8 Prototype tram by the company of CSR with supercapacitors. (Adapted from Nissangenis, 2014, online image available at: https://commons.wikimedia.org/wiki/File:Guangzhou_Haizhu_District_CSR-Zhuzhou_Tram_For_No.05.jpg.)

Table 20.9 Tramway systems which use batteries in combination with Overhead Catenary System (OCS)

Vehicles	Company	Battery type	Length, off wire/total length (km)	City	Power supply
CITADIS	Alstom	Ni-MH (SAFT)	0.91/8.7	Nice, France (2007)	Batteries + OCS
LIBERTY	Brookville	ABB Li-ion nickel manganese cobalt	1.6/2.6	Dallas, USA (2015)	Batteries + OCS
FORCITY CLASSIC	Skoda	CATFREE Nano lithium-titanium	1.8/21	Konya, Turkey (2015)	Batteries + OCS
FLEXITY	Bombardier	PRIMOVE Li-ion	7.5/8.3	Nanjing, China	Batteries + OCS
LIBERTY	Brookville	ABB Li-ion nickel	3.1/5.1	(2014)	Batteries + OCS
TRIO	Inekon	manganese cobalt	4.0 /4.0	Detroit, USA (2016)	Batteries + OCS
TRAMLINK V4	Vossloh	Li-ion (SAFT) ABB Li-titanate	0.4/1.4	Seattle, USA (2016) Santos, Brazil (2016)	Batteries + OCS

Source: Adapted from Akiyama, S., Tsutsumi, K., and Matsuki S., 2008, The development of low floor battery-driven LRV 'SWIMO', Kawasaki Heavy Industries, Ltd., Kobe, Japan, May 2008; Vuchic, V. R. 2007, *Urban Transit Systems and Technology*, John Wiley & Sons, ISBN: 978-0-471-75823-5, p. 624; Railway Technology, n.d., b, *Shenyang Tramway, China*, available online: http://www.railway-technology.com/projects/shenyang-tramway/ (accessed 21 August 2014); Railly News, 2013, Bombardier partner CSR Puzhen to supply catenary-free trams to Nanjing, available online at: http:// www.raillynews.com/2013/bombardier-partner-csr-puzhen-to-supply-catenary-free-trams-to-nanjing/ (accessed 17 August 2014); Marotte B., 2011, *Building a New Canada – Bombardier's transit vision pairs efficiency with wireless technology*, available online: http://www.theglobeandmail.com/news/national/time-to-lead/bombardiers-transit-vision-pairs-efficiency-with-wireless-technology/article4181172/ (accessed 17 August 2014).

vehicles are charged automatically, through a device that is placed between the rails, which is activated as vehicles pass over the track (Railway Gazette, 2014b).

Manufacturers have invested in the potential of lithium-ion batteries which offer higher energy storage density than Ni-MH cells and which have achieved significant development in the sector of rolling stock manufacturing.

In Table 20.9, various examples of tramway systems using batteries are presented.

20.6 AUTOMATION OF TRAINS

20.6.1 Definition and Grades of Automation

The term 'automation' generally refers to the ability of a system to fulfil a specific group of functionalities for a specific purpose in an automated manner, meaning without human intervention.

According to the International Electro-Technical Commission and the IEC 62290-1 standard (EC, 2014), there is a clear classification of the Grades of Automation (GoAs) of railway systems. The GoAs are classified into four categories (see also Section 5.2.2):

- GoA1: Operation with a driver.
- GoA2: Semi-automatic Train Operation.
- GoA3: Driverless Train Operation.
- GoA4: Unattended Train Operation.

The GoA is dependent to a large extent on the driving support systems (ATP, ATO, DTO) with which the infrastructure and the rolling stock are equipped. Generally, automation is considered to be achieved when the train does not have a driver (GoA4 and GoA3).

Trains with a GoA3 are characterised as autonomous, while the ones with GoA4 as fully autonomous.

In metros and monorails, all four Grades of Automation have been developed and implemented. Specifically, automation was introduced as a technic for the first time in 1962 in the New York Metropolitan Railway (Wang et al., 2016), while in 1983 in France in the Lille metro, GoA4 is implemented for the first time.

On the contrary, on conventional railways, only GoA1 has been extensively and systematically been implemented, while for the remaining three, the applications have so far been limited. Specifically (Tomson and Page, 2020; Goverde, 2020):

- In Australia, the Rio Tinto iron ore railways were fully automated recently, allowing trains to operate with GoA4 on a 1,700 km route (Autohall system).
- The Czech railways have been operating with GoA2 Semi-automatic Train Operation (STO) since 1991.
- In the UK, the Thameslink has been operating with STO over ETCS Level 2 since 2019, with plans to upgrade to GoA4.
- The Netherlands is also currently running GoA 2 trials for both passenger and freight trains, building upon the work developed within the Shift2Rail programme.
- Similary, in France, the French National Railway Company (SNCF) recently completed a test run of a remotely controlled autonomous train. The trial was conducted as part of an initiative to develop GoA4 prototypes by 2023. The locomotive-hauled train travelled a distance of 4 km using satellite and 4G technology.
- In Germany, German Rail (DB) is currently working with technology company Siemens to develop a fully automated S-Bahn line in Hamburg by 2021. German Rail plans to automate the entire Hamburg S-Bahn network.

Large part of this work is linked to or built on the ATO research and innovation activities performed in the Shift2Rail programme, which expands also to rail freight ATO with tests run during the last part of 2020. Test at GoA4 is expected to take place during 2021 in Czech Republic.

According to a recent report published by the Japan Transport and Tourism Research Institute (JTTRI), there are currently 64 fully automated train lines (both light and heavy rail) in the world, with more than half located in Asia (Goverde, 2019).

Although there are no GoA4 trains currently operating in Europe, prototypes are being developed and we may see fully automated trials being carried out as early as 2023.

In recent years, automation has started to be developed for all modes of transport as it constitutes an important factor for the enhancement of competition between land-based transport modes. Regarding railways, the goal is to implement automation on intercity and regional railway networks of conventional and high speed for both passenger and freight trains and to generally incorporate it in mixed traffic composition railway networks.

The difficulties that need to be overcome for the adoption of automation in the networks of conventional and high-speed railway systems, as opposed to metros, are attributed to their different characteristics presented in Table 20.10 (Goverde, 2019).

20.6.2 Implementing automation

The Communications-Based Train Control (CBTC) and European Rail Traffic Management System (ERTMS) are currently the two most widely known railway communication systems. The CBTC is implemented in metros and monorails and uses ATP. It is based on the

Table 20.10 Different characteristics between metros and conventional-/high-speed railway systems that hinder the implementation of automation on the latter

Metro	Conventional- and high-speed railway systems
Dedicated traffic (passenger trains)	Mixed traffic (Passengers + freight trains)
Trainsets with similar technical and operational characteristics	Trainsets with different technical and operational characteristics
Short train headway	Longer train headway
Short boarding and alighting time	Much longer boarding and alighting time
Short distance between intermediate stations/stops	Much longer distance between intermediate station/stops
Regular timetables	Not always regular timetables
Track maintenance during night hours	Track maintenance also during operation (at the track side)
Simple track superstructure (usually slab track)	Different configurations of track superstructure
Underground or elevated system	Usually at grade system
Simple organisational structure	Many stakeholders
Single stakeholder	Differentiation of the management of the infrastructure and of the operation
Short time needed for the installation of the suitable equipment onboard and on the track	Long time needed for the installation of the suitable equipment onboard and on the track
Narrow and predefined track occupation width	Larger and modified track occupation width
The rolling stock is used to the same infrastructure	The rolling stock is used to different lines
Exclusive corridor with no LRCs and usually without fencing	Corridor with LRCs and fencing

concept of a moving block and allows for the movement of trains with shorted headway, thus allowing for an increased track capacity.

Levels 1 and 2 of ERTMS/ETCS (European Train Control System – which is the signalling and control component of the ERTMS) – are installed on the tracks of an increased number of conventional-/high-speed railway lines. On these tracks, the ETCS adopts the use of ATP as part of the architecture of a European system for railway traffic management. Shift2Rail and other research is underway regarding ETCS level 3 which will introduce the moving block concept.

These two systems (CBTC and ERTMS) are incompatible regarding the infrastructure and radiocommunication technologies on which they are based. Urban railway networks (specifically metros and monorails) have relatively shorter lengths, are characterised by the use of exclusively passenger trains of unified construction and operational characteristics, and require increased track capacity and thus short headways. For these reasons, CBTC is the optimal choice. On the contrary, conventional and high-speed railway lines are characterised by increased lengths, higher speeds, and mixed traffic composition. For these reasons, ERTMS is the optimal choice for signalling. The use of Wi-Fi as a technology implies a significant cost for installation and maintenance due to the high number of required Access Points. That disadvantage in combination with the susceptibility to interference, and the high speeds involved make this technology unsuitable for use in conventional and high-speed networks (Critical Software, 2014).

In order to ensure interoperability between networks that are equipped with CBTC and ERTMS, a new trend is to use hybrid equipment trains. One such project is the London 'Crossrail'. However, since the onboard and trackside equipment for these two systems is vastly different, the cost for such a combined use is high.

Nevertheless, the introduction of advanced communication technologies, such as 5G, will bring a convergence of CBTC and ERTMS, as one of the key objectives is to increase the capacity of existing lines to create new services and business models; a performance based regulatory framework would most probably pave the way for such convergence.

In this respect, attempts are being made to ensure the maximum possible gain from automation on conventional and high-speed lines, which implies the ability to utilise ATO over ETCS.

Such a use must fulfil the following functions (Goverde, 2019):

- Driving functions.
- Timing point management.
- Train door operation.
- Dwell time management (countdown).
- Add/skip stopping point (updating Journey Profile).
- Hold train at stopping point (updating Journey Profile).
- Low adhesion management.
- Reporting management (status information for monitoring).

Regarding driving functions, the following must be fulfilled:

- TimeTable Speed Management (TTSM): Optimal speeds to stop or pass on time in the most energy efficient way.
- Supervised Speed Envelope Management (SSEM): Maximum speed respecting the ETCS speed limits.
- Automatic Train Stopping Management (ATSM): Speed profile to stop the train accurately at the stopping points.
- ATO Traction/Brake Control: Output commands to drive the train.

For the use of an ATO system over ETCS, the following equipment is required:

ATO Onboard: Calculate constantly the optimal speed profile based on available data from the infrastructure, the track, and the timetables. Controls the traction and braking systems (Siemens, n.d.).

ATO Trackside: Accumulates static and dynamic data from the infrastructure, the track, and the timetable from the Traffic Management System (TMS) and passes them on to the ATO Onboard.

In terms of research, importance is given to the creation of obstacle detection systems.

Automation is considered one of the four basic components that are required for the digitalisation of railway systems (wireless communication and signalling, automatic train operation, intelligent traffic management, automatic timetabling) (Goverde, 2019).

20.6.3 The advantages and disadvantages of automation

Automation optimised most functions of a railway system since, among others, it minimises the possibility of incidents caused by human error. At the same time, it increases track capacity and reduces energy consumption. The results of the limited, so far, use are encouraging (Chatzikonstantinou, 2018). Indicatively:

- The energy consumption in passenger trains operating under the CBTC system is calculated to be half (per passenger per kilometre) in relation to conventional trains. Shift2Rail demonstration activities reached a reduction of up to 30% of energy consumption depending on the operational conditions.
- No time is consumed at terminal stations for drivers changing trains or for breaks and the time delays attributed to a driver's reaction speed are also alleviated. This results in increased track capacity. At the same time, the optimised speed profiles allow for shorter headway and thus increased track capacity.

An increase, in the GoA of an existing system has however also drawbacks (Keevill, 2016):

- Usually, existing depot installations are not adequate and require expansion. The same is true for power supply systems, ventilation systems, and signalling.
- Modifications on the existing rolling stock or the acquisition of new rolling stock are required.
- New communication systems are required to be installed to allow for the dialogue between passenger and control staff and for the increased supervision of passengers via CCTV.
- Obstacle detection must be achieved at a greater distance than when compared with systems with a driver.
- Platform screen doors are required.

Table 20.11 attempts a SWOT (Strengths, Weaknesses, Opportunities, Threats) analysis for the case of automation in railway systems.

Table 20.11 SWOT analysis for automation in railway systems

	Strengths	*Weaknesses*
SWOT ANALYSIS	• Increased track capacity • Increased timetable accuracy • Increased timetable frequency • Increased dynamic comfort for passengers • Lower energy consumption • Lower maintenance costs • Lower operation costs • Optimised driving profile • Conditions that better favour competition	• Higher facility, infrastructure, and rolling stock cost • Increased time required for the development of new systems • Different rolling stock characteristics • Presence of level crossings • Need for the cooperation of infrastructure managers and operators • Increased requirements for communication systems • Lack of appropriate/required legislation
	Opportunities	*Threats*
	• Creation of new job positions • Increased interoperability • Development of moving block through ETCS Level 3 • Concurrent use of ETCS and CBTC	• Lack of trust from users • Long transition period • Lack of experience from previous applications • Overconfidence with regard to the capabilities of new technologies • Lack of standardisation of ATO systems on conventional and high-speed networks • Lack of qualified personnel • Lack of specialised companies to take on required works in all countries • Acceptance of new reality by drivers and need for transition to other positions within an organisation

One of the key elements for the successful progressive introduction of automation will be to put the human at the centre of the digital transformation of rail. In fact, the human/ machine interfaces will have to be built in such a manner, on the one hand, to ensure the correct transition processes and, on the other hand, to complement ATO with remotely controlled activities by operators.

REFERENCES

Akiyama, S., Tsutsumi, K. and Matsuki S. 2008, *The Development of Low Floor Battery-Driven LRV 'SWIMO'*, Kawasaki Heavy Industries, Ltd., Kobe, Japan, May 2008.

ANSALDO STS. n.d., *TramWave Ground-Level Power Supply System (No Overhead Lines)*, Italy, available online at: http://www.ansaldo-sts.com/sites/ansaldosts.message-asp.com/files/imce/tramwave_eng.pdf (accessed 11 April 2015).

Bonsor, K. 2010, *How Smart Windows Work*, available online at: www.tlc.howstuffworks.com/home/smart-window.htm (accessed 7 April 2015).

Brecher, A. and Arthur, D. 2014, Review and evaluation of wireless power transfer (WPT) for electric transit applications, *Federal Transit Administration Report No. 0060*, U. S. Department of Transportation, Volpe National Transportation Systems Center, August.

Brown, M. 2014, Metro bosses bring in new 24 hr a day technology in battle against leaves on the line, *Chronicle Live*, available online at: http://www.chroniclelive.co.uk (accessed 7 April 2015).

CCM. 2014, *Flywheel Technology*, available online at: http://www.ccm.nl/en/ccm/vliegwiel-techn ologie/flywheel-technology.html (accessed 24 August 2014).

Chatzikonstantinou, A. 2018, Η αυτοματοποίηση της κίνησης των τρένων και οι επιπτώσεις της στο σιδηροδρομικό σύστημα, Θεσσαλονίκη: Διπλωματική εργασία, μεταπτυχιακό πρόγραμμα "Διοίκηση και Διαχείριση Τεχνικών Έργων", Τμήμα Πολιτικών Μηχανικών, Πολυτεχνική Σχολή, Α.Π.Θ., 2008.

Cooper Bussmann. 2012, *Third Rail Fuse Link Application Guide*, available online at: http://www 1.cooperbussmann.com/pdf/7c3f8580-423b-4c28-8a78-c90da28faffe.pdf (accessed 11 April 2015).

Critical Software. 2014, ERTMS and CBTC side by side. The main difference and state-of-the art, February 2014, available online at: https://www.criticalsoftware.com/multimedia/critical/en/q5X4azTKP-CSW-Railway-WhitePaper-ERTMSandCBTC-Side-by-Side_1.pdf

Daily Mail. 2010, Runway train overshoots station by more than two miles ... because of leaves on the line, *Mail Online*, available online at: http://www.dailymail.co.uk (accessed 7 April 2015).

Duffin, R. 2020, *6 Important Rail Industry Innovation Ideas for 2020*, available online at: https ://www.wifispark.com/blog/important-rail-industry-innovation-ideas-for-2020 (accessed 8 January 2021).

Dutta, S. 2020, *Five Innovations that Could Shape the Future of Rail Travel*, available online at: https://theconversation.com/five-innovations-that-could-shape-the-future-of-rail-travel-147962 (accessed 8 January 2021).

EC. 2014, EC 62290-1:2014, *Railway Applications – Urban Guided Transport Management and Command/Control Systems – Part 1: System Principles and Fundamental Concepts*.EC Brussels

Finn, B. 2007, Tram System in Bordeaux Report on the Tram System and Underground Power Supply (APS) for Dublin City Business Association, December 2007, available online at: http://www .dcba.ie/wp-content/uploads/2012/04/Bordeaux-Report-07.12.07.pdf (accessed 11 April 2015).

Global Mass Rapid Transport. 2014, *Catenary-Free Trams: Technology and Recent Developments*, April 1.

Gonzalez-Gil, A., Palacin, R. and Batty, P. 2013, Sustainable urban rail systems: Strategies and tech-nologies for optimal management of regenerative braking energy, *Elsevier Energy Conversion and Management*, Vol. 75, pp. 374–388.

Goverde, R. 2019, Railway operations research and the development of digital railway traffic systems, *8th International Conference on Railway Operations Modelling and Analysis*, Norrköping, June 17–20, 2019.

Goverde, R. 2020, Trends and developments in the automation of heavy rail operations, *Global Railway Review*, available online at:https://www.globalrailwayreview.com/article/97734/trend s-developments-automation-heavy-rail/(accessed 8 January 2021).

Gray, R. 2014, End of the line for autumn train delays? Dutch begin trials to zap leaf litter from the tracks with lasers, *Mail Online*, available online at: http://www.dailymail.co.uk (accessed 11 April 2015).

Guerrieri, M. 2019, Catenary-free tramway systems: Functional and cost-benefit analysis for a metropolitan area, *Urban Rail Transit*, Vol. 4, pp. 289–309, available online at: https://doi.org/10 .1007/s40864-019-00118-y / (accessed 8 January 2021).

https://shift2rail.org/

https://webstore.iec.ch/publication/6777

Horizon. 2020, Work programme 2014–2015, Extract from Part 19 – Commission Decision C(2014)4995, 'General Annexes, G. Technology readiness levels (TRL)', available online at: https://ec.europa.eu/research/participants/data/ref/h2020/wp/2014_2015/annexes/h2020-wp1415-annex-g-trl_en.pdf./(accessed 8 January 2021).

Katanbafnasab, M. and Abu-Hijleh, B. 2013, Assessment of the energy impact of using building integrated photovoltaic and electrochromic glazing in office building in UAE, *Engineering*, Vol. 5, pp. 56–61.

Keevill, D. 2016, *Increasing Levels of Automation with CBTC, IRSE Seminar 2016 – CBTC and Beyond.*

Light Rail Now. n.d., Available online at: http://www.lightrailnow.org/ (accessed 20 August 2014).

Marotte, B. 2011, *Building a New Canada – Bombardier's Transit Vision Pairs Efficiency with Wireless Technology*, available online at: http://www.theglobeandmail.com/news/national/ time-to-lead/bombardiers-transit-vision-pairs-efficiency-with-wireless-technology/article4 181172/ (accessed 17 August 2014).

Meuser, U. 2019, How is the future railway system going to look like?, ERRAC rail 2050 vision ERRAC 2030 R&I priorities Shift2Rail contribution for freight rail, *KTH Railway Group Seminar*, Stockholm, 2019, available online at: https://www.kth.se/polopoly_fs/1.921728. 1600689126!/2019_05_23KTHv02.pd.

Mott MacDonald Group. 2008, *Power Distribution for Trams and Electric*, Technical Note No. 7, July 2008, Surface Transport Master Plan, Connecting Abu Dhabi 2030, Department of Transport, available online at: http://dot.abudhabi.ae/en/content/download?file=TN07.pdf&loc =forms (accessed 11 April 2015).

Nissangenis, 2014, Online image, available online at: https://commons.wikimedia.org/wiki/File :Guangzhou_Haizhu_District_CSR-Zhuzhou_Tram_For_No.05.jpg.

Novales, M. 2010, Overhead wires free light rail systems, *90th TRB Annual Meeting 2011*, Washington, DC, 23–27 January, available online at: http://assets.conferencespot.org/fileserver /file/32564/filename/16668k.pdf) (accessed 11 April 2015).

Online image, available online at: http://www.greenpacks.org/2010/03/15/smart-energy-glasses-ge nerate-energy-stay-smart/ (accessed 7 April 2015).

Pline. 2008, Online image, available online at: https://commons.wikimedia.org/wiki/File:XDSC _7576-tramway-Bordeaux-ligne-B-place-des-Quinconces.jpg (accessed 8 August 2015).

Railly News. 2013, Bombardier partner CSR Puzhen to supply catenary-free trams to Nanjing, available online at: http://www.raillynews.com/2013/bombardier-partner-csr-puzhen-to-supply -catenary-free-trams-to-nanjing/ (accessed 17 August 2014).

Rail Strategic Research and Innovation Agenda. 2020, Available online at: https://errac.org/publica tions/rail-strategic-research-and-innovation-agenda-december-2020/ (accessed 8 January 2021).

Railway Gazette. 2014a, Dimmable tram windows, available online at: http://www.railwaygazette .com (accessed 7 April 2015).

Railway Gazette. 2014b, Guangzhou supercapacitor tram unveiled, available online at: http://www .railwaygazette.com/news/urban/single-view/view/guangzhou-supercapacitor-tram-unveiled .html, 2014 (accessed 11 April 2015).

Railway Gazette. n.d., Available online at: http://www.railwaygazette.com/ (accessed 12 July 2014).

Railway Technology. n.d., a, *Nice Tramway*, France, available online at: http://www.railway-techno logy.com/projects/nice-trams/ (accessed 20 August 2014).

Railway Technology. n.d., b, *Shenyang Tramway*, China, available online at: http://www.railway-t echnology.com/projects/shenyang-tramway/ (accessed 21 August 2014).

Research Frontiers. n.d., available online at: http://www.refr-spd.com (accessed 7 April 2015).

Shift2Rail. 2015, *Shift2Rail Joint Undertaking, Multi-Annual Action Plan*, Brussels, November 2015, available online at: https://www.shift2rail.org/wp-content/uploads/2013/07/MAAP-f inal_final.pdf/ (accessed 8 January 2021).

Shift2Rail. 2019a, *Annual Activity Report 2018*, June 2019, available online at: https://shift2rail.org /wp-content/uploads/2019/07/S2R-JU-Annual-Activity-Report-2018.pdf/ (accessed 8 January 2021).

Shift2Rail. 2019b, *Catalogue of Solutions*.Brussels

Shift2Rail. 2020, *Annual Activity Report 2019*, June 2020, available online at https://shift2rail.o rg/wp-content/uploads/2020/06/GB-Decision-05-2020_AAR-2019_Annex_AAR-2019.pdf/ (accessed 8 January 2021).

Siemens. 2012, *Qatar Foundation to Get Turnkey Tram System from Siemens*, available online at: http://www.siemens.com/press/en/feature/2012/infrastructure-cities/rail-systems/2012-08-av enio.php?content[]=IC&content[]=ICRL (accessed 24 August 2014).

Siemens. n.d., Automatic train operation: Solutions for automated driving for high-and low-density mainline, freight and regional traffic, *Siemens Mobility GmbH*, Article No. MOMM-B10208-00-7600, German, available online at: https://assets.new.siemens.com/siem ens/assets/api/uuid:0b9b1530-f878-4713-9e01-4891d44bde16/brochure-automatic-train-op eration-e.pdf. (accessed 06-04-2021)

Structon Rail. n.d., *Laser Railhead Cleaner Tests*, available online at: http://www.struktonrail.com/ about-us/laser-railhead-cleaner-tests/ (accessed 11 April 2015).

SYSTRA. 2012, *Feasibility of Alternative Power Supply Systems for the LUAS BXD*, April 2012.

Thomson, J. and Page, I. 2020, Autonomous trains – Driverless trains in Europe: Heavy rail automa-tion explained, *Intelligent Mobility Xpierence*, available online at: https://www.intelligent-m obility-xperience.com/amp/driverless-trains-in-europe-heavy-rail-automation-explained-a-95 5881/?cmp=go-ta-art-trf-IMX_DSA-20200217&gclid=EAIaIQobChMIzr2O7ImJ7gIVFuJ3 Ch0wHg (accessed 8 January 2021).

Vuchic, V.R. 2007, *Urban Transit Systems and Technology*, John Wiley & Sons, New Jersey, ISBN: 978-0-471-75823-5, 624pp.

Wang, Y., Zhang, M., Ma, I. and Zhou, X. 2016, Survey on driverless train operation for Urban Rail Transit Systems, *Urban Rail Transit*, Vol. 2(3), pp. 1–8, available online at: https://www.res earchgate.net/publication/311734230_Survey_on_Driverless(Accessed 06-04-2021_Train_Ope ration_for_Urban_Rail_Transit_Systems.

Wikipedia. 2015a, *Fenbahn*, available online at: http://de.wikipedia.org/wiki/Stra%C3%9Fenbahn (accessed 15 July 2014)

Wikipedia. 2015b, *Tramway*, available online at: http://fr.wikipedia.org/wiki/Tramway (accessed 22 July 2014).

Young, C. 2020, *Radical Railways: 15 Technologies that Could Drive the Future of Trains*, avail-able online at: https://interestingengineering.com/radical-railways-15-technologies-that-could -drive-the-future-of-trains (accessed 8 January 2021).

Chapter 21

Applicability verification

A supporting tool for the conduction of feasibility studies of urban mass railway transportation systems

21.1 APPLICABILITY VERIFICATION – DEFINITION AND THE NEED FOR ITS INTEGRATION IN URBAN RAILWAY PROJECT STUDIES

For the construction and operation of a railway transportation system a number of studies have to be undertaken: land use, urban, and traffic transportation planning studies, feasibility studies, design/construction studies, and operation and maintenance studies (Figure 21.1).

The feasibility study constitutes a special study category since it is required by almost all financing organisations, in order to fund a specific project. Taking into account that the final objective of a feasibility study is the financial and socioeconomic evaluation of the assessed project, a definition of the project's constructional and operational characteristics, a prediction of future transport demand, an evaluation of constructional and operational costs, and an estimate of the expected benefits of the project's implementation must have taken precedence. Based on the level of detail and precision with which the above elements of a feasibility study are approached, the study is characterised as preliminary or final. Depending on the nature of the project, the feasibility study may precede the preliminary studies, it may be conducted after a number of preliminary studies have been completed (e.g., alignment of the railway line) or after a number of the pre-studies have been completed. The last option allows for a more accurate estimation of the cost of the project (e.g., a project that requires the construction of very costly civil engineering works). Finally, in a few cases, the feasibility study may be attributed to the same contractor that is undertaking the preliminary track alignment study.

The stages, computational methods, and more importantly the mathematical tools used for conducting a feasibility study differ amongst designers. The European Union in an effort to homogenise and harmonise this type of studies, especially in the case of large investments/projects that are proposed to be funded by it, has issued a Directive regarding the cost-benefit analysis of an investment, that is to be conducted within the scope of a feasibility study (European Commission Directorate-General for regional and urban policy, 2014).

In the case of an urban/suburban railway system (tramway, metro, monorail, suburban), many of the required studies may be characterised as specialised. The integration of the railway infrastructure within the urban environment poses significant design, construction, and operational issues to the designers. These problems are often attributed to the lack of available space, the difficulty in finding a suitable and economic location for the depot, the need for coexistence of the railway system with other transport modes, the high level of service that is required to be assured for users, the integration with existing urban layout and architectures; and the more intense need for environmental protection. All the above

DOI: 10.1201/9781003046073-21

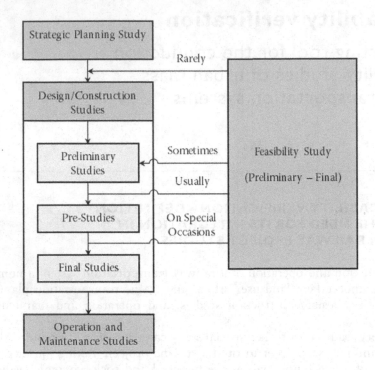

Figure 21.1 Categories and sequence of studies that are required for the construction and operation of railway transportation systems.

make imperative the need for alternative alignments that are evaluated within the scope of a feasibility study to be feasible.

The construction of a new, urban railway system is technically difficult and expensive and has impacts on the road and pedestrian traffic, both at the stage of construction and at the stage of the system's operation.

In this frame, it is necessary for competent bodies and decision-makers to be aware of the feasibility and the implementation cost of the new transport mean, before proceeding with costly studies and an even more expensive construction.

In this context, it is proposed that, before the financial and socioeconomic evaluation of the construction of an urban railway project is conducted, a thorough study of all the issues that pertain to the technical and operational feasibility of the project is undertaken. This particular approach, without changing the number and sequence of the aforementioned four basic study categories, includes in the whole process, as a 'supporting' tool, the 'technical and operational applicability verification' (Pyrgidis, 2003; Pyrgidis, 2008; Chatziparaskeva, 2018, Barbagli and Pyrgidis, 2018; Chatziparaskeva and Pyrgidis, 2018; Poirazis, 2018; Pyrgidis, 2019). This verification may be integrated within the feasibility study and consti-tute its initial stage or it may be conducted independently. In any case, it takes data from and provides data to the preliminary studies and mainly to the preliminary alignment study. At the same time, it generates the required data for the undertaking of the preliminary and final feasibility study (Figure 21.2).

At this stage, it should be noted that before conducting the applicability verifications, one must have concluded the relevant land use, urban, and traffic transportation planning studies.

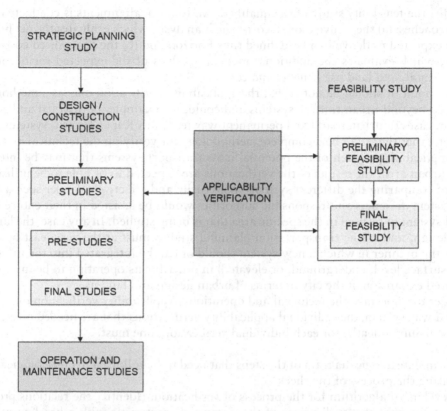

Figure 21.2 Categories and sequence of studies that are required for the construction and operation of urban railway transportation systems.

In this initial stage of studies, alternative alignments are, firstly, identified and preselected, according to the objectives and prerequisites set by the contracting authority, such as the starting and ending points of the alignment, the areas to be served, and the accomplishment of the priorities for future sustainable mobility. The number of these preselected alignments is usually high (more than five alternative alignments). Subsequently, the aforementioned alternative alignments are evaluated and those with the highest potential (usually two or three of them) qualify for the subsequent stages of study. The attained evaluation process can be performed by means of a multicriteria analysis (Pyrgidis et al., 2015).

Secondly, the applicability verification of these two to three alignments is undertaken.

The whole process includes individual verifications regarding the characteristics of the transport system (technical and operation characteristics, performances, environmental impacts, etc.), where all verifications must be satisfied. The number and the type of the individual verifications vary, depending on the category of the studied urban system. Each individual verification is essentially a comparison between the 'real' values of the system's characteristics and the 'reference' values. The 'real' values result from the choices made for the specific project, from in situ measurements and from calculations, while the 'reference' values are determined based on the current international practice. If the 'reference' values are satisfied, then the verification is positive. Otherwise, either the initial features of the system must be changed or the alignment must be changed or the specific alternative alignment should not be selected or the feasibility study should be the final criterion for accepting or not accepting the project.

Thirdly, the feasibility study of the qualified two to three alignments is conducted, using two approaches: (a) the 'conventional' cost-benefit analysis, which evaluates the net financial benefits expected to derive in a predefined time horizon and (b) the generalised cost-benefit analysis, which evaluates the additional monetary values of the expected socioeconomic, environmental, and land use benefits and costs.

At this point, it should be noted that the applicability verification may as a methodology be applied beyond urban railway systems and could, for example, be adapted and adopted for use in cases of urban road fixed permanent way (e.g., Bus Rapid Transit) systems. Based on the fact that there will be a complete methodology for verifying the technical and operational applicability of each of the potential mass transport systems that may be integrated into an urban area, the results of the verifications are expected to be able to be utilised as a means of comparing the different systems more easily and effectively. Under such a framework, system designers and responsible authorities would be assisted in their choice for the optimal system to be used in the specific area that is being studied. In any case, the land use, urban design, and traffic transportation planning studies must at the very least be able to indicate the manner in which a new system should or can be integrated into the urban area (i.e., at surface level, underground, or elevated) in order for its operation to be in sync with the desired expansion of the city in terms of urban design and land use.

In order to undertake the technical and operational applicability verification of a specific urban railway system, the individual applicability verifications that are needed must, firstly, be defined. Subsequently, for each individual verification, one must:

- Formulate a logic diagram of the steps that need to be followed in chronological order during the process of the check.
- Develop an algorithm for the process of verification. (Identify the relations/processes that compare the 'real' values of the systems characteristic with a baseline of 'reference' values.)
- Identify the data that need to be collected/selected/calculated by the ones undertaking the verification. (Identification of all the data that will be fed into the algorithm and selection of their acquisition/gathering process.)
- Define/develop the 'scientific tools' for the application of the algorithm (mathematical relations, empirical tables, simulation models, questionnaires, recording methods, etc.).

Under the framework of this chapter, four (tram, suburban, monorail, and metro) urban railway systems are examined. For trams and suburban railways, all the aforementioned processes/steps that need to be followed for undertaking the applicability verifications (both the individual ones and as a whole) are given. For monorails and metros, only the individual verifications that need to be undertaken are defined.

21.2 APPLICABILITY VERIFICATION OF A TRAMWAY LINE

21.2.1 Individual required verifications

The geometric integration of a tramway system in the urban environment, throughout or in part of its route, is technically difficult, while its operation creates a new traffic situation in the city (Chatziparaskeva and Pyrgidis, 2015; Chatziparaskeva, 2018; Barbagli and Pyrgidis, 2018; Chatziparaskeva and Pyrgidis, 2018). Specifically, the existing roads should be wide enough to accommodate the tramway track alignment. However, the most important problem of the integration of a tramway system lies in finding the right location for the construction and the operation of the depot (Chatziparaskeva et al., 2015; Pyrgidis et al., 2016; Pyrgidis et al., 2015). In addition, due to the lack of road arteries with sufficient width, the integration of a tramway system often leads to a reduction of the number of lanes used by other means of transport or even to the need of reorganising the flow in some roads. In many cases, these interventions make it difficult for the other means of transport to circulate and lead to the loss of parking spaces along the road. In even more unfavourable cases, it is necessary to reduce the width of the side-walks on both sides of the roads along the tramway alignment, and especially in the area of the tram stops, creating problems in the smooth and safe pedestrian traffic. Also, it is required to completely restructure the itineraries of the other public transport modes operating in its area of influence. Finally, the environmental impact of a tramway system, both during its construction phase and during its operation phase, should not be overlooked, with the most important being noise pollution, vibration, and visual annoyance (Pyrgidis et al., 2018; Pyrgidis et al., 2020).

In the case of the implementation of a new tramway system, it is necessary that the selected alternative alignments satisfy the following conditions:

- Is the geometrical insertion of the alignments feasible?
- Do they ensure an adequate level of service to the users (low run time, frequent service, acceptable transport capacity)?
- Can their possible negative impacts on the other transport modes be overcome in an effective and low-cost way?
- Do they create negative environmental impacts on the area of influence, which may be prohibitive for the implementation of the tramway project?
- Is there available space for the construction and efficient operation of rolling stock parking, maintenance, and repair facilities (depot)?
- Does the preliminary estimation of the cost of the tramway project in accordance with the international practice (average cost per track-kilometre)?

The satisfaction of the above conditions pre-requires, for each alternative alignment, the individual applicability verifications illustrated in Figure 21.3 which must be satisfied simultaneously.

21.2.2 Verification of track alignment and geometric integration

While carrying out this individual verification two partial checks are conducted in chronological order, namely (a) track alignment check and (b) geometric integration check (see also Chapter 4).

21.2.2.1 Track alignment check

Algorithms/reference values

- Minimum horizontal curve radii of the network: $R_{cmin} \geq R_c = 25$ m (Clark,1984; TRB, 2012; Prescott, 2012; Commissions of Railway Regulations, 2008).
- Maximum longitudinal gradient of the network: $i_{max} < i$, whereas (TRB, 2012).

$$i = 7\%, \text{when } L_i \leq 150\,m$$

$$i = 6\%, \text{when } L_i \leq 750\,m$$

$$i = 4\%, \text{when } L_i > 750\,m$$

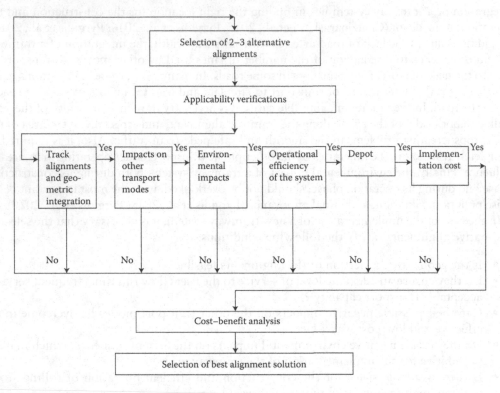

Figure 21.3 Individual applicability verifications applied in case of a tramway project.

where:
   L_i: Length of inclined sections of the track.
   • Available height clearance under the structures: h > 3.90 m (D.D.E, 1993).

Data needed: R_{cmin}, i_{max}, h, i_{max} = f (L_i) along the whole line. Measurements in situ and at the office.
   Scientific tools: Slope meters, AutoCAD.

21.2.2.2 Geometric integration check

Algorithms/reference values

   • Total Tramway Right-Of-Way in Straight sections: TTROWS < available road width.
   • Total Tramway Right-Of-Way in Stops: TTROWST < available road width.
   • In curves: Widths b1 and b2 of the two intersecting roads 1 and 2 should have the available width (provided from tables, see below scientific tools).

Data needed

   • Preselection of the category of tramway corridor for each road artery (A, B, C, D, E).
   • Preselection of the integration type of the tramway tracks (a single track per direction at two opposite sides of the road, double track on one side of the road, and central alignment).
   • Preselection of the vehicle height h_v and length L_v.

- Preselection of the integration type of tramway stops (island platform, laterally staggered platforms, laterally opposed platforms).
- Selection of the power supply system.
- Selection of the width (static g_v and dynamic g_{dv}) of the vehicle.
- Data needed for the calculation of TTROWS and TTROWST (width of separators b_{sw}, width needed for the installation of electrification masts b_{em}, width of platforms b_{lp}, b_{cp}).
- Data needed for the design of current cross sections and of the cross sections after the integration of the tramway (width of the roads and of the sidewalks).

Scientific tools

1. Calculation of TTROWS and TTROWST: Equations 4.1–4.11.
2. In curves, a design simulation for all possible combinations regarding the integration type of the tramway track before and after the intersection and for intersection angles that vary between $\varphi_o = 90°$ and $\varphi_o = 170°$ (see Section 4.4.2.2, Table 4.8).
3. Design (a) of current cross sections and (b) of cross sections after the integration of the tramway, at strategic locations of the alignment ('open line' and stops). Recording of the data showed in Figure 21.4 (indicatively).

Road width	Sidewalk width	Current layout	TTROW	Remaining road width	New layout
14.30 m	3.16 + 3.16 m	4 × 3.75 m ↓	6.40 m	7.90 m	1TT + 2 × 3.50 m ↓+1TT

where 1TT: Single Tramway Track
↓: Direction of road traffic

Figure 21.4 Example of the current layout and the new layout for a road before and after the integration of a tramway system – a single track per direction at two opposite sides of the road – Queen Olga Street, Thessaloniki, Greece. (Adapted from Thessaloniki Public Transport Authority, 2014, Investigation of the implementation of a tramway system in Thessaloniki, Thessaloniki.)

If the new situation is not acceptable (e.g., inadequate width of sidewalks and inadequate width or inadequate number of traffic lanes), then intervention is required for the modification of one or more of the above initial selections in order to achieve an acceptable geometric integration.

21.2.3 Applicability verification of operational efficiency

While carrying out this individual verification two partial checks are conducted in chronological order, namely (a) check of the commercial speed and (b) check of the passenger transport volume.

21.2.3.1 Check of the commercial speed

Algorithm/reference values

Commercial speed: $V_c > 18–19$ km/h or $t_{AB} \leq t$ (target value) (Pyrgidis et al., 2013; Chatziparaskeva, 20018; ERRAC, 2012).

where:
t_{AB} : Run time from point A (origin) to point B (destination).
t : Run time (target value).

Data needed

- S_A, S_B, S_C, S_D, S_E.
- Route length S.
- Data needed for the calculation of the commercial speed via simulation models: Track alignment geometric characteristics, distance between stops, rolling stock performances (power, traction effort, acceleration, etc.).
- Run time t (target value).

Scientific tools

1a. Calculation of the run time (and then of the commercial speed) via empirical formula (Equations 21.1 and 21.2) (Bieber, 1986).

$$t = \frac{S_A}{V_{cA}} + \frac{S_B}{V_{cB}} + \frac{S_C}{V_{cC}} + \frac{S_D}{V_{cD}} + \frac{S_E}{V_{cE}} \tag{21.1}$$

$$V_c = \frac{S}{t} \tag{21.2}$$

where:
V_c: Commercial speed.
S: Total route length.
S_A, S_B, S_C, S_D, S_E: Tramway corridor length for corridor categories A, B, C, D, E, respectively.
$V_{cA}, V_{cB}, V_{cC}, V_{cD}, V_{cE}$: Commercial speed of tramways running on corridor categories A, B, C, D, E, respectively.

This approach takes into consideration the category of tramway corridor along every road artery, the total length of each corridor category and needs to preselect (a) the number of intersections with roads where the tram will have priority at traffic signals; (b) an average distance between successive stops n_s = 500 m; (c) an average waiting time at each stop t_s = 20 s.

The following values are then applied for V_c *(see also Table 4.1)*:

Tramway systems without priority to the traffic signals

- Corridor category E: V_{cE} = 12–15 km/h.
- Corridor category D: V_{cD} = 16–20 km/h.
- Corridor category C: V_{cC} = 18–20 km/h.
- Corridor category B: V_{cB} = 20 km/h.
- Corridor category A: V_{cA} = 30 km/h.

Tramway systems with priority to the traffic signals

Up to 25% increase in commercial speeds in D and B tram category corridors.

1b. Alternatively, calculation of the commercial speed via simulation models existing in the market.

If the resulting commercial speed is not within acceptable limits (typically 18–25 km/h) or a run time that was initially set as a target is not met, the check is repeated after modifying one or more of the above options (e.g., priority to the tram at all traffic signals, change of the tramway corridor category). For example, if priority is given to the tram at all signalised intersections, and if it is considered that this causes an increase in the commercial speed by 25%, Equation 21.1 applies where V_{cB} = 25 km/h and V_{cD} = 22.5 km/h.

21.2.3.2 Check of the passenger transport volume

21.2.3.2.1 Check of the estimated passenger transport volume

Algorithm/reference values

P_{dyear} / S $\geq$ X_{rv} (Chatziparaskeva, 2018; Consultancy HTM, 2003)

$$X_{rv} = \begin{cases} 945{,}714 \text{ passengers / S} & \text{for ppl} < 300{,}000 \\ 1{,}176{,}536 \text{ passengers / S} & \text{for } 300{,}000 \leq \text{ppl} \leq 600{,}000 \\ 1{,}318{,}416 \text{ passengers / S} & \text{for } 600{,}000 \leq \text{ppl} \leq 900{,}000 \\ 1{,}626{,}065 \text{ passengers / S} & \text{for } 900{,}000 \leq \text{ppl} \leq 1{,}200{,}000 \\ 1{,}797{,}063 \text{ passengers / S} & \text{for } 1{,}500{,}000 \leq \text{ppl} \leq 1{,}800{,}000 \end{cases}$$

where:

X_{rv}: Reference values of yearly passenger transport volumes of tramway systems.
ppl: Population of a city.
S: Network length.
P_{dyear}: Yearly predicted passenger transport volume.
Data needed: ppl, S, P_{dyear}

Scientific tools

1. P_{dyear} can be predicted with the aid of traffic forecasting models existing in the market.
2. X_{rv} is defined after an extensive bibliography research (see Table 4.3) (Chatziparaskeva, 2018).

21.2.3.2.2 Check of the transport capacity of the system

Algorithm/reference values

Total daily transport capacity for both directions of the system: $P'_{dday} \geq$ Predicted daily transport load: P_{dday}.

Data needed

- Transport capacity C_v of tramway vehicles in peak hours C_{vph} and non-peak hours C_{vnph}.
- Number of peak hours (n_{ph}) and non-peak hours (n_{nph}) within the 24-h day.
- Headway (f) of trams in peak hours (f_{ph}) and non-peak hours (f_{nph}).
- Duration of the network operating hours within the 24-h day.
- P_{dday}.

Scientific tools

Nonspecific tool is needed for the calculation of P'_{dday}. The following formula is applied:

$$P'_{dday} = \left[n_{ph} \times c_{vph} \times (60/f_{ph}) + n_{nph} \times c_{vnph} \times (60/f_{nph}) \right] \times 2 \qquad (21.3)$$

CASE STUDY

RELEVANT DATA

Total predicted passenger volume per day per direction: P_{dday} = 25,000 passengers.

Predicted daily passenger volume per direction during peak hours: P_{ddayph} = 4,000 passengers.

Route length (AB): S = 10 km.

Run time: t_{AB} = 0.5 h = 30 min.

Commercial speed: V_c = 20 km/h.

Waiting time of trams at each terminal station: t_{ts} = 4 min.

Train transport capacity: C_v = 200 passengers (150 standing and 50 seated – density 4 passengers/m² – during non-peak hours).

Train transport capacity: C_{vph} = 275 passengers (225 standing and 50 seated – density 6 passengers/m² – during peak hours).

The tram is considered operational from 05:30 to 00:30, that is, for a total of 19 h while four of these operating hours are considered as peak hours.

Train headway during the peak hours: f_{ph} = 7 min.

Train headway during the non-peak hours: f_{nph} = 10 min.

Based on the above values:

Two-way route run time (round trip + waiting time at the two terminal stations): t_{ABA} = 2 × 30 + 2 × 4 = 68 min.

The number of vehicles F_1 required to ensure the daily service of the line is 68 / f_{ph} = 68 / 7 = 9.7 = 10 'vehicles'. This figure should be increased by:

- One replacement vehicle at the terminal or the tramway depot for the replacement of any vehicle that is damaged during operation.
- Twelve per cent (percentage of immobilised vehicles based on experience) of the estimated initial number of vehicles, namely one vehicle intended to replace any vehicle that is immobilised for repair or maintenance purposes at the tramway depot (Baumgartner, 2001).

Therefore, the total required fleet of vehicles for the operation of the line (included reserves) is equal to F_{lt} = 13 vehicles. As regards the transport volume, the tramway system can carry in total ([200 passengers × 60 min] / 10 min) × 15 h = 18,000 passengers per direction during the 15 non-peak hours of network operation and ([275 passengers × 60 min] / 7 min) × 4 h = 9,428 passengers per direction during the 4 peak hours of network operation. This results in a total of 54,856 passengers during the operating hours of the tramway network. Therefore, the predicted passenger volume for the optimistic scenario P_{dday} = 25,000 × 2 = 50,000 passengers is satisfied (< P'_{dday} = 54,856).

21.2.4 Applicability verification of a tramway depot

While carrying out this individual verification, four partial checks are conducted in chronological order, namely (a) check of the required and the available tramway ground plan area; (b) check of the distance of the tramway depot from the nearest terminal station of the main line of the network ('dead mileage'); (c) check of the landscape; (d) check of the ability to acquire the land and locating of the tramway depot.

21.2.4.1 Check of the required and the available tramway ground plan area

Algorithm/reference values

Required ground plan area for the depot E_d ≤ available ground plan area for the depot E_a (Chatziparaskeva et al., 2015; Chatziparaskeva, 2018).

Data needed

- Data needed for the calculation of the size of the fleet F_l and F_{lt}: Waiting time t_{ts} of the trams at both terminals A and B of a route AB, route length S, headway of trams in non-peak hours f_{nph}, commercial speed V_c, and percentage of immobilised vehicles.

where F_l: Number of vehicles required to ensure the daily service of the line.
  F_{lt}: Total required fleet of vehicles for the operation of the line (included reserves).
- E_a, L_v.

Scientific tools

1. Calculation of the size of the fleet of the vehicles. Nonspecific tool is needed. The following relations are applied:

$$
\begin{aligned}
1. \quad & t_{AB} = S/V_c \\
& t_{ABA} = 2 \times t_{AB} + 2 \times t_{ts} F_l \\
& t_{ABA}/f_{ph} \\
& F_{lt} = 1.12(F_l + 1)
\end{aligned}
\tag{21.4}
$$

2. Calculation of the required area E_d of the tramway depot's ground plan. For the estimation of the required ground plan area, the methodology that was described in Section 4.6.3 can be applied (Chatziparaskeva et al., 2015).

The designers then compare the size of the area that is proposed for the location of the depot (available land) to the estimated required area.

21.2.4.2 Check of the distance of the tramway depot from the tramway main network

Algorithm/reference values

$$D_m \leq D_{mmax} = 2km \quad \text{(Chatziparaskeva, 2018)}$$

where:
  D_m: 'Dead mileage' (distance between the entrance of the depot and the nearest terminal station of the tramway line).
  D_{mmax}: Maximum permissible value for 'dead mileage'.

  Data needed: D_m.
  Scientific tools: Nonspecific tool is needed.

21.2.4.3 Check of the landscape

Algorithm/reference values

Maximum longitudinal gradient of the tracks in depot $i_{dmax} < i_d = 2.5\%$.
  It should be noted that construction-wise it is feasible to achieve the desired value i_d; however, this can significantly increase the construction cost of the depot.
  Data needed: Topography of the terrain and approximate calculation of the construction cost of the depot.

21.2.4.4 Check of the ability to acquire the land and locating of the tramway depot

21.2.4.4.1 In this check, only qualitative parameters are examined.

Algorithm

- Is it possible to obtain an area for the construction of the tramway depot? Yes/No.
- Generally, areas that require the lowest cost of expropriation should be selected.
- Is there compatibility with adjacent land uses? Yes/No.
- Is the integration to the environment acceptable? Yes/No.

21.2.5 Applicability verification of the implementation cost

While carrying out this individual verification, one check is conducted: Check of the total implementation cost (infrastructure cost + purchase cost of rolling stock).

Algorithm/reference values

Implementation cost per track-km ≤ average implementation cost per track-km of similar tramway systems operating in the same wider geographical area (e.g., Europe, the United States, and North Africa).

The cost of the project must be similar to the cost which is referred to in the international practice. The average construction cost of a tramway line (infrastructure and rolling stock) is calculated to €20–€25 M per track-km (2014 data) (Chatziparaskeva, 2018).

Data needed

- Cost of different specific works/activities.
- Cost of acquiring the rolling stock.
- Fleet size Flt.

Scientific tools

1. Mathematical equations (by assuming unit prices for each work, material, and equipment) have been developed that calculate the construction cost, in relation to the length of the alignment, the length of civil engineering works, the kind of the power supply system, and the length of each type of tramway corridors (Chatziparaskeva, 2018).
2. Statistical analysis of data from existing tramway systems to estimate the cost of acquiring the rolling stock. The average cost is: €2,500,000, for trams with length 30–35 m (data 2014).

21.2.6 Applicability verification of the environmental impacts

While carrying out this individual verification five partial checks are conducted: (a) check of noise pollution; (b) check of visual annoyance; (c) check of safety issues; (d) check on impact on the urban space; (e) check on impact during construction.

21.2.6.1 Check of noise pollution and vibrations

21.2.6.1.1 Algorithm

1. Evaluation of environmental impacts, regarding noise pollution and vibrations, as: (a) limited and (b) unacceptable. In case of unacceptable impacts, designers should make interventions in the alignment.
2. Calculation of the cost of measures to face noise pollution and vibrations. Consideration of the above extra costs in the total implementation cost of a tramway system.

21.2.6.2 Check of visual annoyance

Algorithm/reference values

Visual Nuisance (VN) < 2.52 (Visual Nuisance Qualitative Category I or II) (see below – scientific tools) (Pyrgidis et al., 2018; Pyrgidis et al., 2020)

Data needed

Data needed for the calculation of VNP_i (see Section 19.5.1).
  where:
  VNP_i: Visual Nuisance Points of every structural element i of the tramway.

Scientific tools

A relatively objective and numerical method for evaluating the visual nuisance VN caused by the tramway is developed in Chapter 19, Section 19.5.1.

21.3 APPLICABILITY VERIFICATION OF A SUBURBAN LINE

21.3.1 Individual required verifications

The integration of a new suburban railway service in an already built-up urban area is always of concern to the designers as it falls under many parameters that are either imposed by the technical and operational specifications of the suburban railway itself or by the characteristics of the railway system transit and service area. The existence of satisfactory and steady passenger transport, the securing of rolling stock maintenance, repair, and storage sites, the environmental impacts during the construction and operation of the system, and the implementation cost are the substantial parameters that must be verified.

When routing suburban service trains on an existing infrastructure, more applicability verifications are required (as the track already exists) but the implementation cost is comparatively much lower. Also, in such a case, the existence of satisfactory and steady passenger transport constitutes the basic parameter to examine. In parallel, the presence of other trains on the track confers much more gravity to the track capacity check.

As in the case of the tramway, this process imposes for each of the alternative alignments, the individual verifications of applicability that are given in the logical diagram of Figure 21.5, and which must be simultaneously satisfied (Pyrgidis, 2008).

21.3.2 Operation of suburban trains on existing infrastructure

In the case where the operation is scheduled to take place on existing railway infrastructure, all the issues cited in Section 8.4 must be satisfied.

21.3.2.1 Applicability verification of constructional features of the railway infrastructure

This applicability verification examines issues 1, 2, 4, 8, 9, and 10 of Section 8.4.

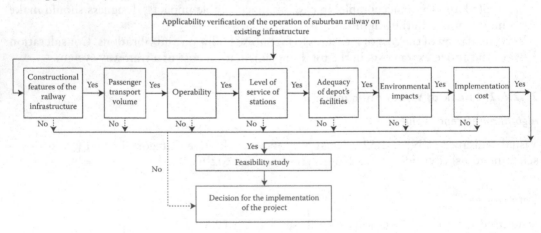

Figure 21.5 Operation of suburban railway on existing infrastructure −individual applicability verifications applied.

It includes the following partial checks:

1. Connection length.
2. Number of tracks.
3. Average distance between intermediate stops.
4. Track geometry alignment design (alignment radii, longitudinal gradients).
5. Components of the track superstructure.
6. Signalling system.
7. Traction system.
8. Fencing.
9. Average running speed.

The following steps are applied to check the average running speed (V_{ar}):

- Definition of the desired run time t (target time) for the connection length S (taking into account the travel time and particularly the travel time that is required by the competitive means of transportation).
- Definition of the number of intermediate stops n_s and the halt time t_s at each stop.
- Calculation of the average running speed of trains V_{ar} that is required in order to achieve the target time, by using Equation 1.3 (see Section 1.4.1).
- Recording of the permitted running speed of trains along the existing line V_{maxtr} and calculation of the average permitted speed V_{amaxtr} for the total connection length.
- Comparison between V_{ar} and V_{amaxtr}.

The check is considered positive when $V_{ar} \leq V_{amaxtr}$. Otherwise, necessary interventions must be made to the existing railway superstructure so as to secure the V_{ar} (e.g., by reducing the number of intermediate stops, by increasing the quality level of maintenance of the track, and by improving the geometry of the track alignment).

21.3.2.2 Applicability verification of the passenger transport volume

This verification examines issues 3 and 14 in Section 8.4, and it also provides data that helps in conducting other verifications.

In order to conduct this verification, the following are required:

- Daily ridership measurements including all transportation modes along the route. These measurements should cover both peak and off-peak periods.
- A questionnaire survey addressing the users of the existing transportation modes along the connection. This survey must use well-structured questionnaires.
- Recording of information on the passenger flow characteristics and, particularly, origin-destination, trip purpose, and trip frequency.
- Recording of users' opinions concerning the development of a suburban railway system and their intention to use this system if it shall operate with the foreseen design characteristics.

A combination of the two above-mentioned checks provides an initial estimate of the number of transported passengers that are expected to use the suburban railway service (daily ridership, passenger/h/direction), as well as an indication of whether there will be seasonal fluctuation of the transport volume, and whether there will be adequate transport demand between the intermediate stops of the connection.

At this point, it should be clarified that the measurements and the questionnaire surveys conducted in this phase do not substitute those to be conducted in the phase of the comprehensive feasibility study.

In such case, by suitable processing of the data acquired during the measurements and the questionnaire surveys, and by making some simplified assumptions, the potential transport volume of the suburban railway service (total number of passengers P_d carried per hour and daily) can be estimated.

Concerning the ridership, values in the range of 5,000 passengers/h/direction are considered acceptable for the suburban railway, while in the case of a regional railway, the respective values are lower (Batisse, 1999).

21.3.2.3 Applicability verification of system operability

This verification concerns the issues 5 and 7 in Section 8.4. It includes four partial checks.

21.3.2.3.1 Check of the system's passenger transport capacity

During this check:

a. Knowing the desired run time (one-way trip).
b. Having selected:
   • The type and formation of the trains.
   • The passenger capacity of the trains (this estimation depends on the number and the transport capacity of the vehicles in relation to the acceptable passenger density).
   • The waiting time of the trains at the two terminal stations.
   • The network operation hours during the day (total of 24 h).
c. And having considered various scenarios related to the headway of trains at peak and off-peak hours
   • The passenger transport capacity of the system P'_d is calculated (passenger/h/direction and total per day).

In the case where the resulting total daily passenger transport capacity P'_d is smaller than the potential transport volume of the suburban railway service P_d, one can then adjust one of the parameters (b) or (c) in an effort to satisfy the anticipated demand. The choice of double-deck trains is always a solution. Yet, in this case, a check of the structure gauge must also be conducted.

21.3.2.3.2 Track capacity check

The track capacity must allow the operation of additional trains. Its calculation can be performed by using one of the existing capacity calculation methods (Kontaxi and Ricci, 2009).

21.3.2.3.3 Check of required rolling stock

The theoretically required number of trains is calculated by dividing the two-way route run time (round trip plus waiting time at the two terminal stations) by the headway of trains at peak hours and rounding it up to the nearest integral values. The finally required rolling stock should allow for reserves for maintenance and reserves in the event of a breakdown (Baumgartner, 2001).

21.3.2.3.4 Check of availability and adequacy of rolling stock

This partial check comprises the following procedures:

- Definition of the basic specifications to be met by the rolling stock in order for the above verifications to have positive results (rolling stock design speed, traction characteristics, train type, train formation and passenger transport capacity, compatibility between the vehicle's floor height and the platform's height, and compatibility between the vehicle's dynamic gauge (width) and its gap from the platforms).
- Inventory of the available rolling stock which satisfies the aforementioned standards with such features. At this point, it should be considered that, in the target year, all railway vehicles in service will have an age of 15 years or less (30 years usually being the useful lifespan of a railway vehicle).

The necessary fleet is estimated for the target year by subtracting the available rolling stock in the target year from the finally required rolling stock.

21.3.2.4 Applicability verification of the station service level

This verification examines issue 6 of Section 8.4. It includes the following partial checks:

1. Check of connectivity with other transport systems and, in general, of park and ride service facilities.
2. Check of location of the stations/stops.
3. Check of the security and of the services that are provided in the stations/stops.

21.3.2.5 Applicability verification of the availability of the depot facilities

This verification deals with issue 11 of Section 8.4.

When a relevant infrastructure exists, this verification can also examine whether those can meet the needs of the new railway service.

21.3.2.6 Applicability verification of the environmental impacts

This verification examines issue 12 of Section 8.4.

When trains operate on existing infrastructure, the main possible impact is the increase of noise pollution and vibrations due to the increased train traffic. It is estimated that these problems can often be addressed with financially accepted solutions unless the length of the line where interventions are needed is considerably extensive (see Chapter 19).

21.3.2.7 Applicability verification of the implementation cost

This verification examines issue 13 of Section 8.4.

When trains operate on existing infrastructure, limits cannot be placed with regard to the acceptable implementation cost. In general, in such cases, the cost is significantly lower than the respective cost of contemplating the construction of new infrastructure.

21.3.3 Operation of suburban trains on new infrastructure

As already mentioned in Section 8.4, when the construction of new infrastructure is contemplated, the issues to be examined are 1, 2, 3, 11, 12, 13, and 14, while all the others must be fulfilled by the project-tendering specifications.

In any case, the selected track design speed V_d must ensure the desired run time (issue 4). In this context, the commonly applied practice is to calculate the average running speed V_{ar} by applying Equation 1.4 (see Section 1.4.1).

The value of V_d is calculated by rounding up the result of Equation 1.4 and by selecting values of V_d equal to 120, 140, 150, 160 km/h.

21.3.3.1 Applicability verification of the constructional features of the railway infrastructure

This verification deals with issues 1 and 2 of Section 8.4.

It includes two partial checks:

- Connection length.
- Average distance between intermediate stops.

21.3.3.2 Applicability verification of the passenger transport volume

As in the case of the existing infrastructure (Section 21.3.2.2).

21.3.3.3 Applicability verification of the location, construction, and operation of the depot facilities

This verification examines issue 11 of Section 8.4.

It is a prerequisite that the functional connection of the main traffic line with an area should be where the following activities can be performed rationally:

- Maintenance, repair, and washing of rolling stock.
- Parking of rolling stock.

The above sites are mainly sought out at the end of the lines and (or) at places where empty runs are minimised. Their dimensioning depends on the size of the fleet of the rolling stock, the length of the trains, and, mainly, the future network extension plans.

21.3.3.4 Applicability verification of the environmental impacts

This assessment examines issue 12 of Section 8.4.

When the construction of new infrastructure is contemplated, problems may arise during both the construction and the system operation phases. The most common impacts relate to noise and visual annoyance (see Chapter 19).

The object of environmental applicability verification is to examine to which extent the environmental engineer considers such consequences minor in relation to the functionality of the network or the capability of generating a radical reorganisation of the project's design.

21.3.3.5 Applicability verification of the implementation cost

This verification deals with issue 13 of Section 8.4.

The implementation cost includes the cost of construction of the infrastructure and the cost of purchase of the rolling stock. The verification requires a first approximate evaluation of the cost of the project and the comparison of the investment amount with an average value that is considered to characterise similar projects. An amount of €15 M per track-km can be adopted as such a value. This price comprises the track, the signalling/electrificatio n/telecommunication systems, the station's facilities, and the electromechanical equipment (that is to say, without taking the cost of rolling stock purchase into account).

21.4 APPLICABILITY VERIFICATION OF A MONORAIL LINE

The verification of technical and operational applicability as a process could also be proven to be effective for monorail systems. In fact, due to monorail systems' elevated integration in the urban environment (usually along the entirety of their route), issues that have to be immediately addressed might arise. At the same time a monorail project at both the constructional and operational stage creates new traffic arrangements within a city which, despite being widely less impacting in comparison with an at-grade solution like tramway, still need to be dealt with. Eventually, issues that deal with safety and optical annoyance are in many cases of major importance for the effective operation of the system.

The applicability verification of a monorail system has to answer the following questions:

1. Is the integration of the system into the urban area that it is meant to serve geometrically possible?
2. Is the spatial integration of the elevated infrastructure on the ground of the urban area possible?
3. Is the construction of the stations possible?
4. Can the consequence on the rest of the traffic (road vehicles, pedestrians) for the duration of the construction and operation of the project be addressed effectively with a relatively small cost?
5. Is there a possibility for feeding the monorail by other transport modes and what changes will the construction of the monorail create on the choices and habits of the citizens concerning their daily transport?
6. Is the proper integration of the monorail with the other existing and planned transport means, private and public, possible?
7. Can the environmental consequences of the construction and operation of the project, in the area of the alignment, as well as the wider area, constitute a deterrent for the project's implementation?
8. Is there the possibility to ensure the necessary space for the construction and effective operation of the depot for this particular monorail project?
9. Is the provided level of service (run times, frequency etc.) acceptable based on international practice?
10. Will the transport capacity of the system be adequate for predicted demand?
11. Are the characteristics of the rolling stock completely harmonised with the needs of the alignment geometry and the predicted transport capacity of the system?
12. Is the technology that is to be selected and adopted, available in the industry market?
13. Can the safety of the system be effectively assured with a relatively small cost?
14. Does the cost of the project, as it is formulated from an initial evaluation, range within the limits set by the current international practice?
15. Is the whole project acceptable to the citizens/users of the serviced area?

The answer to the aforementioned questions requires, respectively, for each of the alternative alignment solutions, applicability verifications that are presented in Figure 21.6 and which have to be satisfied simultaneously. These individual verifications may lead designers to make simple adjustments in the design, structural, and operational elements of the

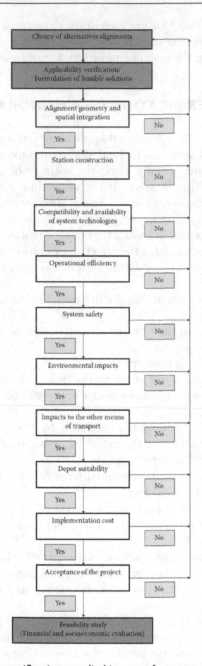

Figure 21.6 Individual applicability verifications applied in case of a monorail project.

monorail system that they are suggesting. They may however also lead to a change in the alignment.

The verification of the alignment geometry and spatial integration of the infrastructure provides answers to questions 1 and 2. The verification for the feasibility of the construction of the stations is directly associated with questions 1, 2, and 3. The operability verification of the system is linked with questions 9, 10, and 11, the compatibility and availability of the technology verification with questions 11 and 12, the environmental consequences verification with question 7, the traffic consequences verification with questions 4, 5, and 6, the depot feasibility verification with question 8, the system safety verification with question 13, the implementation cost verification with question 14, and finally the acceptance of the project with question 15.

21.5 APPLICABILITY VERIFICATION OF A METRO LINE

The construction of a metro system is both technically difficult and costly and has significant impacts on the traffic of a city both during the construction and operation phases. Problems arise with the spatial integration of its infrastructure underground (geotechnical characteristics, available space, land uses), the required construction expertise, the insurance of the necessary expropriations for the construction of stations, the possible existence of antiquities, the repositioning of utility pipelines, the complementarity with existing transport modes, the high level of service that is required to be attained for users (short run times, frequent stops, safety), and the need for environmental protection (noise, vibrations).

At this stage, it is deemed necessary for responsible parties and decision-makers to be aware of the degree of applicability and construction cost of the new transport mode before the move on to costly studies and even costlier construction.

In any case, the applicability verification of a metro system is required to give an answer to the following questions:

1. Is the geometrical alignment acceptable?
2. Is the special integration under the urban area that is to be serviced possible?
3. Is the construction of the stations and tunnels possible with the use of conventional methods?
4. Can the effects on the rest of the traffic (road vehicles, pedestrians) during the construction of the project be addressed effectively and at a relatively low cost?
5. Is there a possibility for the complementarity of the metro with the rest of the transport modes and what changes does the construction of a metro have on the preferences and habits of inhabitants in their daily movements?
6. Can the environmental impacts during the construction of the project in the alignment zone and the wider area constitute an inhibitory factor in its realisation?
7. Is there a capability to acquire the necessary space for the construction and proper operation of facilities for the maintenance, repair, and parking of the rolling stock of this particular metro network?
8. Is the offered level of service (run times, frequency, etc.) acceptable based on international practice?
9. Is the transport capacity of the system adequate based on the expected demand?
10. Are the construction characteristics of the rolling stock in sync with the needs of the alignment, the cross sections of the tunnels, the platforms, and the required transport capacity of the system?
11. Does the construction of the track superstructure ensure the necessary levels of noise reduction as well as the required speed, safety, and dynamic comfort of the passengers? Does it concurrently ensure a low maintenance cost?
12. Does the cost of the project, as this is derived from an initial estimation, range within what is accepted by international practice?

The answers to the aforementioned questions require for each alignment alternative the respective applicability verifications that are presented in Figure 21.7, which are required

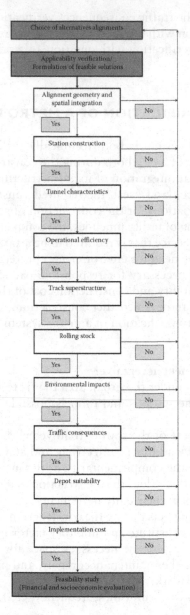

Figure 21.7 Individual applicability verifications applied in case of a metro project (Poirazis, 2018).

to be concurrently satisfied (Poirazis, 2018). These individual verifications may lead designers to simple adjustments in the design, construction, and operational characteristics of the metro system that is being suggested or may lead to a different alignment.

Indicatively, a correlation of the aforementioned questions with individual verifications is presented in order to make their use more apparent. The geometrical alignment and infrastructure special integration verification provide answers to questions 1 and 2. The station construction applicability verification is in direct relation to questions 1, 2, and 3. Similarly, the verification for the characteristics and construction of the tunnels answers questions 1 and 2. The system functionality verification is related to questions 8 and 9, while the superstructure elements verification is related to question 11, the rolling stock verification to question 10, the environmental consequences verification to question 6, the consequences on traffic verification to questions 4 and 5, the depot verification to question 7, and finally the cost verification to question 12.

REFERENCES

Batisse, F. 1999, Les transports urbains et suburbains un défi pour le rail ou un marché, *Rail International*, April 1999, Brussels, pp. 2–11.

Baumgartner, J.P. 2001, *Prices and Costs in the Railway Sector*, École Polytechnique Fédérale de Lausanne, Lausanne, January 2001.

Barbagli, M. and Pyrgidis, C. 2018, Applicability verification of the implementation of a monorail system, *10th International Monorail Association Annual Conference (Monorailex 2018)*, Berlin, Germany, 13–15 September 2018.

Bieber, C.A. 1986, *Les choix techniques pour les transports collectifs*, Ecole Nationale des Ponts et Chaussées, Paris.

Chatziparaskeva, M. and Pyrgidis, C. 2018. A decision-making tool for the applicability verification of tramway projects, *8th International Symposium on Speed up and Sustainable Technology for Railway and Maglev Systems*, Barcelona, Sidges, Spain, September 2018.

Chatziparaskeva, M., Christogiannis, E., Kidikoudis, C. and Pyrgidis, C. 2015, Estimation of required ground plan area for a tram depot, *Journal of Rail Rapid Transit, Proceeding of institution of Mechanical Engineer, PART F*, Vol. 230, pp. 946–960.

Chatziparaskeva, M. and Pyrgidis, C. 2015, Integration of a tramway line alignment in the urban transport system towards sustainability, *2nd International Conference 'Changing Cities: Spatial, Design, Landscape & Socio-economic Dimensions'*, Porto Heli, Peloponnesus, Greece, 22–26 June 2015.

Chatziparaskeva, M. 2018, *Methodology and Mathematical Tools for the Applicability Verification of the Implementation of a Tramway Line*, Phd Dissertation, School of Civil Engineering, Aristotle University of Thessaloniki.

Clark, R. 1984, *Guidelines for the Design of Light Rail Transit Facilities in Edmonton*. http://www.trolleycoalition.org/pdf/lrtreport.pdf (accessed date 06-04-2021)

Transportation Research Board of the National Academies. 2012, Transit cooperative research program: Report 155, *Track Design Handbook for Light Rail Transit: Second Edition*.The National Academies Press, Danvers.

Commission for Railway Regulations. 2008, *Guidelines for the Design of Railway Infrastructure And Rolling Stock, Section 7: Tramways*, August, RSC-G-008-B.

Consultancy HTM. 2003, *Light Rail Project Bergen: Review of Light Rail Systems in the World and Analysis of Comparable Cities with Bergen*, August 2003.

D.D.E. 1993, *Projet de rocade tramway en site propre entre Saint Denis et Bobigny: schéma de principe*, RATP, Paris, France, Février

European Commission, Directorate- General for Regional and Urban Policy. 2014, *Guide to Cost – Benefit Analysis of Investment Projects. Economic Appraisal Tool for Cohesion Policy 2014–2020*. EC, Brussels.

European Rail Research Advisory Council. 2012, *UITP, Metro, Light Rail and Tram Systems in Europe*. ERRAC, Brussels.

Kontaxi, E. and Ricci, S. 2009, Techniques and methodologies for railway capacity analysis: Comparative studies and integration perspectives, *RailZurich 2009, 3rd International Seminar on Railway Operations Modeling and Analysis*.Zurich February 2009.

Poirazis, C. 2018. *Applicability Verification of the Implementation of a Metro Project*, Thesis of final studies, Aristotle University of Thessaloniki, School of Civil Engineering, Thessaloniki, Greece.

Prescott, T. 2012, *Modern Light Rail Vehicles: Fit for Purpose? The Design and Technology of Modern Light Rail Vehicles*, Transit Australia Publishing, Rockdale.

Pyrgidis, C. 2003, Tramway integration in urban areas, alignment applicability control methodology, *Rail Engineering International*, Vol. 1, pp. 12–16.

Pyrgidis, C. 2008, Integration of railway systems into suburban areas – Proposal of a methodology to verify project applicability, *International Congress 'Transportation Decision Making: Issues, Tools, Models and Case Studies'*, 14/11/2008, Venice.

Pyrgidis, C., Chatziparaskeva, M. and Politis, I. 2013, Investigation of parameters affecting the travel time at tramway systems: The case of Athens, Greece, *3rd International Conference on Recent Advances in Railway Engineering (ICRARE 2013)*, Tehran, Iran, 30/4–1/5/2013.

Pyrgidis, C., Chatziparaskeva, M. and Siokis, P. 2015, Design and operation of tramway and metro depots – Similarities and differences, under acceptance for presentation to the *International Conference 'Railway Engineering-2015'*, 30/6-1/7/2015, Edinburgh, Scotland.

Pyrgidis, C., Chatziparaskeva, M. and Siokis, P. 2016, Tram and metro depots: A comparison of design and operation characteristics, *Rail Engineering International*, No 4, pp. 7–10.

Pyrgidis, C., Lagarias, A. and Dolianitis, A. 2018, The aesthetic integration of a tramway system in the urban landscape – Evaluation of the visual nuisance, *4th Conference on Sustainable Urban Mobility – CSUM2018*, Skiathos Island, Greece, 24–25 May 2018.

Pyrgidis,C. 2019, Applicability verification – A supporting tool for undertaking feasibility studies on urban railway systems, *9th International Congress ICTR 2019*, 24–25 October 2019, Athens.

Pyrgidis, C., Lagarias, A., Garefallakis, I., Spithakis, I. and Barbagli, M. 2020, Evaluation of the aesthetic impact of urban mass transportation systems, *5th Conference on Sustainable Urban Mobility – CSUM2018*, Skiathos Island, Greece, 17–19 June 2020.

Thessaloniki Public Transport Authority (THEPTA). 2014, *Investigation of the Implementation of a Tramway System in Thessaloniki*, Thessaloniki.

Index

Printed in the United States
by Baker & Taylor Publisher Services

Printed in the United States
by Baker & Taylor Publisher Services